W9-CSD-269

WATER

A COMPREHENSIVE TREATISE

Volume 1

The Physics and Physical Chemistry of Water

WATER
A COMPREHENSIVE TREATISE

Edited by Felix Franks

Volume 1	The Physics and Physical Chemistry of Water
Volume 2	Water in Crystalline Hydrates; Aqueous Solutions of Simple Nonelectrolytes
Volume 3	Aqueous Solutions of Simple Electrolytes
Volume 4	Aqueous Solutions of Amphiphiles and Macromolecules
Volume 5	Water in Disperse Systems

WATER

A COMPREHENSIVE TREATISE

Edited by Felix Franks

Unilever Research Laboratory
Sharnbrook, Bedford, England

Volume 1
The Physics and Physical Chemistry of Water

PLENUM PRESS • NEW YORK—LONDON • 1972

First Printing – July 1972
Second Printing – July 1974

Library of Congress Catalog Card Number 78-165694
ISBN 0-306-37181-2

A Division of Plenum Publishing Corporation
227 West 17th Street, New York, N.Y. 10011

United Kingdom edition published by Plenum Press, London
A Division of Plenum Publishing Company, Ltd.
Davis House (4th Floor), 8 Scrubs Lane, Harlesden, London, NW10 6SE, England

Printed in the United States of America

αριστον μεν ὕδωρ.
PINDAR

Preface

Although the Greek philosophers and poets were acutely aware that water plays a very special role in Nature, the attitude among modern scientists has been along the lines of "familiarity breeds contempt." With a very few notable exceptions, the scientific community has taken this liquid very much for granted, and active interest on a significant scale only dates from the early 1960's. Since then progress has been slow, but it has been finite. However, for a triatomic molecule water has presented physicists, chemists, biologists, and technologists with a large number of practical and conceptual problems. The renewed interest in their solution during the past decade has been greatly assisted by the active support of the U. S. Office of Saline Water, but even acknowledging all that has been achieved, we have barely scratched the surface. Perhaps the most significant achievement is that we are now beginning to understand the nature of the problem confronting us.

It is against this background that this volume should be approached, at a time when even simple liquids still present us with many unsolved mysteries. In the various chapters of this book a range of properties of water is discussed and it is in the nature of the general problem that conclusions based on the application of any one experimental technique cannot always be easily reconciled with results derived from the use of another technique. Indeed, some of the interpretations advanced for observed phenomena are by no means universally accepted by the scientific "water community"—this became very evident during the first Gordon Research Conference on Water and Aqueous Systems, held in 1970.

The aims in publishing this volume at this particular time are threefold: to review the physical and theoretical techniques which have been and are being applied to a study of the intermolecular nature of water; to discuss critically the type of information obtainable by these techniques, hopefully to arrive at some temporary consensus model or models; and to present reliable physical data pertaining to water under a range of conditions, i.e., "Dorsey revisited," albeit on a less ambitious scale.

I should like to acknowledge a debt of gratitude to several of my colleagues, to Prof. D. J. G. Ives and Prof. Robert L. Kay for valuable guidance and active encouragement, to the contributors to this volume for their willing cooperation, and to my wife and daughters for the understanding shown to a husband and father who hid in his study for many an evening. My very special thanks go to Mrs. Joyce Johnson, who did all the correspondence and much of the arduous editorial work with her usual cheerful efficiency.

F. Franks

Biophysics Division
Unilever Research Laboratory Colworth/Welwyn
Colworth House, Sharnbrook, Bedford
March 1972

Contents

Chapter 5

Raman and Infrared Spectral Investigations of Water Structure

G. E. Walrafen

Chapter 6

Nuclear Magnetic Resonance Studies on Water and Ice

Jay A. Glasel

Chapter 7

Liquid Water: Dielectric Properties

J. B. Hasted

Chapter 8

Liquid Water: Scattering of X-Rays

A. H. Narten and H. A. Levy

Chapter 9

The Scattering of Neutrons by Liquid Water

D. I. Page

Chapter 12

Liquid Water—Acoustic Properties: Absorption and Relaxation

Charles M. Davis, Jr., and Jacek Jarzynski

Chapter 13

Water at High Temperatures and Pressures

Klaus Tödheide

Chapter 14

Structural Models

Henry S. Frank

Contents of Volume 2: **Water in Crystalline Hydrates; Aqueous Solutions of Simple Nonelectrolytes**

Contents of Volume 3: Aqueous Solutions of Simple Electrolytes

Contents of Volume 4: Aqueous Solutions of Amphiphiles and Macromolecules

Contents of Volume 5: Water in Disperse Systems

WATER

A COMPREHENSIVE TREATISE

Volume 1

The Physics and Physical Chemistry of Water

CHAPTER 1

Introduction—Water, The Unique Chemical

F. Franks
Unilever Research Laboratory Colworth/Welwyn
Colworth House
Sharnbrook, Bedford, England

1. INTRODUCTION

The physical nature of water, its role in life processes, and its occurrence, distribution, and turnover present a wide spectrum of problems to physical and life scientists and engineers. Although each specialist is normally conversant with such issues as can be formulated in the particular terminology of his own discipline, few people have a full appreciation of the influence which water has had, and will have, on ecological, sociological, and industrial developments.

It is not the function of this volume to survey in depth subjects such as water conservation, purification, or pollution, or technologies based on changing the state of water in a variety of natural and synthetic products. Rather, it is our aim to review the present state of knowledge concerning the nature of the molecular interactions in the liquid, and to suggest areas where more information is yet required. At the same time it must be borne in mind that there are vast areas where it is hoped that any success achieved by physicists and physical chemists can be translated into applications and it is mainly for this reason that this introductory chapter contains a brief summary of the manner in which our total water supplies are distributed, utilized, and turned over. Similarly, a short factual account of those life processes in which water plays an important role suggests that a better understanding of the nature of water–solute interactions at the molecular level may prove to be of considerable help in the elucidation of problems such as enzyme catalysis, biological mass transport, and the formation of biological structures from molecules and molecular aggregates.

Probably because of the ubiquity of water in our environment, the general realization that this liquid had many unique features worth closer study came relatively recently and this is highlighted in Section 4 dealing with the history of the scientific study of water. However much interest is now shown in problems of water structure, we can only hope to advance our knowledge of this subject at the same rate at which progress is made in the general understanding of the liquid state. The final section of this chapter therefore discusses briefly the concepts and techniques which have been applied to the characterization of the liquid state in terms of bulk properties and molecular correlations and potentials. A comparison of the thermodynamic and transport properties of water in the solid and liquid states with those of a number of simple substances suggests that, in many ways, the behavior of water is more or less what might be expected from an assembly of H_2O molecules. Differences between water and simple fluids are often subtle, but all the same they have far-reaching implications in regard to the conditioning of our terrestrial environment.

2. THE OCCURRENCE AND DISTRIBUTION OF WATER ON THE EARTH

The available water is distributed over the face of the earth in a very uneven manner. Thus, less than 0.027% of the total water is fresh and immediately available. In fact, most of the fresh water is locked up in the Arctic and Antarctic ice caps and, with the world demand for fresh water constantly increasing, it has been suggested that the towing of icebergs to the temperate zones is quite feasible and, in terms of economics, compares favorably with all presently known desalination processes. Thus, it has been calculated[20] that of an iceberg measuring $2700 \times 2700 \times 250$ m, towed at a speed of half a knot from the Amery ice shelf to Australia, 30% would arrive intact. The water would be worth \$ 5.5 million, i.e., about 10% of the cost of a similar quantity of desalinated water, compared with a towing cost of \$ 1 million.

Ninety seven percent of the water available on the earth is found in the large oceans. Thus, the oceans cover an area of 3.6×10^8 km^2 and contain 13×10^8 km^3 of water. Compared to this, the lower seven miles of the earth's atmosphere only contain 13×10^3 km^3, i.e., 0.000053% of the total water. The main process of the hydrological cycle consists of evaporation from the oceans and subsequent precipitation and runoff back into the oceans. The annual turnover of water amounts to 3.5×10^5 km^3 (3.5×10^{14} tons). The Antarctic ice cap, which covers 1.5×10^7 km^2, makes up the largest volume of fresh water (2.5–2.9×10^7 km^3). If melted, it could supply all the earth's rivers for 830 years. By comparison, the Greenland ice cap is quite insignificant; it contains 2.6×10^6 km^3 of water. However, in terms of the available fresh water this is still a very large volume since, if it were melted, it could supply river systems such as the Amazon or Mississippi for 4000–5000 years. The total water locked up in glaciers amounts to only 2.1×10^5 km^3. Just as glaciers can be compared with rivers, so the ice caps can be compared with lakes, in that they cover the landscape and flow radially outward.

The world's great lakes contribute a minor but important amount of the available surface fresh water, totaling 1.2×10^5 km^3. More than half of this volume comes from the four largest lakes: Baikal (26,000), Tanganyika (20,000), Nyasa (13,000), and Superior (12,000). A volume of water equal to that in the fresh-water lakes also exists in saline lakes, of which the Caspian Sea is by far the most important (78,000 km^3). On the continents of Asia, America, and Africa, 75% of the fresh surface water is accounted for by lakes. The corresponding figure for Europe is only 2%. This is one of the reasons why, in spite of the high density of population, the problems connected with pollution and the adequate supply of fresh water are not as pressing in Europe as they are in America.

Although the amount of fresh water present in all the rivers of the world is only 1200 km^3, the annual runoff into the oceans amounts to 34,400 km^3. Of this, the three largest river systems, the Amazon, Congo, and Mississippi, respectively, contribute 5100 km^3. The fact that the rivers of the United States discharge fresh water into the sea at the rate of 4.9×10^4 m^3 sec^{-1} dramatically illustrates the large turnover of water.

The underground reservoirs provide further supplies of fresh water. The topmost layer is a saturated zone in which the liquid is held by capillary forces. The upper boundary is referred to as the water table and this may lie at the land surface, e.g., in swamps, or several hundred meters below, as in deserts. Further down is an unsaturated zone from which water percolates to the water table. This zone contains about 41,000 km^3 of water, which is not extractable but which recharges the ground-water reservoirs from which the water can be extracted. Altogether, 4.1×10^6 km^3 of fresh ground water extends down to a depth of 1 km. Further down, large reservoirs of highly mineralized water exist, but the extraction of these is not economically feasible.

The period for which water remains underground varies from a few hours to several hundred years, and at very great depths the water may remain for up to ten thousand years. The amount of water in the top layer of the earth's crust is equivalent to 4000 times the water in all the earth's rivers. Underground water is, however, not wholly self-renewing, and it is for this reason that water conservation is now of great importance. It has been calculated that for irrigation schemes in arid regions the level of the underground water supply may be reduced through pumping by 60 cm per year, whereas it is only replenished at the rate of 0.5 cm per year.

Finally, probably the most important source of fresh water is rain, the distribution of which over the earth is quite nonuniform. As a result of rainfall and percolation from the water table to the topsoil, the total moisture content of the soil in the world is 25,000 km^3. Plants normally grow on what is considered to be "dry" land, and it is not generally realized that even "dry" dust contains up to 15% of water. It appears that plant growth requires extractable water. Thus, an ordinary tree withdraws and transpires about 190 liters per day.

The average annual rainfall in the United States is 75 cm, 53 cm of which is returned to the atmosphere and only 7.5 cm is used by man. The remainder goes to replenish the underground water reservoirs. Many attempts have been made to increase man's control over the rate of precipitation and to retard losses by evaporation, since the problems associated with the supply of fresh water are becoming more urgent all the time. Urbanization always leads to increases in water usage which far outstrip the rate at which the population increases. In some countries, the use and supply of water are already subject to government legislation, and as problems associated with water pollution become more urgent and receive wider publicity, so more economical methods of cleaning waste water and of using second-rate water in industrial processes will be developed.

3. WATER AND LIFE

Even a superficial study of liquid water, and to some extent of ice, must suggest that life on this planet has been conditioned by its abnormal properties, since water was present on this planet long before the evolution of life. It is well known that water forms a necessary constituent of the cells of all animal and plant tissues and that life cannot exist, even for a limited period, in the absence of water, so that we have the somewhat strange position that the only naturally occurring inorganic liquid is essential for the maintenance of organic life. Bearing in mind also that natural processes are characterized by the economy with which energy (matter) is utilized, it seems permissible to conclude that in organisms which consist of up to 95% water, this liquid fulfills a function other than that of an inert substrate. It is, of course, very much harder to elucidate the exact role of water in life processes,[1043] although biochemical and medical studies have yielded some useful data. Apart from acting as a proton-exchange medium, water moves through living organisms and functions as a lubricant in the form of surface films and viscous juices, e.g., dilute solutions of mucopolysaccharides. Nothing is yet known about the manner in which water acts in the formation of organized biological structures at the subcellular, cellular, and multicellular levels, and, at the molecular level, the role of water in the stabilization of native conformations of biopolymers has only recently been receiving some attention.[248,748,1043] The almost complete disregard of the role of the solvent in tertiary and quaternary structure phenomena is an interesting example of how established experimental findings are sometimes ignored because they cannot be reconciled with existing concepts. Thus, some time before X-ray techniques were successfully applied to establish the structure of DNA, it was well known that the polymer required some 30% of water to maintain its native conformation in the crystalline state, and that partial dehydration led to denaturation. Available X-ray techniques cannot "see" the water in biopolymers because of its relatively high mobility and therefore when the double helix structure was confirmed it was claimed that it owed its stability to intramolecular hydrogen bonds, van der Waals-type interactions between purine and pyrimidine bases, and electrostatic interactions between sugar phosphate groups. Clearly this cannot be the whole story, and further studies will reveal the function of water in the stabilization of the double helix. At present the most useful method for observing small displacements of molecules or segments of molecules in solution is undoubtedly NMR spectroscopy, and the application of this technique to the study of water in biopolymer systems shows some promise.[184,372,394]

Although the contribution of water to life processes at a molecular level is almost completely unexplored, a considerable amount of data exists on the distribution, synthesis, and turnover of water at a more complex level.[64,343,582] Thus, the water content of living organisms varies between the extremes of 96–97% in some marine invertebrates to less than 50% in bacterial spores. The adult human has a water content of 65–70%, but the water is unevenly distributed; nervous tissue contains 84%, liver 73%, muscle 77%, skin 71%, connective tissue 60%, and adipose tissue 30%. The water content of biological fluids such as plasma, saliva, and gastric juices is between 90–99.5%. Approximately 45–50% of the organism is made up of intracellular water, 5% of plasma water, 30–35% of nonaqueous matter, and the remainder can be termed interstitial or extracellular water. The hydration of an organism changes during its development. A human embryo during its first month has a water content of 93% and, as a child develops to maturity, so the intracellular water content increases at the cost of the extracellular liquid. These levels are maintained constant until old age, when the process is reversed.

Water is the solvent which promotes biological hydrolysis (digestion) in which proteins and carbohydrates are broken down; lipids, although not actually modified chemically, are solubilized in the aqueous medium. On the other hand, biosynthesis of water results from condensation polymerization, examples being the production of glycogen from glucose and the formation of proteins from amino acids. Thus, the energy required for biosynthesis derives partially from the energy of formation of water.

The study of the properties and functions of water in biological systems is complicated by the nature of the medium, which contains polymers and colloidally dispersed particles or may be in a gel state. Important fields of study include the origin of resistance toward freezing and dehydration shown by most animal and plant tissues. This raises the question of the nature of "bound" water which is currently receiving some attention[22,148,231,958,1154] and can be characterized by such diverse techniques as osmometry, spectroscopy, and adsorption measurements. Another important function of water is the thermal regulation of living organisms; its large heat capacity coupled with the high water content (45 kg in an adult human) are responsible for maintaining isothermal conditions and the high thermal conductivity of water prevents serious local temperature fluctuations. The high latent heat of evaporation permits large losses of heat: the average adult eliminates water at a daily rate of 300–400 g by respiration and 600–800 g by cutaneous evaporation, with an associated loss of heat amounting to 20% of the total heat produced in a day.

It is well known that living organisms cannot survive without a minimum supply of water, although the tolerance toward dehydration varies widely throughout the animal and plant kingdoms. The average daily intake of water by an adult is 2.5 liters in the form of drink and solid food. Table I shows that of an average daily intake of 1.5 kg of "solid" food, 57% is actually water. In addition, the average adult produces water endogenously by the combustion of food at a daily rate of 350 g, accompanied by a heat liberation of 1.31 kcal. This endogenous production of water is surprisingly constant under different physiological conditions and always

TABLE I. Average Daily Intake of Food and Its Water Content

	Weight of food, g	Corresponding water intake, g
Bread	300	100
Milk	200	175
Lean meat	100	76
Potatoes	300	225
Vegetable	150	133
Fruit	50	40
Cheese	60	21
Fish	60	49
Meat products (sausage, etc.)	80	9.5
Fat	40	0
Sugar	40	0

takes place via a series of coupled reactions involving cytochrome and cytochrome oxidase in which eventually hydrogen and oxygen become available in a form in which they can combine to yield water.

Most of the water present in living organisms acts as an irrigant, distributing nutrients and removing waste products. The internal circulation of water proceeds via intestinal absorption, hemodynamic flow, and diuresis. Most diseases connected with water result from irregularity in the rate of blood flow, the composition of the interstitial and cellular aqueous media, and partial dehydration or hyperhydration. The uptake and endogenous production of water constitute one part of the biological water cycle which is completed by the processes which involve loss of water, namely surface transport, excretion, and vapor loss by respiration; the last of these also regulates the supply of carbon dioxide. The water vapor thus lost enters the hydrological cycle via the atmosphere, whereas liquid waste water eventually replenishes the natural water reservoir. The hydrological cycles of the biosphere are shown in Fig. 1. The subcycle relating to plants is completed by the process of photosynthesis, in which water vapor and carbon dioxide are assimilated. Isotope tracer studies have shown that the oxygen liberated is eventually reconverted to water. The importance of this subcycle is demonstrated by the annual turnover of 6.5×10^{11} tons of water by photosynthesis of green plants and marine organisms.

The above short summary of the different ways in which water is bound up with life processes indicates clearly that water, acting as solvent and dispersing and lubricating medium, is also a versatile reactant and that

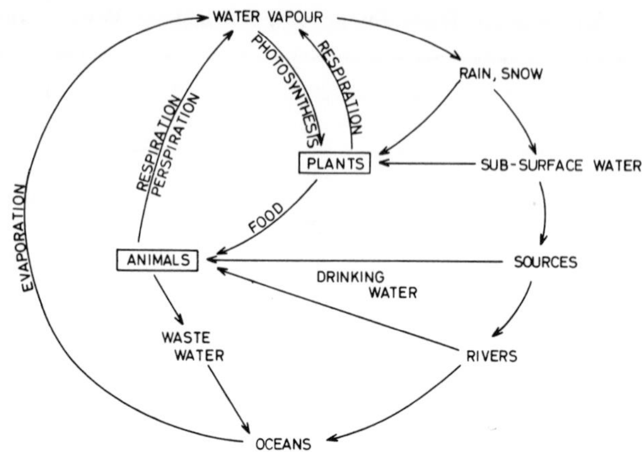

Fig. 1. The hydrological cycle, including the biological subcycles.

morphologically and functionally, life and water are inseparable. It is therefore hardly surprising that living organisms are sensitively attuned to the properties of water. This can be demonstrated by changing these properties "slightly," e.g., by raising the temperature. It appears that for each species there is a definite temperature above which it can exist for only very limited periods; for mammals this temperature is 40°C. Another method for altering the properties of water is by changing its isotopic composition. Although from a chemical standpoint H_2O and D_2O closely resemble each other, the rates of many chemical, and especially of biochemical processes are extremely sensitive to the nature of the aqueous substrate, and life processes can be retarded by replacing H_2O in the body fluids by D_2O. Until recently it was believed that D_2O could not support life, but Katz[576] has shown that algae, bacteria, yeasts, fungi, and protozoa can be induced to grow in 99.8% D_2O, and higher organisms in lower concentrations. In fact, completely deuterated biopolymers have thus been biosynthesized. How such adaptation is brought about and what are the effects of deuteration on the functional properties of biopolymers is not yet fully understood, but investigations in this field clearly indicate that much can be learned by tracing the path of hydrogen, rather than carbon, in biological processes.

4. THE SCIENTIFIC STUDY OF WATER—A SHORT HISTORY

In order to place the subject matter contained in this volume in its right perspective, it is helpful to outline briefly the historical development of the scientific study of water in all its various aspects. The realization that water is a unique substance goes back a long way in history, and already at the time of the famous Greek philosophers it was realized that water played an important role in supporting life. Aristotle included water among the four elements, alongside earth, air, and fire. Not until the 18th century was it established that earth and air were mixtures, and that fire was simply the manifestation of chemical change. In 1781 Priestley synthesized the last remaining one of Aristotle's elements, and shortly afterward Lavoisier and Cavendish succeeded in decomposing liquid water into "ordinary air" (oxygen) and "inflammable air" (hydrogen). Although these classical experiments established the fact that water cannot be regarded as an element (in the sense now accepted), it nevertheless plays a unique role among chemical compounds. It is therefore somewhat surprising that some of the famous 19th century scientists did not enquire more deeply into its nature. The explanation may lie in the fact that on this planet water is ubiquitous and has, for a long time, been taken for granted. Looking back now at the volume of experimental data on aqueous solutions of electrolytes and nonelectrolytes assembled over the past hundred years, it is easy to see that physical and organic chemists would not have encountered so many difficulties if the most readily available solvent had not been water. Thus, the observed phenomena associated with equilibrium and kinetic processes were usually interpreted without allowing for the fact that the solvent itself was a highly complex one. Probably one of the best examples is provided by the development of the Debye–Hückel theory of electrolytes, based almost completely on observations derived from studies on aqueous solutions.[(914)] Yet the only physical property ascribed to the solvent was a dielectric constant, and although in its limiting-law form the theory is of great value, the various extensions which have, from time to time, been proposed to account for the concentration dependence of activity coefficients are, at best, semiempirical.

Thus, in the main, scientists did not pay much attention to any possible functions of the solvent in physicochemical processes, although an increasing number of the properties of water were accurately measured and found to differ from those of other liquids or chemically related compounds. Probably the earliest suggestion that liquid water might contain "solid particles" was made in 1884, and in 1891 Vernon[(1108)] postulated the aggregation of water molecules in order to account for the phenomenon of maximum density. A year later Röntgen[(916)] qualitatively explained several of the apparently anomalous properties of water. However, none of these early attempts to draw attention to the peculiar nature of water was followed up. This is also true for the now famous monograph by Dorsey,[(269)] ironically entitled "Properties of Ordinary Water Substance," which contains many examples of the physical anomalies which had been observed over the previous 50 years. Even the classic paper by Bernal and Fowler,[71] in which they advanced the first plausible model for liquid water, failed to make an immediate impact, although it has since been realized that it laid the foundation for subsequent work, and we are even now still in the era of Bernal and Fowler. Later in the 1930's came the first X-ray[761] and infrared[720] studies on liquid water, and Butler[158,159] and Eley[304] investigated the origin of negative partial entropies exhibited by a variety of simple solutes in aqueous solution. These studies culminated in the famous paper by Frank and Evans,[359] which introduced the concept of "icebergs" induced in water by solute molecules. At the same time Samoilov[940] was investigating the nature of the interactions between water and ions in solution. In 1948 Hall,[444] studying acoustic relaxation in water, advanced the first detailed mixture model, i.e., a model based on the concept of water as a mixture of two distinguishable species. Shortly afterward Wang[1144–1146] published his extensive studies on the self-diffusion behavior of water of different isotopic compositions. The next major step forward was made by Pople,[855] resulting from his earlier studies into the properties of the hydrogen bond. Basing his calculations on the earlier model of Bernal and Fowler, he suggested the possibility of hydrogen bond bending, rather than breaking, to account for the temperature dependence of the dielectric constant and some other properties of water.

In the 1950's physicochemical studies of water and its interactions with solutes gradually gathered momentum and several molecular models were proposed, but the subject of water structure only attained the status of scientific respectability in the 1960's. Frank's reminder that the existence of long-lived structures in liquid water was most unlikely and that a useful description might involve "flickering clusters"[361] of hydrogen-bonded molecules was very timely, whilst Kauzmann's outstanding discussion of the possible role of water in determining protein conformation and denaturation behavior[579] helped to make biochemists aware of the peculiar nature of the solvent medium in which life processes take place. In fact, this review was to prove instrumental in promoting a veritable flood of publications describing various aspects of the phenomenon of "hydrophobic bonding" in biopolymer systems.

In 1962 Nemethy and Scheraga published a series of papers[787–789] in which they attempted to apply statistical mechanics to the development of a molecular description of water and of aqueous solutions of apolar solutes and proteins. Although the mathematical treatment has subsequently been criticized,[1082] and the numerical parameters employed may well be unrealistic, nevertheless these papers constitute the first attempt to put on a quantitative basis what had, up to that time, been based on empiricism, and as such they constitute a significant development.

Comparisons of various bulk and microscopic transport processes (e.g., viscosity, self-diffusion, and dielectric and NMR spin–lattice relaxation) in water had shown that they could all be associated with an identical energy of activation, and this led to speculations regarding the nature of molecular motions in water. High-precision Raman studies[1134] of the intermolecular (hydrogen bonding) modes in liquid water followed, and recent infrared and laser Raman techniques have been applied to the investigation of the uncoupled intramolecular OH and OD stretching modes in H_2O and D_2O.[962,1138,1140] The spectral contours have been carefully analyzed in an effort to adduce evidence for or against models which treat water as a mixture. At the time of writing, however, this question is still unresolved, although mixture (on a suitable time scale) models are at present the more favored ones. In fact, the observed effects of solutes on the physical properties of water cannot readily be accounted for by a continuum model.

Until recently all the model calculations had to be based on experimental results covering only the normal liquid range, but during the last five years some of the physical properties of water have been studied up to and beyond the critical point.[354,703] so that much more rigorous tests can now be applied to the molecular models.

The development of sophisticated computing techniques has led to attempts to study water–water interactions by *ab initio* quantum mechanical methods. Thus, water dimers have received attention[250,764] and work is currently in progress to extend these techniques to trimers and larger aggregates.[251] It is hoped that this approach will provide detailed information about the nature of the hydrogen bond in liquid water and, in particular, whether hydrogen-bonding processes in liquid water involve a cooperative element. Computer simulation is also being applied to the study of water, although experiments reported so far have been confined to Monte Carlo calculations of thermodynamic properties based on a rather simple pair potential.[51]

Although a rigorous treatment of simple liquids, let alone water, in terms of realistic potential functions may not become available for some time (see Chapter 3), nevertheless, the manner in which solute molecules interact with water and with one another in aqueous solutions can be profitably studied, and the last five years have seen some useful progress in this field.[66,362,363,633] There is, however, an urgent need for reliable experimental data on very dilute aqueous solutions of a large range of solutes, and with very few notable exceptions, such data are only now finding their way into the literature. Even precise experimental data on aqueous nonelectrolytes, covering the whole composition range, are still scarce. The present position is put in perspective by reference to standard texts on liquids and solutions,[505,873,928] in which little space is devoted to aqueous solutions.

One interesting issue which still awaits resolution concerns the existence, or nonexistence, of thermal anomalies (also referred to as "kinks" and "thermal discontinuities") in the physical properties of water. The chief protagonist of their existence has been Drost-Hansen,[278] whose claims have been supported by some, and hotly contested by others. Essentially the suggestion is that higher-order structural transitions occur in water at, or near, 15, 30, 45, and 60° and that these transitions give rise to discontinuities in the second or higher temperature derivatives of physical properties. Although at one time Drost-Hansen believed that such transitions occurred in pure water,[277] subsequent experimental measurements at closely spaced temperature intervals and analyses of previous work have led him and others to cast doubt on this hypothesis. However, in some biological systems, especially those involving membrane transport phenomena,[1063,1064] these discontinuities have been clearly established and even in relatively simple aqueous systems, especially those with large area/volume ratios, reliable experimental studies[887] have shown that temperature derivatives of physical properties can adopt complex forms, with distinct evidence for the anomalies postulated by Drost-Hansen. Unfortunately, objective discussion of this interesting issue, based on experimental evidence, has often been obscured by dogmatic statements and bitter polemics, so that the problem is still unresolved.

In the investigations of water the most recent chapter, and one of the most intriguing, must surely be the story of "polywater." Since 1961 Deryagin, Fedyakin, and their collaborators[257,328] have studied the properties of water and other liquids in thin films and in capillaries. That water in confined spaces has some very interesting properties (e.g., it does not freeze at 0°C) had been known for many years. In the course of his experiments, Deryagin reported the condensation of a liquid from an atmosphere of unsaturated water vapor in narrow (<20 μm) capillaries. On cooling, this liquid separated into two phases, one of which appeared to be ordinary water and could be removed by distillation. The remainder did not freeze at 0°C but showed a complex freeze–melt behavior. It was claimed to possess a density of 1.3 g cm^{-3} and a very high viscosity. Although much of this information had been published previously in Soviet journals, the first publication in an English journal was not until 1966,[256] and even then the significance of the discoveries was largely overlooked for another two years, during which Western scientists adopted a highly sceptical attitude toward Deryagin's experiments. However, a group of British scientists managed to substantiate most of Deryagin's fairly modest claims,[1172] and during 1969 the "polywater" bandwagon began to roll. Since then many claims, much more sweeping than those initially made by Deryagin, have been advanced, based on very much less concrete evidence.[61,691] Speculations about "a brilliant future in industry" have been made,[314] and various industrial uses proposed. However, at the time of writing few serious full-length papers on the subject have yet appeared in the scientific literature. Instead there are numerous short notes,[313,821,836] sometimes based on scant experimental effort, and at present it is not at all certain whether the effects observed are not due to impurities.[644,923] A climax was provided by a warning to scientists against experimenting with polywater as this might affect the oceans, turning them solid.[266] Although the unlikelihood of such a disaster was quickly pointed out,[70,320] several daily and weekly newspapers featured the story and "polywater" has since been the subject of several articles in newspapers[24,46,863] and popular scientific publications.[19,21]

With all the effort which has, over the past fifty years, been devoted to the study of water, we are still a long way from understanding the exact nature of the molecular interactions in the liquid. It is therefore improbable that with the present very sketchy and controversial experimental evidence the various suggestions advanced for the structure of polywater[10,268,313,402,687,744,757] can be taken seriously, although the future may yet see interesting developments, if it can be established that the substance isolated and described by Deryagin is indeed a new structural modification of H_2O.

5. THE PLACE OF WATER AMONG LIQUIDS

Classical thermodynamics can give no direct information about the microscopic behavior of liquids, but if bulk thermodynamic properties are used in conjunction with a simple molecular theory of matter, then some useful observations can be made. The simplest comparison that can be made is between the thermodynamic and transport properties of materials under similar conditions. Table II shows some relevant properties of argon, benzene, water, and sodium for the condensed phases. The most striking

TABLE II

	Argon		Benzene		Water		Sodium	
	Solid	Liquid	Solid	Liquid	Solid	Liquid	Solid	Liquid
Density, g cm^{-3}	1.636	1.407	1.0	0.899	0.920	0.997	0.951	0.927
Latent heat of fusion, cal mol^{-1}	1880		8300		1430		26,200	
Latent heat of evaporation, cal mol^{-1}	1600		6000		9700		25,600	
Heat capacity, cal (mol deg)$^{-1}$	6.2	5.4	2.7	3.1	9	18	6.8	7.7
Melting point, °K	84.1		278.8		373.2		371.1	
Liquid range, deg	3.5		75		100		794	
Compressibility (isothermal), 10^5 bar^{-1}	1	20	8.1	8.7	2	4.9	1.7	1.9
Expansibility, 10^3 deg^{-1}	2	4.4	0.8	1	0.11	0.3	0.21	0.25
Surface tension, dyn cm^{-1}	13		28.9		72		190	
Viscosity, 10^3 P	2.8		8.0		8.9		6.8	
Self-diffusion coefficient, cm^2 sec^{-1}	10^{-9}	1.6×10^{-5}	10^{-9}	1.7×10^{-5}	10^{-10}	2.2×10^{-5}	1.9×10^{-7}	4.3×10^{-5}
Thermal conductivity, cal (sec cm deg)$^{-1}$	7.1×10^{-4}	2.9×10^{-4}	6.5×10^{-4}	3.7×10^{-4}	5×10^{-3}	1.4×10^{-3}	0.32	0.20

feature is the similarity between the liquid and solid states; the small increases in volume, the small differences between the latent heats of fusion and evaporation, and the similarity of the specific heats for the simple substances all indicate that liquids are highly condensed, with strong intermolecular cohesive forces. The thermal motions are presumably small-amplitude vibrations, as large vibrations would not be possible in the solid state. None of the bulk properties has a unique value, or range of values, for a given state, except, of course, surface tension. These observations give weight to the suggestion by Frenkel[366] that liquids and solids are similar in the arrangement of their molecules, at least near the melting point. When comparing liquids among themselves, similar gross observations are possible. The increase of intermolecular bonding in the series argon to sodium is reflected to a greater or lesser extent in all the thermodynamic properties. The most notable of these are the temperature ranges of the liquid state, the melting points, and the surface tensions. The heat capacities are exceptional, as here the intramolecular vibrations in the complex molecular liquids upset the general trends.

The transport properties can exhibit quite dramatic differences. The viscosity shows the largest change in any property found for a change of state (frequently by a factor of 10^{10}–10^{15}). The phenomena of viscosity and self-diffusion reflect the breakdown of long-range order in the liquid state, resulting in the enhanced mobility of the molecules. Thermal conductivities do not show quite such dramatic changes, indicating that, although the molecules are free to move, the heat conduction process is relatively unaffected.

A more quantitative approach can be made using equations of state. Many such equations have been proposed, some of which adequately account for the properties of the liquid. They are generally arranged in a virial form and, in certain cases, can be related to the theoretical molecular potential functions. Theoretically, the potential function $U(r)$ is a complete description of a liquid, and all other properties may be derived from it. Unfortunately, direct measurements of $U(r)$ are impossible on any but the simplest systems, e.g., low-pressure gases where interactions between pairs of molecules only need be considered. The calculation of $U(r)$ is just as difficult, but in a few cases, e.g., argon, theoretical calculations have produced results which are said to be "better than experimental values."

As it is impossible to calculate or measure $U(r)$ for most liquids, spectroscopic and other physical data are usually expressed in simpler terms. The time-dependent correlation function[301] provides a convenient general method, where the basic measured quantity is expressed as an explicit function of the microscopic molecular motion. One early use of these functions was by Dirac,[264] who postulated a time-dependent perturbation theory for the probability of a molecule being excited to another quantum state by the absorption of radiation. The space–time correlation function[288] is useful in the description of diffusional modes in liquids. In a review of the use of correlation functions, Gordon[414] suggests that a better understanding of $U(r)$ may be gained if several correlation functions could be obtained for a given system.

Correlation functions give a mathematical picture of a liquid which is not readily visualized. For this reason most of the microscopic techniques are interpreted in terms of model systems. For example, the relaxation times found by acoustic or dielectric methods arise from a particular molecular species in a liquid being unable to keep pace with reversals in the direction of an applied field. This period is related to a maximum reorientation rate of the species in the particular environment and is a direct microscopic measurement. Changes in the relaxation time can be measured against changes in the microscopic environment. Unfortunately, the microscopic environment in a liquid is not known and usually a physical model is proposed. Therefore, the results are dependent on the quality of the proposed model. The use of models, in spite of their limitations, is widespread in the interpretation of microscopic measurements.

A prerequisite of any microscopic technique is that it can resolve individual molecules. However, the resolution time of a given technique is also vitally important, as this determines the type of motion which will be studied. The whole range of resolution times is required if a complete picture of a liquid is to be obtained. For the relaxation techniques this range is readily deduced from the frequency range of available equipment, although modern experimental techniques and developments in light scattering have extended the range of ultrasonic absorption measurements. The spectroscopic absorption techniques, as well as X-ray, infrared, Raman, and neutron scattering, rely on the detection of changes in the properties of well-characterized incident beams. These changes are due to interactions between the incident radiation and the molecules in the liquid.

The measured bulk properties of liquids arise from the average of a large number of molecular interactions, and ideally each molecule should be considered separately. This would involve the consideration of all the attractive and repulsive forces experienced by the molecule, followed by a summation of these effects over all the molecules. It would then be theoretically possible to predict any of the bulk properties. Unfortunately, many-body potential functions are insoluble, and in any case the computations involved are beyond our present resources. One major simplification is to reduce $U(r)$ to a simple, averaged two-body potential which can be considered as being typical of the forces experienced by a molecule. Such pair potentials are widely used and the success of a given method depends to a large extent on the quality of $U(r)$, whether it is defined or implicit.[929]

A formal approach has been developed in which the radial distribution function is calculated from $U(r)$, and hence, using the virial theorem, an equation of state is deduced. Unfortunately, approximations in the first step of the calculation are necessary. This method has been applied with reasonable success to theoretical hard-sphere fluids, Lennard-Jones fluids, and a few simple liquids.[610] To avoid the rather laborious integrations and assumptions involved in this formal approach, other methods have been developed. Bernal[68] made elegant use of ball-and-wire models and other devices to demonstrate the properties of liquids. His models are particularly useful when applied to liquids near their melting points, and they predict a radial distribution function $g(r)$ which agrees well with experimental values for simple liquids. Many conceptual, rather than physical, models have been proposed.[288] They have the advantage of computational simplicity and, if successful, they give a visual picture of the liquid state. Most of these models yield partition functions from which the Helmholtz free energy and other thermodynamic properties may be calculated and compared with experimental values.

Most liquids expand up to about 10% on melting. This expansion is not accompanied by an increase in nearest-neighbor distances, but rather by a reduction in the number of nearest neighbors.[637] Most models are concerned with the spaces thus created, and in 1932 Frenkel proposed a "hole" theory for the perturbation of order in a simple crystal. He extended this idea to include the melting of a crystal when a critical number of holes was reached. Eyring[321] extended the hole theory to liquids and, with Cernushi,[172] suggested a quasilattice liquid.

Early work on the cell model, in which a mobile central molecule is situated in a cell comprised of stationary hard-sphere molecules, gave poor results. Lennard-Jones and Devonshire[668] modified the hard-sphere potential and obtained better results. A more popular theory is the free-volume theory, where the excess volume is seen as holes or vacancies. These vacancies qualitatively account for the mobility, density, and entropy of liquids. This approach has been modified by several workers, and Henderson[489] obtained good agreement with experimental second virial coefficients. In an attempt to describe a more disordered lattice system, Barker[50] proposed a tunnel theory in which the liquid is seen as being divided into lines of molecules moving in one-dimensional tunnels, and the partition function is calculated from the longitudinal and transverse motions in the tunnel.

Most of these theories are capable of predicting some of the bulk properties of simple liquids with reasonable accuracy, i.e., better than 5%. However, for associated liquids, or liquid polymers, agreement is not generally as good and additional parameters are sometimes required to obtain reasonable results.

Of all known liquids, water is probably the most studied and least understood, although many of its physical properties are accepted as international standards, e.g., its triple point, density, mass, and viscosity. The wealth of accurate data available imposes severe tests on any model proposed for water, and possible anomalies in the data themselves have led to several generalized observations on the microscopic nature of water. Reference to Table II shows that, as a first approximation, water can be regarded as a liquid with bonding characteristics intermediate between benzene and sodium. The bulk properties certainly give this impression, e.g., melting point, boiling point, liquid range, heat of sublimation, and surface tension. If, however, water is compared with other hydrides,[550] then a very different picture emerges. In any case, not too much significance can be attached to the fusion and evaporation behavior of series of compounds. It must be borne in mind that melting points depend mainly on the structural features of the lattice. This is demonstrated in Fig. 2, in which the melting point trends are plotted for series of hydrides and metallic and nonmetallic halides,[625] showing that the behavior ascribed to hydrogen

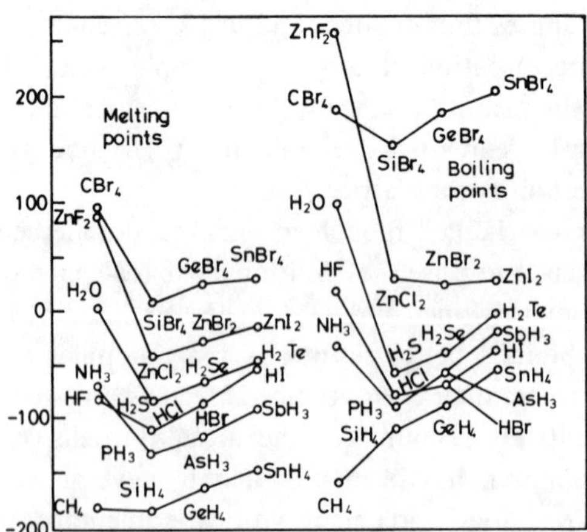

Fig. 2. Melting points and boiling points of series of related hydrides and halides, showing common trends.

bonding in the case of water is by no means unique. The one feature which is peculiar to the water molecule is its capability of forming four hydrogen bonds, with the possible promotion of three-dimensional structures. However, the postulate of association by hydrogen bonding, although accounting for some of the mysteries, also raises several new problems, because the nature of the hydrogen bond is as yet only imperfectly understood, and there are no realistic potential functions to describe this particular type of interaction.

A closer inspection of the thermodynamics of water, and in particular of its *PVT* properties, does reveal a behavior which is very different from that of benzene or sodium. Thus, the density increases on melting, and the temperature and pressure derivatives of volume show extrema, indicating that in addition to unusual bonding arrangements, structural equilibria may also exist, giving rise to higher-order phase transitions. The higher pressure and temperature derivatives of the volume, when plotted against temperature, reveal an even more complex situation, which has not yet been fully explored. The transport properties are also consistent with the concept of a "structured" liquid, the viscosity being noticeably high, as would be expected.

The physical and chemical anomalies of water are well known and documented, but their significance in quantitative terms is not always realized. The large heat capacity of water reflects its ability to store considerable quantities of thermal energy. The oceans are thus able to act as vast thermostats from which the heat energy is carried to the cooler regions by currents such as the Gulf stream. To obtain a quantitative estimate of the heat-carrying capacity of the ocean currents, the following calculation may serve as a useful illustration: if one mile^3 of water flows from a region of high temperature to a distant region where the temperature is 20 C deg lower, then the heat transfer amounts to 4.6×10^{13} kcal, which is equivalent to the heat produced by 7 million tons of coal. If a current 100 miles wide and $\frac{1}{4}$ mile deep flows at a rate of 1 mph, then 25 mile^3 of water are transported every hour, and the heat energy thus transferred to the cooler region is equivalent to that produced by 175 million tons of coal. All the coal mined in the world in one year would supply heat at this rate for only 12 hr.

Another property intimately connected with the movement of water by ocean currents is the density maximum near 4°C, the origins of which were already discussed in 1667 by the Florentine academicians of Cimento. As is generally known, this density maximum just above the freezing point of water results in the freezing of rivers and lakes from the surface downward. If the oceans in the cooler regions of the earth would freeze upward from the seabed, only the surface layers would thaw during the summer, with the eventual result that the vast masses of water transported by the warm currents could no longer find their way to the temperate zones, so that these would gradually freeze up.

In this connection the compressibility of water is also of interest. The coefficient of compressibility is only 5×10^{-5} bar^{-1} at ordinary temperatures and pressures, but if water were completely incompressible, then the level of the sea would be higher by 40 m and the total land area reduced by 5%. The expansion of the liquid on freezing, although not unique to water, is another property which has had a major influence on the nature of the earth's surface. Water, as a result of its very high surface tension, easily penetrates into rock crevices, where, on freezing, it fragments the rock substance, a process which has led to the formation of soil.

Other remarkable properties, although not producing results quite as spectacular, include the almost universal solvent action of liquid water, making its rigorous purification extremely difficult. Nearly all known chemical compounds will dissolve in water to a slight, but detectable extent, but no instance is known of a thermodynamically ideal solution in which one of the components is water. Apart from being the universal solvent, water is also one of the most reactive chemicals, in that it readily interacts with ions and molecules. It is in fact one of the most corrosive substances known, but in spite of this it is, unlike other hydrides, physiologically innocuous.

Strictly speaking, water is a mixture containing various combinations of the isotopes ^1H, ^2H, ^3H, ^{16}O, ^{17}O, ^{18}O in the form of molecules, hydrogen ions, and hydroxyl ions. The mean isotopic ratios of hydrogen and oxygen in natural waters are $^2H : {}^1H = 1 : 69{,}000$, $^{17}O : {}^{18}O : {}^{16}O = 1 : 5 : 2500$. These ratios appear to be the same whatever the source of the water. Tritium ($^3H : {}^1H = 1 : 10^{18}$) occurs in rain and snow, and since it has a half life of 12.5 years, water can be dated by its tritium content.

As will be shown in this series of volumes, the presence of molecules, ions, microorganisms, and dispersed particles generally in liquid water appears to give rise to subtle changes in the water–water interactions which in turn can markedly affect solute–solute correlations. The particular type of solute hydration envelope appears to be very sensitive to the nature of the dissolved or dispersed particles. The control of many technological and biological processes involves the modification of particle hydration shells by heating, cooling, agitation, changes in *p*H and ionic strength, or by the addition of organic molecules, surface-active agents, or polymers. It is therefore to be expected that a better insight into the molecular interactions in pure water and in aqueous solutions and dispersions may lead to improvements in several areas of technology and medicine.

CHAPTER 2

The Water Molecule

C. W. Kern

Battelle Memorial Institute and The Ohio State University
Department of Chemistry
Columbus, Ohio

and

M. Karplus

Harvard University
Department of Chemistry
Cambridge, Massachusetts

1. INTRODUCTION

The behavior of water in any of its phases depends ultimately on the structure and properties of the isolated water molecule. Since quantum mechanics provides an accurate description of molecular phenomena, a detailed understanding of the water molecule is available from theory. This implies that it is possible, in principle, to predict the structure and properties of water, or of any other molecule, from a knowledge of the number of particles and the values of their masses, charges, and spins. Because of significant limitations on the extent to which such a theoretical analysis can be carried out in practice, a judicious combination of theory and carefully designed experiments provides the most powerful tool for the investigation of chemically interesting species.

It is the purpose of this chapter to summarize the principles of structure and spectra that are relevant to an understanding of an isolated water molecule, and to review in a selective way the experimental and theoretical data currently available. Since general surveys of the water molecule have appeared recently,[301,337] we shall attempt a complementary discussion that emphasizes the quantitative aspects of the underlying theoretical development.

2. PRINCIPLES OF STRUCTURE AND SPECTRA: THE BORN–OPPENHEIMER SEPARATION

For a water molecule in the absence of external fields or other perturbing interactions, the quantized form of the total Hamiltonian function $\mathscr{H}_t$ is independent of time. As a result, the time dependence of the total wave function can be separated out, and we need concern ourselves only with the time-independent Schrödinger equation

$$\mathscr{H}_t \Psi_n = W_n \Psi_n \tag{1}$$

in which the energies W_n are associated with the stationary eigenstates Ψ_n that depend on the space and spin coordinates of all the particles. The precise form of $\mathscr{H}_t$ for the water molecule is

$$\mathscr{H}_t = T_N + T_e + U \tag{2}$$

where the first two terms represent the kinetic energy of the three nuclei and ten electrons, respectively, and where

$$U(\mathbf{r}, \mathbf{R}) = -\sum_{K=1}^{3} \sum_{k=1}^{10} \frac{Z_K e^2}{|\mathbf{R}_K - \mathbf{r}_k|} + \frac{1}{2} \sum_{K=1}^{3} \sum_{K' \neq K}^{3} \frac{Z_K Z_{K'} e^2}{|\mathbf{R}_K - \mathbf{R}_{K'}|} + \frac{1}{2} \sum_{k=1}^{10} \sum_{k' \neq k}^{10} \frac{e^2}{|\mathbf{r}_k - \mathbf{r}_{k'}|} \tag{3}$$

is the potential energy function composed of electrostatic interactions between all pairs of particles.

Because we are interested primarily in the internal structure of water, it is convenient to perform* a transformation to a center-of-mass, molecule-fixed (gyrating) axis system. If we assume that the center-of-mass motion has been separated out, eqns. (1)–(3) describe the internal motion to high accuracy. The appropriate electronic and nuclear kinetic energy operators can be written†

$$T_e = -(\hbar^2/2m) \sum_{k=1}^{10} \nabla_k^2 \tag{4}$$

* Documentation can be found in numerous references, including the review by Nielsen[795] and the textbook by Wilson, Decius, and Cross.[1173]

† The mass polarization and relativistic corrections are neglected in this chapter.[537]

and

$$T_N = \tfrac{1}{2} \sum_{\alpha=1}^{3} \sum_{\beta=1}^{3} \left(P_\alpha + \sum_{s=1}^{3} A_s^{(\alpha)} p_s\right) \mu_{\alpha\beta} \left(P_\beta + \sum_{s=1}^{3} A_s^{(\beta)} p_s\right) + \tfrac{1}{2} \sum_{s=1}^{3} p_s^2 - (\hbar^2/8) \sum_{\alpha=1}^{3} \mu_{\alpha\alpha} \tag{5}$$

respectively. Equation (4) can be simply interpreted as the kinetic energy of the electrons referred to nuclei that are fixed with respect to their center of mass. The kinetic energy operator* as written in eqn. (5) is the one to use for studying the rotational and vibrational motions of the water molecule when the atoms are arranged in a nonlinear configuration. The first term, for example, represents the three components ($\alpha, \beta = 1, 2, 3$ for X, Y, Z) of the pure rotational energy through the quantity $\mu_{\alpha\beta} P_\alpha P_\beta$, where P_α is the α component of the total angular momentum of the molecule relative to a rotating-body–fixed-axis system whose origin coincides with the center of mass and the elements of $\mu_{\alpha\beta}$ are related to the effective moments of inertia. The first term also includes Coriolis coupling to the three normal coordinates (Q_s) of vibration through the conjugate momentum operator $p_s = -i\hbar\, \partial/\partial Q_s$. The second term contains the pure vibrational kinetic energy, and the third term acts as a mass-dependent contribution to the potential energy U. The lowest-order terms in eqn. (5) approximate the nuclear kinetic energy as the energy of a rigid, rotating body plus the vibrational kinetic energy of a nonrotating molecule (see Section 4). The symbols $A_s^{(\alpha)}$ and $\mu_{\alpha\beta}$ appearing in T_N are defined in the appendices.

Following Born and Oppenheimer,[103] we now note the large difference in mass between the ten electrons (each of mass m) and the three nuclei, the ratio λ being of the order of $m/m_p \simeq 0.0005$ in the least favorable case of a proton (of mass m_p). Since this implies that the ratio of nuclear to electronic kinetic energies are typically on the order of $\lambda^{1/2} \simeq 0.02$ or less, the electrons can be assumed to move, as a good first approximation, in a force field of fixed nuclei. An approximate solution Φ_n of the Schrödinger equation (1) for the internal motion is therefore expected to be given by the *electronic* Schrödinger equation

$$\mathscr{H}_e \Phi_n(\mathbf{r}; \mathbf{R}) = E_n(\mathbf{R}) \Phi_n(\mathbf{r}; \mathbf{R}) \tag{6}$$

where

$$\mathscr{H}_e = -(\hbar^2/2m) \sum_{k=1}^{10} \nabla_k^2 + U(\mathbf{r}; \mathbf{R}) \tag{7}$$

* The form of the kinetic energy operator given here results from a simplification, due to Watson,[1152] of the vibrational–rotational Hamiltonian derived classically by Wilson and Howard[1174] and quantum mechanically by Darling and Dennison.[232]

is formed from the full Hamiltonian in eqn. (2) by neglecting the nuclear kinetic energy contribution. Since the electronic wave function $\Phi_n(\mathbf{r}; \mathbf{R})$ depends explicitly on the electron coordinates but implicitly on the nuclear geometry (as indicated in a symbolic fashion by $\mathbf{r}; \mathbf{R}$), it is clear that the electronic energies, $E_n(\mathbf{R})$ for the eigenstate n, generate an energy *surface*, each point on the surface corresponding to a different spatial arrangement $(\mathbf{R})$ of the nuclei. Pursuing this line of reasoning a step further, we argue that the *nuclear* eigenvalue equation is of the form

$$[T_N + E_n(\mathbf{R})]\tilde{\chi}_{n\nu}(\mathbf{R}) = \tilde{W}_{n\nu}\tilde{\chi}_{n\nu}(\mathbf{R}) \tag{8}$$

the electronic energy $E_n(\mathbf{R})$ serving as the potential energy function for the nuclear motion. In the limit of essentially complete separation of electronic and nuclear motion, the total wave function is thus given by

$$\tilde{\Psi}_{n\nu}(\mathbf{r}, \mathbf{R}) = \tilde{\chi}_{n\nu}(\mathbf{R})\Phi_n(\mathbf{r}; \mathbf{R}) \tag{9}$$

where the indices n and ν label the electronic and nuclear energy levels, respectively.

The above argument can be made more rigorous by applying the Rayleigh–Ritz variational principle, which is the basis of a powerful technique for solving the Schrödinger equation. It states that exact solutions to eqn. (1) have the property that the average energy

$$W_n = \langle \Psi_n \mid \mathscr{H}_t \mid \Psi_n \rangle / \langle \Psi_n \mid \Psi_n \rangle \tag{10}$$

is stable with respect to variations in Ψ_n (i.e., $\delta W_n = 0$ for arbitrary $\delta\Psi_n$), and further that the energy functional $\tilde{W}_n$ corresponding to any arbitrary but mathematically "well-behaved" trial function $\tilde{\Psi}_n$ will provide an upper bound to the true energy; that is,

$$\tilde{W}_n = \langle \tilde{\Psi}_n \mid \mathscr{H}_t \mid \tilde{\Psi}_n \rangle / \langle \tilde{\Psi}_n \mid \tilde{\Psi}_n \rangle \geq W_n \tag{11}$$

By a suitable choice of the trial function $\tilde{\Psi}_n$, we can evidently approach the exact limits of W_n and Ψ_n as closely as we wish.*

Let us now use this principle to obtain the best possible solution to eqn. (1), with $\tilde{\Psi}_{n\nu}(\mathbf{r}; \mathbf{R})$ restricted to the form given in eqn. (9). The electronic Schrödinger equation (6) is assumed to have been solved exactly (see Section 3), and both its solution $\Phi_n(\mathbf{r}; \mathbf{R})$ and the nuclear trial function

* The conditions on the general validity of eqn. (11) are discussed in various places, such as Shull and Löwdin.[977]

$\tilde{\chi}_{n\nu}(\mathbf{R})$ are taken, for simplicity, to be real and normalized. Applying the calculus of variations to eqn. (11), and using eqns. (2) and (7), we find

$$\int d\mathbf{R}\,(\delta\tilde{\chi}_{n\nu})\Big[\int d\mathbf{r}\,\Phi_n\mathscr{H}_e\Phi_n\Big]\tilde{\chi}_{n\nu} + \int d\mathbf{R}\,(\delta\tilde{\chi}_{n\nu})\Big[\int d\mathbf{r}\,\Phi_n T_N(\Phi_n\tilde{\chi}_{n\nu})\Big]$$
$$= \tilde{W}_{n\nu}\int d\mathbf{R}\,(\delta\tilde{\chi}_{n\nu})\tilde{\chi}_{n\nu} \tag{12}$$

We now observe that the integral over electronic coordinates in the first term on the l.h.s. is just $E_n(\mathbf{R})$ by virtue of eqn. (6), and that the corresponding integral in the second term reduces to

$$\int d\mathbf{r}\,\Phi_n T_N(\Phi_n\tilde{\chi}_{n\nu}) = (T_N + \int d\mathbf{r}\,\Phi_n T_N \Phi_n)\tilde{\chi}_{n\nu} \tag{13}$$

with integrals of the type $\int d\mathbf{r}\,\Phi_n\,\nabla\Phi_n$ vanishing because of the identity

$$\nabla\int\Phi_n{}^2\,d\mathbf{r} = 2\int\Phi_n\,\nabla\Phi_n\,d\mathbf{r} = 0 \tag{14}$$

Combining eqns. (12) and (13), we are led to

$$\int d\mathbf{R}\,(\delta\tilde{\chi}_{n\nu})[T_N + E_n(\mathbf{R}) + \langle\Phi_n \mid T_N \mid \Phi_n\rangle - \tilde{W}_{n\nu}]\tilde{\chi}_{n\nu} = 0 \tag{15}$$

and to the corresponding eigenvalue equation

$$[T_N + E_n(\mathbf{R}) + \langle\Phi_n \mid T_N \mid \Phi_n\rangle]\tilde{\chi}_{n\nu}(\mathbf{R}) = \tilde{W}_{n\nu}\tilde{\chi}_{n\nu}(\mathbf{R}) \tag{16}$$

for independent variations in $\tilde{\chi}_{n\nu}$. Comparing orders of magnitude, we conclude that the two terms $E_n(\mathbf{R})$ and $\langle\Phi_n \mid T_N \mid \Phi_n\rangle$ appearing in the nuclear potential energy function are generally in the ratio of 1 to λ ($\simeq 0.0005$ for a proton), justifying the heuristic argument leading to eqn. (8). Significant exceptions can occur, however, when the electronic state is degenerate (Jahn–Teller[557a] and Renner[899a] effects).

Improvements in the trial function $\tilde{\Psi}_{n\nu}(\mathbf{r}; \mathbf{R})$ can be incorporated into the analysis by introducing additional electronic states; in fact, a formal expansion of $\Psi_n(\mathbf{r}, \mathbf{R})$ in terms of the complete orthonormal set of states $\{\Phi_m(\mathbf{r}; \mathbf{R})\}$, with nuclear wave functions $\chi_m(\mathbf{R})$ as the expansion coefficients, gives an infinite set of coupled equations

$$[T_N + E_n(\mathbf{R}) + \langle\Phi_n \mid T_N \mid \Phi_n\rangle - W_n]\chi_n(\mathbf{R}) = -\sum_{(m\neq n)}\langle\Phi_n \mid T_N \mid \Phi_m\rangle\chi_m(\mathbf{R}) \tag{17}$$

for the $\chi_n(\mathbf{R})$ without approximation. Use of the assumption

$$\langle\Phi_n \mid T_N \mid \Phi_m\rangle = 0 \tag{18}$$

for all n and m reduces eqn. (17) to eqn. (8), which is sometimes referred to as the Born–Oppenheimer approximation. Although eqns. (6) and (17) taken together constitute an exact description for the internal motion, they are completely equivalent to, and as difficult to solve as, the original Schrödinger equation (1).

Approximate solutions to eqn. (1) in the Born–Oppenheimer approximation are adequate for many purposes and can be determined using techniques such as the variational principle. The way one proceeds operationally is first to obtain solutions to the electronic Schrödinger equation (6) and then to use these for constructing the effective potentials that govern the nuclear motion. Each electronic-state energy surface $\tilde{E}_n(\mathbf{R})$ generates a set of $\tilde{\chi}_{nv}(\mathbf{R})$ that corresponds in the simplest approximation to vibration of the nuclei and to rotation of the entire molecule about its center of mass. Since the vibrational and rotational energies in water can be strongly coupled, it is necessary to include interactions between rotation and vibration. Cross terms of this type are in fact contained in the nuclear kinetic energy operator T_N in eqn. (5), and are treated in Section 4 after we first investigate the form of $E_n(\mathbf{R})$ in Section 3.

3. THE ELECTRONIC MOTION

Within the framework of the Born–Oppenheimer separation of electronic and nuclear motion, our first objective in finding a complete solution to the Schrödinger equation is to determine the electronic eigenvalues and eigenfunctions defined in eqns. (6) and (7). Although exact results do not exist, numerous approximate treatments are available. Most of these make use of the variational principle to approximate the eigenfunctions and to obtain an upper bound on the energy. A common form of trial function that has been applied to different electronic states of water is the Hartree–Fock or self-consistent-field molecular orbital (SCF-MO) function,[918] in which each of the ten electrons is assigned to one-electron molecular spin orbitals (MSO's) ϕ_{ni} that extend over the entire molecule. Since an interchange of coordinates for any two electrons must cause the variational function $\tilde{\Phi}_n(\mathbf{r};\mathbf{R})$ to change sign and thereby satisfy the Pauli exclusion principle, it is convenient to use a *determinantal* product of the form

$$\tilde{\Phi}_n(\mathbf{r};\mathbf{R}) = [1/(10!)^{1/2}]\det[\phi_{n1}(1)\phi_{n2}(2)\cdots\phi_{n,10}(10)] \qquad (19)$$

The inverse $(10!)^{1/2}$ factor guarantees that $\tilde{\Phi}_n(\mathbf{r};\mathbf{R})$ is normalized for orthonormal ϕ_{ni}. The subscripts ni on ϕ_{ni} represent a set of quantum numbers that characterize the spatial distribution, symmetry properties, and spin factor of the function ϕ_{ni} in this approximation to the wave function $\Phi_n(\mathbf{r}; \mathbf{R})$.

If we use $\tilde{\Phi}_n(\mathbf{r}; \mathbf{R})$ in the variational energy functional corresponding to eqn. (11) specialized for the electronic problem at fixed $\mathbf{R}$, we find that the *best* ϕ_i's (dropping the index n for convenience in notation) satisfy the so-called Hartree–Fock equation

$$F(k)\phi_i(k) = \varepsilon_i \phi_i(k) \tag{20}$$

where ε_i is the energy of orbital i, and where $F(k) = H(k) + G(k)$ is the Hartree–Fock operator for electron k. It consists of the Hamiltonian for an electron in the field of the nuclei,

$$H(k) = -\frac{\hbar^2}{2m}\nabla_k^2 - \sum_{K=1}^{3} \frac{Z_K e^2}{|\mathbf{R}_K - \mathbf{r}_k|} \tag{21}$$

and a contribution

$$G(k) = e^2 \sum_{j=1}^{10} \left[\int d\mathbf{r}_l\, \phi_j^*(l)\phi_j(l) \frac{1}{|\mathbf{r}_k - \mathbf{r}_l|} - \phi_j(k) \int d\mathbf{r}_l\, \phi_j^*(l) \frac{P_{kl}}{|\mathbf{r}_k - \mathbf{r}_l|} \right] \tag{22}$$

that represents the average field that electron k experiences due to all of the other electrons. The first term in $G(k)$ corresponds to the Coulomb interaction between electrons; the second term is the exchange potential, which involves a permutation operator P_{kl} that interchanges the coordinates of electrons k and l, and is a consequence of the antisymmetry (Pauli-exclusion) requirement on $\tilde{\Phi}(\mathbf{r}; \mathbf{R})$. The resulting Hartree–Fock energy is given by

$$\tilde{E} = \sum_{i=1}^{10} H_{ii} + \tfrac{1}{2} \sum_{i=1}^{10} \sum_{j=1}^{10} (J_{ij} - K_{ij}) + \tfrac{1}{2} \sum_{K=1}^{3} \sum_{K' \neq K}^{3} \frac{Z_K Z_{K'} e^2}{|\mathbf{R}_K - \mathbf{R}_{K'}|} \tag{23}$$

where the first two terms are the expectation values of $H(k)$ and $G(k)$ in eqns. (21) and (22); that is,

$$H_{ii} = \int \phi_i^*(k) H(k) \phi_i(k)\, d\mathbf{r}_k \tag{24}$$

$$J_{ij} = [\phi_i\phi_i \mid \phi_j\phi_j] = \iint \phi_i^*(k)\phi_i(k)(e^2/|\mathbf{r}_k - \mathbf{r}_l|)\phi_j^*(l)\phi_j(l)\, d\mathbf{r}_k\, d\mathbf{r}_l \tag{25}$$

and

$$K_{ij} = [\phi_i\phi_j \mid \phi_j\phi_i] = \iint \phi_i^*(k)\phi_j(k)(e^2/|\mathbf{r}_k - \mathbf{r}_l|)\phi_j^*(l)\phi_i(l)\, d\mathbf{r}_k\, d\mathbf{r}_l. \tag{26}$$

Using eqns. (20)–(22), we can also write the energy expression as

$$\tilde{E} = \sum_{i=1}^{10} \varepsilon_i - \tfrac{1}{2} \sum_{i=1}^{10} \sum_{j=1}^{10} (J_{ij} - K_{ij}) + \tfrac{1}{2} \sum_{K=1}^{3} \sum_{K' \neq K}^{3} \frac{Z_K Z_{K'} e^2}{|\mathbf{R}_K - \mathbf{R}_{K'}|} \tag{27}$$

where [see eqn. (20)]

$$\varepsilon_i = H_{ii} + \sum_{j=1}^{10} (J_{ij} - K_{ij}) \tag{28}$$

is the average (orbital) energy of each electron including the nuclear field (H_{ii}) and that of the other electrons $[\sum_{j=1}^{10}(J_{ij} - K_{ij})]$.

The essence of the molecular orbital picture for water is that it approximates the solution to the ten-electron Schrödinger equation by solutions to ten coupled one-electron Hartree–Fock equations, which are easier, though by no means trivial, to solve by a self-consistent-field procedure. In most cases, it is assumed that in the ground state the ten electrons doubly occupy the five lowest-energy molecular orbitals (MO's), one electron in each orbital having α spin and the other β spin. A basis set of atomic orbitals (AO's) is introduced and each of the five occupied MO's ϕ_i is expanded in terms of them. Two types of AO's are commonly used in molecular calculations. These are Slater-type orbitals (STO's)

$$\varphi_j \equiv \varphi_{nlm} = N_{nlm} r^{n-1} Y_l^m(\hat{r}) \exp(-\zeta r), \quad n \geq 1; \quad 0 \leq l \leq n-1; \quad -l \leq m \leq l \tag{29}$$

and "primitive" Gaussian-type orbitals (GTO's)

$$\begin{aligned} \varphi_j \equiv \varphi_{nlm} &= N_{nlm} r^{n-1} Y_l^m(\hat{r}) \exp(-\zeta r^2) \\ n &\geq 1; \quad l = n-1, n-3, \ldots \geq 0; \quad -l \leq m \leq l \end{aligned} \tag{30}$$

where r represents the radial coordinate from a suitably chosen origin and $\hat{r}$ is the corresponding unit vector. The functions $Y_l^m(\hat{r})$ are spherical harmonics, and N_{nlm} is a normalizing constant that depends on the quantum numbers n, l, and m and on the effective nuclear charge ζ (orbital exponent). Frequently, linear combinations of the $Y_l^m(\hat{r})$ are taken so that the φ_j's are all real. The difficulty of evaluating the integrals that appear in the Hartree–Fock equation will depend on where the AO's are placed and whether STO's or GTO's are used. For example, the AO's can be centered on each nucleus, on only one nucleus (one-center wave function), or at other points in space (e.g., Gaussian-lobe wave function). In each case, the expansion of MO's in terms of AO's is of the form

$$\phi_i = \sum_{j=1}^{[N]} \eta_{ij}\varphi_j \tag{31}$$

where the φ_j can differ from each other in their coordinate origin and in the choice of n, l, m, and ζ. When this expansion is substituted into the Hartree–Fock equation (20), a set of matrix equations is obtained for the coefficients η_{ij}, which are usually determined by iterating in a self-consistent fashion from estimated starting values. In principle, exact solutions to the Hartree–Fock equations result in the limit of a complete basis set ($N \to \infty$). As a practical matter, only a finite number of AO's is used, so that their careful selection is an essential part of any calculation. It is also important to emphasize that neglect or inaccurate computation of even a single integral that is required for the evaluation of eqn. (11) may invalidate the bounding feature that allows one to approach the exact solution from above. Although selection of the basis set and integral evaluation were once considered serious "bottlenecks" in applying quantum mechanics to polyatomic species, the water molecule has been investigated by detailed calculations that used accurate values for the many multicenter integrals and that achieved good control over the AO basis set representation of the MO's.

At the Hartree–Fock limit, which has been approached very closely for the ground state, the total energy and certain other one-electron properties are given to within a few per cent of their observed values (see Section 3.1). Any residual error is due to the constraint that each electron sees only the average rather than the instantaneous field of the other electrons and must be accounted for by correlating the interelectronic motion.

A fully correlated wave function for the water molecule can be obtained, in principle, by diagonalizing the ten-electron Hamiltonian [eqn. (7)] over some complete set of ten-electron functions. One such set consists of all determinants that can be constructed from the complete set of Hartree–Fock MSO's that satisfy eqn. (20). The most important contribution in such an expansion is made by the Hartree–Fock solution $\tilde{\Phi}_n$ [eqn. (19)]; additional configurations (i.e., determinants) are built up from all possible single, double, triple, quadruple, etc., substitutions of "unoccupied" MSO's for the spin orbitals $\phi_{n1}\phi_{n2}\cdots\phi_{n,10}$ that are "occupied" in $\tilde{\Phi}_n$. We can therefore write the expansion of the exact electronic wave function Φ_n in the form

$$\Phi_n = c_{nn}\tilde{\Phi}_n + \sum_{m \neq n}^{\infty} c_{nm}\tilde{\Phi}_m \tag{32}$$

with $c_{nn} \sim 1$ if $\langle \Phi_n \mid \Phi_n \rangle = \langle \tilde{\Phi}_n \mid \tilde{\Phi}_n \rangle = 1$. Constructing the energy functional in eqn. (11), treating the linear coefficients c_{nm} as parameters, and applying the variational principle to them, we obtain the set of homogeneous linear equations

$$\sum_{m=0}^{\infty} c_{nm}[\langle\tilde{\Phi}_n \mid \mathscr{H}_e \mid \tilde{\Phi}_m\rangle - E_n\,\delta_{mn}] = 0 \tag{33}$$

This equation produces successive sets of eigenvectors $\{c_{nm}\}$ and associated energy eigenvalues E_n, the lowest root ($n = 0$) representing the ground state and successive higher roots ($n > 0$) the excited states.

This method of treating electron correlation is called configuration interaction (CI); it consists in a superposition of configurations constructed from one-electron orbitals optimized in the single-determinant approximation. Unfortunately, its apparent simplicity is deceptive, largely because slow convergence requires the calculation of matrices of very large dimension (the current record[968]* is $12{,}077\times 12{,}077$) and of a special character (a dominant main diagonal and otherwise very sparse). Advances in computer technology and in data manipulation, however, provide some hope that techniques can be developed to choose excited configurations more judiciously. A promising variant of conventional CI, called the multiconfiguration self-consistent field (MC-SCF) method,[233] involves the use of a few selected initial configurations in which all of the orbitals, including the unoccupied ones, are variationally determined by solving Hartree–Fock-type equations within the CI framework. Alternatively, the interelectronic coordinates, such as powers of $|\,\mathbf{r}_k - \mathbf{r}_l\,|$, can be introduced into the wave function as in the classic James–Coolidge calculation on H_2. Although these three methods of approaching the exact solution to the electronic Schrödinger equation (6)—the CI method, the MC-SCF method, and the directly correlated method—have not been fully implemented for water (even for the ground electronic state), the progress that has been made is described in Section 3.1.

Certain other extensions beyond the Hartree–Fock limit, such as the separated-pair and the so-called "GI" limits, include some correlation and have been approached more closely. A separated pair function[544] incorporates correlation by means of an appropriately antisymmetrized product of "geminals", which are two-electron analogs of one-electron MO's. In a molecule like water, where geminals correspond naturally to the inner shell, the two OH bonds, and the two lone pairs, a model of this type might be expected to provide a useful and accurate approximation to the exact

* The calculation was performed on the ground state of ethylene.

solution of the electronic Schrödinger equation. The "GI" model, due to Goddard,[398–401,647] is based on projecting from a simple product of spatial orbitals,* such as those forming the diagonal elements of the determinant in eqn. (19), an eigenfunction of S^2 as specified by an appropriate Young tableau. The projected function is then determined by the variational principle to obtain the "best" spatial orbitals subject to the constraints corresponding to the choice of the particular Young tableau. There are two choices which are of particular interest. One generates a type of variational valence-bond wave function (G1 function) in which some electron correlation is included by associating slightly different spatial orbitals ϕ_i with the two electrons in each pair.[699–700,976] The other choice produces a function (the GF function) which is equivalent to letting all the MSO's for the ten electrons be different, projecting out the component of the total wave function with the desired spin properties, and then determining the MSO's by applying the variational principle. The optimum mixture[566a,647] of G1 and GF with the other possible GI spin functions is referred to as the SOGI function; in valence-bond language, SOGI corresponds to the superposition of all possible canonical structures which can be obtained from a particular set of orbitals.

3.1. The Ground Electronic State of Water

We shall now describe the extent to which the various theoretical techniques have been implemented for the determination of the structure of water in its lowest electronic state. Using the symmetry designations of the appropriate C_{2v} point group, one finds that the ground-electron configuration in the single-determinant MO description is

$$(1a_1)^2(2a_1)^2(1b_2)^2(3a_1)^2(1b_1)^2$$

that is, there are five doubly occupied MO's, three belonging to the totally symmetric representation A_1, one belonging to B_1 which changes sign upon reflection through a plane containing the molecule, and one belonging to B_2 which changes sign on reflection in the plane bisecting the HOH angle. On the basis of qualitative arguments,† the arrangement of atoms

* Properties of spatial eigenfunctions are discussed by Kotani *et al.*,[631] Matsen,[736] Harris,[466] and Hameed *et al.*[452]

† A review of Walsh's rules can be found in Herzberg.[500] For more recent developments, see Peyerimhoff *et al.*[837] and Kaufmann *et al.*[578]

in such a state is expected to be nonlinear and symmetric (i.e., equal OH distances) in the equilibrium configuration.

The early theoretical work (see the bibliography in Appendix A) on the ground state of water attempted to approximate the occupied MO's using a "minimal" set of STO's as the functions φ_j in eqn. (29), with orbital exponents chosen by applying Slater's rules. This means that the φ_j are to be identified with the STO's $1s$ ($\zeta = 7.7$), $2s$ ($\zeta = 2.275$), $2p_x, 2p_y, 2p_z$ ($\zeta = 2.275$) on oxygen and $1s$ ($\zeta = 1.0$) on each of the hydrogen atoms. The first SCF-MO calculations of this type were performed by Ellison and Shull[307] in a widely quoted paper published in 1955. Their work, carried out without access to a high-speed digital computer, contained some rather large errors* in the individual integral values which have only recently been fully corrected in the literature by Pitzer and Merrifield.[(849,850)] The corrected wave function, rather crude by present standards, predicts a minimum energy of $-75.6568e^2/a_0$ at an H—O—H angle θ of 100°, with the OH bond distance R fixed at $1.8103a_0 = 0.958$ Å. Variation of R and θ, and of the orbital exponents as a function of both, leads[(849,850,1040a)] to a somewhat lower energy minimum of $-75.7055e^2/a_0$ at $R_e = 1.8711a_0 = 0.990$ Å and $\theta_e = 100.3°$, to be compared with the experimental values[(63)] of $-76.431e^2/a_0$, $1.8089a_0 = 0.9572$ Å, and 104.52°. Binding energies, harmonic force constants, and other one-electron properties corresponding to this best minimal-basis-set SCF function have also been determined by Pitzer and Merrifield.[(849,850)]

Since 1955, a majority of the theoretical investigations on water have been directed toward using an expanded basis set in order to approach the exact Hartree–Fock solution for the ground state. A list of pertinent references can be found in Appendix A. To demonstrate in a quantitative fashion that basis sets of increasing size, both of the STO and GTO types, produce charge distributions that converge to the Hartree–Fock limit, we compare a series of calculations in which the same molecular properties have been evaluated at the same geometry. All of the wave functions in Table I have been calculated with $R = 1.8111a_0$ and $\theta = 104°27'$. This is the experimental geometry[(498)] frequently adopted in *ab initio* calculations. More recent and accurate values are given by Benedict *et al.*[(63)] Six of

* It is important to recognize that some of the theoretical work on water is vitiated because it depends, in one way or another, on the Ellison and Shull results (see Refs. 849 and 850). The studies affected include numerous textbooks and literature references such as Ellison,[(306)] Banyard and March,[(47,48)] Das and Ghose,[(234)] Bersohn,[(74)] McWeeny and Ohno,[(719)] Spiess *et al.*,[(1008)] and Coulson and Deb.[(213)]

the nine functions listed there are constructed from STO's and the remaining three from GTO's. The parenthesis notation $(ns, mp, \ldots / n's, m'p, \ldots)$ specifies that $n, m, \ldots$ and $n', m', \ldots$ s-like, p-like, ... AO's centered on oxygen and each hydrogen, respectively, are used to construct the MO's. In the case of Gaussians, a square bracket $[\cdots/\cdots]$ denotes that "contracted" orbitals are employed. These may be regarded as suitable linear combinations of "primitive" GTO's which are treated as a group; that is, the coefficients of the individual "primitive" GTO's are not varied in the calculation, only that of the group as a whole. The total energies $\tilde{E}_0$ corresponding to these nine basis sets are ordered across the table beginning with the highest, and proceeding to the lowest, energy. Since the exact Hartree–Fock energy is estimated* to be $\tilde{E}_0 = -76.066 \pm 0.002e^2/a_0$, we can see that the best Gaussian function, a carefully chosen $[6s5p2d/3s1p]$ set contracted from a $(11s7p2d/5s1p)$ primitive set,* approaches this idealized limit very closely. The best Slater set $(5s4p1d/3s1p)$ approaches it even more closely.† The least accurate wave function, composed of a $(2s1p/1s)$ "minimal" basis set, has an energy more than 11 eV above the best function. Intermediate energy points between these two extremes are reached by optimization of the orbital exponents and/or by the addition of more basis functions. For example, the $(2s1p/1s)$ STO-optimized set[36] is composed of a minimal set of orbitals with exponents optimized for the molecule ($\zeta_{1s} = 7.66$, $\zeta_{2s} = 2.25$, $\zeta_{2p} = 2.21$ for oxygen and $\zeta_{1s} = 1.27$ for hydrogen); addition of a second $2p$ set of orbitals and full reoptimization produces the next function,[36] a $(2s2p/1s)$ STO basis in which the oxygen exponents become $\zeta_{1s} = 7.65$, $\zeta_{2s} = 2.26$, $\zeta_{2p} = 1.56$, $\zeta_{2p'} = 3.60$ and the hydrogen exponent $\zeta_{1s} = 1.50$. One proceeds in this manner to build up the basis set so that the electrons are ultimately given complete flexibility to adapt to their molecular environment in the Hartree–Fock model.

In addition to the total energy $\tilde{E}_0$ and the orbital energies ε_i of the five doubly occupied orbitals, additional checks on the convergence of the wave function can be obtained by computing expectation values of various operators. The properties that are included in Table I are:

(a) Molecular electric quadrupole moment at the center of mass:

$$\Theta = \tfrac{1}{2}e \sum_{K=1}^{3} Z_K(3\mathbf{R}_K\mathbf{R}_K - R_K{}^2\mathbf{1}) - \tfrac{1}{2}e\langle\tilde{\Phi}_0 | \sum_{k=1}^{10} (3\mathbf{r}_k\mathbf{r}_k - r_k{}^2\mathbf{1}) | \tilde{\Phi}_0\rangle \quad (34)$$

* T. H. Dunning, Jr., private communication.

† R. M. Pitzer, private communication.

TABLE I. Selected Properties of the Ground Electronic State Charge Distribution of the Water Molecule Obtained in the Single-Determinant Approximation[a]

Property	(2s1p/1s) STO-Slater	(2s1p/1s) STO-optimized	(2s2p/1s) STO	(3s3p1d/2s1p) STO	(4s3p/2s) STO	(9s5p2d/4s1p) [4s3p2d/2s1p] GTO	(10s6p2d/5s1p) [5s4p2d/3s1p] GTO	(11s7p2d/5s1p) [6s5p2d/3s1p] GTO	(5s4p1d/3s1p) STO	Experimental values[b]
$\tilde{E}_0$, e^2/a_0	−75.6560	−75.7033	−75.9693	−76.0048	−76.0200	−76.0504	−76.0596	−76.0621	−76.0631	−76.431[c]
$\varepsilon_1(1a_1)$, e^2/a_0	−20.5055	−20.5560	−20.5421	−20.5655	−20.5668	−20.5563	−20.5607	−20.5627	−−20.5608	—
$\varepsilon_2(2a_1)$, e^2/a_0	−1.2987	−1.2850	−1.3534	−1.3393	−1.3671	−1.3443	−1.3494	−1.3513	−1.3498	—
$\varepsilon_3(1b_2)$, e^2/a_0	−0.6377	−0.6243	−0.7099	−0.7283	−0.7218	−0.7099	−0.7146	−0.7166	−0.7155	—
$\varepsilon_4(3a_1)$, e^2/a_0	−0.4728	−0.4660	−0.5638	−0.5950	−0.5724	−0.5775	−0.5818	−0.5836	−0.5830	—
$\varepsilon_5(1b_1)$, e^2/a_0	−0.4257	−0.4026	−0.5077	−0.5211	−0.5105	−0.5027	−0.5061	−0.5077	−0.5067	—
Θ_{xx}, 10^{-26} esu cm^2	0.200	−0.007	−0.123	−0.052	−0.221	−0.080	−0.084	−0.080	−0.070	−0.13±0.03[d]
Θ_{yy}, 10^{-26} esu cm^2	−1.020	−1.485	−2.307	−2.586	−2.449	−2.435	−2.497	−2.515	−2.542	−2.50±0.02[d]
Θ_{zz}, 10^{-26} esu cm^2	0.820	1.492	2.430	2.639	2.670	2.515	2.581	2.595	2.612	2.63±0.02[d]
μ, D ($H_2^+O^-$)	1.435	1.921	2.827	2.035	2.775	2.043	2.075	2.084	2.054	1.884±0.012[e]
σ_D^d, ppm	101.0	103.6	101.7	102.0	101.8	102.5	102.4	102.4	102.4	102.4[d]
σ_O^d, ppm	417.7	415.3	415.8	415.0	416.0	416.1	416.0	416.0	416.1	—
$\mathscr{F}_{H1}(z)$, mdyn	1.03	0.593	0.904	0.013	0.796	−0.033	−0.030	−0.023	−0.069	0

mdyn	−0.442	−0.135	−0.292	0.029	−0.289	0.050	0.054	0.050	0.068	0
$\mathscr{F}_O(x)$, mdyn	25.39	23.65	18.07	4.63	10.62	4.06	2.29	1.24	−0.474	0
$\Sigma_N \mathscr{F}_N(x)$, mdyn	24.51	23.38	17.49	4.69	10.04	4.16	2.40	1.34	−0.339	0
$(eq_{yy}Q/h)_D$, kHz	−204.6	−193.5	−214.7	−204.0	−203.6	−187.9	−185.0	−185.1	−192.2	−174.78±0.20[f]
$(eq_{\perp}Q/h)_D$, kHz	−159.8	−151.1	−175.4	−158.6	−161.3	−142.3	−140.0	−140.2	−144.6	−133.13±0.14[f]
$(eq_{\parallel}Q/h)_D$, kHz	364.4	344.6	390.1	362.6	364.8	330.2	325.0	325.3	336.8	307.95±0.14[f]
φ	2°8′	2°22′	2°44′	0°56′	2°37′	0°49′	0°51′	0°51′	0°46′	−1°16′[f]
$(eq_{xx}Q/h)_O$, MHz	−1.51	−1.66	−1.20	−0.87	−1.07	−1.11	−1.12	−1.12	−1.12	−1.28±0.09[f]
$(eq_{yy}Q/h)_O$, MHz	17.47	14.76	11.82	10.35	11.63	10.84	10.65	10.63	10.77	10.17±0.07[f]
$(eq_{zz}Q/h)_O$, MHz	−15.96	−13.10	−10.62	−9.48	−10.56	−9.73	−9.53	−9.51	−9.65	−8.89±0.03[f]

[a] The STO and GTO calculations cited here were preformed by Pitzer and co-workers[36] and by T. H. Dunning, Jr. (Refs. 281a, b, c, and private communication), respectively, at a geometry of $R = 1.8111a_0$ and $\theta = 104° 27'$. See eqns. (34)–(38) of the text for definitions of molecular properties.

[b] All the experimental values refer to H_2O except the quadrupole coupling constants and $\sigma_D{}^d$, which correspond to $HD^{17}O$ and D_2O, respectively.

[c] Nonrelativistic energy excluding zero-point vibrations. The Hartree–Fock energy is estimated to be -76.066 ± 0.002 au (T. H. Dunning, Jr., private communication).

[d] Reference 1106. See also H. Taft and B. P. Dailey, *J. Chem. Phys.* **51**, 1002 (1969) for different but apparently less accurate experimental values of θ. The theoretical values of θ are referred to the observed H_2O center of mass shown in Fig. 1.

[e] Reference 683.

[f] References 1105 and 1107. The principal deuteron field gradients $q_{\perp}$ and $q_{\parallel}$ are directed approximately perpendicular and parallel to the OD bond line but rotated at an angle of φ to it, such that $q_{\parallel}$ crosses the OH bond line. Theoretical values based on $Q_D = 0.002796$ b and $Q_o = -0.024$ b. A somewhat larger value for Q_o ($\cong -0.026$ b) is reported by H. P. Kelly, *Phys. Rev.* **180**, 55 (1969) and by H. F. Schaefer, R. A. Klemm, and F. E. Harris, *Phys. Rev.* **176**, 49 (1968).

(b) Molecular electric dipole moment:

$$\boldsymbol{\mu} = e \sum_{K=1}^{3} Z_K \mathbf{R}_K - e\langle \tilde{\Phi}_0 | \sum_{k=1}^{10} \mathbf{r}_k | \tilde{\Phi}_0 \rangle \tag{35}$$

(c) Average diamagnetic shielding constant:

$$\sigma_N{}^d = (e^2/3mc^2)\langle \tilde{\Phi}_0 | \sum_{k=1}^{10} (1/r_{Nk}) | \tilde{\Phi}_0 \rangle \tag{36}$$

(d) Hellmann–Feynman force:

$$\mathscr{F}_N = -Z_N e^2 \left[\sum_{K \neq N}^{3} (Z_K \mathbf{R}_{NK}/R_{NK}^3) - \langle \tilde{\Phi}_0 | \sum_{k=1}^{10} (\mathbf{r}_{Nk}/r_{Nk}^3) | \tilde{\Phi}_0 \rangle \right] \tag{37}$$

(e) Nuclear quadrupole coupling constant:

$$(e\mathsf{q}_N Q/h) = eQ \left\{ \sum_{K \neq N}^{3} [Z_K(3\mathbf{R}_{NK}\mathbf{R}_{NK} - R_{NK}^2 \mathsf{1})/R_{NK}^5] \right.$$
$$\left. - \langle \tilde{\Phi}_0 | \sum_{k=1}^{10} [(3\mathbf{r}_{Nk}\mathbf{r}_{Nk} - r_{Nk}^2 \mathsf{1})/r_{Nk}^5] | \tilde{\Phi}_0 \rangle \right\} \tag{38}$$

The coordinate system is defined in Fig. 1. Each successive property (a)–(e) depends on smaller powers of the nuclear and electron position vectors $\mathbf{R}_K$ and $\mathbf{r}_k$. These five properties, taken together, therefore sample the charge density over different regions and provide complementary measures of the accuracy of each wave function.

Except for the Hellmann–Feynman force, the *absolute* accuracy of the various properties can only be assessed by comparison with experiment. That the force constitutes an exception to this can be seen in the following way.[599] In the Born–Oppenheimer approximation, the force on one of the three nuclei (say N) in water can be obtained from the potential energy surface for the nuclear motion, which we write in the form

$$E = E(\mathbf{R}_{12}, \mathbf{R}_{13}, \mathbf{R}_{23}) \tag{39}$$

where $(\mathbf{R}_{12}, \mathbf{R}_{13}, \mathbf{R}_{23})$ represents the set of interatomic position vectors for the nuclei 1, 2, 3. Since the force on nucleus N is given by

$$\mathbf{F}_N = -\boldsymbol{\nabla}_N E \tag{40}$$

we consider the quantity

$$\sum_{N=1}^{3} \mathbf{F}_N = -\sum_{N=1}^{3} \sum_{M \neq N}^{3} (\partial E/\partial R_{NM}) \boldsymbol{\nabla}_N R_{NM} \tag{41}$$

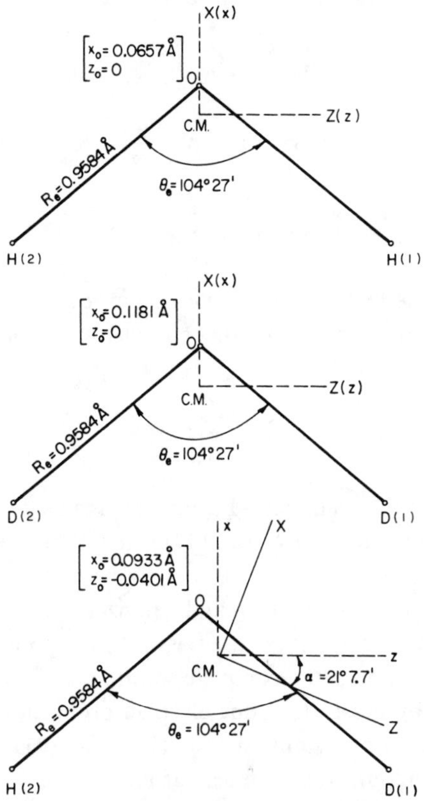

Fig. 1. Equilibrium geometry and principal axis system for H_2O, D_2O, and HDO.

where R_{NM} is the distance between nuclei N and M. Summing eqn. (41) over N and $M(N \neq M)$ and making use of the fact that

$$\nabla_N R_{NM} = -\nabla_M R_{NM} \tag{42}$$

we obtain

$$\sum_{N=1}^{3} \mathbf{F}_N = \mathbf{0} \tag{43}$$

that is, the sum of the forces on all the nuclei is equal to zero. Although eqn. (43) may be regarded as "intuitively obvious," it is of some interest because it leads to a simple, purely theoretical *necessary condition* that an *exact* Hartree–Fock function or an exact wave function must satisfy at *every nuclear configuration*. Writing $\mathbf{F}_N$ as the sum of the two types of terms that contribute to the energy derivatives, we have for the wave function Φ

$$\mathbf{F}_N = \mathscr{F}_N + \Delta_N \tag{44}$$

where

$$\mathscr{F}_N = -\langle \Phi \mid \boldsymbol{\nabla}_N \mathscr{H}_e \mid \Phi \rangle \tag{45}$$

and

$$\Delta_N = -\langle \boldsymbol{\nabla}_N \Phi \mid \mathscr{H}_e \mid \Phi \rangle - \langle \Phi \mid \mathscr{H}_e \mid \boldsymbol{\nabla}_N \Phi \rangle \tag{46}$$

It can be demonstrated[599] that for either the exact Hartree–Fock function or the exact wave function (e.g., the complete CI solution), eqn. (43) reduces to the expression

$$\sum_{N=1}^{3} \mathscr{F}_N = \mathbf{0} \tag{47}$$

Here, $\mathscr{F}_N$ represents the Hellmann–Feynman force as defined in eqn. (37). For approximate wave functions, eqn. (47) can still be used to estimate the actual force sum, but the equality will not be necessarily satisfied because the contribution from the term Δ_N may be nonzero. Thus, deviations from eqn. (47) mean that the function used is only an approximation to the Hartree–Fock function or to the complete solution. This condition contrasts with that provided by most other one-electron properties, for which it has not been possible to obtain any simple Hartree–Fock or configuration-interaction limits on purely theoretical grounds.

Referring again to Table I, we see that the error in most of the calculated properties can be large until the Hartree–Fock solution is approached rather closely. This is illustrated in Fig. 2, where the Hellmann–Feynman force sum in eqn. (47) is plotted as a function of the improvement in the total energy. The force is apparently a sensitive indicator of wave-function quality. For the other properties the available experimental data are given for comparison in the last column of Table I. Agreement is quite good for most properties, although even the best function shows significant errors; for example, the calculated dipole moment is about 9% too large. Some of the deviations arise in part from the fact that the experimental quantities correspond to averages over the molecular vibrations (see Section 4). However, there are significant contributions to these properties from the effects of electron correlation which are not included in the Hartree–Fock model. Also, it should be noted that certain of the calculated properties (e.g., the dipole moment) do not appear to converge in any simple manner with improvement in the total energy. This demonstrates that some caution

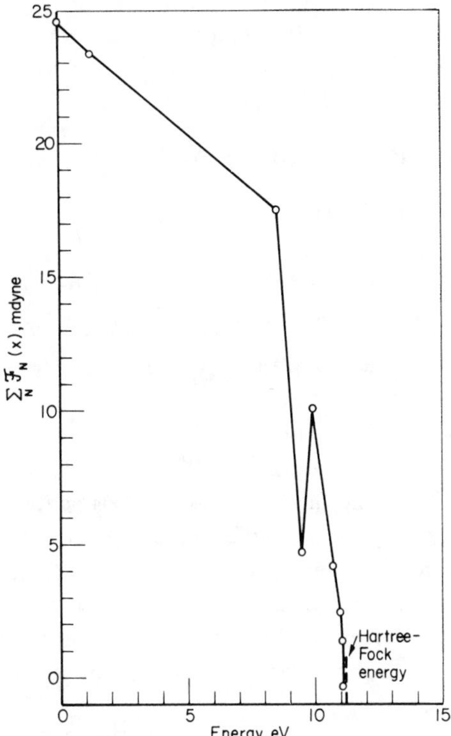

Fig. 2. Plot of the Hellmann–Feynman force sum $\Sigma_N \mathscr{F}_N(x)$ in eqn. (47) vs. energy improvement as the Hartree–Fock limit is approached. The data are obtained from Table I.

is necessary when one speaks of improving a wave function; it is generally necessary to specify in what way it is being improved.

An investigation of the Hartree–Fock potential energy surface $\tilde{E}_0(\mathbf{R})$ has also been performed in the neighborhood of the energy minimum with the [4s3p2d/2s1p] GTO basis cited in Table I. Using a judiciously chosen set of 30 symmetric (equal OH bond lengths) and 15 asymmetric (unequal OH bond lengths) configurations of nuclei, it is found[315] that the theoretical minimum in the energy occurs at $R_e = 1.779a_0$ and $\theta_e = 106.1°$ and that the shape of the energy surface can be represented by a Taylor series expansion of the form

$$\tilde{E}_0(\mathbf{R}) = \tilde{E}_0(\mathbf{R}_e) + \tilde{E}_0(\mathbf{R}'^2) + \tilde{E}_0(\mathbf{R}'^3) + \tilde{E}_0(\mathbf{R}'^4) + \cdots \tag{48a}$$

Here $\mathbf{R}'$ is a displacement coordinate referring collectively to the two bond lengths $R_1' = R_1 - R_e$, $R_2' = R_2 - R_e$, and to the bond angle $\theta' = \theta - \theta_e$.

The first term $\tilde{E}_0(\mathbf{R}_e) = -76.0510e^2/a_0$ gives the energy at equilibrium, the next term (not shown) $\tilde{E}_0(\mathbf{R}')$ must vanish by the definition of equilibrium, the harmonic and cubic terms are

$$\tilde{E}_0(\mathbf{R}'^2) = \tfrac{1}{2}K_r(R_1'^2 + R_2'^2) + \tfrac{1}{2}K_\theta R_e{}^2\theta'^2 + K_r'R_1'R_2' + K_{r\theta}R_e\theta'(R_1' + R_2') \tag{48b}$$

and

$$\begin{aligned}\tilde{E}_0(\mathbf{R}'^3) = (1/R_e)[&K_{rrr}(R_1'^3 + R_2'^3) + K_{rrr}'(R_1' + R_2')R_1'R_2' \\ &+ K_{rr\theta}(R_1'^2 + R_2'^2)R_e\theta' + K_{rr\theta}'R_eR_1'R_2'\theta' \\ &+ K_{r\theta\theta}(R_1' + R_2')R_e{}^2\theta'^2 + K_{\theta\theta\theta}R_e{}^3\theta'^3]\end{aligned} \tag{48c}$$

respectively, and so on for the higher terms. The computed force constants are compared with the experimental data[(641)] in Table II. Taking into account the uncertainties in both the theoretical and experimental constants, we see that the changes in energy near equilibrium are predicted rather well at this level of approximation. However, for displacements far from equilibrium, the Hartree–Fock energy is not adequate because the wave function does not dissociate to the correct limits. Thus, to obtain a complete potential surface, some configuration interaction or alternative refinement in the wave function is essential.

The discussion so far suggests that the Hartree–Fock theory is reasonably satisfactory in accounting for many observed features of the water

TABLE II. Comparisons of Calculated and Observed Force Constants[a]

Force constant	Computed value[b]	Observed value[c]
K_r	9.876 ± 0.001	8.454
K_θ	0.881 ± 0.001	0.761
K_r'	-0.079 ± 0.001	-0.101
$K_{r\theta}$	0.258 ± 0.001	0.228
K_{rrr}	-10.51 ± 0.01	-9.55 ± 0.03
K_{rrr}'	-0.0004 ± 0.004	-0.32 ± 0.08
$K_{rr\theta}$	-0.033 ± 0.006	$+0.16 \pm 0.02$
$K_{rr\theta}'$	-0.53 ± 0.01	-0.66 ± 0.01
$K_{r\theta\theta}$	-0.149 ± 0.002	$+0.15 \pm 0.20$
$K_{\theta\theta\theta}$	-0.145 ± 0.001	-0.14 ± 0.01

[a] All values in units of 10^5 dyn cm^{-1} referred to eqn. (48).
[b] Reference 315. The errors reflect the uncertainty in fitting the computed surface to the function in eqn. (48).
[c] Reference 641.

molecule. An examination of these SCF-MO's reveals, however, that information about chemical bonding in terms of traditional concepts (e.g., bonds, lone-pairs) is masked by the delocalized nature of the MO's themselves. Although such a delocalized description is very useful for some purposes (e.g., definition of symmetry species, discussion of electronic transitions), it can be largely eliminated(286,671) by subjecting the MO's to a transformation such that the intraorbital Coulomb energy J_{ii} is maximized, or equivalently, the interorbital Coulomb J_{ij} and exchange K_{ij} energies are minimized. Liang(682) and Taylor(1056) applied this energy localization criterion to the SCF-MO's expressed in terms of an optimized, minimal set of STO's, with the results shown in Table III, part A. The total energy $\tilde{E}_0$ is invariant to this transformation. The new expansion coefficients with respect to the STO basis demonstrate that the localized MO's (LMO's) may be classified into an oxygen inner-shell (is) orbital (consisting mostly of the $1s$-oxygen STO), two equivalent OH bond orbitals, and two equivalent oxygen lone-pair (lp) orbitals, each accommodating two electrons with opposed spins. The angle between the radial maxima in the two lone-pairs is about 120°. The second part of Table III, part A, relates the LMO's directly to the delocalized SCF-MO's; that is, it gives the linear coefficient of each SCF-MO in each LMO.

The LMO interpretation of the electronic structure of water is consistent with the traditional view of bonding derived from valence theory in which the occupied orbitals are constructed from four normalized, locally orthogonal, hybrid orbitals centered on oxygen, two of which combine to form bond orbitals with the hydrogen $1s$ functions. To make a comparison with the LMO results, let us define these four hybrids such that two of them have their maxima directed along the OH bond lines in the xz plane; that is,

$$h_{\mathrm{OH}}^{\pm} = 0.447 \mid 2s_{\mathrm{O}}^{*}\rangle - 0.548 \mid 2px_{\mathrm{O}}\rangle \pm 0.707 \mid 2pz_{\mathrm{O}}\rangle \qquad (49a)$$

The other two hybrids,

$$h_{\mathrm{lp}}^{\pm} = 0.548 \mid 2s_{\mathrm{O}}^{*}\rangle + 0.447 \mid 2px_{\mathrm{O}}\rangle \pm 0.707 \mid 2py_{\mathrm{O}}\rangle \qquad (49b)$$

are made orthogonal to $h_{\mathrm{OH}}^{\pm}$ by a similar construction in the xy plane. Here $\mid 2px_{\mathrm{O}}\rangle$, $\mid 2py_{\mathrm{O}}\rangle$, and $\mid 2pz_{\mathrm{O}}\rangle$ are the optimized oxygen STO's in the coordinate system of Fig. 1. The hydrogen-like $\mid 2s_{\mathrm{O}}^{*}\rangle$ orbital is orthogonalized to the $\mid 1s_{\mathrm{O}}\rangle$ orbital. Using $h_{\mathrm{OH}}^{\pm}$, $h_{lp}^{\pm}$, and the remaining basis functions $\mid 1s_{\mathrm{O}}\rangle$ and $\mid 1s_{\mathrm{H}}\rangle$, we can now form five doubly occupied orbitals according to the rules of directed valence: (i) the $1s$ oxygen AO is assumed to constitute, by itself, one of the five ϕ's; (ii) two additional ϕ's are identified

TABLE III. Localized Molecular Orbitals for H_2O[a]

Orbital	O, is	OH_1	OH_2	O, lp_1	O, lp_2
A. LMO's[b]			$\tilde{E}_0 = -75.7033e^2/a_0$		
H_1, 1*s*	−0.00668	0.49662	−0.10216	−0.08723	−0.08723
H_2, 1*s*	−0.00668	−0.10216	0.49662	−0.08723	−0.08723
O, 1*s*	1.02022	−0.04126	−0.04126	−0.06059	−0.06059
O, 2*s*	−0.11826	0.25990	0.25990	0.64325	0.64325
O, 2*px*	−0.03299	−0.40663	−0.40663	0.39051	0.39051
O, 2*pz*	0	0.44136	−0.44136	0	0
O, 2*py*	0	0	0	0.70711	−0.70711
$1a_1$	0.99076	0.04503	0.04503	−0.08469	−0.08468
$2a_1$	−0.12239	0.56587	0.56587	−0.41509	−0.41509
$3a_1$	0.05847	0.42162	0.42162	0.56615	0.56615
$1b_2$	0	0.70711	−0.70711	0	0
$1b_1$	0	0	0	0.70711	−0.70711
B. OBO's[c]			$\tilde{E}_0 = -75.6396e^2/a_0$		
H_1, 1*s*	−0.00565	0.47855	−0.02904	−0.01531	−0.01531
H_2, 1*s*	−0.00565	−0.02904	0.47855	−0.01531	−0.01531
O, 1*s*	1.00178	−0.06902	−0.06902	−0.12545	−0.12545
O, 2*s*	−0.00573	0.24685	0.24685	0.54776	0.54776
O, 2*px*	0.00961	−0.36676	−0.36676	0.47290	0.47290
O, 2*pz*	0	0.49234	−0.49234	0	0
O, 2*py*	0	0	0	0.70711	−0.70711
C. NBO's[c]			$\tilde{E}_0 = -75.2104e^2/a_0$		
H_1, 1*s*	0	0.47408	0	0	0
H_2, 1*s*	0	0	0.47408	0	0
O, 1*s*	1.0	−0.06909	−0.06909	−0.13028	−0.13028
O, 2*s*	0	0.29871	0.29871	0.56327	0.56327
O, 2*px*	0	−0.35637	−0.35637	0.44688	0.44688
O, 2*pz*	0	0.45984	−0.45984	0	0
O, 2*py*	0	0	0	0.70711	−0.70711

[a] Based on the STO minimal-basis-set, optimized-exponent wave function in Ref. 36 at $R = 1.8111a_0$, $\theta = 104°27'$.

[b] The LMO's are orthonormal and are obtained by the method of Refs. 682 and 1056.

[c] Ref. 105a.

with $h_{lp}^{\pm}$ and may be regarded as the lone-pair orbitals; (iii) the two OH bond orbitals arise from linear combinations of the form $h_{OH}^{\pm} + \lambda \mid 1s_H \rangle$, where λ is a variable mixing coefficient. These five ϕ's, when properly normalized, can then be used to construct a Hartree type of electrostatic model for the water molecule in which a product function instead of the determinant [eqn. (19)] is employed. We refer to these ϕ's as normalized bond orbitals (NBO's). When the NBO's are orthogonalized (OBO), however, each orbital mixes with every other orbital to a small extent, and this mixing introduces exactly the amount of change that is needed to satisfy the determinantal (Pauli-exclusion) requirement; that is, the charge distribution obtained from summing over the five OBO's is exactly the same as that resulting from using the NBO's in a determinantal function [eqn. (19)] with the proper normalization. For the optimized-exponent, minimal STO basis set, the best value of λ is found[105a] to be 0.729 using the variation principle. The corresponding NBO and orthogonalized bond functions (OBO's) are given in parts C and B of Table III, respectively. Comparing these functions with the full variational self-consistent-field treatment, we see that the NBO's, OBO's, and LMO's are similar in form. This shows that there is relatively little delocalization in the water molecule other than that required to satisfy the Pauli-exclusion principle. Moreover, it is clear that sophisticated *ab initio* calculations support, qualitatively, the usual valence theory description of water in terms of separate inner-shell, lone-pair, and bonding electrons.

A decomposition of the integrated charge distribution $\sum_{i=1}^{10} \int \mid \phi_i \mid^2$ into orbital populations can be obtained by expanding it in terms of AO's and dividing all overlap terms equally among them. The application of this population analysis[773] to the optimized minimal basis-set SCF function described above yields the STO occupation numbers

$$1s_{\mathrm{O}}^{2.00} 2s_{\mathrm{O}}^{1.82} 2px_{\mathrm{O}}^{1.50} 2pz_{\mathrm{O}}^{1.12} 2py_{\mathrm{O}}^{2.00} 1s_{\mathrm{H1}}^{0.78} 1s_{\mathrm{H2}}^{0.78}$$

for the ground state at the calculated equilibrium geometry. Summing over the orbitals on the different atoms, we can assign a total electronic charge of 8.44 to the oxygen atom and 0.78 to each hydrogen, so that oxygen has a net excess negative charge of 0.44 electrons, and each hydrogen an excess positive charge of half that amount. This result is consistent with the usual chemical notion that the bond polarity is in the direction O^-H^+. From these atomic charges, a dipole moment (1.24 D) considerably smaller than the experimental result (1.88 D) is obtained. Although other electrostatic models[161] can be devised which give a more accurate value for the dipole moment, these models yield quadrupole and higher moments that differ drastically from experiment. As far as the "origin" of the dipole moment is concerned, it is found that the OH bonds subtract about twice as much as the lone pairs add to the nuclear contribution when the radius vector is measured from the oxygen atom; that is, $\mu = 5.64$ (nuclear) $+2.99$ (lone pairs) $- 6.72$ (OH bonds) $+ 0.01$ (inner shell) $= 1.92$ D for the LMO function in Table III, part A. The qualitative features of the decomposition are unchanged for the OBO and NBO functions, although their net respective dipole moments of 2.50 and 2.15 D are somewhat larger.

A more detailed analysis of the charge distribution can be made by plotting electron density maps for individual LMO's and for their sum and differences. Figures 3–7 illustrate the flexibility and level of detail which can be extracted from this approach. Figure 3a,b shows the three-dimensional electron density for the molecule as a whole. The contrast between the OH bond and lone-pair distributions can be seen in Figs. 4a,b and 5a,b. Difference maps in Figs. 6 and 7 show, respectively, the change in the OH bond on going from the OBO to the LMO model and the effect on the overall density due to molecular formation (i.e., the difference between the superposed atoms and the charge distribution). It is evident that chemical binding is a complex process in terms of its effects on the electron density.*

The ground-state correlation energy, defined as the difference between the exact nonrelativistic electronic and Hartree–Fock energies, is found to be about 10 eV, or 1 eV per electron, which is a typical order of magnitude for this quantity (see below). Although this correlation correction is only about $\frac{1}{2}\%$ of the total energy of the molecule, it is nearly equal to 100% of the observed binding energy (10.1 eV). The binding energy in the Hartree–Fock approximation is about 7 eV (i.e., the difference in energy between Hartree–Fock calculations for the molecule and for the separated atoms). For chemistry, it is clear that a proper determination of the correlation energy represents a significant part of the problem. Furthermore, as already mentioned above, a Hartree–Fock function for the ground state does not dissociate properly to $O(^3P) + 2H(^2S)$, so that a correct description of adiabatic dissociation also requires the inclusion of some correlation.

The SOGI and the separated-pair (SP) models described above do go beyond the Hartree–Fock limit and the former almost always dissociates correctly to atoms for molecules in their ground states. It is of interest

* An excellent film loop, "Atoms to Molecules" by A. C. Wahl and U. Blukis, is available from the McGraw-Hill Book Co. It shows the formation of H_2O from H and OH in terms of electron density maps.

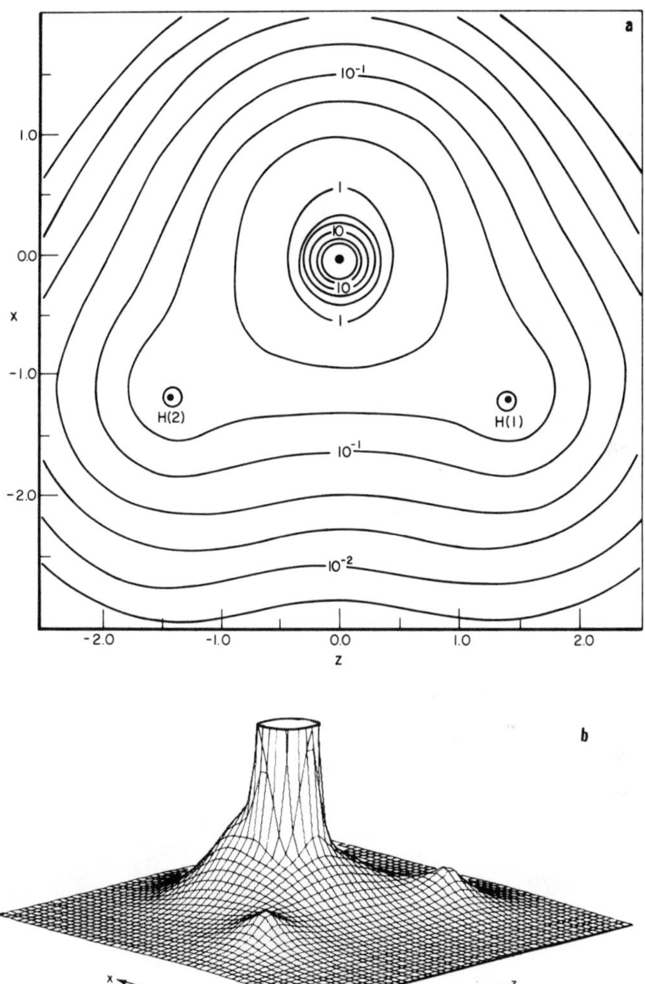

Fig. 3. (a) The total electron density ϱ of the ground state of water in the molecular (xz) plane in units of electrons per $a_0{}^3$. Densities are computed with the optimized, minimal STO basis-set ($2s1p/1s$) function in Table I. The large solid dot in the center represents the oxygen nucleus ($\varrho \simeq 300$) at $(x, y, z) = (0, 0, 0)$ and the smaller circles the protons ($\varrho \simeq 0.5$). The density gradient is $(0.1)^{1/3}$ per contour. Distances along the axes are measured in units of a_0. (b) A three-dimensional plot of the total electron density shown in (a). The density is plotted along the axis perpendicular to the molecular (xz) plane. The viewer is oriented 45° clockwise from the line bisecting the HOH angle and 15° above the plane.

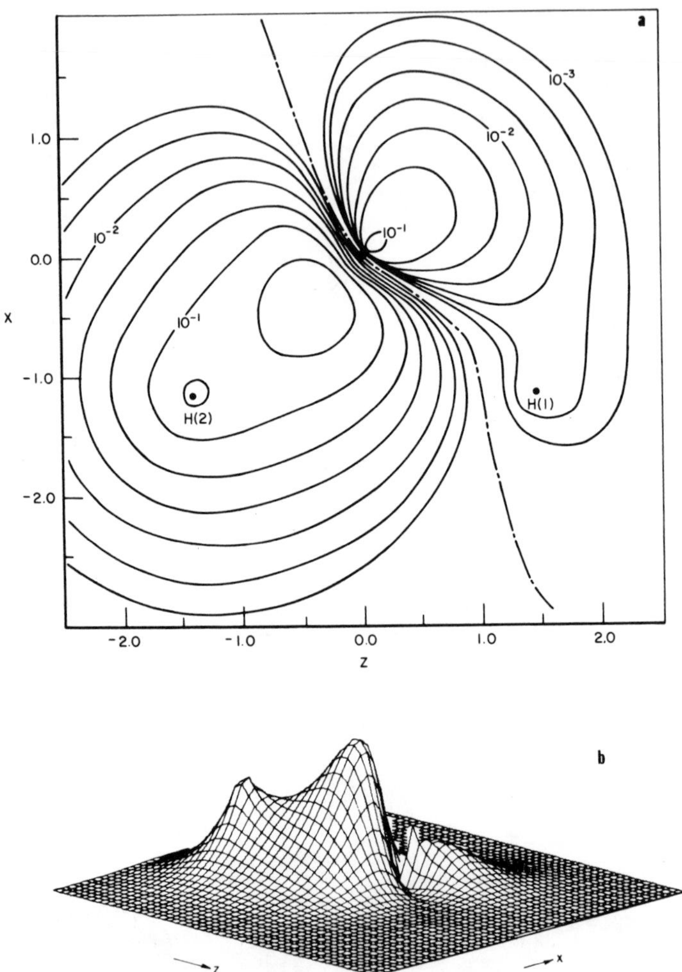

Fig. 4. (a) Density for one electron in the LMO bonding orbital OH_2 of the water molecule for the wave function in Table III, part A. Plot is in the *xz* plane. The densities at oxygen $[(x, z) = (0, 0)]$, hydrogen H_1, and hydrogen H_2 are 0.02, 0.004, and 0.22 electrons/a_0^3, respectively. The dot-dash line represents a node. See caption to Fig. 3a for other units. (b) A three-dimensional plot of the LMO bonding orbital density shown in (a). The density is plotted along the axis perpendicular to the molecular (*xz*) plane. The viewer is oriented 45° counterclockwise from the line bisecting the HOH angle and 15° above the plane. The density scale is the same as in Fig. 5b and five times larger than that in Fig. 3b.

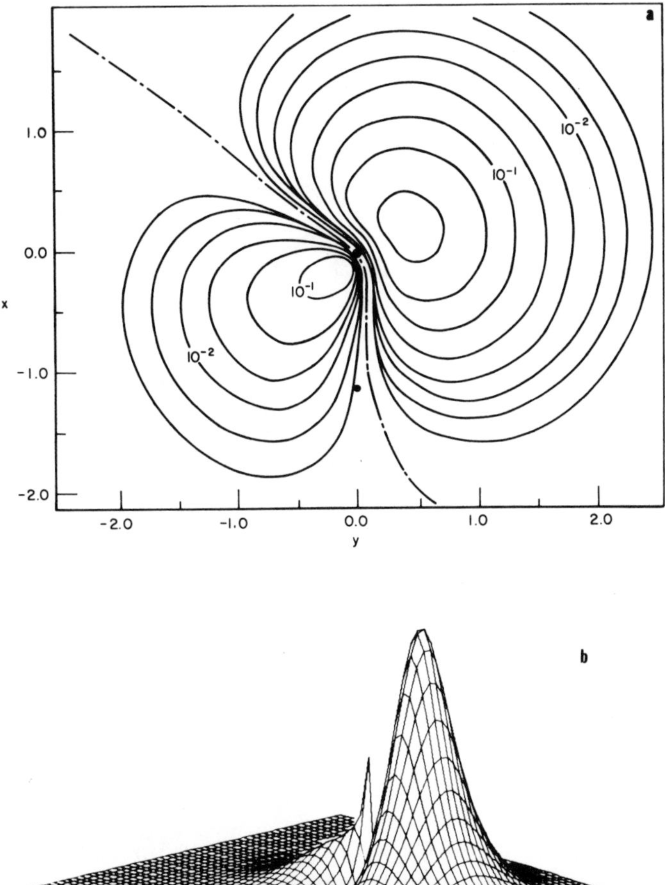

Fig. 5. (a) Density for one electron in the LMO lone-pair orbital lp_1 of the water molecule for the wave function in Table III, part A. Plot is in the xy plane. The densities at oxygen $[(x, y) = (0, 0)]$ and the projected hydrogen atom position (small dot along $y = 0$) are 0.55 and 0.001 electrons/a_0^3, respectively. The dot-dash line represents a node. See caption to Fig. 3a for other units. (b) A three-dimensional plot of the LMO lone-pair orbital density shown in (a). The density is plotted along the axis perpendicular to the xy plane. The viewer is oriented 45° above the molecular xz plane and 15° above the xy plane. The density scale is five times larger than that in Fig. 3b.

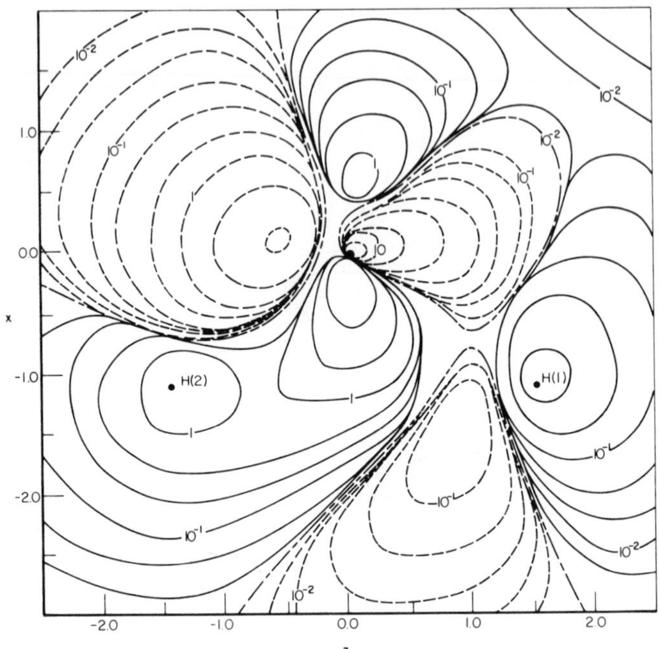

Fig. 6. Density difference map per electron of the LMO minus OBO bonding orbitals OH_2 for the wave function in Table III, parts A and B. Plot is in the *xz* plane. Solid, dash, and dot-dash lines represent positive, negative, and nodal contours, respectively. See caption to Fig. 3a for units.

to know what fraction of the total correlation energy they recover at the equilibrium geometry. Using basis sets that yield energies in the neighborhood of $-76.035e^2/a_0$ in the Hartree–Fock approximation {i.e., approximately midway between the $(4s3p/2s)$ STO and $(5s4p1d/3s1p)$ STO sets in Table I}, Guberman and Goddard[430] and Scarzafava[948] have obtained GF[430], G1[948] with strong orthogonality, and SP[948] wave functions with energies equal to -76.078, -76.110, and $-76.120e^2/a_0$, respectively.* These values, relative to the SCF energy obtained from the same basis, correspond to 12%, 20%, and 22% of the correlation energy; that is, only a small fraction of the Hartree–Fock energy error is accounted for in these calculations. Although the inclusion of additional basis functions would

* Separated-pair calculations have also been performed by W. J. Hunt, P. J. Hay, and W. A. Goddard, III (private communication), and by W. J. Stevens, Ph. D. Thesis, Indiana University (1971).

certainly lead to some improvement, it is impossible from results of this type alone to interpret the correlation energy. Nevertheless, Scarzafava[(948)] has attempted to resolve the correlation correction into interorbital and intraorbital pair contributions by incorporating CI into the SP framework, by making certain assumptions and definitions, and by using data from other sources. He surmises that the oxygen inner-shell pair, the two lone-pairs, and the two bond-pair orbitals contribute about 10% (1 eV), 20% (2 eV), and 22% (2.2 eV), respectively, and that the remaining 48% (4.8 eV) comes from the six types of interorbital interactions. The "extra" correlation energy due solely to molecular formation (i.e., the difference in correlation between the atoms and the molecule) is computed[(519)] to be ~2.5 eV and comes mainly from the two OH bonds. More definitive calculations are needed to test these estimates.

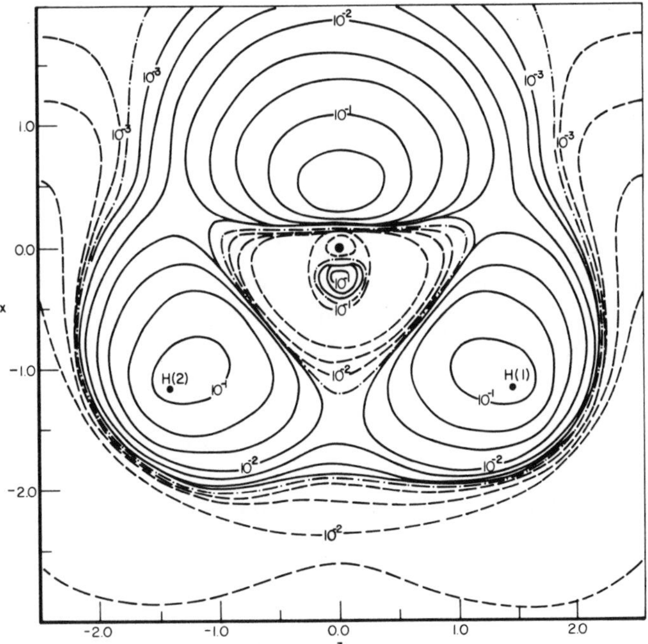

Fig. 7. Electron density difference map of the total molecular minus sum of atomic densities in the *xz* plane. Molecular densities are computed with the optimized, minimal STO basis set (2*s*1*p*/1*s*) in Table I. An optimized (2*s*1*p*) STO function is used for the oxygen atom with the 2*py* orbital doubly occupied. Solid, dash, and dot-dash lines represent positive, negative, and nodal contours, respectively. See caption to Fig. 3 for units.

The most accurate calculations that have been reported to date on the ground electronic state of water are the very recent CI studies by Shavitt and co-workers,[535] in which as many as 6779 configurations were employed with comparable STO and GTO basis sets of the type [$4s2p/2s$], by Schaefer and Bender,[949] who combined fewer configurations (1027) with a more accurate [$4s2p1d/2s1p$] GTO starting function, and by Meyer[745a] who employed a ($11s7p4d1f/5s1p$) GTO basis with about 240 *nonorthogonal* configurations and with an approximate treatment of products of double substitutions. Their best calculations, respectively, yield total energies of -76.148, -76.242, and $-76.383e^2/a_0$, corresponding to 38%, 53%, and 87% of the correlation energy at the equilibrium geometry. Many-body perturbation theory has also been applied to the water molecule; the most accurate study by Miller and Kelly[750a] reports an energy of $-76.48 \pm 0.07e^2/a_0$, which corresponds to $100 \pm 20\%$ of the correlation energy. Analysis of these results should provide additional insight into the effects of electron correlation on the various molecular properties.[28]

3.2. The Excited Electronic States of Water

Although the excited electronic states of water are of great interest for spectroscopy and for a variety of collision processes, much less is known about their properties. In fact, most of the known excited states are probably Rydberg states which arise, in the MO description, from promotions of electrons from the outer $3a_1$ and $1b_1$ MO's into the *ns*-, *np*-, *nd*- like orbitals of the united atom. For example, the configuration

$$(1a_1)^2(2a_1)^2(1b_2)^2(3a_1)^2(1b_1)(nsa_1)$$

resulting from the excitation $1b_1 \rightarrow nsa_1$, gives rise to a $^1B_1$ and a $^3B_1$ state for the singlet and triplet coupling of the unpaired spins, respectively. Successive states of this type, corresponding to $3sa_1$, $4sa_1$, $5sa_1$, ... in the outer shell nsa_1 orbital, approach the first ionization limit at 12.618 eV above the ground $\tilde{X}\,^1A_1$ state. For the *np* orbitals, there are three Rydberg series (npa_1, npb_1, npb_2); for the *nd* orbitals, there are five; and so on. Similar sets of energy levels also arise from excitations of the $3a_1$ level. Excitations from the inner orbitals $1a_1$, $2a_1$, and $1b_2$ lie much higher in energy and have not been investigated. States corresponding to the simultaneous promotion of two or more electrons also occur, but again these lie rather far above the ground state and have not been observed. A good review of these results is given by Herzberg.[500]

TABLE IV. Vertical Electronic Excitation Energies of the Water Molecule[a]

State	Observed[b]	IVO[d]	CI[e]	SCF-I[f]	SCF-II[g]	INDO[h]
$^1B_1(1b_1 \rightarrow 3sa_1)$	7.49	7.30	—	7.90	6.50	7.43
$^1A_1(3a_1 \rightarrow 3sa_1)$	9.75	9.60	—	9.87	—	10.85
$^3A_1(1b_1 \rightarrow 3pb_1)$	(9.81)[c]	9.71	—	8.99	—	8.67
$^1B_1(1b_1 \rightarrow 3pa_1)$	10.00	10.04	10.75	—	—	—
$^1A_1(1b_1 \rightarrow 3pb_1)$	10.17	10.17	11.26	—	—	—

[a] All values in eV.
[b] References 998a, 1071a, and 1151.
[c] The 9.81 eV state, observed in electron impact, is not assigned in Ref. 1071a. The suggested assignment is based on theoretical calculations.[543]
[d] Improved virtual orbital (IVO) calculation with correlation energy included as the difference between the experimental and computed ionization potential.[543]
[e] Configuration interaction (CI) calculation with a one-center basis and $^1B_1(1b_1 \rightarrow 3pa_1)$ and $^1A_1(1b_1 \rightarrow 3pb_1)$ energies referred to the experimental values of $^1B_1(1b_1 \rightarrow 3sa_1)$ and $^1A_1(3a_1 \rightarrow 3sa_1)$, respectively.[86]
[f] Self-consistent-field (SCF) calculation.[533]
[g] Self-consistent-field (SCF) calculation.[751]
[h] Intermediate neglect of differential overlap (INDO) calculation.[189]

For only one of these states, the $\tilde{C}^1B_1$ state arising from the $1b_1 \rightarrow 3pa_1$ excitation, has the rotational structure been analyzed.[561] It suggests that the excited-state geometry is similar to that of the ground state, which is reasonable since the $1b_1$ and $3pa_1$ orbitals involved in the excitation are not sensitive to molecular geometry.[500] By contrast, the excitation $3a_1 \rightarrow 3sa_1$ probably leads[500] to a linear excited state $\tilde{B}\,^1A_1$, in agreement with the calculated angular dependence of the energy of the $3a_1$ orbital. This change in geometry has strong effects on the energy distribution in the photodissociation products from the excited state.[167,179,1160] The geometry of the $\tilde{D}^1A_1$ state ($1b_1 \rightarrow 3pb_1$) has also been estimated from a vibrational analysis.[60a]

Theoretical studies on the excited states of water are in a more primitive stage of development than the ground-state calculations that we have described in Section 3.1. Since much additional work is needed to obtain definitive results, we refer the reader to some of the available treatments[40a,40b,86,86a,189,372a,458,533,534,543,652,684,745a,751,1075a]* and present only a few of the results that have been obtained so far. Table IV lists vertical electronic excitation energies for five states calculated by various *ab initio* and semiempirical approaches, and compares them with the spectroscopic data.

* Also K. Morokuma, L. Pedersen, and M. Karplus, unpublished calculations.

It is evident that several of the treatments yield reasonably satisfactory results. Many states other than those cited here have been investigated as well. For instance, Meyer(745a) has examined certain high-lying states observed in photoelectron spectra.(139a) Also of interest are the more extensive theoretical predictions by Hunt and Goddard,(543) and the very recent synthesis of all the available optical, electron-impact, and photodissociation data by Claydon *et al.*(189)

4. THE NUCLEAR MOTION

In the previous sections, we showed that the solution to the electronic Schrödinger equation (6) for each electronic state of water yields a potential energy surface for the nuclear motion associated with that state. Since relatively more is known both theoretically and experimentally about the properties of the ground electronic state, we shall consider the solution of the nuclear Schrödinger equation (16) for this case. Most of the discussion will revolve around the low-symmetry HDO isotopic variant of water. Corresponding results for the symmetric isotopes (e.g., H_2O, D_2O) can be obtained as limiting cases. Implicit in such a general treatment are the very good approximations that the potential function $E_0(\mathbf{R})$ for the ground electronic state is independent of the isotopic variant and that the kinetic energy operator T_N in eqn. (5) applies to atoms treated as point masses.

We place the molecule in the xz plane as shown in Fig. 1 with the origin of coordinates at the center of mass. We use the symbols m_{O}, m_{H}, and m_{D} for the masses of the oxygen, hydrogen, and deuterium atoms and $M = m_{\mathrm{O}} + m_{\mathrm{H}} + m_{\mathrm{D}}$ for the total mass of the HDO molecule. Designating the principal axes of the molecule as X, Y, and Z, we find that the equilibrium coordinates $X_K^{(\mathrm{e})}$, $Y_K^{(\mathrm{e})}$, $Z_K^{(\mathrm{e})}$ and the equilibrium principal moments of inertia $I_{XX}^{(\mathrm{e})}$, $I_{YY}^{(\mathrm{e})}$, $I_{ZZ}^{(\mathrm{e})}$ are given by

$$X_{\mathrm{O}}^{(\mathrm{e})} = -(1/M)[u(m_{\mathrm{D}} - m_{\mathrm{H}}) \sin\alpha - v(m_{\mathrm{D}} + m_{\mathrm{H}}) \cos\alpha] \tag{50a}$$

$$X_{\mathrm{H}}^{(\mathrm{e})} = -(1/M)[u(2m_{\mathrm{D}} + m_{\mathrm{O}}) \sin\alpha + v m_{\mathrm{O}} \cos\alpha] \tag{50b}$$

$$X_{\mathrm{D}}^{(\mathrm{e})} = (1/M)[u(2m_{\mathrm{H}} + m_{\mathrm{O}}) \sin\alpha - v m_{\mathrm{O}} \cos\alpha] \tag{50c}$$

$$Y_{\mathrm{O}}^{(\mathrm{e})} = Y_{\mathrm{H}}^{(\mathrm{e})} = Y_{\mathrm{D}}^{(\mathrm{e})} = 0 \tag{50d}$$

$$Z_{\mathrm{O}}^{(\mathrm{e})} = -(1/M)[u(m_{\mathrm{D}} - m_{\mathrm{H}}) \cos\alpha + v(m_{\mathrm{D}} + m_{\mathrm{H}}) \sin\alpha] \tag{50e}$$

$$Z_{\mathrm{H}}^{(\mathrm{e})} = -(1/M)[u(2m_{\mathrm{D}} + m_{\mathrm{O}}) \cos\alpha - v m_{\mathrm{O}} \sin\alpha] \tag{50f}$$

$$Z_{\mathrm{D}}^{(\mathrm{e})} = (1/M)[u(2m_{\mathrm{H}} + m_{\mathrm{O}}) \cos\alpha + v m_{\mathrm{O}} \sin\alpha] \tag{50g}$$

and by

$$I_{XX}^{(e)} = \tfrac{1}{2}[I_{YY}^{(e)} + (1/M)(U^2 + V^2)^{1/2}] \tag{51a}$$

$$I_{YY}^{(e)} = (1/M)[V + 2m_{\rm O}(m_{\rm H} + m_{\rm D})v^2] \tag{51b}$$

$$I_{ZZ}^{(e)} = \tfrac{1}{2}[I_{YY}^{(e)} - (1/M)(U^2 + V^2)^{1/2}] \tag{51c}$$

where α is the angle relating the principal axis system to the molecule-fixed system,

$$u = R_e \sin(\theta_e/2); \qquad v = R_e \cos(\theta_e/2) \tag{51d}$$

$$U = 2uvm_{\rm O}(m_{\rm D} - m_{\rm H}) \tag{51e}$$

and

$$V = m_{\rm O}(m_{\rm H} + m_{\rm D})(u^2 - v^2) + 4m_{\rm H}m_{\rm D}u^2 \tag{51f}$$

It is clear that the water molecule has three unequal moments of inertia and is therefore an asymmetric top. As we have indicated, corresponding results for H_2O and D_2O can be obtained from eqns. (50) and (51) by setting $\alpha = 0$ and $m_{\rm H} = m_{\rm D}$.

Numerical values for the structural parameters of the water molecule can be obtained from theory (Section 3) or experiment. Since the zero-point motion cannot be eliminated, the experimental values corresponding to the potential minimum are obtained by effectively extrapolating the appropriate data to a hypothetical "equilibrium" level. The observed equilibrium geometry and corresponding principal axes which have been found[(498)] in this manner are shown in Fig. 1. We adopt these values to maximize consistency with the *ab initio* results of Section 3.1, although somewhat more accurate results are given by Benedict *et al.*[(63)] From the data given in Fig. 1, the moments of inertia can be calculated and the results are listed in Table V. Noting that

$$I_{ZZ}^{(e)} < I_{XX}^{(e)} < I_{YY}^{(e)} \tag{51g}$$

we conclude that Z, X, and Y can be identified with the conventional principal axes of inertia a, b, and c, respectively; this is referred to as the I^r representation by King *et al.*[(605)]

The purely vibrational motion of the molecule can be determined independently of its rotation to a first approximation. To do this, we subject the set of instantaneous position coordinates $X_K' = X_K - X_K^{(e)}$, $Y_K' = Y_K - Y_K^{(e)}$, and $Z_K' = Z_K - Z_K^{(e)}$ for each of the atoms K to certain mathematical conditions[(1173)] which enable the vibrations to be treated as if the

TABLE V. Observed Moments of Inertia of the Water Molecule[a]

	H_2O	D_2O	HDO
$I_{XX}^{(e)}$	1.921	3.839	3.071
$I_{YY}^{(e)}$	2.945	5.680	4.283
$I_{ZZ}^{(e)}$	1.024	1.841	1.212
$I_{XX}^{(0)}$	1.933	3.860	3.080
$I_{YY}^{(0)}$	3.020	5.784	4.371
$I_{ZZ}^{(0)}$	1.009	1.819	1.203

[a] Equilibrium [$I_{\alpha\alpha}^{(e)}$] and ground-state vibrational [$I_{\alpha\alpha}^{(0)}$] moments of inertia in units of 10^{-40} g cm^2 are based[861,862] on the geometry in Fig. 1. More accurate values are given by Benedict *et al.*[63]

molecule were not rotating. Applying these conditions, we find that through first order the purely vibrational kinetic energy contribution to T_N [i.e., the second term in eqn. (5)] can be written as

$$T_{\text{vib}} = \tfrac{1}{2} \sum_{K=1}^{3} m_K(\dot{X}_K'^2 + \dot{Y}_K'^2 + \dot{Z}_K'^2) \tag{52a}$$

Introducing the bond and angular displacement coordinates $R_1' = R_1 - R_e$, $R_2' = R_2 - R_e$, $\theta' = \theta - \theta_e$ used in Eq. (48), we obtain[795,1173]

$$T_{\text{vib}} = \tfrac{1}{2}[G_{11}^{-1}\dot{R}_1'^2 + G_{22}^{-1}\dot{R}_2'^2 + G_{33}^{-1}\dot{\theta}'^2 + 2G_{12}^{-1}\dot{R}_1'\dot{R}_2' + 2G_{13}^{-1}\dot{R}_1'\dot{\theta}' + 2G_{23}^{-1}\dot{R}_2'\dot{\theta}'] \tag{52b}$$

where $\mathbf{G}^{-1}$ is the inverse of the matrix

$$\mathbf{G} = \begin{bmatrix} \dfrac{1}{m_\text{O}} + \dfrac{1}{m_\text{H}} & \dfrac{\cos\theta_e}{m_\text{O}} & -\dfrac{\sin\theta_e}{m_\text{O}R_e} \\ \dfrac{\cos\theta_e}{m_\text{O}} & \dfrac{1}{m_\text{O}} + \dfrac{1}{m_\text{D}} & -\dfrac{\sin\theta_e}{m_\text{O}R_e} \\ -\dfrac{\sin\theta_e}{m_\text{O}R_e} & -\dfrac{\sin\theta_e}{m_\text{O}R_e} & \dfrac{1}{R_e^2}\left(\dfrac{1}{m_\text{H}} + \dfrac{1}{m_\text{D}} + \dfrac{2}{m_\text{O}} - \dfrac{2\cos\theta_e}{m_\text{O}}\right) \end{bmatrix} \tag{52c}$$

The coefficients G_{ij}^{-1} are usually obtained by numerical methods once the values of G_{ij} are specified; here G_{ij}^{-1} is defined as the ij-element of $\mathbf{G}^{-1}$.

To find the normal coordinates Q_s which enter into the Hamiltonian of eqn. (5), we combine the vibrational kinetic energy operator T_{vib} in eqn. (52b) with the quadratic (harmonic) part of the potential energy function in eqn. (48b). The normal coordinate basis is obtained from R_1', R_2', and θ' by the inverse transformation of

$$\begin{pmatrix} R_1' \\ R_2' \\ \theta' \end{pmatrix} = \sum_{s=1}^{3} \begin{pmatrix} \mathscr{C}_{1s} \\ \mathscr{C}_{2s} \\ \mathscr{C}_{3s} \end{pmatrix} Q_s \tag{53}$$

which simultaneously reduces T_{vib} and $E_0(\mathbf{R}'^2)$ to simple sums of squares

$$T_{\text{vib}} = \tfrac{1}{2} \sum_{s=1}^{3} \dot{Q}_s^2, \qquad E_0(\mathbf{R}'^2) = \tfrac{1}{2} \sum_{s=1}^{3} \omega_s^2 Q_s^2 \tag{54}$$

over the coordinates Q_s, $s = 1, 2, 3$. Thus, the normal coordinate analysis consists in finding coefficients $\mathscr{C}_{1s}$, $\mathscr{C}_{2s}$, $\mathscr{C}_{3s}$ such that the motion of the molecule can be described as a superposition of simple harmonic motions characterized by the angular frequencies ω_s. Once the equilibrium geometry and force constants are specified, this analysis can be carried out by well-known methods (such as Wilson's *GF*-matrix method[1173]), with the results given in Table VI for the isotopes H_2O, D_2O, and HDO. An examination of the $\mathscr{C}$ transformation matrix shows that the normal motions for H_2O and D_2O are accurately described as symmetric and asymmetric stretching of the two OH (or OD) bonds and a bending of the HOH (or DOD) angle; for HDO, the stretching normal motions correspond more closely to compression or elongation of the individual O—H and O—D bonds, as illustrated in Fig. 8. The harmonic frequencies show that the zero-point bending motion involves much less energy than bond stretching. The O—H and O—D bond stretching frequencies (3890 cm^{-1}, 2824 cm^{-1}) in HDO are very similar to the observed values (3735 cm^{-1}, 2721 cm^{-1}) for the diatomic molecules OH and OD,[499] respectively. This correspondence of the potential functions is in agreement with the similarity of the different charge distributions.*

Having determined the normal frequencies and coordinates, we return to the more general problem of analyzing the full nuclear Hamiltonian in eqn. (5). We have already indicated [eqn. (48)] the form for the Hartree–Fock potential energy $\tilde{E}_0(\mathbf{R})$; analogously, we have

$$E_0(\mathbf{R}) = E_0(\mathbf{R}_e) + \tfrac{1}{2} \sum_{s} \omega_s^2 Q_s^2 + \sum_{ss's''} k'_{ss's''} Q_s Q_{s'} Q_{s''} + \sum_{ss's''s'''} k'_{ss's''s'''} Q_s Q_{s'} Q_{s''} Q_{s'''} + \cdots \tag{55}$$

* The overlap of H_2O and OH^- bond geminals is 0.9937 for basis sets of comparable quality.[8a]

TABLE VI. Transformation Matrix $\mathscr{C}$ Relating Bond Length and Bend Displacements to Normal Coordinates[a]

	H_2O		
	Symmetric stretch ($\omega = 3832\ cm^{-1}$)	Bend ($\omega = 1648\ cm^{-1}$)	Asymmetric stretch ($\omega = 3943\ cm^{-1}$)
R_1'	0.0169	−0.0004	0.0171
R_2'	0.0169	−0.0004	−0.0171
θ'	−0.0007	0.0189	0
	D_2O		
	Symmetric stretch ($\omega = 2764\ cm^{-1}$)	Bend ($\omega = 1206\ cm^{-1}$)	Asymmetric stretch ($\omega = 2889\ cm^{-1}$)
R_1'	0.0122	−0.0001	0.0125
R_2'	0.0122	−0.0001	−0.0125
θ'	−0.0014	0.0138	0
	HDO		
	O—D stretch ($\omega = 2824\ cm^{-1}$)	Bend ($\omega = 1440\ cm^{-1}$)	O—H stretch ($\omega = 3890\ cm^{-1}$)
R_1'	0.0174	−0.0002	−0.0010
R_2'	0.0008	−0.0003	0.0240
θ'	−0.0009	0.0166	−0.0005

[a] All values referred to eqn. (53) of the text and re-expressed in atomic units ($e = \hbar = m = 1$). The $\mathscr{C}$ matrix is based on experimental quadratic force constants which are slightly different from those in Table II. See Ref. 600.

employing the normal coordinates. The summations over the cubic and quartic force constants $k'_{ss's''}$ and $k'_{ss's''s'''}$ are restricted by $s \leq s' \leq s'' \leq s'''$. The kinetic energy contribution is treated by expanding T_N as a power series in the normal coordinates. To do this, we first expand the tensor $\mu_{\alpha\beta}$ appearing in eqn. (5) in terms of the Q_s, obtaining

$$\mu_{\alpha\beta} = \frac{1}{I_{\alpha\alpha}^{(e)} I_{\beta\beta}^{(e)}} \left[I_{\alpha\alpha}^{(e)} \delta_{\alpha\beta} - \sum_{s=1}^{3} a_s^{(\alpha\beta)} Q_s - \sum_{s,s'=1}^{3} G_{ss'}^{(\alpha\beta)} Q_s Q_{s'} + \cdots \right] \qquad (56)$$

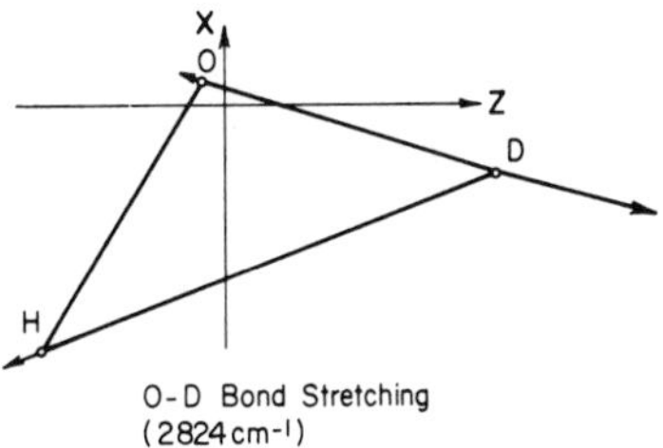

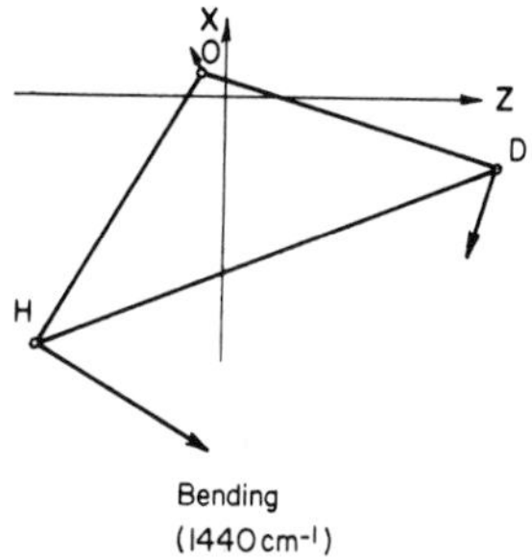

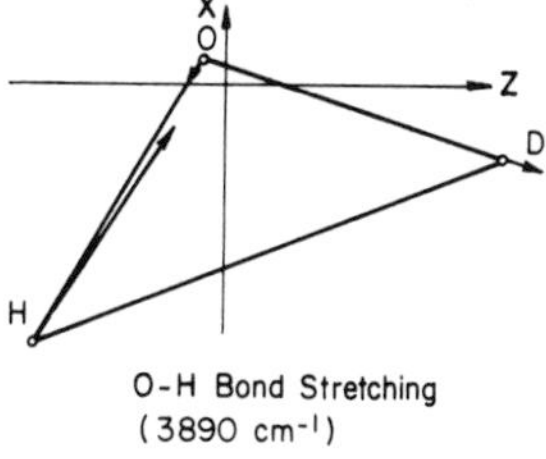

Fig. 8. Normal vibrations in the HDO molecule.

from eqns. (B5)–(B8) of Appendix B. It can then be shown by substituting this expression for $\mu_{\alpha\beta}$ into eqn. (5), combining the result with eqn. (55), and omitting constant terms [i.e., $E_0(\mathbf{R}_e)$, $(\hbar^2/8)\sum_\alpha \mu_{\alpha\alpha}$] which contribute only an overall energy level shift, that the vibration–rotation Hamiltonian

$$\mathscr{H}_N = T_N + E_0(\mathbf{R}) + \langle \Phi_0 \mid T_N \mid \Phi_0 \rangle \tag{57}$$

can be written

$$\mathscr{H}_N = \mathscr{H}_N^{(0)} + \mathscr{H}_N^{(1)} + \mathscr{H}_N^{(2)} + \cdots \tag{58}$$

where the successive terms[795,861]

$$\mathscr{H}_N^{(0)} = \tfrac{1}{2}\hbar \sum_{s=1}^{3} \omega_s(p_s'^2 + q_s^2) + \tfrac{1}{2} \sum_{\alpha=1}^{3} [P_\alpha^2/I_{\alpha\alpha}^{(e)}] \tag{59a}$$

$$\mathscr{H}_N^{(1)} = \sum_{ss's''} k_{ss's''} q_s q_{s'} q_{s''} - \sum_\alpha \frac{P_\alpha \hat{p}_\alpha}{I_{\alpha\alpha}^{(e)}} - \frac{\hbar^{1/2}}{2} \sum_{\alpha\beta} \sum_s \frac{P_\alpha P_\beta}{I_{\alpha\alpha}^{(e)} I_{\beta\beta}^{(e)}} \frac{a_s^{(\alpha\beta)}}{\omega^{1/2}} q_s$$
$$+ \frac{\hbar^{1/2}}{2} \sum_{\alpha\beta} \sum_s \frac{P_\alpha}{I_{\alpha\alpha}^{(e)} I_{\beta\beta}^{(e)}} \frac{a_s^{(\alpha\beta)}}{\omega_s^{1/2}} (\hat{p}_\beta q_s + q_s \hat{p}_\beta) \tag{59b}$$

$$\mathscr{H}_N^{(2)} = \sum_{ss's''s'''} k_{ss's''s'''} q_s q_{s'} q_{s''} q_{s'''} + \tfrac{1}{2} \sum_\alpha \frac{\hat{p}_\alpha^2}{I_{\alpha\alpha}^{(e)}}$$
$$- \frac{\hbar}{2} \sum_{\alpha\beta} \sum_{ss'} \frac{P_\alpha P_\beta}{I_{\alpha\alpha}^{(e)} I_{\beta\beta}^{(e)}} \frac{G_{ss'}^{(\alpha\beta)}}{(\omega_s \omega_{s'})^{1/2}} q_s q_{s'} \tag{59c}$$

are given in terms of the dimensionless coordinates

$$q_s = (\omega_s/\hbar)^{1/2} Q_s; \qquad p_s' = p_s/(\hbar\omega_s)^{1/2}, \qquad s = 1, 2, 3 \tag{60}$$

and

$$\hat{p}_\alpha = \hbar \sum_{ss'} (\omega_{s'}/\omega_s)^{1/2} \zeta_{ss'}^{(\alpha)} q_s p_{s'}', \qquad \alpha = X, Y, Z \tag{61}$$

Notice that we have also introduced new cubic and quartic potential constants k (without the primes), and that we have used the commutator $[P_\alpha, \hat{p}_\alpha] = 0$.

Recognizing that the two terms in $\mathscr{H}_N^{(0)}$ [Eq. (59a)] represent the harmonic oscillator and rigid-rotator model limits, which are known[795,1173] to be a good starting point for the description of vibrational and rotational motion, respectively, we use perturbation theory to introduce corrections for anharmonicity, Coriolis interaction, and centrifugal distortion. A standard procedure for solving the perturbation problem is to diagonalize the energy matrix by applying contact transformations[965] to successively higher-order terms in $\mathscr{H}_N$. Since this method is somewhat involved if carried to high order,[177,964,965,965a,1188] we employ conventional low-order Rayleigh–Schrödinger perturbation theory to gain some insight into the results. We choose the usual zeroth-order representation of $\mathscr{H}_N^{(0)}$ to be diagonal in the harmonic oscillator vibrational quantum numbers v_s and in the symmetric-top rotational quantum numbers J and K. In this representation, the nonzero harmonic oscillator matrix elements of q_s and

p_s' are given by

$$\langle v_s \mid q_s \mid v_s \pm 1 \rangle = \pm i \langle v_s \mid p_s' \mid v_s \pm 1 \rangle = [\tfrac{1}{2}(v_s + \tfrac{1}{2} \pm \tfrac{1}{2})]^{1/2} \tag{62a}$$

and the rigid-rotor elements by

$$\mp i \langle J, K \mid P_X \mid J, K \pm 1 \rangle = \langle J, K \mid P_Y \mid J, K \pm 1 \rangle = \tfrac{1}{2}\hbar[J(J+1) - K(K \pm 1)]^{1/2} \tag{62b}$$

and

$$\langle J, K \mid P_Z \mid J, K \rangle = \hbar K \tag{62c}$$

$$\langle J, K \mid P^2 \mid J, K \rangle = \hbar^2 J(J+1) \tag{62d}$$

We can truncate the perturbation expression

$$\begin{aligned} \langle E_0 \rangle_v = {} & \langle v_1 v_2 v_3 \mid \mathscr{H}_N^{(0)} \mid v_1 v_2 v_3 \rangle \\ & + \langle v_1 v_2 v_3 \mid \mathscr{H}_N^{(1)} \mid v_1 v_2 v_3 \rangle + \langle v_1 v_2 v_3 \mid \mathscr{H}_N^{(2)} \mid v_1 v_2 v_3 \rangle \\ & + \sum_{v_1' v_2' v_3'}^{\infty}{}' \left[\frac{ | \langle v_1 v_2 v_3 \mid \mathscr{H}_N^{(1)} \mid v_1' v_2' v_3' \rangle |^2 }{ (E^{(0)}_{v_1 v_2 v_3} - E^{(0)}_{v_1' v_2' v_3'}) } \right] \end{aligned} \tag{63}$$

at second order and determine the element diagonal in v_s, $s = 1, 2, 3$. Introducing eqn. (59), we obtain

$$\langle E_0 \rangle_v = E_{\text{vib}} + \mathscr{H}^v_{\text{rot}} \tag{64a}$$

by direct calculation, where

$$E_{\text{vib}} = \hbar \sum_s \omega_s (v_s + \tfrac{1}{2}) + hc \sum_s \sum_{s' \geq s} x_{ss'} (v_s + \tfrac{1}{2})(v_{s'} + \tfrac{1}{2}) \tag{64b}$$

is the pure vibrational part that depends on v alone* and where

$$\mathscr{H}^v_{\text{rot}} = \tfrac{1}{2} \sum_{\alpha\beta} \sigma^v_{\alpha\beta} P_\alpha P_\beta + \tfrac{1}{4} \sum_{\alpha\beta\gamma\delta} \tau_{\alpha\beta\gamma\delta} P_\alpha P_\beta P_\gamma P_\delta \tag{64c}$$

includes the pure rotational plus interaction terms. The coefficients $x_{ss'}$, $\sigma^v_{\alpha\beta}$, and $\tau_{\alpha\beta\gamma\delta}$ are related to the force constants, moments of inertia, and other parameters in eqn. (59), as defined in Appendix C. An examination of eqn. (64b) shows that the first term corresponds to the harmonic oscillator energy $E^{(0)}_{v_1 v_2 v_3}$ $[= \hbar \sum_s \omega_s (v_s + \tfrac{1}{2})]$, and that corrections to this zeroth-order

* A term that is independent of v has been omitted [cf., M. Wolfsberg, A. A. Massa, and J. W. Pyper, *J. Chem. Phys.* **53**, 3138 (1970)].

approximation arise from the first two terms in $\mathscr{H}_N^{(2)}$ [eqn. (59c)] and from the coupling of excited (and deexcited) states $(v_1'v_2'v_3')$ with the unperturbed state $(v_1v_2v_3)$ through $\mathscr{H}_N^{(1)}$ [eqn. (59)b]. The first-order vibrational contribution from $\mathscr{H}_N^{(1)}$ is zero. The leading term in eqn. (64c) contains the rigid-rotator energy, and succeeding terms represent appropriate corrections. Thus, to this order of approximation, we can proceed to calculate the rotational matrix elements of vibrational state v from eqn. (64c) using the results in eqn. (62). After tedious but straightforward manipulation, it is possible to obtain expressions for matrix elements diagonal in K and connecting rotational state K with adjacent states. The results are[795,861,862]

$$\langle J, K \,|\, \mathscr{H}^v_{\text{rot}} \,|\, J, K\rangle = \mathscr{R}_0 + \mathscr{R}_2 K^2 - D_K K^4 \tag{65a}$$

$$\begin{aligned}\langle J, K \,|\, \mathscr{H}^v_{\text{rot}} \,|\, J, K \pm 1\rangle = &\pm [J(J+1) - K(K\pm 1)]^{1/2}(2K \pm 1)\\ &\times \{\tfrac{5}{2}\mathscr{R}_7 + \mathscr{R}_8 J(J+1) + \mathscr{R}_9[K^2 + (K\pm 1)^2]\}\end{aligned} \tag{65b}$$

$$\begin{aligned}\langle J, K \,|\, \mathscr{H}^v_{\text{rot}} \,|\, J, K \pm 2\rangle = &[J(J+1) - K(K\pm 1)]^{1/2}[J(J+1)\\ &- K(K\pm 2)]^{1/2}\{\mathscr{R}_4 - \mathscr{R}_5[K^2 + (K\pm 2)^2]\}\end{aligned} \tag{65c}$$

$$\begin{aligned}\langle J, K \,|\, \mathscr{H}^v_{\text{rot}} \,|\, J, K \pm 3\rangle = &\pm [J(J+1) - K(K\pm 1)]^{1/2}[J(J+1)\\ &- K(K\pm 2)]^{1/2}[J(J+1) - K(K\pm 3)]^{1/2}(2K\pm 3)\mathscr{R}_7\end{aligned} \tag{65d}$$

and

$$\begin{aligned}\langle J, K \,|\, \mathscr{H}^v_{\text{rot}} \,|\, J, K \pm 4\rangle = &[J(J+1) - K(K\pm 1)]^{1/2}[J(J+1)\\ &- K(K\pm 2)]^{1/2}[J(J+1) - K(K\pm 3)]^{1/2}[J(J+1)\\ &- K(K\pm 4)]^{1/2}\mathscr{R}_6\end{aligned} \tag{65e}$$

These expressions depend on the rotational quantum numbers J and K, and on certain constants that characterize the potential energy surface of a particular electronic and vibrational state. These constants are explicitly defined in Appendix D. To obtain the rotational energy levels, the matrix corresponding to eqn. (65) must be diagonalized by numerical methods.

In the absence of vibrational degeneracies, the theory which we have outlined provides an adequate basis for understanding most of the important effects of nuclear motion known at the present time. In particular, it is possible to deduce numerical values for the various constants appearing in eqns. (55)–(65) and to show that they are consistent with the spectra. Although this has been done to some extent (see below) by purely theoretical means starting, for example, from the near-Hartree–Fock potential energy surface in eqn. (48), most of the constants have been obtained by experimental investigations in the near-infrared and at longer wavelengths.[498] In the region between 10^3 and 10^4 cm^{-1}, absorption bands corresponding to excitations of the normal modes are observed. Within each band are series of closely spaced lines representing transitions between different rotational levels. Pure rotational transitions, without any accompanying vibrational changes, can also be observed in the far-infrared and microwave regions. The most accurate experimental data currently available* for water vapor have been obtained by Benedict *et al.*[63,63a] (D_2O, HDO) and by Rao and coworkers[346,347] (H_2O, $H_2^{18}O$) in the near-infrared region, by Hall and Dowling[445,446] (H_2O) in the far-infrared region, and by Lichtenstein *et al.*[683] (H_2O), Powell and Johnson[865] ($H_2^{18}O$), Verhoeven *et al.*[1105,1106,1107] (D_2O, $HD^{17}O$), and Gordy *et al.*[253a] (HDO) in the microwave region. Verhoeven *et al.* have also investigated the hyperfine structure of various rotational transitions (see Section 5).

All of the spectroscopic data concerning the ground electronic state of water are consistent with each other and lead to energy level diagrams of the type shown in Fig. 9 for H_2O, D_2O, and HDO. The vibrational states are designated by the triple of quantum numbers $[\nu_1\nu_2\nu_3]$, where the subscripts 1, 2, and 3 refer, respectively, to the symmetric stretching, bending, and asymmetric stretching modes for H_2O and D_2O and to the OD bond stretching and bending and OH bond stretching for HDO. The rotational states are denoted by the symbol $J_{K_{-1}K_1}$, where the two subscripts K_{-1} and K_1, respectively, represent the K values of the limiting prolate and oblate symmetric tops. We see from Fig. 9 that the amount of energy required to singly excite any zero-point mode to a "fundamental" level ($\nu_s = 0$ to $\nu_s = 1$) is on the order of several thousand cm^{-1}. Some of the observed rotational levels within the [000] bands are also shown. Since the average spacing between the rotational J levels is on the order of 10 cm^{-1}, we have arbitrarily chosen to include selected ones at intervals of 200 cm^{-1}. Energy level assignments for the $J = 1$ states of [000] are given in Table VII. These and other values for higher vibrational states with $\sum_s \nu_s \leq 3$ and their associated rotational manifolds can be found in the literature.[63,346,347,445,446,683,865,1105–1107]

Numerical values for all of the constants which appear in eqns. (55)–(65) have been obtained. Table VIII gives the transformation matrix relating

* Investigations of the absorption spectrum of water vapor were actually carried out before the turn of the century (e.g., Paschen[825]). However, it was not until quantum theory was developed that the spectra were understood in a quantitative fashion. The papers by Shaffer and Nielsen[965] and by Darling and Dennison[232] played important roles in this connection.

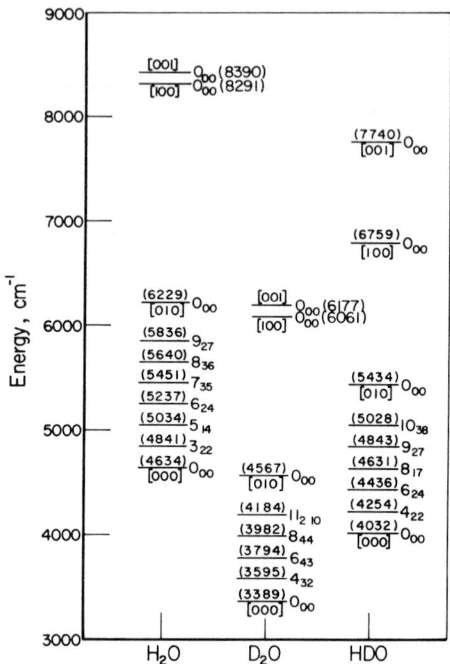

Fig. 9. Energy level diagram of the lower-lying vibrational and rotational states of H_2O, D_2O, and HDO relative to the minimum of the potential energy surface. The vibrational states are designated as $[\nu_1, \nu_2, \nu_3]$ for the three normal modes, where ν_1, ν_2, and ν_3 refer, respectively, to quantum numbers for the symmetric stretch, the bend, and the asymmetric stretch of H_2O and D_2O and for the O—D bond stretch, the bend, and the O—H bond stretch of HDO. The rotational states within the [000] band are arbitrarily shown at approximately 200 cm^{-1} intervals.

principal axis displacements to normal coordinates, Table IX lists the observed constants characterizing the anharmonic vibrational motion, and Table X includes the corresponding constants for the rotational motion of of the [000] vibrational state. In these tabulations, we have made extensive use of the experimental data and analyses in Refs. 63, 861, and 862. Qualitatively, the results show that vibration and rotation cause rather large deviations from the harmonic oscillator, rigid-rotator approximation, and that these must be taken into account in any detailed interpretation of the spectra.

TABLE VII. Observed Rotational Energy Levels (cm^{-1}) for $J = 1$ of the [000] Bands of Water[a]

$J_{K_{-1}K_1}$	H_2O	D_2O	HDO
0_{00}	0	0	0
1_{01}	23.79	12.11	15.51
1_{11}	37.14	20.23	29.87
1_{10}	42.36	22.64	32.55

[a] The 0_{00} states of H_2O, D_2O, and HDO are located, respectively, at 4634, 3387, and 4032 cm^{-1} with respect to $E_0(R_e, \theta_e)$ [cf. eqn. (48)].

In contrast to many polyatomic molecules, the distortion effects observable in water spectra are so large that higher-order theory is required. For the lowest-lying vibrational and rotational states, the comparatively simple second-order theory which we have outlined here is fairly adequate; most lines are fitted to within 0.1 cm^{-1}. However, highly excited states require that correspondingly higher corrections be added to the Hamiltonian.(63,140,177,964,1151a,1188)

To discuss further the significance of the vibrational–rotational Hamiltonian and its fit to the observed spectra, we examine for $J_{K_{-1}K_1} = 0_{00}$ the effect of the zero-point vibrational motion on the bond lengths and angle. This can be done by averaging the coordinates R_1', R_2', and θ' over the vibrational wave function χ_{nv}. Making use of the normal coordinate transformation [eqn. (53)], we have

$$\langle \chi_{nv} \mid \begin{pmatrix} R_1' \\ R_2' \\ \theta' \end{pmatrix} \mid \chi_{nv} \rangle = \sum_{s=1}^{3} \begin{pmatrix} \mathscr{C}_{1s} \\ \mathscr{C}_{2s} \\ \mathscr{C}_{3s} \end{pmatrix} \langle \chi_{nv} \mid Q_s \mid \chi_{nv} \rangle \tag{66}$$

For the ground vibrational ($v = 0$) states, it can be shown that for $\mathscr{C}$ in atomic units(600,640)

$$\langle \chi_{n0} \mid \begin{pmatrix} R_1' \\ R_2' \\ \theta' \end{pmatrix} \mid \chi_{n0} \rangle = -\frac{1}{2\hbar} \sum_{s} \frac{1}{\omega_s} \begin{pmatrix} \mathscr{C}_{1s} \\ \mathscr{C}_{2s} \\ \mathscr{C}_{3s} \end{pmatrix} \left[3k_{sss} + \sum_{s'>s} (k_{ss's'} + k_{sss'}) \right] \tag{67}$$

to an order of approximation in which the vibrational wave function χ_{n0} is corrected for anharmonicity through $\mathscr{H}_N^{(1)}$. A similar expression can be derived for $(R_1'^2)^{1/2}$, $(R_2'^2)^{1/2}$, and $(\theta'^2)^{1/2}$. The results obtained (for the ground electronic state $n = 0$) from experimental data(223a,640) and from the near-Hartree–Fock basis-set calculations(315) described in Section 3 are given in

TABLE VIII. Transformation Matrix $L^{(\alpha)}$ Relating Principal Axis Displacements to Normal Coordinates[a]

	H_2O		
	Symmetric stretch (ω = 3832 cm^{-1})	Bend (ω = 1648 cm^{-1})	Asymmetric stretch (ω = 3943 cm^{-1})
X_O'	0.1958	0.2713	0
X_{H1}'	−0.3899	−0.5404	0.4169
X_{H2}'	−0.3899	−0.5404	−0.4169
Z_O'	0	0	0.2702
Z_{H1}'	0.5734	−0.4138	−0.5382
Z_{H2}'	−0.5734	0.4138	−0.5382
	D_2O		
	Symmetric stretch (ω = 2764 cm^{-1})	Bend (ω = 1206 cm^{-1})	Asymmetric stretch (ω = 2889 cm^{-1})
X_O'	0.2759	0.3537	0
X_{D1}'	−0.3887	−0.4983	0.4025
X_{D2}'	−0.3887	−0.4938	−0.4025
Z_O'	0	0	0.3688
Z_{D1}'	0.5576	−0.4349	−0.5196
Z_{D2}'	−0.5576	0.4349	−0.5196
	HDO		
	O—D stretch (ω = 2824 cm^{-1})	Bend (ω = 1440 cm^{-1})	O—H stretch (ω = 3890 cm^{-1})
X_O'	0.0953	0.2844	−0.1941
X_H'	−0.0356	−0.5229	0.8158
X_D'	−0.2434	−0.3940	−0.0301
Z_O'	−0.3067	−0.1076	−0.1437
Z_H'	−0.0675	0.6707	0.5235
Z_D'	0.9120	−0.1712	0.0346

[a] All values refer to eqn. (B4) of Appendix B in dimensionless units. The $L^{(\alpha)}$ matrix is based on experimental quadratic force constants which are slightly different from those in Table II (see Ref. 861).

TABLE IX. Observed Constants Characterizing Anharmonic Vibrational Motion[a]

	H_2O	D_2O	HDO
x_{11}	−42.58	−22.58	−43.36
x_{22}	−16.81	−9.18	−11.77
x_{33}	−47.57	−26.15	−82.88
x_{12}	−15.93	−7.58	−8.60
x_{13}	−165.82	−87.15	−13.14
x_{23}	−20.33	−10.61	−20.08
k_{111}	−319.4 ± 3.3	−193.3 ± 2.1	−276.5
k_{112}	39.6 ± 2.4	7.6 ± 1.6	36.3
k_{113}	0	0	24.7
k_{122}	255.4 ± 31.4	191.1 ± 24.8	10.7
k_{222}	−61.9 ± 2.5	−33.4 ± 2.1	−41.4
k_{223}	0	0	203.1
k_{133}	−921.7 ± 1.5	−632.1 ± 18.0	−91.1
k_{233}	147.3 ± 0.4	93.7 ± 2.1	64.6
k_{333}	0	0	−442.9
k_{123}	0	0	−87.7
k_{1111}	38.5 ± 1.4	18.8 ± 0.7	37.8
k_{1122}	−140.4 ± 15.9	−96.8 ± 13.6	−25.6
k_{2222}	11.2 ± 4.1	9.9 ± 3.6	13.2
k_{1133}	209.2 ± 2.5	135.7 ± 6.5	3.76
k_{2233}	−122.2 ± 7.6	−74.3 ± 5.8	−132.7
k_{3333}	35.0 ± 0.2	26.0 ± 2.3	71.2

[a] All values in cm^{-1}. The subscripts 1, 2, 3 refer, respectively, to the symmetric stretching, bending, and asymmetric stretching modes for H_2O and D_2O, and to OD bond stretching and bending and OH bond stretching in HDO. The anharmonic constants $x_{ss'}$ are taken from the data of Benedict *et al.*[63] and are referred to in eqn. (64b) of the text. The reduced cubic force constants [eqn. (59b)] are based on the experimental cubic constants in Table II. The H_2O and D_2O values are taken from Ref. 641 and the HDO values from the "SUBFF" model of D. Papoušek and J. Pliva, *Collection Czech. Chem. Commun.* **29**, 1973 (1964). Experimental values for certain other quartic constants (k_{1112}, k_{1222}, k_{1233}) are not known.

Table XI for the three isotopic species H_2O, D_2O, and HDO. We see that although the mean bond distance and angle shifted by only 1% or less, the root-mean-square amplitudes are nearly 10% of the equilibrium values (see Fig. 1).

The good agreement between these experimental and theoretical results has prompted additional investigations of the zero-point vibrational cor-

TABLE X. Observed Structural Constants Characterizing Rotational Motion for the [000] Vibrational State[a]

	H_2O	D_2O	HDO
$\zeta_{12}^{(Y)}$	0	0	0.5451
$\zeta_{13}^{(Y)}$	0.0055	−0.0568	−0.0178
$\zeta_{23}^{(Y)}$	-0.9999_8	−0.9984	−0.8382
$a_1^{(XX)}$	2.2479	3.0900	3.1496
$a_2^{(XX)}$	−1.6222	−2.4103	−1.4342
$a_3^{(XX)}$	0	0	−0.5545
$a_1^{(YY)}$	3.4322	4.7589	3.4695
$a_2^{(YY)}$	0.0189	−0.2707	−0.0736
$a_3^{(YY)}$	0	0	−2.2563
$a_1^{(ZZ)}$	1.1843	1.6689	0.3199
$a_2^{(ZZ)}$	1.6411	2.1395	1.3606
$a_3^{(ZZ)}$	0	0	−1.7018
$a_1^{(ZX)}$	0	0	−0.7719
$a_2^{(ZX)}$	0	0	−1.2520
$a_3^{(ZX)}$	−1.6346	−2.2309	−1.1461
$A_{11}^{(XX)}$	0.6576	0.6217	0.9304
$A_{22}^{(XX)}$	0.3424	0.3783	0.4907
$A_{33}^{(XX)}$	0.6523	0.6759	0.2959
$A_{11}^{(ZZ)}$	0.3424	0.3783	0.0696
$A_{22}^{(ZZ)}$	0.6576	0.6217	0.5093
$A_{33}^{(ZZ)}$	0.3477	0.3241	0.7041
$A_{11}^{(ZX)}$	0	0	−0.2488
$A_{22}^{(ZX)}$	0	0	−0.2051
$A_{33}^{(ZX)}$	0	0	0.4539
$b_1^{(XX)}$	0.0107	0.0044	0.0068
$b_2^{(XX)}$	−0.0067	−0.0027	−0.0060
$b_3^{(XX)}$	0.0021	0.0009	0.0009
$b_1^{(YY)}$	0.0070	0.0027	0.0038
$b_2^{(YY)}$	0.0045	0.0017	0.0024
$b_3^{(YY)}$	0.0051	0.0018	0.0030
$b_1^{(ZZ)}$	0.0203	0.0058	0.0100
$b_2^{(ZZ)}$	−0.0959	−0.0380	−0.0601
$b_3^{(ZZ)}$	0.0460	0.0190	0.0362

Table X. (*Continued*)

	H_2O	D_2O	HDO
τ_{XXXX}	−1.3524	−0.3386	−0.3637
τ_{YYYY}	−0.1508	−0.0410	−0.0777
τ_{ZZZZ}	−13.8495	−4.2864	−7.1449
τ_{XXYY}	−0.2271	−0.0705	−0.1225
τ_{YYZZ}	−0.4480	−0.0836	−0.1815
τ_{ZZXX}	2.8815	0.8021	0.7842
τ_{ZXZX}	−0.6281	−0.1689	−0.8923
τ_{ZXXX}	0	0	−0.2428
τ_{ZXYY}	0	0	0.0246
τ_{ZXZZ}	0	0	1.7900
D_{XX}	−0.0844	−0.0221	−0.1019
D_{YY}	0.0901	0.0248	0.1460
D_{ZZ}	−0.0480	−0.0138	−0.1061

[a] Based on the analysis of Ref. 861, 862, the assignment of modes to the subscripts 1, 2, 3 is given in the footnote to Table IX. The following units are used for dimensioned quantities: $a_s^{(\alpha\beta)}$, 10^{-20} $g^{1/2}$ cm; $b_s^{\alpha\beta}$, 10^{40} g^{-1} cm^{-2}; $\tau_{\alpha\beta\gamma\delta}$, 10^{90} g^{-3} cm^{-6} sec^2; $D_{\alpha\beta}$, 10^{37} g^{-1} cm^{-2}.

TABLE XI. Zero-Point Vibrational Corrections to the Equilibrium Structure of the Water Molecule[a]

	H_2O		D_2O		HDO	
	Expt.[b]	Theory[c]	Expt.[b]	Theory[c]	Expt.[b]	Theory[c]
R_1'	0.0140	0.0147	0.0102	0.0107	—	0.0108
R_2'	0.0140	0.0147	0.0102	0.0107	—	0.0148
θ'	0.18	0.12	0.12	0.09	—	0.11
$(R_1'^2)^{1/2}$	0.0677	0.0651	0.0578	0.0556	0.0578	0.0556
$(R_2'^2)^{1/2}$	0.0677	0.0651	0.0578	0.0556	0.0677	0.0651
$(\theta'^2)^{1/2}$	8.72	8.68	7.49	7.43	—	8.13

[a] Bond lengths and angles are in angstroms and degrees, respectively.
[b] References 640 ("Calc. I") and 223a. Also see ref. 639a.
[c] Reference 315. The theoretically computed angles reported in Table VII of this reference are all somewhat too large due to an inaccurate conversion from radians to degrees.

rections to various one-electron properties. Any electronic property $\mathscr{P}$ can be expanded in reduced normal coordinates as

$$\mathscr{P} = \mathscr{P}_e + \sum_s \mathscr{A}_s q_s + \sum_{ss'} \mathscr{B}_{ss'} q_s q_{s'} + \sum_{ss's''} \mathscr{C}_{ss's''} q_s q_{s'} q_{s''} + \cdots \tag{68}$$

where $\mathscr{P}_e$ is the value of $\mathscr{P}$ at the theoretical equilibrium geometry. The expansion coefficients $\mathscr{A}$, $\mathscr{B}$, and $\mathscr{C}$, like those for the energy in eqn. (55), are obtained by calculating $\mathscr{P}$ for a suitable set of geometries and fitting eqn. (68) to the resulting values. Averaging the q_s over the ground-state vibrational function χ_{00}, we find that the zero-point vibrational correction to $\mathscr{P}$ can be written in the form[315]

$$\begin{aligned}\langle \chi_{00} | \mathscr{P} | \chi_{00} \rangle - \mathscr{P}_e = {} & \tfrac{1}{2} \sum_s \mathscr{B}_{ss} - \frac{3}{2\hbar} \sum_{ss'} (\mathscr{A}_s k_{ss's'}/\omega_s) \\ & - \frac{1}{4\hbar} \sum_s (\mathscr{C}_{sss}/\omega_s)(11 k_{sss} + 9 \sum_{s' \neq s} k_{ss's'}) \\ & - \frac{9}{4\hbar} \sum_{s \neq s', s''} (\mathscr{C}_{sss'}/\omega_{s'}) k_{s's''s''} \\ & - \frac{9}{2\hbar} \sum_{s \neq s'} [\mathscr{C}_{sss'} k_{sss'}/(2\omega_s + \omega_{s'})] \\ & - \frac{1}{4\hbar} \sum_{s \neq s' \neq s''} [\mathscr{C}_{ss's''} k_{ss's''}/(\omega_s + \omega_{s'} + \omega_{s''})] + \cdots \end{aligned} \tag{69}$$

For simplicity, only the zeroth- and first-order terms arising from $\mathscr{H}_N^{(0)}$ and $\mathscr{H}_N^{(1)}$ are shown in the form of sums which are unrestricted except as indicated.*

Calculations based on eqns. (68) and (69) have been carried out[315] for the electronic properties cited in Table I. A [$4s3p2d/2s1p$] GTO basis set was used to determine the property values at the same points that were required for the *ab initio* force constant calculations described in Section 3. These were fitted to an expansion in R_1', R_2', θ', and then transformed to reduced normal coordinates [eqn. (68)]. Averaging as in eqn. (69) yielded the results in Table XII for the isotopes H_2O, D_2O, and HDO. The two values of $\mathscr{P}_e$ listed there correspond to the unaveraged result for the theoretical [$\mathscr{P}_e$ (calc. min.)] and for the experimental [$\mathscr{P}_e$ (expt. min.)] minimum-energy geometries. The zero-point vibrational corrections are purely theoretical in that they are relative to the computed minimum, though it is likely that the exact values are similar. The corrections are

* This differs from the convention that we have used elsewhere for sums over normal modes (see page 56).

TABLE XII. Zero-Point Vibrational Corrections to Selected Electronic Properties[a]

Property	$\mathscr{P}_e$ (calc. min.)	$\mathscr{P}_e$ (expt. min.)	Vibrational corrections		
			H_2O	D_2O	HDO
Θ_{xx} (O), 10^{-26} esu cm^2	0.0811	0.1884	0.009	0.007	0.007
Θ_{yy} (O), 10^{-26} esu cm^2	−2.5101	−2.5690	−0.062	−0.044	−0.052
Θ_{zz} (O), 10^{-26} esu cm^2	2.4291	2.3806	0.053	0.037	0.045
μ, D ($H_2{}^+O^-$)	1.9975	2.0427	0.0061	0.0046	0.0051
$\sigma_H{}^d$, ppm	103.983	102.486	0.115	0.086	0.112
$\sigma_O{}^d$, ppm	416.515	416.101	0.019	0.014	0.016
$(eq_{yy}Q/h)_D$, kHz	−208.837	−187.902	—	2.73	2.76
$(eq_{\perp}Q/h)_D$, kHz	−161.717	−142.306	—	2.78	2.99
$(eq_{\parallel}Q/h)_D$, kHz	370.555	330.208	—	−5.51	−5.75
φ	0°39′	0°49′	—	0°8.5′	0°9.4′
$(eq_{xx}Q/h)_O$, MHz	−0.727	−1.106	−0.165	−0.122	−0.143
$(eq_{yy}Q/h)_O$, MHz	10.475	10.839	0.207	0.151	0.178
$(eq_{zz}Q/h)_O$, MHz	−9.748	−9.733	−0.042	−0.029	−0.035

[a] Based on a [4s3p2d/2s1p] GTO basis set.[(315)] All vibrational corrections are for the ^{16}O isotopic variant except for $(eqQ/h)_D[HD^{17}O]$ and $(eqQ/h)_O[H_2^{17}O,\ D_2^{17}O,\ HD^{17}O]$. Definitions of properties are given in eqns. (34)–(38) of the text and in footnotes to Table I, except that here the quadrupole moments refer to oxygen as origin.

generally seen to be on the order of 1 to 10%, and to improve agreement with experiment in most cases. The correction to $(eq_{xx}Q/h)_O$ is the largest effect, being 23% of $\mathscr{P}_e$ at the calculated potential minimum for H_2O. In situations of this type, it is probable that one obtains the most accurate results by combining $\mathscr{P}_e$ at the experimental minimum with the calculated vibrational correction.

The results in Table XII suggest that for some properties the effects of nuclear motion on the near-Hartree–Fock charge distribution may be larger than the neglected correlation effects (Section 3). Thus, it would be worthwhile to perform corresponding calculations which included configuration interaction in the electronic wave function in addition to rotational terms in the nuclear Hamiltonian, so as to obtain rigorous *ab initio* values for molecular properties as a function of the vibrational and rotational states. Such an approach is necessary for a strict comparison between theoretical calculations and the beam-maser data[(1105–1107)] in Table I. Of particular interest as a test of the generalized treatment are the isotope shifts, which can be estimated from differences between the vibrational corrections in Table XII. Accurate measurements of these shifts are needed.

5. EXTERNAL-FIELD EFFECTS

In previous sections, we have considered the Hamiltonian for the electronic and nuclear motion of a water molecule in field-free space. It is also of importance to develop a formulation for a water molecule in the presence of external fields. These could arise from the surrounding medium or in a laboratory apparatus as part of the measurement process. The effect of such fields can be introduced by adding to the Hamiltonian the terms that describe the interaction between the fields and the molecule. These terms are often sufficiently small that they do not cause substantive modifications of the wave functions or the total energy. Thus, they permit us to probe the electronic and nuclear motions without greatly affecting them. Moreover, since the various terms depend upon different aspects of the charge distribution, a given wave function can be sampled in several ways by choice of the appropriate fields.

We illustrate the effect of external-field perturbations by considering first the interaction[891,1091] between a freely rotating water molecule in an electronic state of zero orbital and spin angular momentum (e.g., the $^1A_1$ ground state), and an external magnetic field with induction **B**. The applied field interacts with the magnetic moments of the three nuclei and with the moment induced in the electronic charge cloud by molecular rotation. This is the well-known (linear) Zeeman effect; it gives rise to an energy contribution $-\boldsymbol{\mu}_m \cdot \mathbf{B}$, where $\boldsymbol{\mu}_m$ is one of the magnetic moments. In addition, the external field perturbs the electron distribution itself, producing an induced molecular moment which is directly proportional to **B**, the constant of proportionality being the magnetic susceptibility χ. This induced field can also interact with the nuclear moments and produce nuclear shielding σ which gives rise to the chemical shift. The application of a uniform electric field $\mathscr{E}$ introduces a term $-\boldsymbol{\mu}_e \cdot \mathscr{E}$, where $\boldsymbol{\mu}_e$ is the electric dipole moment. It also induces a moment proportional to $\mathscr{E}$ in the charge cloud, and the corresponding proportionality constant is called the static polarizibility α. Although the theoretical evaluation of χ, σ, and α is rather complicated, they are of considerable interest for our understanding of the molecular electronic structure. Experimentally, accurate values can be obtained for these quantities by a variety of techniques.

In addition to the external-field interactions, any nuclear moments in the molecule can couple to internal fields and produce hyperfine structure in the molecular spectrum; this can be measured by microwave or beam-resonance techniques. In rough ascending order of magnitude, these interactions arise from nuclear spin–spin coupling, nuclear spin–(molecular) rotation coupling, and nuclear electric-quadrupole coupling. The first of these is the well-known interaction between two magnetic dipoles; it involves either permanent direct dipole–dipole interactions or indirect interactions. The direct effect can be estimated from a knowledge of the nuclear moments and the molecular geometry; the indirect terms depend upon the electrons to couple one nucleus to another. The nuclear spin–rotation interaction is related to the chemical shift. It arises from the coupling of molecular rotation with the nuclear magnetic moment in the absence of an applied field. Finally, the largest effect comes about from the coupling between the electric-field gradient q at a given nucleus and its electric quadrupole moment Q; it is responsible for the nuclear quadrupole interaction ($e\mathsf{q}Q/h$) that we have already defined in eqn. (38).

To show how some of these interactions are related to the molecular electronic structure, we examine expressions for the magnetic susceptibility χ, the nuclear shielding σ, and the polarizibility α of water in its ground electronic state. These quantities, all of which are second-rank tensors, have the form

$$\chi = -(e^2/4mc^2)\langle\Phi_0|\sum_{k=1}^{10}(r_k^2\mathsf{1} - \mathbf{r}_k\mathbf{r}_k)|\Phi_0\rangle + (e^2/2m^2c^2)\sum_{n=1}^{\infty}\left[|\langle\Phi_0|\sum_{k=1}^{10}\mathbf{M}_k|\Phi_n\rangle|^2 \Big/ (E_n - E_0)\right] \tag{70}$$

$$\sigma = (e^2/2mc^2)\langle\Phi_0|\sum_{k=1}^{10}(1/r_k^3)(r_k^2\mathsf{1} - \mathbf{r}_k\mathbf{r}_k)|\Phi_0\rangle - (e^2/2m^2c^2) \times \sum_{n=1}^{\infty}\frac{\left\{\left[\langle\Phi_0|\sum_{k=1}^{10}\mathbf{M}_k|\Phi_n\rangle\langle\Phi_n|\sum_{k=1}^{10}(\mathbf{M}_k/r_k^3)|\Phi_0\rangle\right] + [\text{c.c.}]\right\}}{(E_n - E_0)} \tag{71}$$

(where c.c. indicates complex conjugate) and

$$\alpha = 2e^2\sum_{n=1}^{\infty}\left[|\langle\Phi_0|\sum_{k=1}^{10}\mathbf{r}_k|\Phi_n\rangle|^2 \Big/ (E_n - E_0)\right] \tag{72}$$

The first two terms of χ and σ depend only on the ground state Φ_0; they are referred to as the diamagnetic contributions ($\chi^{\mathrm{d}}, \sigma^{\mathrm{d}}$) and describe the effect expected from a homogeneous magnetic field if the electron cloud were as free to respond as in a closed-shell atom (e.g., He). The second terms of χ and σ correspond to a change in the molecular response due to asphericity (see Fig. 3), and are designated as the paramagnetic contributions χ^{p} and σ^{p}, respectively. They both depend on the angular momentum operator

$\mathbf{M}_k = \mathbf{r}_k \times \mathbf{p}_k$ for electron k and on the complete spectrum of excited states, including those in the continuum. Since we only have a reasonably satisfactory description of the ground state (Section 3), indirect procedures are needed to evaluate these excited-state sums, which make a significant contribution in the water molecule (see Table XIII).

Corresponding to the operators appearing in eqns. (70)–(72), we modify the Hamiltonian $\mathscr{H}_e$ in eqn. (7) to include contributions from the applied magnetic and electric fields $\mathbf{B}$ and $\mathscr{E}$, and from the magnetic dipole moment of a nucleus N, $\boldsymbol{\mu}_N$; additional nuclei are introduced by summing over N. The appropriate form is

$$\begin{aligned}\mathscr{H}(\mathbf{B}, \boldsymbol{\mu}_N, \mathscr{E}) = \mathscr{H}_e &+ (e/2mc)\mathbf{B} \cdot \sum_{k=1}^{10} \mathbf{M}_k + (e/mc)\boldsymbol{\mu}_N \cdot \sum_{k=1}^{10} (\mathbf{M}_k/r_k^3) \\ &+ e\mathscr{E} \cdot \sum_{k=1}^{10} \mathbf{r}_k + (e^2/8mc^2)\mathbf{B} \cdot \left[\sum_{k=1}^{10} (r_k^2\mathbf{1} - \mathbf{r}_k\mathbf{r}_k)\right] \cdot \mathbf{B} \\ &+ (e^2/2mc^2)\boldsymbol{\mu}_N \cdot \left[\sum_{k=1}^{10} (1/r_k^3)(r_k^2\mathbf{1} - \mathbf{r}_k\mathbf{r}_k)\right] \cdot \mathbf{B} + \cdots \end{aligned} \tag{73}$$

Because the field-dependent terms add a very small amount of energy to $\mathscr{H}_e$, it is natural to describe their effects on the molecular charge distribution by using perturbation theory. Accordingly, we expand the total wave function as

$$\Phi(\mathbf{B}, \boldsymbol{\mu}_N, \mathscr{E}) = \Phi_0 + (e/2mc)\mathbf{B}\Phi_0^{(\mathbf{B})} + (e/mc)\boldsymbol{\mu}_N\Phi_0^{(\mu)} + e\mathscr{E}\Phi_0^{(\mathscr{E})} + \cdots \tag{74}$$

with successive terms specifying the amount of change in the unperturbed wave function (Φ_0) that results from adding the respective field contributions to $\mathscr{H}_e$. If $\Phi(\mathbf{B}, \boldsymbol{\mu}_N, \mathscr{E})$ is normalized, the total energy including the field perturbations is given by

$$E(\mathbf{B}, \boldsymbol{\mu}_N, \mathscr{E}) = \langle \Phi(\mathbf{B}, \boldsymbol{\mu}_N, \mathscr{E}) \,|\, \mathscr{H}(\mathbf{B}, \boldsymbol{\mu}_N, \mathscr{E}) \,|\, \Phi(\mathbf{B}, \boldsymbol{\mu}_N, \mathscr{E})\rangle \tag{75}$$

in the Born–Oppenheimer approximation.

Substitution of eqns. (73) and (74) into eqn. (75) and comparison with the definitions of χ, σ, and α in eqns. (70)–(72) show that

$$\chi = -(\partial^2 E/\partial \mathbf{B}^2)_{\mathbf{B}=\boldsymbol{\mu}_N=\mathscr{E}=0} \tag{76a}$$

$$\sigma(N) = (\partial^2 E/\partial \boldsymbol{\mu}_N \partial \mathbf{B})_{\mathbf{B}=\boldsymbol{\mu}_N=\mathscr{E}=0} \tag{76b}$$

and

$$\alpha = -(\partial^2 E/\partial \mathscr{E}^2)_{\mathbf{B}=\boldsymbol{\mu}_N=\mathscr{E}=0} \tag{76c}$$

TABLE XIII. Comparisons of Calculated and Experimental Values for Selected Electric and Magnetic Properties of the Water Molecule[a]

Property	*xx* component		*yy* component		*zz* component	
	Calculated	Experimental	Calculated	Experimental	Calculated	Experimental
α_{ii}, Å^3	1.162	—	1.069	—	1.279	—
χ_{ii}^{d}, ppm cgs mol^{-1}	−15.581	−14.5 ± 2.0	−16.411	−15.6 ± 2.0	−14.762	−13.6 ± 2.0
$\chi_{ii}^{d} - \bar{\chi}^{d}$, ppm cgs mol^{-1}	0.004	0.147 ± 0.016	−0.826	−1.073 ± 0.012	0.883	0.929 ± 0.008
χ_{ii}^{p}, ppm cgs mol^{-1}	1.034	1.38	2.061	2.57	0.540	0.810
χ_{ii}, ppm cgs mol^{-1}	−14.547	−13.1	−14.350	−13.0	−14.222	−12.8
$\chi_{ii} - \bar{\chi}$, ppm cgs mol^{-1}	−0.174	−0.066 ± 0.016	0.023	−0.089 ± 0.001	0.151	0.155 ± 0.008
σ_{ii}^{d}(D), ppm	102.8	—	130.2	—	75.3	—
σ_{ii}^{p}(D), ppm	−52.6	−71.3	−79.2	−108.3	−25.8	−37.3
σ_{ii}(D), ppm	50.2	—	51.0	—	49.5	—
σ_{ii}^{d}(O), ppm	415.2	—	414.8	—	413.4	—
σ_{ii}^{p}(O), ppm	−108.2	—	−91.1	—	−53.6	—
σ_{ii}(O), ppm	307.0	—	323.7	—	359.8	—

[a] All experimental results taken from Ref. 1106. The calculated values(28,33,348) are based on the $(5s3p1d/2s1p)$ STO function of Arrighini *et al.*,(29) which has an energy of $-76.0384e^2/a_0$ at $R = 1.8103a_0$ and $\theta = 105°$. The origin of the vector potential is at the H_2O center of mass for χ, at $(x, y, z) = (-0.0056$ Å, 0, 0) for σ(O), and at deuterium for σ(D). The experimental data are with reference to D_2O.

if we identify $\Phi_0^{(\mathbf{B})}$, $\Phi_0^{(\mu)}$, and $\Phi_0^{(\mathscr{E})}$ with the usual first-order perturbation theory expressions; that is,

$$\Phi_0^{(\mathbf{B})} = -\sum_{n=1}^{\infty}\left[\langle\Phi_n \mid \sum_{k=1}^{10} \mathbf{M}_k \mid \Phi_0\rangle \Big/ (E_n - E_0)\right]\Phi_n \tag{77a}$$

$$\Phi_0^{(\mu)} = -\sum_{n=1}^{\infty}\left[\langle\Phi_n \mid \sum_{k=1}^{10} (\mathbf{M}_k/r_k^{3}) \mid \Phi_0\rangle \Big/ (E_n - E_0)\right]\Phi_n \tag{77b}$$

and

$$\Phi_0^{(\mathscr{E})} = -\sum_{n=1}^{\infty}\left[\langle\Phi_n \mid \sum_{k=1}^{10} \mathbf{r}_k \mid \Phi_0\rangle \Big/ (E_n - E_0)\right]\Phi_n \tag{77c}$$

These results suggest that, rather than using the formal summations in eqns. (70)–(72) or eqns. (77) to evaluate χ, σ, and α, it may be possible to construct the perturbed wave functions by an alternative, more practical procedure based on the derivative relationships in eqn. (76). This can be done by generalizing the argument which led to the Hellmann–Feynman force condition in eqn. (47). The essential point to notice is that $\mathbf{\Delta}_N$ defined in eqn. (46) and its derivatives are equal to zero for either Hartree–Fock or exact wave functions. Applications of this condition to field-dependent forces lead to necessary conditions on the perturbed wave functions $\Phi_0^{(\mathbf{B})}$, $\Phi_0^{(\mu)}$, $\Phi_0^{(\mathscr{E})}$, and to more tractable procedures for the determination of their associated properties.

5.1. Perturbed Hartree–Fock Method

In Section 3 we described the Hartree–Fock solution $\tilde{\Phi}_0$ for the ground state of the water molecule in the absence of external fields. We wrote this approximate solution of the Schrödinger equation in determinantal form [eqn. (19)], and stated that it satisfied an eigenvalue equation, eqn. (20). If the SCF-MO's ϕ_i appearing in $\tilde{\Phi}_0$ are now taken to be field-dependent, eqns. (19)–(28) apply except that each quantity in them depends on $\mathbf{B}$, $\boldsymbol{\mu}_N$, and $\mathscr{E}$. Since $\tilde{\Phi}_0$ is not an eigenfunction of the exact Hamiltonian, it is appropriate to introduce an effective Hamiltonian $\mathscr{H}$ such that $\tilde{\Phi}_0$ satisfies the equation

$$\mathscr{H}(\mathbf{B}, \boldsymbol{\mu}_N, \mathscr{E})\tilde{\Phi}_0(\mathbf{B}, \boldsymbol{\mu}_N, \mathscr{E}) = \tilde{E}_0(\mathbf{B}, \boldsymbol{\mu}_N, \mathscr{E})\tilde{\Phi}_0(\mathbf{B}, \boldsymbol{\mu}_N, \mathscr{E}) \tag{78}$$

where

$$\begin{aligned}\tilde{E}_0(\mathbf{B}, \boldsymbol{\mu}_N, \mathscr{E}) &= \langle\tilde{\Phi}_0(\mathbf{B}, \boldsymbol{\mu}_N, \mathscr{E}) \mid \mathscr{H}(\mathbf{B}, \boldsymbol{\mu}_N, \mathscr{E}) \mid \tilde{\Phi}_0(\mathbf{B}, \boldsymbol{\mu}_N, \mathscr{E})\rangle \\ &= \langle\tilde{\Phi}_0(\mathbf{B}, \boldsymbol{\mu}_N, \mathscr{E}) \mid \mathscr{H}(\mathbf{B}, \boldsymbol{\mu}_N, \mathscr{E}) \mid \tilde{\Phi}_0(\mathbf{B}, \boldsymbol{\mu}_N, \mathscr{E})\rangle\end{aligned} \tag{79}$$

For the field-dependent single-determinant function of eqn. (19), the effective Hamiltonian $\tilde{\mathscr{H}}$ must have the form (omitting the constant nuclear repulsion term)

$$\tilde{\mathscr{H}}(\mathbf{B}, \boldsymbol{\mu}_N, \mathscr{E}) = \sum_{i=1}^{10} F(i; \mathbf{B}, \boldsymbol{\mu}_N, \mathscr{E}) - \tfrac{1}{2}\sum_{i=1}^{10}\sum_{j=1}^{10} [J_{ij}(\mathbf{B}, \boldsymbol{\mu}_N, \mathscr{E}) - K_{ij}(\mathbf{B}, \boldsymbol{\mu}_N, \mathscr{E})] \qquad (80)$$

in order to be consistent with eqn. (27).

Introduction of the many-electron Hamiltonian $\tilde{\mathscr{H}}$ corresponds to the investigation of a model system whose Hamiltonian depends upon the occupied one-electron functions, $\phi_i(\mathbf{B}, \boldsymbol{\mu}_N, \mathscr{E})$. Since $\tilde{\Phi}_0$ is an exact eigenfunction of $\tilde{\mathscr{H}}$, generalized forms of eqns. (44)–(46) are applicable. Denoting the applied field and moment vectors such as $\mathbf{B}$, $\boldsymbol{\mu}_N$, and $\mathscr{E}$ by $\boldsymbol{\lambda}_1, \boldsymbol{\lambda}_2, \boldsymbol{\lambda}_3, \ldots, \boldsymbol{\lambda}_t$ and remembering that all the relevant operators and wave functions are dependent on them, we can write

$$\tilde{\mathbf{F}}_{t\ldots1} = \tilde{\mathscr{F}}_{t\ldots1} + \tilde{\boldsymbol{\Delta}}_{t\ldots1} \qquad (81)$$

where

$$\tilde{\mathscr{F}}_{t\ldots1} = (\partial/\partial\boldsymbol{\lambda}_t)\cdots(\partial/\partial\boldsymbol{\lambda}_2)\langle\tilde{\Phi}_0 \mid \partial\tilde{\mathscr{H}}/\partial\boldsymbol{\lambda}_1 \mid \tilde{\Phi}_0\rangle \qquad (82)$$

and

$$\tilde{\boldsymbol{\Delta}}_{t\ldots1} = (\partial/\partial\boldsymbol{\lambda}_t)\cdots(\partial/\partial\boldsymbol{\lambda}_2)\{[\langle\partial\tilde{\Phi}_0/\partial\boldsymbol{\lambda}_1 \mid \tilde{\mathscr{H}} \mid \tilde{\Phi}_0\rangle] + [\text{complex conjugate}]\} \qquad (83)$$

The derivatives with respect to $\boldsymbol{\lambda}_1$ of $\tilde{\mathscr{H}}$ and $\tilde{\Phi}_0$ can be expanded, by introducing the form of $\tilde{\mathscr{H}}$ [eqn. (80)] and the determinantal form of $\tilde{\Phi}_0$, to yield

$$\partial\tilde{\mathscr{H}}/\partial\boldsymbol{\lambda}_1 = \sum_{i=1}^{10} [\partial F(i)/\partial\boldsymbol{\lambda}_1] - \sum_{i=1}^{10}\sum_{j=1}^{10} \{([\phi_i\phi_i \mid \boldsymbol{\Lambda}_j^1\phi_j] - [\phi_i\phi_j \mid \boldsymbol{\Lambda}_j^1\phi_i]) + (\text{complex conjugate})\} \qquad (84)$$

and

$$\partial\tilde{\Phi}_0/\partial\boldsymbol{\lambda}_1 = [(10)!]^{-1/2} \sum_{i=1}^{10} \det[\phi_1(1)\cdots\boldsymbol{\Lambda}_i^1(i)\cdots\phi_{10}(10)] \qquad (85)$$

where

$$\boldsymbol{\Lambda}_i^{t\ldots1} = \partial^t\phi_i/\partial\boldsymbol{\lambda}_t\cdots\partial\boldsymbol{\lambda}_1 \qquad (86)$$

From the definition of $F(i; \boldsymbol{\lambda}_1, \ldots, \boldsymbol{\lambda}_t)$ [eqns. (20)–(22)], we have

$$\frac{\partial F}{\partial \boldsymbol{\lambda}_1} = \frac{\partial H}{\partial \boldsymbol{\lambda}_1} + e^2 \Big\{ \sum_{j=1}^{10} \Big[\Big[\int d\mathbf{r}_l\, \Lambda_j^{1*}(l)\phi_j(l) \frac{1}{|\mathbf{r}_k - \mathbf{r}_l|} - \phi_j(k) \int d\mathbf{r}_l\, \Lambda_j^{1*}(l) \frac{P_{kl}}{|\mathbf{r}_k - \mathbf{r}_l|} \Big] \Big\} + \{\text{complex conjugate}\} \tag{87}$$

Combining eqns. (84) and (87) and substituting into eqn. (82), we obtain

$$\tilde{\mathscr{F}}_{t\ldots 1} = (\partial/\partial\boldsymbol{\lambda}_t)\cdots(\partial/\partial\boldsymbol{\lambda}_2) \sum_{i=1}^{10} \langle \phi_i \,|\, \partial H/\partial\boldsymbol{\lambda}_1 \,|\, \phi_i \rangle \tag{88}$$

Correspondingly,

$$\tilde{\boldsymbol{\Delta}}_{t\ldots 1} = \frac{\partial}{\partial \boldsymbol{\lambda}_t} \cdots \frac{\partial}{\partial \boldsymbol{\lambda}_2} \Big\{ \sum_{i=1}^{10} \Big[\langle \boldsymbol{\Lambda}_i^1 \,|\, H \,|\, \phi_i \rangle + \sum_{j=1}^{10} ([\phi_j\phi_j \,|\, \boldsymbol{\Lambda}_i^1 \phi_i] - [\phi_j\phi_i \,|\, \boldsymbol{\Lambda}_i^1 \phi_j]) \Big] \Big\} + \text{complex conjugate} = 0 \tag{89}$$

The term in braces vanishes and $\tilde{\mathbf{F}}_{t\ldots 1} = \tilde{\mathscr{F}}_{t\ldots 1}$ when $\tilde{\Phi}_0$ is the *exact* Hartree–Fock solution, as indicated in eqn. (47). This can be shown explicitly from the form of eqn. (89) by use of the fact that $\boldsymbol{\Lambda}_i^1$ can be expanded in a complete set of Hartree–Fock orbitals. There results an expression involving the sum of off-diagonal elements of the Hartree–Fock matrix, which are equal to zero. For an approximate Hartree–Fock solution obtained using a finite basis-set expansion, eqn. (89) is only satisfied within the subspace of these functions.

Equation (89) provides a condition on the MSO derivatives $\boldsymbol{\Lambda}_i^{t\ldots 1}$ and thereby a means for obtaining the perturbed Hartree–Fock wave function, which can then be used to evaluate $\tilde{\mathscr{F}}_{t\ldots 1}$ from eqn. (88). To demonstrate this, we consider the important case $t = 2$ that is required in eqn. (76) for χ, σ, and α; that is, for these respective second-order properties, we make the identification $\boldsymbol{\lambda}_1 = \boldsymbol{\lambda}_2 = \mathbf{B}$; $\boldsymbol{\lambda}_1 = \mathbf{B}, \boldsymbol{\lambda}_2 = \boldsymbol{\mu}_N$; $\boldsymbol{\lambda}_1 = \boldsymbol{\lambda}_2 = \mathscr{E}$. From eqn. (89), we find for any pair of these vectors that

$$\begin{aligned}\tilde{\boldsymbol{\Delta}}_{21} = \Big[&\langle \boldsymbol{\Lambda}_i^{21} \,|\, H \,|\, \phi_i \rangle + \sum_{j=1}^{10} ([\phi_j\phi_j \,|\, \boldsymbol{\Lambda}_i^{21}\phi_i] - [\phi_j\phi_i \,|\, \boldsymbol{\Lambda}_i^{21}\phi_j]) \\ &+ \langle \boldsymbol{\Lambda}_i^1 \,|\, H \,|\, \boldsymbol{\Lambda}_i^2 \rangle + \sum_{j=1}^{10} ([\phi_j\phi_j \,|\, \boldsymbol{\Lambda}_i^1\boldsymbol{\Lambda}_i^2] - [\phi_j\boldsymbol{\Lambda}_i^2 \,|\, \boldsymbol{\Lambda}_i^1\phi_j]) \\ &+ \langle \boldsymbol{\Lambda}_i^1 \,|\, \partial H/\partial\boldsymbol{\lambda}_2 \,|\, \phi_i \rangle + \sum_{j=1}^{10} ([\phi_j\boldsymbol{\Lambda}_j^2 \,|\, \boldsymbol{\Lambda}_i^1\phi_i] - [\phi_j\phi_i \,|\, \boldsymbol{\Lambda}_i^1\boldsymbol{\Lambda}_j^2]) \\ &+ \sum_{j=1}^{10} ([\boldsymbol{\Lambda}_j^2\phi_j \,|\, \boldsymbol{\Lambda}_i^1\phi_i] - [\boldsymbol{\Lambda}_j^2\phi_i \,|\, \boldsymbol{\Lambda}_i^1\phi_j]) \Big] + [\text{complex conjugate}] = 0\end{aligned} \tag{90}$$

The function $\boldsymbol{\Lambda}_i^{21}$ in eqn. (90) can be eliminated by the orthogonality condition, which requires that

$$(\langle\phi_i \mid \boldsymbol{\Lambda}_i^{21}\rangle + \langle\boldsymbol{\Lambda}_i^{2} \mid \boldsymbol{\Lambda}_i^{1}\rangle) + (\text{complex conjugate}) = 0 \tag{91}$$

Using the expansions

$$\boldsymbol{\Lambda}_i^{r} = \sum_p^{\infty} \mathbf{a}_{pi}^{(r)}\phi_p, \qquad r = 1, 2 \tag{92}$$

and

$$\boldsymbol{\Lambda}_i^{21} = \sum_p^{\infty} \mathbf{b}_{pi}^{(21)}\phi_p \tag{93}$$

together with eqns. (90) and (91), we obtain

$$\begin{aligned}(\varepsilon_p - \varepsilon_q)\mathbf{a}_{pq}^{(2)} &+ \langle\phi_p \mid \partial H/\partial\boldsymbol{\lambda}_2 \mid \phi_q\rangle + \sum_{i=1}^{10}\sum_{j=i+1}^{\infty} [([\phi_i\phi_j \mid \phi_p\phi_q] \\ &- [\phi_i\phi_q \mid \phi_p\phi_j])\mathbf{a}_{ji}^{(2)} + ([\phi_j\phi_i \mid \phi_p\phi_q] - [\phi_j\phi_q \mid \phi_p\phi_i])\mathbf{a}_{ji}^{(2)*}] \\ &+ \text{complex conjugate} = 0\end{aligned} \tag{94}$$

The set of linear inhomogeneous equations can be solved for the $\mathbf{a}_{pq}^{(2)}$ which are needed to determine $\tilde{\mathscr{F}}_{21}$. A corresponding equation for the $\mathbf{a}_{pq}^{(1)}$ can be obtained by interchanging the superscripts 1 and 2 on the $\boldsymbol{\Lambda}_i^{t\cdots 1}$ in eqn. (90).

Equation (94) is identical to what is usually referred to as the fully coupled form of the perturbed Hartree–Fock equation.[(1021)]* Although it has been used rather extensively[(1017–1020)] to study magnetic susceptibilities, chemical shielding, and polarizabilities in diatomic molecules, applications to the water molecule have been reported only recently. The theoretical chemistry group at Pisa[(28–33,348)] has been the most active in implementing the procedure for water and for a few other small polyatomic molecules (NH_3, CH_4, CH_3F). Some of their H_2O results are summarized and compared to the available experimental data in Table XIII. A ($5s3p1d/2s1p$) STO basis set is used in their most accurate calculations (see Table I for a comparison with other functions); its unperturbed energy is[29] $-76.0384e^2/a_0$ near the equilibrium geometry ($R = 1.8103a_0$, $\theta = 105°$). This means that 29 MO's generated from the same number of AO's in the basis set (counting all components of the p and d orbitals), are included in the expansion summations in eqns. (92) and (93). The level of agreement between the computed and observed values in Table XIII suggests that for properties such

* Langhoff *et al.*[(651)] discuss approximations to the fully coupled procedure.

as the chemical shielding constant of the proton, the choice of basis-set polarization functions and orbital exponents is probably not optimum. It is apparent from Table I that this is true for the unperturbed wave function (e.g., a $(5s4p1d/3s1p)$ STO function is seen to yield a lower energy); and it is probably true for the perturbed wave functions as well. Additional calculations with improved basis sets over a range of geometries (incorporating zero-point vibrational corrections as described in Section 4) are needed to make definitive comparisons with the available measurements.

5.2. Perturbed Configuration Interaction Method

Although the perturbed Hartree–Fock model limit has proved generally to be a reliable tool for evaluating second-order electric and magnetic field properties, it may be important to include electron correlation in the wave function for some purposes. For example, we see from Table XIII that the diamagnetic and paramagnetic contributions to σ are of the same order of magnitude but opposite in sign. Correlation could affect these two terms differently, and this difference may be especially critical when taking second differences to determine shifts among the various isotopic species of the water molecule.

A perturbation scheme that includes electron correlation can be formulated in the configuration interaction approach, which expresses the solution of the Schrödinger equation as a complete set of determinants [eqn. (32)]. We consider the perturbed electronic Hamiltonian

$$\mathscr{H} = \mathscr{H}_e + \sum_{k=1}^{10} h_k \tag{95}$$

where the operator h_k represents a one-electron perturbation such as those in eqn. (73). In contrast to the Hartree–Fock Hamiltonian $\tilde{\mathscr{H}}$, $\mathscr{H}$ is independent of the wave function, although it depends on the $\{\boldsymbol{\lambda}_t\}$ through h_k.

Expanding the derivatives of Φ_0 in the same complete set of functions which were employed for Φ_0 itself [eqn. (32)],

$$\frac{\partial \Phi_0}{\partial \boldsymbol{\lambda}_r} = \sum_{p}^{\infty} d_p^{(r)} \tilde{\Phi}_p \tag{96}$$

and using eqns. (32) and (33) with the index n suppressed, we obtain

$$\tilde{\boldsymbol{\Delta}}_{t\ldots 1} = \frac{\partial}{\partial \boldsymbol{\lambda}_t} \cdots \frac{\partial}{\partial \boldsymbol{\lambda}_2} \left[E_0 \Big(\sum_{m}^{\infty} c_m{}^* d_m^{(1)} + c_m d_m^{(1)*} \Big) \right] = 0 \tag{97}$$

as the analog of eqn. (83). Just as $\boldsymbol{\Lambda}_i^r$ in eqn. (92) can be calculated from the first-order Hartree–Fock equations, so the coefficients $d_p^{(1)}$ in eqn. (97) specialized for $t = 2$ can be evaluated from

$$\boldsymbol{\Delta}_{21} = \boldsymbol{\Delta}_{12} = \left(\left\langle \Phi_0 \mid \mathscr{H} \mid \frac{\partial^2 \Phi_0}{\partial \boldsymbol{\lambda}_1 \, \partial \boldsymbol{\lambda}_2} \right\rangle + \left\langle \Phi_0 \left| \frac{\partial \mathscr{H}}{\partial \boldsymbol{\lambda}_1} \right| \frac{\partial \Phi_0}{\partial \boldsymbol{\lambda}_2} \right\rangle \right.$$
$$\left. + \left\langle \frac{\partial \Phi_0}{\partial \boldsymbol{\lambda}_1} \mid \mathscr{H} \mid \frac{\partial \Phi_0}{\partial \boldsymbol{\lambda}_2} \right\rangle \right) + (\text{c.c.}) = 0 \qquad (98)$$

where c.c. indicates complex conjugate. Proceeding in a manner similar to the Hartree–Fock treatment [eqns. (90)–(94)], we find the set of coupled equations

$$\left(\sum_m^{\infty} c_m{}^* \langle \tilde{\Phi}_m \mid \partial \mathscr{H} / \partial \boldsymbol{\lambda}_1 \mid \tilde{\Phi}_q \rangle \right.$$
$$\left. + \sum_p^{\infty} d_p^{(1)*} \langle \tilde{\Phi}_p \mid \mathscr{H} \mid \tilde{\Phi}_q \rangle - \langle \Phi_0 \mid \mathscr{H} \mid \Phi_0 \rangle d_q^{(1)*} \right) + (\text{c.c.}) = 0 \qquad (99)$$

for the $d_p^{(1)*}$ coefficients. The linear equations for $d_p^{(2)*}$ can be obtained from eqn. (99) by replacing $\boldsymbol{\lambda}_1$ and $d_\alpha^{(1)*}$, $\alpha = p, q$, by $\boldsymbol{\lambda}_2$ and $d_\alpha^{(2)*}$, $\alpha = p, q$. Similar conditions can be found for higher t values.

Although the solution of the infinite set of equations given by (99) would yield the exact first-order perturbed wave function, we wish to emphasize that in practice only a limited set of starting functions $\{\tilde{\Phi}_p\}$ is available to span the space of $\partial \Phi_0 / \partial \boldsymbol{\lambda}_r$. The result is that the upper limits on the infinite summations in (99) become finite and lead to a finite set of linear inhomogeneous equations. When the $\tilde{\Phi}_p$ are Hartree–Fock solutions, the number of $\tilde{\Phi}_p$ which can be constructed depends upon the size of the one-electron basis-set expansion in eqn. (31). Once the exact or approximate $d_p^{(1)}$ and $d_p^{(1)*}$ coefficients are known, equations of the form

$$\mathscr{F}_{12} = \sum_p^{\infty} \left(d_p^{(1)} \langle \Phi_0 \mid \partial \mathscr{H} / \partial \boldsymbol{\lambda}_2 \mid \tilde{\Phi}_p \rangle + d_p^{(1)*} \langle \tilde{\Phi}_p \mid \partial \mathscr{H} / \partial \boldsymbol{\lambda}_2 \mid \Phi_0 \rangle \right) \qquad (100)$$

can be used to evaluate χ, σ, α, or other second-order properties. Implementation of this procedure for water would be very worthwhile. These formulations can also be generalized[515] to include time-dependent perturbations such as those required for the frequency-dependent polarizability; however, neither theoretical treatments nor experimental data on these interactions are currently available for the isolated water molecule.

6. CONCLUSION

In this chapter, we have sought to show how quantum mechanical theory and spectroscopic measurements combine to provide quantitative information on the electronic structure and the nuclear potential function of the water molecule. It appears to us that, although more is known about the electronic ground state of water than any other nonlinear polyatomic molecule, much remains to be learned. However, rapid theoretical and experimental advances are being made to increase our knowledge. It is our hope that the theoretical background included here will serve the reader as a summary of the present information, as well as an introduction to the subsequent chapters of this treatise on interactions of water molecules in gas, liquid, and solid phases.

ACKNOWLEDGMENTS

Many colleagues have helped us in significant ways to compile this survey. We are most indebted to T. H. Dunning, Jr. and R. M. Pitzer for performing many of the calculations in Table I and for a large number of other favors. We are very grateful to Dr. Dunning and Dr. N. W. Winter for preparing the electron density maps in Figs. 3–7. The following also kindly supplied data in advance of publication and/or special assistance: J. D. Allen, D. M. Bishop, M. J. T. Bowers, C. R. Claydon, C. A. Coulson, W. C. Ermler, F. D. Glasser, W. A. Goddard, III, W. J. Hunt, B. J. Krohn, C. W. Mathews, W. Meyer, W. G. Richards, P. J. Rossky, I. Shavitt, O. J. Sovers, W. J. Taylor, S. Trajmar, J. Verhoeven, A. C. Wahl, J. K. G. Watson, and L. H. Weller, Jr.

APPENDIX A. BIBLIOGRAPHY OF THEORETICAL CALCULATIONS ON THE ELECTRONIC STRUCTURE OF THE WATER MOLECULE (1925–1970)

This bibliography of calculations on the water molecule complements existing tabulations [M. Krauss, Compendium of *Ab Initio* Calculations of Molecular Energies and Properties, NBS Technical Note 438 (1967): L. C. Allen, *Ann. Rev. Phys. Chem.* **20**, 315 (1969); R. G. Clark and E. T. Stewart, *Quart. Rev.* **24**, 95 (1970); W. G. Richards, T. E. H. Walker, and R. K. Hinkley, "Bibliography of *Ab Initio* Molecular Wave Functions,"

Clarendon Press, Oxford (1970)] by the inclusion of both semiempirical and *ab initio* studies during the period 1925–1970. References to work on water polymers are arbitrarily excluded, even though they usually contain, as a preliminary, results on a single water molecule. With this constraint, our compilation pretends to be complete; however, the vast literature on water almost certainly makes this a hope rather than a claim.

1925

F. Hund, Die Gestalt Mehratomiger Polarer Molekeln. I., *Z. Physik* **31**, 81 (1925); II. Molekeln, die aus einem Negativen Ion und aus Wasserstoffkernen Bestehen, *Z. Physik* **32**, 1 (1925).

1931

L. Pauling, The Nature of the Chemical Bond. Applications of Results Obtained from the Quantum Mechanics and from a Theory of Paramagnetic Susceptibility to the Structure of Molecules, *J. Am. Chem. Soc.* **53**, 1367 (1931).

J. C. Slater, Molecular Energy Levels and Valence Bonds, *Phys. Rev.* **38**, 1109 (1931).

1932

A. S. Coolidge, A Quantum Mechanics Treatment of the Water Molecule, *Phys. Rev.* **42**, 189 (1932).

R. S. Mulliken, Electronic Structures of Polyatomic Molecules and Valence, *Phys. Rev.* **40**, 55 (1932).

1933

R. S. Mulliken, Electronic Structures of Polyatomic Molecules and Valence. V. Molecules RX_n, *J. Chem. Phys.* **1**, 492 (1933).

J. H. Van Vleck and P. C. Cross, A Calculation of the Vibration Frequencies and Other Constants of the H_2O Molecule, *J. Chem. Phys.* **1**, 357 (1933).

1935

R. S. Mulliken, Electronic Structures of Polyatomic Molecules. VII. Ammonia and Water-Type Molecules and Their Derivatives, *J. Chem. Phys.* **3**, 506 (1935); XI. Electroaffinity, Molecular Orbitals, and Dipole Moments, *J. Chem. Phys.* **3**, 573 (1935); XII. Electroaffinity and Molecular Orbitals, Polyatomic Applications, *J. Chem. Phys.* **3**, 586 (1935).

1941

C. A. Coulson, Quantum Theory of the Chemical Bond, *Proc. Roy. Soc.* (*Edinburgh*) **A61**, 115 (1941).

1948

D. F. Heath and J. W. Linnett, Molecular Force Fields. I. The Structure of the Water Molecule, *Trans. Faraday Soc.* **44**, 556 (1948).

1949

C. A. Coulson, Liaisons Localisées et Non-Localisées, *J. Chim. Phys.* **46**, 198 (1949).

1950

J. A. Pople, The Molecular Orbital Theory of Chemical Valency. V. The Structure of Water and Similar Molecules, *Proc. Roy. Soc.* (*London*) **A202**, 323 (1950).

1951

C. A. Coulson, Critical Survey of the Method of Ionic-Homopolar Resonance, *Proc. Roy. Soc.* (*London*) **A207**, 63 (1951).

1953

J. W. Linnett and A. J. Poe, Directed Valency in Elements of the First Short Period, *Trans. Faraday Soc.* **49**, 217 (1953).

A. B. F. Duncan and J. A. Pople, The Structure of Some Simple Molecules with Lone-Pair Electrons, *Trans. Faraday Soc.* **49**, 217 (1953).

F. O. Ellison and H. Shull, An LCAO MO Self-Consistent Field Calculation of the Ground State of H_2O, *J. Chem. Phys.* **21**, 1420 (1953).

J. A. Pople, The Electronic Structure and Polarity of the Water Molecule, *J. Chem. Phys.* **21**, 2234 (1953).

A. D. Walsh, The Electronic Orbitals, Shapes, and Spectra of Polyatomic Molecules. Part I. AH_2 Molecules, *J. Chem. Soc.* 2260 (1953).

1955

F. O. Ellison and H. Shull, Molecular Calculations. I. LCAO MO Self-Consistent Field Treatment of the Ground State of H_2O, *J. Chem. Phys.* **23**, 2348 (1955).

F. O. Ellison, Molecular Calculations. II. Methods of Approximation of Molecular Integrals, *J. Chem. Phys.* **23**, 2358 (1955).

R. S. Mulliken, Bond Angles in Water-Type and Ammonia-Type Molecules and Their Derivatives, *J. Am. Chem. Soc.* **77**, 887 (1955).

R. S. Mulliken, Electronic Population Analysis on LCAO MO Molecular Wave Functions, *J. Chem. Phys.* **23**, 1833 (1955).

1956

K. E. Banyard and N. H. March, X-Ray Scattering by "Neon-like" Molecules, *Acta Cryst.* **9**, 385 (1956).

S. F. Boys, G. B. Cook, C. M. Reeves, and I. Shavitt, Automatic Fundamental Calculations of Molecular Structure, *Nature* **178**, 1207 (1956).

S. Smith and J. W. Linnett, Molecular Force Fields. XVI. Force Constants in Some Non-linear Triatomic Molecules, *Trans. Faraday Soc.* **52**, 891 (1956).

1957

K. E. Banyard and N. H. March, Distribution of Electrons in the Water Molecule, *J. Chem. Phys.* **26**, 1416 (1957).

K. E. Banyard and N. H. March, Central-Field Approach for NH_3 and H_2O, *J. Chem. Phys.* **27**, 977 (1957).

K. Funabashi and J. L. Magee, Central-Field Approximation for the Electronic Wave Functions of Simple Molecules, *J. Chem. Phys.* **26**, 407 (1957).

J. A. Pople, The Theory of Chemical Shifts in Nuclear Magnetic Resonance. II. Interpretation of Proton Shifts, *Proc. Roy. Soc.* (*London*) **A239**, 550 (1957).

1959

T. P. Das and T. Ghose, Magnetic Properties of Water Molecule, *J. Chem. Phys.* **31**, 42 (1959).

R. Gaspar and I. Tamassy-Lentei, United Atom Model for the Hydride Molecules. HO, HO^-, and H_2O, *Acta Phys. Hung.* **10**, 149 (1959).

1960

K. E. Banyard, Diamagnetism as a Test of Wave Functions for Some Simple Molecules, *J. Chem. Phys.* **33**, 832 (1960).

R. Bersohn, Field Gradients at the Deuteron in Molecules, *J. Chem. Phys.* **32**, 85 (1960).

R. McWeeny and K. A. Ohno, A Quantum-Mechanical Study of the Water Molecule, *Proc. Roy. Soc.* (*London*) **A255**, 367 (1960).

1962

R. Gaspar, I. Tamassy-Lentei, and Y. Kruglyak, United-Atom Model for Molecules of the Type XH_n, *J. Chem. Phys.* **36**, 740 (1962).

D. P. Merrifield, A Configuration Interaction Study of the Electronic States of the Water Molecule, MIT-SSMTG Quart. Progr. Rept. No. 43, p. 27 (1962).

R. Moccia, One-Center Expansion Self-Consistent Field Molecular Orbital Electronic Wave Functions for XH_n Molecules, *J. Chem. Phys.* **37**, 910 (1962).

1963

M. Allavena and S. Bratŏz, Calcul Electronique des Constantes de Force et du Spectre Infrarouge des Molecules H_2O et D_2O, *J. Chim. Phys.* **60,** 1199 (1963).

R. F. W. Bader and G. A. Jones, The Electron Density Distributions in Hydride Molecules. I. The Water Molecule, *Can. J. Chem.* **41,** 586 (1963).

D. M. Bishop, J. R. Hoyland, and R. G. Parr, Simple One-Center Calculation of Breathing Force Constants and Equilibrium Internuclear Distances for NH_3, H_2O, and HF, *Mol. Phys.* **6,** 467 (1963).

1964

M. Krauss, Calculation of the Geometrical Structure of Some AH_n Molecules, *J. Res. Natl. Bur. Std.* **68A,** 635 (1964).

S. R. La Paglia, Theory of Rydberg Series in Polyatomic Molecules: H_2O, *J. Chem. Phys.* **41,** 1427 (1964).

R. Moccia, One-Center Basis Set SCF MO's. III. H_2O, H_2S, and CH_4, *J. Chem. Phys.* **40,** 2186 (1964).

J. A. Pople and D. P. Santry, Molecular Orbital Theory of Nuclear Spin Coupling Constants, *Mol. Phys.* **8,** 1 (1964).

J. L. J. Rosenfeld, Analysis of Calculations on Some Oxygen Hydrides, *J. Chem. Phys.* **40,** 384 (1964).

L. Zulicke, Über eine quantenmechanische Berechnung des Wassermoleküls, *Z. Naturforsch.* **19a,** 1016 (1964).

1965

D. M. Bishop, Use of Fock–Petrashen and Hydrogenic Orbitals in Single-Center Wave Functions, *J. Chem. Phys.* **43,** 3052 (1965).

L. C. Cusachs, Semi-Empirical Molecular Orbitals for General Polyatomic Molecules. II. One-Electron Model Prediction of the H–O–H Angle, *J. Chem. Phys.* **43,** S157 (1965).

R. M. Glaeser and C. A. Coulson, Multipole Moments of the Water Molecule, *Trans. Faraday Soc.* **61,** 389 (1965).

M. Klessinger, Self-Consistent Group Calculations on Polyatomic Molecules. II. Hybridization and Optimum Orbitals in Water, *J. Chem. Phys.* **43,** S117 (1965).

B. Kockel, D. Hamel, and K. Ruckelshausen, Eine Berechnung des Wassermoleküls, *Z. Naturforsch.* **20a,** 26 (1965).

J. W. Moskowitz and M. C. Harrison, Gaussian Wave Functions for the 10-Electron Systems. III. OH^-, H_2O, H_2O^+, *J. Chem. Phys.* **43,** 3550 (1965); Erratum: *Ibid.* **46,** 2019 (1967).

J. A. Pople and G. A. Segal, Approximate Self-Consistent Molecular Orbital Theory. II. Calculations with Complete Neglect of Differential Overlap, *J. Chem. Phys.* **43,** S136 (1965).

1966

D. M. Bishop and M. Randic, *Ab Initio* Calculation of Harmonic Force Constants, *J. Chem. Phys.* **44**, 2480 (1966).

D. M. Bishop and M. Randic, A Theoretical Investigation of the Water Molecule, *Mol. Phys.* **10**, 517 (1966).

D. G. Carroll, A. T. Armstrong, and S. P. McGlynn, Semi-Empirical Molecular Orbital Calculations. I. The Electronic Structure of H_2O and H_2S, *J. Chem. Phys.* **44**, 1865 (1966).

D. E. Ellis and A. Sambles, Tabulated and used by C. Edmiston and K. Ruedenberg, "Quantum Theory of Atoms, Molecules, and the Solid State," Academic Press, New York (1966), p. 263.

J. A. Pople and G. A. Segal, Approximate Self-Consistent Molecular Orbital Theory. III. CNDO Results for AB_2 and AB_3 Systems, *J. Chem. Phys.* **44**, 3289 (1966).

1967

G. P. Arrighini, M. Maestro, and R. Moccia, Electric Polarizability of Polyatomic Molecules, *Chem. Phys. Letters* **1**, 242 (1967).

R. N. Dixon, Approximate Self-Consistent Field Molecular Orbital Calculations for Valence Shell Electronic States, *Mol. Phys.* **12**, 83 (1967).

A. A. Frost, B. H. Prentice, III, and R. A. Rouse, A Simple Floating Localized Orbital Model of Molecular Structure, *J. Am. Chem. Soc.* **89**, 3064 (1967).

D. Hager, E. Hess, and L. Zulicke, Berechnungen des Wassermoleküls mit Hartree–Fock-Atomorbitalen, *Z. Naturforsch.* **22a**, 1282 (1967).

D. Hamel, Eine Berechnung des H_2O Moleküls, *Z. Naturforsch.* **22a**, 176 (1967).

J. F. Harrison, Some One-Electron Properties of H_2O and NH_3, *J. Chem. Phys.* **47**, 2990 (1967).

R. Moccia, Perturbed SCF MO Calculations. Electrical Polarizability and Magnetic Susceptibility of HF, H_2O, NH_3, and CH_4, *Theor. Chim. Acta* **8**, 192 (1967).

J. A. Pople, D. L. Beveridge, and P. A. Dobosh, Approximate Self-Consistent Molecular-Orbital Theory. V. Intermediate Neglect of Differential Overlap, *J. Chem. Phys.* **47**, 2026 (1967).

P. Pyykkö, Electric Field Gradient Calculations with One-Center Expansion Wave Functions, *Proc. Phys. Soc.* **92**, 841 (1967).

C. D. Ritchie and H. F. King, Gaussian Basis SCF Calculations for OH^-, H_2O, NH_3, and CH_4, *J. Chem. Phys.* **47**, 564 (1967).

L. C. Snyder, Heats of Reaction from Hartree–Fock Energies of Closed-Shell Molecules, *J. Chem. Phys.* **46**, 3602 (1967).

J. G. Stamper and N. Trinajstic, Localized Orbitals for Some Simple Molecules, *J. Chem. Soc.* 782 (1967).

1968

G. P. Arrighini, M. Maestro, and R. Moccia, Magnetic Properties of Polyatomic Molecules. I. Magnetic Susceptibility of H_2O, NH_3, CH_4, H_2O_2, *J. Chem. Phys.* **49**, 882 (1968).

G. P. Arrighini, M. Maestro, and R. Moccia, Calculation of Dipole Hyperpolarizabilities of H_2O, NH_3, CH_4 and CH_3F, *Symp. Faraday Soc.* **2**, 48 (1968).

S. Aung, R. M. Pitzer, and S. I. Chan, Approximate Hartree–Fock Wave Functions, One-Electron Properties, and Electronic Structure of the Water Molecule, *J. Chem. Phys.* **49**, 2071 (1968).

M. J. Cooper, M. Roux, M. Cornille, and B. Tsapline, The Compton Profile of Water, *Phil. Mag.* **18**, 309 (1968).

A. A. Frost, A Floating Spherical Gaussian Orbital Model of Molecular Structure. III. First-Row Atom Hydrides, *J. Phys. Chem.* **72**, 1289 (1968).

C. Guidotti and O. Salvetti, Double Orbital Exponent SCF Functions for H_2O, NH_3, CH_4, *Theor. Chim. Acta* **10**, 454 (1968).

Y. Harada and J. N. Murrell, The Correlation Between Molecular and Atomic Rydberg Levels. Part I. An Analysis of the Rydberg States of Water, *Mol. Phys.* **14**, 153 (1968).

J. F. Harrison, Electric-Dipole Polarizability of H_2O and NH_3, *J. Chem. Phys.* **49**, 3321 (1968).

W. J. Hehre and J. A. Pople, Atomic Electron Populations for Some Simple Molecules, *Chem. Phys. Letters* **2**, 379 (1968).

A. C. Hopkinson, N. K. Holbrook, K. Yates, and I. G. Csizmadia, Theoretical Study on the Proton Affinity of Small Molecules Using Gaussian Basis Sets in the LCAO-MO-SCF Framework, *J. Chem. Phys.* **49**, 3596 (1968).

C. W. Kern and R. L. Matcha, Nuclear Corrections to Electronic Expectation Values: Zero-Point Vibrational Effects in the Water Molecule, *J. Chem. Phys.* **49**, 2081 (1968).

H. Kim and R. G. Parr, One-Center Perturbation Approach to Molecular Electronic Energies. IV. 10-Electron Molecules of Type MH_k, *J. Chem. Phys.* **49**, 3071 (1968).

J. H. Letcher and T. H. Dunning, Jr., Localized Orbitals. I. σ Bonds, *J. Chem. Phys.* **48**, 4538 (1968).

T. F. Lin and A. B. F. Duncan, Calculations on Rydberg Terms of the Water Molecule, *J. Chem. Phys.* **48**, 866 (1968).

D. Neumann and J. W. Moskowitz, One-Electron Properties of Near-Hartree–Fock Wave Functions. I. Water, *J. Chem. Phys.* **49**, 2056 (1968).

J. A. Pople, J. W. McIver, Jr., and N. S. Ostlund, Self-Consistent Perturbation Theory. II. Nuclear-Spin Coupling Constants, *J. Chem. Phys.* **49**, 2965 (1968).

1969

J. Andriessen, A Calculation of the Ground State of H_2O Using a Minimal Basis Set, *Chem. Phys. Letters* **3**, 257 (1969).

D. B. Cook and P. Palmieri, Approximate *Ab Initio* Calculations on Polyatomic Molecules. II., *Mol. Phys.* **17**, 271 (1969).

P. F. Franchini and C. Vergani, GF Calculation with Minimal and Extended Basis Sets for H_2O, NH_3, CH_4, and H_2O_2, *Theor. Chim. Acta* **13**, 46 (1969).

I. H. Hillier and V. R. Saunders, *Ab Initio* Calculations of d-Orbital Participation in Some Sulfur Compounds, *Chem. Phys. Letters* **4**, 163 (1969).

J. A. Horsley and W. H. Fink, *Ab Initio* Calculation of Some Lower-Lying Excited States of H_2O, *J. Chem. Phys.* **50**, 750 (1969).

W. J. Hunt and W. A. Goddard, III, Excited States of H_2O Using Improved Virtual Orbitals, *Chem. Phys. Letters* **3**, 414 (1969).

W. J. Hunt, T. H. Dunning, Jr., and W. A. Goddard, III, The Orthogonality Constrained Basis Set Expansion Method for Treating Off-Diagonal Lagrange Multipliers in Calculations of Electronic Wave Functions, *Chem. Phys. Letters* **3**, 606 (1969).

M. Klessinger, Bond-Angle Deformation and Hybridization in H_2O, *Chem. Phys. Letters* **4**, 144 (1969).

K. J. Miller, S. R. Mielczarek, and M. Krauss, Energy Surface and Generalized Oscillator Strength of the $^1A''$ Rydberg State of H_2O, *J. Chem. Phys.* **51**, 26 (1969).

M. D. Newton, W. A. Lathan, W. J. Hehre, and J. A. Pople, Self-Consistent Molecular-Orbital Methods. III. Comparison of Gaussian Expansion and PDDO Methods Using Minimal STO Basis Sets, *J. Chem. Phys.* **51**, 3927 (1969).

J. D. Petke and J. L. Whitten, *Ab Initio* Studies of Orbital Hybridization in Polyatomic Molecules, *J. Chem. Phys.* **51**, 3166 (1969).

H. Preuss and R. Janoschek, Wave-Mechanical Calculations on Molecules Taking All Electrons Into Account, *J. Mol. Structure* **3**, 423 (1969).

E. Switkes, R. M. Stevens, and W. N. Lipscomb, Polyatomic SCF Calculations Utilizing Anisotropic Basis Sets of Slater-Type Orbitals, *J. Chem. Phys.* **51**, 5229 (1969).

1970

G. P. Arrighini and C. Guidotti, Experimental Data and *Ab Initio* Calculations of Some One-Electron Properties of the H_2O Molecule: A Comparison, *Chem. Phys. Letters* **6**, 435 (1970).

G. P. Arrighini, C. Guidotti, and O. Salvetti, SCF MO's and Molecular Properties of H_2O, *J. Chem. Phys.* **52**, 1037 (1970).

G. P. Arrighini, M. Maestro, and R. Moccia, Magnetic Properties of Polyatomic Molecules. II. Proton Magnetic Shielding Constants in H_2O, NH_3, CH_4, and CH_3F, *J. Chem. Phys.* **52**, 6411 (1970).

G. P. Arrighini, M. Maestro, and R. Moccia, Magnetic Properties of Polyatomic Molecules. III. Magnetic Shielding Constants of Heavy Nuclei in $H_2{}^{17}O$, $^{14}NH_3$, $^{13}CH_4$, and $^{13}CH_3F$, *Chem. Phys. Letters* **7**, 351 (1970).

R. F. W. Bader and R. A. Gangi, The Lowest Singlet and Triplet Potential Surfaces of H_2O, *Chem. Phys. Letters* **6**, 312 (1970).

M. Cornille, M. Roux, and B. Tsapline, Some Molecular Compton Profiles, *Acta Cryst.* **A26**, 105 (1970).

T. H. Dunning, Jr., Gaussian Basis Functions for Use in Molecular Calculations. I. Contraction of $(9s5p)$ Atomic Basis Sets for the First-Row Atoms, *J. Chem. Phys.* **53**, 2823 (1970).

B. Ford, G. G. Hall, and J. C. Packer, Molecular Modeling with Spherical Gaussians, *Internat. J. Quantum Chem.* **4**, 533 (1970).

P. F. Franchini, R. Moccia, and M. Zandomeneghi, Extended Group Function Calculations for H_2O, NH_3, and CH_4, *Internat. J. Quantum Chem.* **4**, 487 (1970).

S. L. Guberman and W. A. Goddard, III, Spin-Generalized SCF Wave Functions for H_2O, OH, and O, *J. Chem. Phys.* **53**, 1803 (1970).

J. A. Horsley and F. Flouquet, The Dissociation of NH_3 and H_2O in Excited States, *Chem. Phys. Letters* **5**, 165 (1970).

R. P. Hosteny, R. R. Gilman, T. H. Dunning, Jr., A. Pipano, and I. Shavitt, Comparison of Slater and Contracted Gaussian Basis Sets in SCF and CI Calculations on H_2O, *Chem. Phys. Letters* **7**, 325 (1970).

H. E. Montgomery, B. L. Bruner, and R. E. Knight, Applications of Variation-Perturbation Theory to Molecules. I. The Water Molecule, *J. Chem. Phys.* **52**, 4407 (1970).

Y.-C. Pan and H. F. Hameka, Calculation of the Diamagnetic Susceptibility of the Water Molecule, *J. Chem. Phys.* **53**, 1265 (1970).

R. M. Pitzer and D. P. Merrifield, Minimum Basis Wave Functions for Water, *J. Chem. Phys.* **52**, 4782 (1970).

P. PuLay, *Ab Initio* Calculation of Force Constants and Equilibrium Geometries in Polyatomic Molecules. II. Force Constants of Water, *Mol. Phys.* **18**, 473 (1970).

I. L. Thomas and H. W. Joy, Protonic Structure of Molecules. II. Methodology, Center-of-Mass Transformation, and the Structure of Methane, Ammonia, and Water, *Phys. Rev.* **A2**, 1200 (1970).

APPENDIX B. DEFINITION OF THE SYMBOLS APPEARING IN T_N

For a model in which certain electronic effects are neglected and in which the atoms are replaced by point masses that vibrate about an equilibrium geometry, the symbols $A_s^{(\alpha)}$ and $\mu_{\alpha\beta}$ appearing in the kinetic energy operator T_N of eqn. (5) are defined in terms of the normal coordinate Q_s and its conjugate momentum p_s through the following relationships. The quantity $A_s^{(\alpha)}$ is given by

$$A_s^{(\alpha)} = \sum_{s'} \zeta_{ss'}^{(\alpha)} Q_{s'} \tag{B1}$$

where

$$\zeta_{ss'}^{(\alpha)} = \sum_K \sum_\beta \sum_\gamma \varepsilon_{\alpha\beta\gamma} L_{Ks}^{(\beta)} L_{Ks'}^{(\gamma)} \tag{B2}$$

$$\varepsilon_{\alpha\beta\gamma} = \begin{cases} + \ (-) & \text{if } \alpha, \beta, \gamma \text{ is an even (odd) permutation of } X, Y, Z \\ 0 & \text{otherwise} \end{cases} \tag{B3}$$

and

$$\alpha_K' = m_K^{-1/2} \sum_s L_{Ks}^{(\alpha)} Q_s \tag{B4}$$

For the water molecule, all of the sums in the above equations go from 1 to 3, and α, β, or γ designate one of the principal axes of inertia X, Y, or Z (Fig. 1). The prime on α' in eqn. (B4) designates a displacement coordinate from equilibrium (Section 4).

The effective moment of inertia $\mu_{\alpha\beta}$ is defined by

$$\mu_{\alpha\beta} = [1/I_{\alpha\alpha}^{(e)} I_{\beta\beta}^{(e)}]\left[I_{\alpha\alpha}^{(e)}\delta_{\alpha\beta} - \sum_{s=1}^{3} a_s^{(\alpha\beta)} Q_s - \sum_{s,s'=1}^{3} G_{ss'}^{(\alpha\beta)} Q_s Q_{s'} + \cdots\right] \quad \text{(B5)}$$

to second order in Q_s. This expression contains the coefficients

$$a_s^{(\alpha\beta)} = \sum_K m_K^{1/2}\left[2\delta_{\alpha\beta}\sum_\gamma \gamma_K^{(e)} L_{Ks}^{(\gamma)} - \alpha_K^{(e)} L_{Ks}^{(\beta)} - \beta_K^{(e)} L_{Ks}^{(\alpha)}\right] \quad \text{(B6)}$$

and

$$G_{ss'}^{(\alpha\beta)} = A_{ss'}^{(\alpha\beta)} - \sum_{s''} \zeta_{ss''}^{(\alpha)} \zeta_{s's''}^{(\beta)} - \sum_\gamma [a_s^{(\alpha\gamma)} a_{s'}^{(\beta\gamma)}/I_{\gamma\gamma}^{(e)}] \quad \text{(B7)}$$

where

$$A_{ss'}^{(\alpha\beta)} = \delta_{ss'}\delta_{\alpha\beta} - \sum_K L_{Ks}^{(\alpha)} L_{Ks'}^{(\beta)} \quad \text{(B8)}$$

The superscript (e) on the various quantities (e.g., the moments of inertia $I_{\alpha\alpha}^{(e)}$) indicates that they refer to the equilibrium structure.

APPENDIX C. DEFINITION OF THE SYMBOLS APPEARING IN E_{vib} AND $\mathscr{H}_{\text{rot}}^{v}$

The expansion coefficients $x_{ss'}$ in eqn. (64b) are given, according to Nielsen,[795] by the following expressions referred to dimensionless normal coordinates:

$$x_{ss} = \frac{1}{4}\left[6k_{ssss} - \frac{15k_{sss}^2}{\omega_s} - \sum_{s'} \frac{(8\omega_s^2 - 3\omega_{s'}^2)k_{sss'}^2}{\omega_{s'}(4\omega_s^2 - \omega_{s'}^2)}\right] \quad \text{(C1)}$$

$$x_{ss'} = \frac{1}{2}\left[k_{sss's'} - \frac{6k_{sss}k_{ss's'}}{\omega_s} - \frac{4\omega_s k_{sss'}^2}{(4\omega_s^2 - \omega_{s'}^2)} - \sum_{s''}\left(\frac{k_{sss''}k_{s''s's'}}{\omega_{s''}}\right.\right.$$

$$\left. + \frac{\omega_{s''}(\omega_{s''}^2 - \omega_s^2 - \omega_{s'}^2)k_{ss's''}^2}{2(\omega_s+\omega_{s'}+\omega_{s''})(\omega_s+\omega_{s'}-\omega_{s''})(\omega_s-\omega_{s'}+\omega_{s''})(\omega_s-\omega_{s'}-\omega_{s''})}\right)$$

$$\left. + \frac{\hbar}{2\pi c}\sum_\alpha \frac{\omega_{s'}(\zeta_{ss'}^{(\alpha)})^2}{\omega_s I_{\alpha\alpha}^{(e)}}\right] \quad (s \neq s') \quad \text{(C2)}$$

These results are general in that they can be applied to any polyatomic molecule for which the vibrational states are nondegenerate. When degeneracy is present, additional terms need to be introduced. For HDO, eqns. (C1) and (C2) do not simplify to any great degree.[(965a)] However, in the case of symmetric molecules such as H_2O or D_2O, the expressions for $x_{ss'}$ are somewhat more compact.[(965)] Equations of this type appear to have been obtained first by Bonner.[(100a)]

The coefficients for the rotational Hamiltonian $\mathscr{H}^v_{\text{rot}}$ in eqn. (64c) are defined by[861,862]

$$\sigma^v_{\alpha\beta} = (\delta_{\alpha\beta}/I^{(e)}_{\alpha\beta}) - \sum_s (v_s + \tfrac{1}{2}) b_s^{(\alpha\beta)} \tag{C3}$$

$$\tau_{\alpha\beta\gamma\delta} = -(1/2 I^{(e)}_{\alpha\alpha} I^{(e)}_{\beta\beta} I^{(e)}_{\gamma\gamma} I^{(e)}_{\delta\delta}) \sum_s (a_s^{(\alpha\beta)} a_s^{(\gamma\delta)}/\omega_s^2) \tag{C4}$$

where

$$b_s^{(\alpha\beta)} = \frac{\hbar}{\omega_s I^{(e)}_{\alpha\alpha} I^{(e)}_{\beta\beta}} \left[A_{ss}^{(\alpha\beta)} - \sum_\gamma \frac{a_s^{(\alpha\gamma)} a_s^{(\beta\gamma)}}{I^{(e)}_{\gamma\gamma}} - 4 \sum_{s'} \frac{\zeta^{(\alpha)}_{ss'} \zeta^{(\beta)}_{ss'} \omega_s^2}{\omega_s^2 - \omega^2_{s'}} \right.$$
$$\left. - 3\omega_s \sum_{s'} \frac{k_{sss'} a_{s'}^{(\alpha\beta)}}{(\hbar\omega_{s'})^{3/2}} \right] \tag{C5}$$

The constants $a_s^{(\alpha\beta)}$, $A_{ss}^{(\alpha\beta)}$, and $\zeta^{(\alpha)}_{ss'}$ are defined in eqns. (B6), (B8), and (B2) of Appendix B, respectively. In the treatment leading to these equations, certain terms which are small for the ground-vibrational state have been neglected.

APPENDIX D. DEFINITION OF THE SYMBOLS APPEARING IN THE ROTATIONAL MATRIX ELEMENTS

The matrix elements in eqn. (65) are defined in terms of the following interrelated quantities[861,862]:

$$\mathscr{R}_0 = \tfrac{1}{2}(B^v_{YY} + B^v_{ZZ})J(J+1) - \tfrac{1}{4}(B^v_{YY} - B^v_{ZZ})(\varkappa - 1)J(J+1)$$
$$- D_J J^2 (J+1)^2 \tag{D1}$$

$$\mathscr{R}_2 = \tfrac{1}{4}(B^v_{YY} - B^v_{ZZ})(\varkappa - 3) - D_{JK} J(J+1) \tag{D2}$$

$$\mathscr{R}_4 = \tfrac{1}{4}(B^v_{YY} - B^v_{ZZ})(\varkappa + 1) + \delta_J J(J+1) \tag{D3}$$

$$\varkappa = (2B^v_{XX} - B^v_{YY} - B^v_{ZZ})/(B^v_{ZZ} - B^v_{YY}) \tag{D4}$$

$$B^v_{\alpha\beta} = \tfrac{1}{2}(\sigma^v_{\alpha\beta} - D_{\alpha\beta}) \tag{D5}$$

$$D_{XX} = \tfrac{1}{8}\hbar^2(\tau_{XXXX} + \tau_{YYYY} - 2\tau_{XXYY} + 4\tau_{XYXY} - 12\tau_{YZYZ} + 8\tau_{ZXZX}) \quad \text{(D6)}$$

$$D_{YY} = \tfrac{1}{8}\hbar^2(\tau_{XXXX} + \tau_{YYYY} - 2\tau_{XXYY} + 4\tau_{XYXY} + 8\tau_{YZYZ} - 12\tau_{ZXZX}) \quad \text{(D7)}$$

$$D_{ZZ} = -\tfrac{3}{4}(D_{XX} + D_{YY}) + \tfrac{5}{8}\hbar^2(\tau_{YZYZ} + \tau_{ZXZX}) \quad \text{(D8)}$$

$$D_{XY} = \tfrac{5}{2}\hbar^2\tau_{YZZX} \quad \text{(D9)}$$

$$D_{YZ} = -\tfrac{5}{8}\hbar^2(\tau_{YZXX} - \tau_{YZYY} + 2\tau_{XYZX}) \quad \text{(D10)}$$

$$D_{ZX} = \tfrac{5}{8}\hbar^2(\tau_{ZXXX} - \tau_{ZXYY} + 2\tau_{XYYZ}) \quad \text{(D11)}$$

$$D_J = -\tfrac{1}{32}\hbar^4(3\tau_{XXXX} + 3\tau_{YYYY} + 2\tau_{XXYY} + 4\tau_{XYXY}) \quad \text{(D12)}$$

$$D_K = D_J - \tfrac{1}{4}\hbar^4(\tau_{ZZZZ} - \tau_{ZZXX} - \tau_{YYZZ} - 2\tau_{YZYZ} - 2\tau_{ZXZX}) \quad \text{(D13)}$$

$$D_{JK} = -D_J - D_K - \tfrac{1}{4}\hbar^4\tau_{ZZZZ} \quad \text{(D14)}$$

$$\delta_J = -\tfrac{1}{16}\hbar^4(\tau_{XXXX} - \tau_{YYYY}) \quad \text{(D15)}$$

$$\mathscr{R}_5 = -\tfrac{1}{32}\hbar^4(\tau_{XXXX} - \tau_{YYYY} + 2\tau_{YYZZ} - 2\tau_{ZZXX} + 4\tau_{YZYZ} - 4\tau_{ZXZX}) \quad \text{(D16)}$$

$$\mathscr{R}_6 = \tfrac{1}{64}\hbar^4(\tau_{XXXX} + \tau_{YYYY} - 2\tau_{XXYY} - 4\tau_{XYXY}) \quad \text{(D17)}$$

$$\mathscr{R}_7 = -\tfrac{1}{16}\hbar^4(\tau_{ZXXX} - \tau_{ZXYY} - 2\tau_{XYYZ}) \quad \text{(D18)}$$

$$\mathscr{R}_8 = \tfrac{1}{16}\hbar^4(3\tau_{ZXXX} + \tau_{ZXYY} + 2\tau_{XYYZ}) \quad \text{(D19)}$$

$$\mathscr{R}_9 = -\mathscr{R}_8 + \tfrac{1}{8}\hbar^4\tau_{ZXZZ} \quad \text{(D20)}$$

Expressions for $\sigma^v_{\alpha\beta}$ and $\tau_{\alpha\beta\gamma\delta}$ are given in Appendix C.

CHAPTER 3

Theory of Hydrogen Bonding in Water

C. N. R. Rao

Department of Chemistry
Indian Institute of Technology
Kanpur, India

1. INTRODUCTION

Among the various types of intermolecular interactions, the hydrogen bond plays the most significant role in determining the structure and properties of molecular systems of importance to chemistry and biology. The phenomenon of hydrogen bonding has been investigated extensively by a variety of spectroscopic and other physical methods and the subject has been reviewed by several authors.[442,453,777,844] Electronic theories of hydrogen bonding were discussed by Bratoz[112] in 1966 and some of the recent molecular orbital studies have been reviewed by Murthy and Rao.[779] Advances in theory of the hydrogen bond in the last few years have mainly been due to: (a) the availability of all-valence-electron semiempirical molecular orbital theories, which have rendered calculations on molecular systems up to 50 atoms fairly straightforward, and (b) the new generation of computers, which have made possible nonempirical (*ab initio*) self-consistent field calculations on relatively large systems. In this chapter we shall examine how recent theoretical studies are able to predict various properties of hydrogen bonds in water and throw light on the mechanism of hydrogen bonding. Some of the important aspects we shall be interested in are: dissociation energy, charge distribution, geometry, proton potential function and spectroscopic properties. Before specifically discussing water, we shall survey the early theoretical treatments of the hydrogen bond and briefly describe the recent theoretical methods.

2. EARLY THEORETICAL STUDIES OF THE HYDROGEN BOND

Pauling[828] first suggested that in a hydrogen bond X—H···Y the hydrogen 1*s* orbital can form only one covalent bond with X and so the interaction with Y (electron donor) is largely electrostatic in character. Several properties of hydrogen bonds have been interpreted on the basis of the electrostatic model.[442,844] The electrostatic nature of the hydrogen bond was subsequently stressed by a few other authors. Thus, Lennard-Jones and Pople[672] and Pople[855] postulated the point-charge model for the tetrahedral water molecule and then considered the effect of distortion from the tetrahedral angle on the properties of hydrogen bonds in water. On the basis of the electrostatic model, Pople calculated the energy of the hydrogen bond in the linear water dimer to be about 6 kcal mol^{-1}. According to this model, a hydrogen bond is undistorted when the O—H bond and the lone-pair of the donor oxygen atom are along the O···O bond. When the direction of the O—H bond or the lone-pair makes an angle ϕ with the O···O axis, the energy of the system is increased by $k_\phi(1 - \cos\phi)$, where k_ϕ is the hydrogen bond-bending force constant. Based on this model, Pople[855] could explain the X-ray radial distribution curve of liquid water. Features of this distorted hydrogen bond model of water have been discussed by Eisenberg and Kauzmann[301] and also elsewhere in this volume.

It was realized that the early electrostatic theory was unsatisfactory when Coulson[209,211] showed that the major contribution to the dipole moment of water comes from the lone-pair electrons occupying the asymmetric (sp^3) orbitals of the oxygen atom; the OH bond is thus much less ionic than assumed earlier. Based on spectroscopic and theoretical considerations, Cannon[162] lists the following three criteria for the formation of a hydrogen bond X—H···Y: (i) The X—H bond must be partially ionic in character; the X atom must therefore have a high electronegativity, leading to $^{-}$X···$^{+}$H, where the 1*s* orbital of the hydrogen atom is not fully utilized in the formation of the covalent X—H bond and is therefore available for overlap with the lone-pair orbital of Y. (ii) The Y atom must have lone-pair electrons in an asymmetric orbital. (iii) For maximum interaction, and, therefore, for maximum hydrogen bond energy, the X—H bond and the axis of the lone-pair orbital must be collinear.

The word "electrostatic" is generally used in the sense that different interacting species can be brought together without deformation of any charge cloud or any electron exchange. We would expect, however, that when two molecules come close together they are bound to polarize each other and there should be some distortion of charge clouds. The large-scale (permanent) distortion of these clouds gives rise to a *delocalization term.* The small-scale (coordinated) motion of the electrons gives rise to a *dispersion term.* A *repulsive term* comes in due to the violation of the Pauli exclusion principle.

If the electrostatic contribution alone were the major controlling factor in determining hydrogen bond energy, there should be a correlation between the strength of the hydrogen bond and the dipole moment of the electron donor. A ketone should be a better donor than an ether and a nitrile better than a tertiary amine. Infrared spectroscopic studies show that the electron-donating ability of a ketone and an ether are about the same, while a tertiary amine is a better donor than a nitrile.[777] The increase in intensity of the O—H stretching band on hydrogen bond formation is sometimes more than the increase expected on purely electrostatic grounds. Coulson[209, 211] has interpreted this behavior as due to a charge movement (charge transfer) to and from the donor during the hydrogen vibration, thus implying electron delocalization. It seems, therefore, that covalent contributions are very important in a hydrogen bond.

It is now clear that the hydrogen bond has contributions from the following forces: (1) electrostatic, (2) delocalization energy (charge transfer), (3) dispersion, and (4) exchange repulsion. The success of the electrostatic model in predicting the correct hydrogen bond energies in certain systems is possibly due to the fact that other terms are not significant. Coulson and Danielsson[212] as well as Tsubomura[1076] carried out rigorous variational calculations on the O—H···O hydrogen bond and found that the contribution from delocalization energy was appreciable. The following wave functions were considered by Tsubomura[1076] in treating the fictitious O—H···O unit:

O_X—H O_Y	covalent O_X—H bond (ψ_1)
O_X^- $H^+ \cdots O_Y$	ionic (ψ_2)
O_X^- H——O_Y^+	fully covalent O_Y—H bond with charge transfer (ψ_3)
O_X^+ $H^- \cdots O_Y$	ionic (ψ_4)
O_X H^- O_Y^+ (with O_X and O_Y linked)	covalent O_X—O_Y bond with charge transfer (ψ_5)

The calculation of Coulson and Danielsson[212] including only the first three structures gives the following weightings: ψ_1, 65%; ψ_2, 31% and ψ_3, 4%. Tsubomura's calculations, however, yield different weightings; as expected, the highest contribution is from ψ_1, but the weightings of the other functions are not satisfactory. The contribution from ψ_1, ψ_2, or ψ_3 varies with O_X—H distance and $O_X\cdots O_Y$ distance.[212] The covalent contribution to the hydrogen bond increases as the O_X—H bond length increases. The energy of the hydrogen bond varies linearly with the weight of the covalent contribution.

3. POTENTIAL FUNCTION FOR THE HYDROGEN BOND

Lippincott and Schroeder[689,690,957] have developed a model for the potential function of the hydrogen bond, based on the Morse function for diatomic molecules. This potential function describes the complete potential curve for the proton within the hydrogen bond, from which the barrier height for the proton transfer can be obtained. The Lippincott–Schroeder potential function for a hydrogen bond X—H$\cdots$Y is made up of four terms:

$$V = V_1 + V_2 + V_3 + V_4$$

Here, V_1 is the potential function for the X—H bond, V_2 the potential function for the H$\cdots$Y bond, V_3 the van der Waals repulsion between the two oxygen atoms, and V_4 the electrostatic attraction between the oxygens:

$$V_1 = D\{1 - \exp[-n(r - r_0)^2/2r]\}$$
$$V_2 = -D^* \exp[-n^*(R - r^* - r_0^*)^2/2(R - r^*)]$$
$$V_3 = Ae^{-bR}$$
$$V_4 = BR^{-m}$$

A, B, b, m are constants, D and D^* are the strengths of the X—H and H$\cdots$Y bonds, n and n^* are related to the ionization potentials of the atoms, r_0 and r_0^* are the internuclear distances in the absence of the hydrogen bond, r and r^* are the distances on formation of the hydrogen bond, R is the X$\cdots$Y distance, and R_0 is the X$\cdots$Y distance at the minimum in the potential function.

Lippincott and Schroeder[690,957] have applied this potential function with fair success to several linear and bent hydrogen bonds. For O—H$\cdots$O bonds successful predictions have been made regarding the dependence of the O—H distance and the O—H stretching frequency on the O$\cdots$O distance. Reid[896] has used a slightly modified Lippincott–Schroeder potential to interpret changes in infrared spectra and the changes in the lattice dimensions of hydrogen-bonded crystals on deuteration. Sokolov[1005] feels that the repulsive and attractive terms (V_3 and V_4) are overemphasized in the Lippincott–Schroeder treatment and suggests that a quantum mechanical description with less emphasis on these terms would explain some of the spectroscopic properties of the hydrogen bond more satisfactorily.

Moulton and Kromhout[770] have modified the Lippincott–Schroeder potential function by multiplying the factor $\exp[-n(r - r_0)^2/2r]$ by $\cos^2 \theta'$ and the factor $\exp[-n^*(R - r^* - r_0^*)^2/2(R - r^*)]$ by $\cos^2 \theta$, where θ' is the angle between X—H bond and some normal bonding direction for the atom X, and θ is the angle between the H···Y bond and some normal bonding direction for the atom Y. Scott and Scheraga[959] neglect the first term (V_1) in their calculations on molecules of biological interest, but introduce the Moulton–Kromhout modification into the second term (V_2), assuming that θ is the angle which the N—H bond makes in the direction of the lone-pair orbital on the oxygen atom. Two terms similar to the second term have been included in the equation, since there are two lone-pairs. Bent hydrogen bonds have been treated by Chidambaram and Sikka[183] by replacing $(R - r)$ in the expression for V_2 by d (increase in the X···Y distance).

Quantum mechanical treatment of the double minimum potential has been discussed by McKinney and Barrow[714] and Pshenichnov and Sokolov.[875] McKinney and Barrow[714] have described the O—H···O hydrogen bond by means of one-dimensional counterparts of the atomic orbitals and calculated the potential energy of the system for various positions of the hydrogen atom. Their calculations indicate the presence of a second minimum; the presence of a double minimum potential has been established experimentally from vibrational spectroscopy.[60] Pshenichnov and Sokolov[875] have investigated quantum effects in a double potential well by a quasiclassical method applied in a region near the top of the potential barrier. This model is helpful in interpreting certain regularities in the spectra of hydrogen-bonded systems. Somarjai and Hornig[1006] have pointed out that double minimum potentials may be asymmetric and that the asymmetry can be best examined by a study of infrared band intensities as well as frequencies at all levels up to and above the barrier.

One can view the double minimum potential with a large barrier applicable to moderately weak, asymmetric hydrogen bonds as one extreme case, and the single minimum potential applicable to the strong symmetric hydrogen bonds as the other extreme case. In between these two extrema we can have a variety of hydrogen bonds with different strengths and, therefore, with different shapes of the potential curves.

4. RECENT THEORETICAL METHODS

In contrast to the early theoretical work, which employed the valence bond method, most of the recent theoretical studies of the hydrogen bond which attempt to predict various properties of hydrogen-bonded systems generally employ the molecular orbital method. Molecular orbital calculations provide accurate charge distributions in monomers as well as in hydrogen-bonded dimers and polymers. We shall now briefly describe the molecular orbital methods used to study single molecules and their hydrogen-bonded complexes.

In order to carry out an *ab initio* molecular orbital calculation, we assume fixed nuclei in the well-known Born–Oppenheimer approximation and employ the exact nonrelativistic Hamiltonian in solving the time-independent Schrödinger equation for the electronic ground state; further, we assume a single determinantal form of the wave function (the Hartree–Fock approximation). The orthogonal molecular orbitals (in the determinant) are linear combinations of the atomic orbitals. The self-consistent field approach[9,843,918] is employed to solve for the set of molecular orbitals which yield the minimum total energy for the system. The neglect of relativistic effects (for atoms of low atomic number) and nuclear electronic coupling does not significantly affect the results of such calculations. There are, however, two serious shortcomings in these *ab initio* molecular orbital calculations: (i) The use of a single determinantal form of the wave function (neglect of electron correlation) is a drastic approximation. Dissociation of molecules into atoms is not at all properly pictured within the Hartree–Fock approximation; thus, two fluorine atoms are predicted to be more stable than the F_2 molecule. Properties such as inversion barriers, rotational barriers, molecular geometries, and energies of hydrogen-bond formation are, however, well represented within the Hartree–Fock approximation. Vibrational frequencies and dipole moments are generally overstimated by 5–20%. (ii) Within the Hartree–Fock approximation, there is a great deal of choice for the atomic orbitals employed as the basis set. This does not mean that the ability of the wave function to predict molecular properties improves with the improvement in the atomic basis used. For example, the dipole moment of water molecule found with a single Slater orbital basis[251] ($E = -75.500$) is 1.82 D; with a near-Hartree–Fock atomic basis[468] ($E = -76.002$) it is 2.57 D; and with the best extended basis[791] ($E = -76.059$) the value is 1.99 D. In order to obtain agreement with the experimental dipole moment (1.85 D), we shall have to include configuration interactions in the calculations with the best extended basis.[36]

In recent years, much progress has been made in developing semi-empirical molecular orbital methods: (i) In one-electron theory, commonly known as the extended Hückel theory,[518] the diagonal elements (Coulomb integrals) of the Hamiltonian are equated to the negative values of the ionization potential of the corresponding valence states of the atoms. The off-diagonal elements are taken to be proportional to the diagonal elements and the overlap between the different atomic orbitals. The Wolfsberg–Helmholtz proportionality factor of 1.75 proposed by Hoffmann[518] is calibrated to reproduce the rotational barrier in ethane. The molecular orbitals are found by diagonalizing the Hamiltonian matrix and the electrons are placed in these orbitals. The total energy E_t is equal to the sum of the energies of all the occupied orbitals. (ii) The two-electron theory, where electron–electron repulsions are considered, was developed by Klopman[615] and Pople and co-workers.[857] The two-electron theory has many forms (the most commonly used ones being CNDO[857] and INDO[856]) which differ in subtle ways. The common features of these methods are: (a) neglect of all three- and four-center electron repulsion integrals, and (b) neglect of certain differential overlaps (NDO), i.e. $\Phi_i \cdot \Phi_j = O$ for $i \neq j$. The methods differ in the choice of some of the parameters and also in the manner in which they neglect the differential overlap. Thus, in CNDO we have complete neglect of differential overlap, while in NDDO we have no diatomic differential overlap; in INDO we have intermediate neglect of differential overlap. The commonly used version is CNDO/2,[858] whose parameters differ slightly from those in the original CNDO/1 parameterization. In this method, the one-center terms are approximated by the Mulliken electronegatives and the two-center attraction terms are equated to the product of the core charge of the nucleus and the electron repulsion integral between the valence s orbitals on the two centers; the nonzero electron repulsion integrals are then evaluated by using the valence-shell s orbitals. The use of only s orbitals in this method to represent two-electron integrals (ss/pp) provides the rotational invariance of the energy, a feature which was not present in the earlier NDO method of Klopman.[615]

5. WATER DIMER

In the last three years, there have been a number of theoretical investigations of the water dimer, of which six are *ab initio* calculations. We shall now summarize the important findings from these studies.

Based on CNDO/2 calculations, Rao and co-workers[403,778] find that the dissociation energy of the hydrogen bond (D_e) in the linear water dimer (I) is about 6.0 kcal mol^{-1} when the experimental structural parameters of H_2O ($r_{OH} = 0.96$ Å and $\angle HOH = 105°$) are used in the calculations. The value of D_e is, however, sensitive to the O—H bond length (Fig. 1, Table I). When $r_{OH} = 1.03$ Å obtained by energy minimization is employed, D_e increases to 8.4 kcal mol^{-1}. CNDO/2 calculations on the linear dimer have been carried out with different values for the angle ϕ between the O···O bond and the direction of the sp^3 lone-pair orbital of the (electron) donor oxygen. The energy of the dimer and D_e do not vary greatly with ϕ, but there is slight stabilization when $\phi = 54°$ (i.e., when the O—H vector is directed in between the two lone-pair orbitals). Although CNDO/2 calculations underestimate the O···O distance, these calculations do show that the linear structure (I) of water dimer is much more stable than either

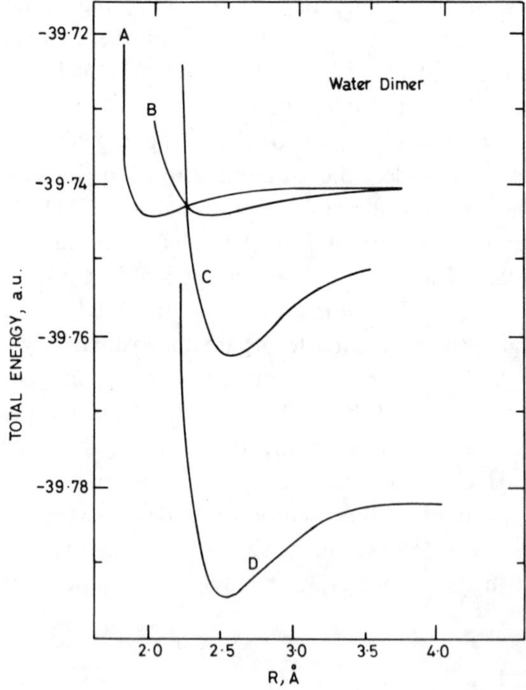

Fig. 1. Plots of the total energy of water dimer (CNDO/2) as a function of the O···O distance R: (A) cyclic dimer (II); (B) bifurcated dimer (III); (C) linear dimer (I), $r_{O\text{-}H} = 0.96$ Å, $r'_{O\text{-}H} = 1.04$ Å, $\phi = 0°$; (D) linear dimer (I), $r_{O\text{-}H} = 1.03$ Å; $r'_{O\text{-}H} = 1.04$ Å, $\phi = 0°$. (From Rao and co-workers.[403])

TABLE I. Calculations on Linear Water Dimer

Basis set	Dimerization energy or D_e, kcal	$R(O\cdots O)$, Å
CNDO (experimental monomer geometry)[403,538,619,778]	5.9	2.55
CNDO (CNDO optimized geometries)[403,538,619,778]	8.4	2.55
(5, 3, 3) Gaussian[763]	12.6	2.68
Hartree–Fock AO contracted Gaussian[620]	5.3	3.00
Hartree–Fock AO "split out" Gaussian[620]	7.9	2.85
Single Slater basis[764]	6.6	2.76
Gaussian fit to Slater basis[251]	6.1	2.73
Extended basis with polarization function[263,455]	4.7	3.00

the cyclic (II) or the bifurcated (III) structures (see Fig. 1).

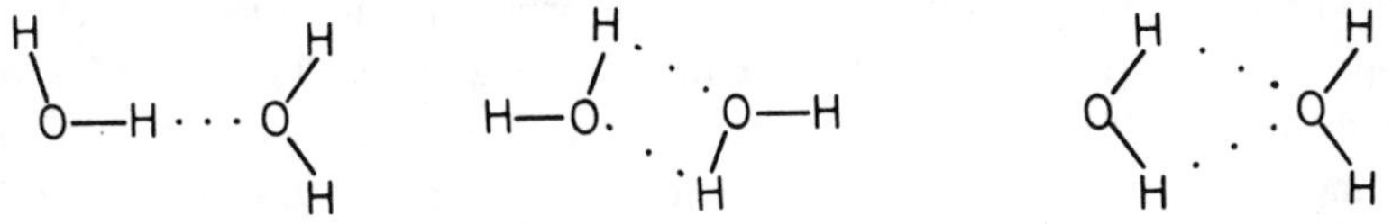

(I) (II) (III)

The value of D_e is 2.4 kcal mol^{-1} in both (II) and (III). Hoyland and Kier,[538] as well as Kollmann and Allen,[619] have reported CNDO/2 results similar to those of Rao and co-workers. Kollmann and Allen,[619] however, find D_e values in cyclic and linear water dimers to be much too high with the NDDO approximation.

The extended Hückel method of Hoffmann is not satisfactory for calculating the properties of the dimer;[778] the method shows no minimum energy configuration when the HOH angle is varied anywhere between 90° and 120°. Further, this method predicts the $H_3O^+OH^-$ configuration of the dimer to be more stable than the nonionic configuration. An iterative extended Hückel calculation of Rein and Harris[897] yields no dimerization energy for the linear dimer.

The first *ab initio* calculation on water dimer was performed by Morakuma and Pedersen,[763] who employed a limited Gaussian basis set with 5*s* and 3*p* orbitals for the oxygen and 3*s* orbitals centered on the hydrogen atom. These workers found the linear configuration to be most stable; next in stability was the bifurcated structure, followed by the cyclic structure. The dissociation energy of the linear dimer was, however, very much overestimated (12.6 kcal mol^{-1}). The study of Kollman and Allen,[620] making use of a contracted Hartree–Fock basis set, also confirmed that the linear dimer was more stable than either the bifurcated or the cyclic structure, but the D_e value was 5.3 kcal mol^{-1}. We may note here that the second virial coefficient data[926] seem to yield 5.0 kcal mol^{-1} for D_e; also, recent infrared investigations favor the linear structure for the water dimer.[1079]

Calculations on the linear dimer by Morokuma and Winick[764] (with a single Slater basis) and Del Bene and Pople[251] (with a more complete study, involving 52 different geometries of the dimer for every O···O distance, using a Gaussian fit to a Slater basis), yield good dimerization energies of 6.6 and 6.1 kcal mol^{-1} respectively. Calculations by Diercksen[263] and Hankins and co-workers[455] with accurate basis set give dimerization energies of 4.7–4.8 kcal mol^{-1}. The results on the linear dimer from the various calculations are summarized in Table I.

The *ab initio* calculations predict the angle ϕ in the linear dimer to be anywhere between 0 and 25°. The fact that this angle is not far from zero points to the validity of the localized orbital picture of an O—H bond approaching a lone-pair (rather than between lone-pairs). The value of ϕ found with the different basis sets varies, probably because of the differences in the dipole moment of the monomer. Just as in the case of the CNDO results, the dimer energy and D_e are not significantly affected by ϕ. All the smaller basis calculations[251,763,764] predict an O···O distance of about 2.76 Å, whereas the larger basis calculations[263,455] predict an O···O distance of 3.0 Å. Bolander and co-workers[99] predict a distance of 2.9 Å using semiempirical statistical mechanical arguments. Infrared studies[1079] on the water dimer in gas matrices seem to indicate a longer O···O distance than in ice I.

The effect of bending the hydrogen bond (IV) on the dimer energy

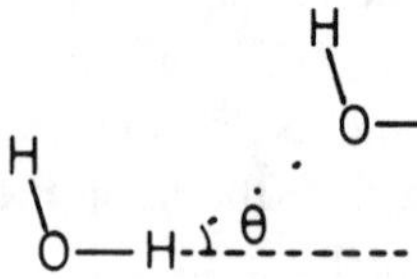

(IV)

has recently been examined by Rao and co-workers[404] by the CNDO/2 method. In (IV), the lone-pair orbital was always directed toward the hydrogen ($\phi = 0$). There appears to be little or no difference in the total dimer energy as well as D_e up to $\theta \approx 25°$; a further increase in θ increases the dimer energy, and markedly decreases D_e (Fig. 2, Table II). The O···O distance

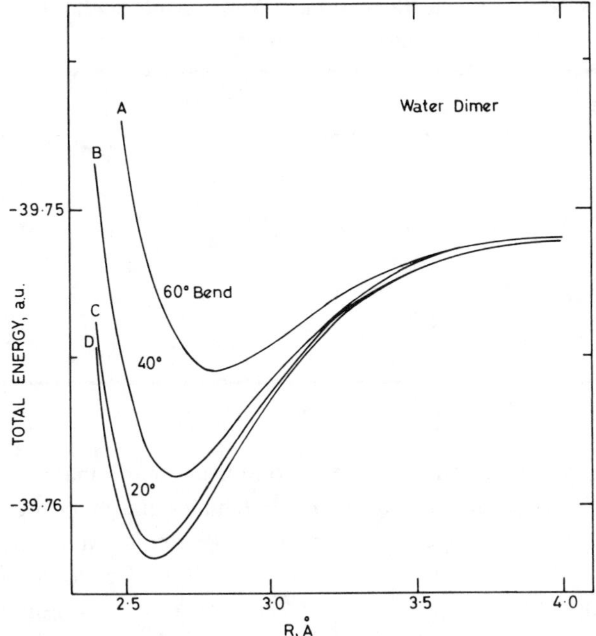

Fig. 2. Plots of the total energy of the linear water dimer as a function of the O···O distance R for different values of the bending angle θ: (A) $\theta = 60°$; (B) $\theta = 40°$; (C) $\theta = 20°$; (D) $\theta = 0°$. (From Rao and co-workers.[404])

also increases markedly beyond 25°. The decrease in D_e as θ becomes greater than 25° is ~100 cal/deg bend. This explains why in many hydrates, bent hydrogen bonds ($\theta \leq 25°$) can be present without significant loss of stability.[267] These results are consistent with the conclusion of Donohue[267] that the maximum allowable bend is 25°.

Considering the present results on the effect of hydrogen bond bending on the stability of the dimer, we feel that bent hydrogen bonds ($\theta \leq 25°$) may possibly be present in liquid water. The results in Table II show that the O···O distance increases by about 0.1 Å when $\theta \approx 30°$. Although the CNDO/2 method is not entirely satisfactory in predicting bond distances in such calculations, it may not be unreasonable to assume that a variation of about 0.1 Å in the O···O distance in bent dimers does not significantly affect the hydrogen bond energy. On the basis of the width of the uncoupled O—H (or O—D) stretching band in the vibrational spectrum of liquid water,[301,1132] it has been inferred that the O···O bond length in water may vary anywhere between 2.75 and 3.00 Å, with an equilibrium value of

TABLE II. CNDO/2 Calculations on the Effect of Hydrogen Bond Bending on the Properties of Water Dimer[a]

θ, deg	R_e, Å	D_e, kcal mol^{-1}	μ, D
0	2.55	7.3	3.62
10	2.60	7.2	3.39
20	2.60	6.8	3.20
30	2.65	6.2	2.97
40	2.70	5.3	2.73
50	2.75	3.4	2.51
60	2.80	3.1	2.28

[a] r_{O-H} = 1.03 Å (distance in the minimum energy configuration); ϕ = 0°.

2.85 Å. The present calculations seem to indicate that a major contribution to the width of the uncoupled O—H stretching band could be due to the presence of bent hydrogen bonds which give rise to different molecular environments. The postulate of a variety of distorted hydrogen bonds in a given associated species provides some support to the continuum approach to liquid water.[952] We may note here that the recent infrared and Raman data[962,1139] on the uncoupled O—H and O—D stretching bands (which suggest the presence of a mixture of species in equilibrium) also show the bands to be very broad, possibly because of the presence of a continuous distribution of distorted hydrogen-bonded species.

6. WATER POLYMERS

Rao and co-workers[403] have carried out CNDO/2 calculations on the various open trimers (V, VI, VII) by minimizing the energy with

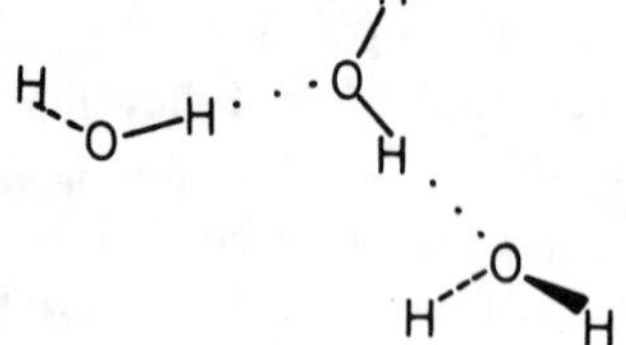

(V)

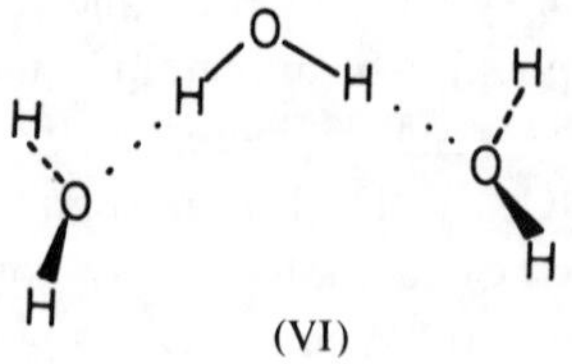

(VI)

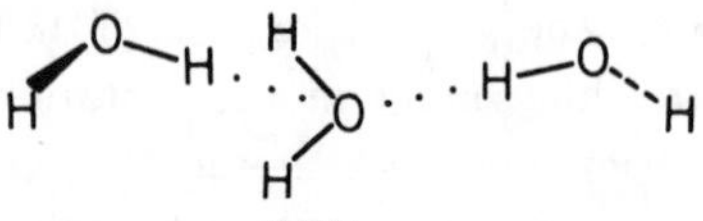

(VII)

respect to the terminal hydrogen bond and find (VI) to be the most stable structure. Hoyland and Kier,[538] however, find (V) to be more stable, which appears more reasonable. The average hydrogen bond energy in the trimer is higher than in the linear dimer.[403,538] Similar observations have been made on the trimer of methanol[776] and formamide.[780] The average hydrogen bond energies in the tetramer (VIII) and the tetrahedral pentamer (IX)

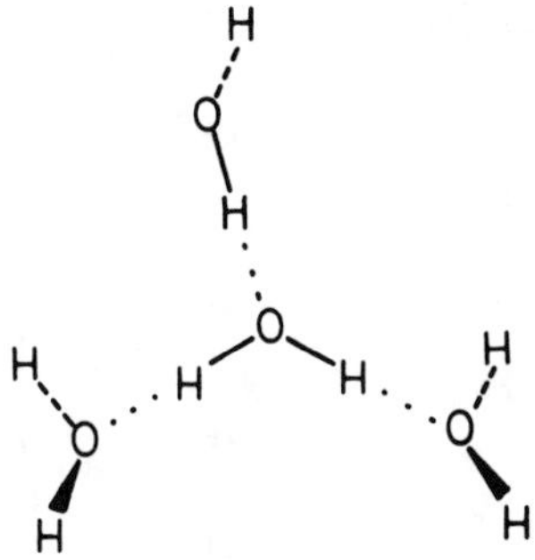

(VIII)

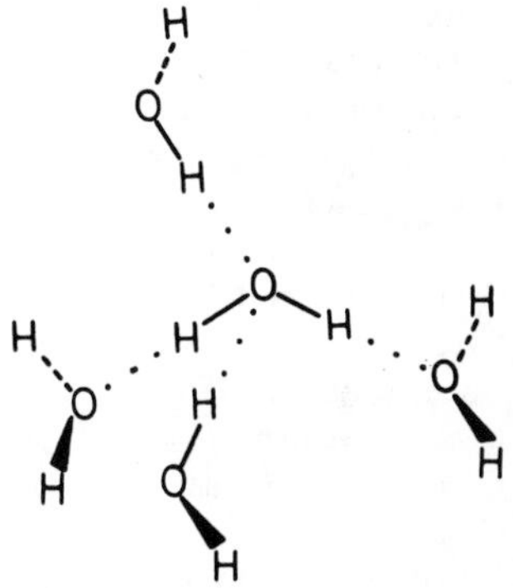

(IX)

appear to be nearly the same as in the stable trimer[403] (Table III). It appears that D_e is not additive up to the trimer, but becomes so beyond the trimer. Kollmann and Allen,[619] as well as Hoyland and Kier,[538] find that in the tetrahedral pentamer, the average D_e is similar to that in a dimer, a result which is somewhat surprising. We may note here that Walrafen[1139] treats the temperature variation of Raman bands of water in terms of an equilibrium between monomers and pentamers (see Chapter 6).

Hankins and co-workers,[455] as well as Del Bene and Pople,[250] have carried out *ab initio* calculations on some open and cyclic polymers of water. These workers find structure (V) for the trimer to be more stable and favor the presence of the OH···OH···OH chain structure in polymers (Table III). *Ab initio* calculations also show that hydrogen bond energies are not additive. Calculations of Del Bene and Pople[250] indicate that cyclic polymers may exist; in fact, the cyclic structures for the trimer and other polymers are more stable than the linear structures. Further, the asymmetric cyclic structures are favored over the symmetric ones. Some of the CNDO/2 calculations[10,402,403] on cyclic polymers yield results similar to those of Del Bene and Pople.[250] CNDO calculations on the asymmetric cyclic trimer show that the open trimers (V) and (VI) are both stable with respect to the cyclic trimer.[403] The asymmetric cyclic pentamer is, however, more stable

TABLE III. CNDO/2 and *ab initio* Calculations on Water Polymers

	CNDO/2[a]		*Ab initio*[b]	
	$R(O\cdots O)$, Å	D_e, kcal	$R(O\cdots O)$, Å	D_e, kcal
Open trimer (V)	2.60	8.8 (6.8)	—	7.3
Open trimer (VI)	2.60	9.5 (5.7)	—	4.7
Open trimer (VII)	2.60	7.3	—	5.1
Cyclic trimer (asymmetric)	2.80	3.0	—	4.9
Cyclic trimer (C_3)	—	—	2.56	6.2
Open tetramer (VIII)	2.6	9.8	—	—
Cyclic tetramer (D_{2h})	—	—	2.85	4.3
Cyclic tetramer (C_4)	—	—	2.47	9.5
Cyclic tetramer (S_4)	2.45	9.5	2.47	10.6
Open pentamer (IX)	2.6	9.9 (5.6)	—	—
Cyclic pentamer (C_5)	—	—	2.45	10.6
Cyclic pentamer (asymmetric)	2.5	8.9	2.45	11.6
Cyclic hexamer (C_6)	—	—	2.44	10.7
Cyclic hexamer (S_6)	2.45	10.8	2.44	12.0

[a] From Refs. 403 and 10; values in parentheses from Ref. 538.
[b] From Refs. 251 and 250.

than the pentamer. The instability of the cyclic trimer may be due to ring strain. The hydrogen bond energies and distances in the polymers of water are summarized in Table III.

While cyclic structures cannot explain the three-dimensional networks believed to exist in liquid water, we cannot exclude the possibility of the presence of isolated cyclic polymers; such cyclic structures have been proposed in the random network model for water.[(301)] Cyclic polymers are indeed believed to be present in alcohols.[(241)] If the proper description for liquid water is likely to be one which includes features of both the continuum and the mixture models, then the different open and cyclic polymer species could still have individual existence, although in each of the species there may be a distribution of distorted hydrogen bonds.

Semiempirical CNDO calculations on possible contributing structures to the much-disputed "anomalous water" (or "polywater") have been carried out by Allen and Kollman (CNDO/2),[(10)] Rao and co-workers[(402)] (CNDO/2), Morokuma[(762)] (CNDO/2), Pederson[(829)] (INDO), Azman and co-workers[(38)] (CNDO/2), and Messmer[(744)] (CNDO/2).[(267)] Allen and Kollman,[(10)] as well as Davis and co-workers,[(241)] have examined many structures of $(H_2O)_n$ by CNDO/2, assuming symmetric hydrogen bonds

($r_{OH} = r_{H \cdots O}$) as suggested by Lippincott and co-workers[691] and find that if anomalous water is composed of $(H_2O)_n$, an sp^2 graphite layer structure is likely, since the other symmetric structures considered have higher energies. There is no *a priori* reason, however, why one should assume the presence of symmetric hydrogen bonds. Cyclic structures with sp^3 water molecules do not provide high stability; they only indicate the presence of asymmetric hydrogen bonds of 10 kcal/H bond. Morokuma[762] considered the planar negative ions proposed by Lippincott and found that, in every case, the asymmetrically bonded structure was more stable than the symmetric bonds. Pederson[829] attempted to rationalize the stability of anomalous water in terms of increased bond orders in cyclic structures, while Azman and co-workers as well as Messmer suggest that the stability of anomalous water may be partially due to the participation of hydrogen $p\pi$ orbitals.

Ab initio calculations on cyclic hexamers yield different results, depending on the basis set employed. Sabin and co-workers[932] show that a (7, 3, 3) basis predicts a cyclic symmetric structure to be more stable than an asymmetric structure, but a Hartree–Fock atomic basis calculation predicts a symmetric structure to be very unstable. The one-electron calculation by Minton[757] also finds the symmetric structure to be unstable with respect to the monomers. We can safely conclude that theory does not really predict the presence of stable "anomalous water" in the form of $(H_2O)_n$.

Ab initio and CNDO/2 calculations carried out on hydrated cations $O_nH_{2n+1}^+$ and anions $O_nH_{2n-1}^-$ show excellent agreement with experimental observations.[255,634] Thus, the energy of formation (36 kcal) of $H_5O_2{}^+$ from H_2O and H_3O^+ is nicely predicted by the *ab initio* calculations. The CNDO/2 method exaggerates the formation energies and predicts chain structures for the hydronium ions.

7. SPECTROSCOPIC PROPERTIES, PROTON POTENTIAL FUNCTIONS, CHARGE DISTRIBUTION, AND RELATED ASPECTS

Besides the study of the geometries and energies of the hydrogen-bonded species of water, there have been a few theoretical attempts to explain various other properties, such as the changes in the vibrational spectra, proton potential functions, NMR chemical shifts, etc. There have also been attempts to decompose the hydrogen bond energy into the different contributions. We shall now briefly review some of these results.

Kollman and Allen[621] have studied the changes in the vibration spectrum of water dimer [as well as $(HF)_2$ and H_2O—HF] and found a very small increase in the length of the O—H bond and a slight decrease in the O—H force constant. Similar results have been reported by Morokuma and Winick.[764] The small shifts predicted by theory are in agreement with the spectroscopic study of the dimer in the gas phase.[1079]

The intensity enhancement observed in infrared spectra is an important property of hydrogen bonds. Kollman and Allen[621] find relative intensity increases of 8 and 20 at $R(\mathrm{O}\cdots\mathrm{O}) = 3.00$ and 2.75 Å, respectively, in the linear water dimer, compared to an increase of 1.9 in $(HF)_2$ at $R = 2.88$ Å. Kollman and Allen propose that intramolecular charge redistribution rather than charge transfer may be the key mechanism in causing the observed intensity enhancement. CNDO/2 results[619] on the O—H force constant and on the intensity enhancement in water dimer are quite similar to those from *ab initio* calculations.

Van Thiel and co-workers[1090] have observed a decrease in infrared intensity of the bending mode upon hydrogen-bond formation. Accordingly, Kollman and Allen[620,621] find that the dipole moment of the dimer decreases from 4.26 D to 4.17 D on increasing the HOH angle of the proton donor molecule from 105° to 120°; in the monomer the dipole moment change is larger (from 2.49 D to 2.20 D) upon increasing the angle by a similar amount. CNDO/2 calculations of Rao and co-workers[402,403] also give similar results.

The O···O stretching force constant for the water dimer has been shown to be compatible with that found experimentally in ice.[621,764] Murthy and Rao[778] report a force constant of 7.6×10^4 dyn cm^{-1} for the linear water dimer. Del Bene and Pople[251] have examined all the intermolecular force constants in the water dimer and find the potential curves to be much steeper for the movement of the proton-donor molecule than for the acceptor molecule. Their force constant for hydrogen bond bending is greater than the value of Kollman and Allen[621] for $(HF)_2$.

Many of the weakly hydrogen-bonded systems do not clearly exhibit the double-minimum potential well for the proton at the equilibrium X···Y distance. At larger X···Y distances, however, a higher energy minimum (corresponding to the H···Y distance) appears. These observations have been made on the basis of *ab initio* as well as CNDO/2 calculations. The extended Hückel method is found to be unsatisfactory for the study of proton potential functions since this method generally predicts the energy at the longer H···Y distance to be lower.[775] Thus, in the linear water dimer, the $H_3O^+ \cdot OH^-$ configuration is predicted to be of lower energy by the

EHT method[778]; the CNDO/2 method appears to predict more realistic proton potential functions.[775,776,778] In Fig. 3, the proton potential functions in the linear water dimer calculated by the EHT and CNDO/2 methods are reproduced for purposes of comparison.

Kraemers and Diercksen[634] have examined the proton potential well for $H_5O_2^+$. This ion has an extremely shallow single-minimum proton well at the minimum-energy O···O distance of 2.38 Å. The energy varies by

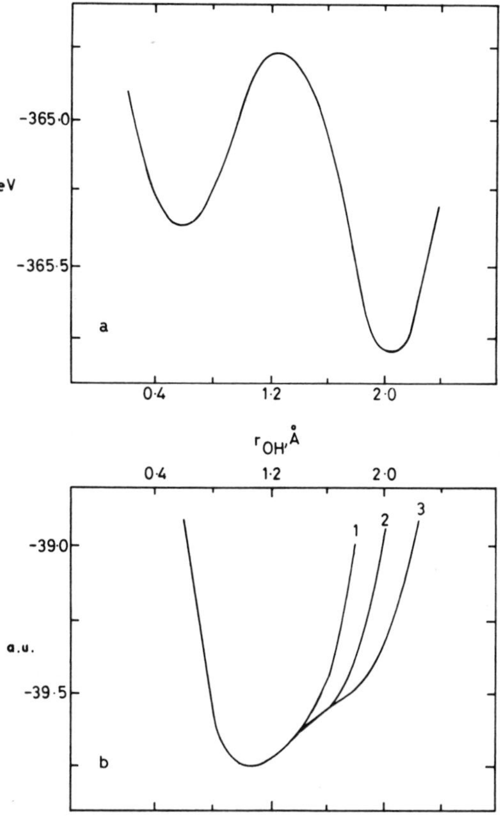

Fig. 3. (a) Total orbital energy (in eV) of the linear water dimer by the EHT method as a function of the position of the hydrogen atom. The O···O distance is 2.7 Å. (b) Total energy of the dimer (in a.u.) by the CNDO/2 method as a function of the position of the hydrogen atom. Curves 1, 2, and 3 correspond to O···O distances of 2.5, 2.7, and 2.9 Å, respectively. The equilibrium O···O distance found by energy minimization is 2.5 Å. (From Murthy and Rao.[778])

only 0.04 kcal when the proton moves from the center of the O···O bond to a position nearer to one of the oxygens. The double well at an O···O distance of 2.48 Å has an energy barrier of 300 cm^{-1}, so that the proton is essentially in a single well at this geometry. At distances greater than 2.5 Å, the proton would become localized in one of the wells and one would have an asymmetric hydrogen bond. We may note here that experimental studies suggest that at O···O distances greater than 2.5 Å, hydrogen bonds are asymmetric.

The studies of Kraemers and Dierksen,[(654)] as well as those of De Paz and co-workers,[(255)] seem to provide some rationalization for the proton mobilities in H_2O. The CNDO/2 studies of the latter group show a difference between the positive hydrated ions and the negative ions. In the positive ions, all the protons in the cluster are nearly equally positive, with the charge evently distributed among the fragments. In the negative ions, the central protons are quite positive, whereas the external hydrogens have a slight negative charge; De Paz and co-workers view these species as OH^- groups held together by protons.

It is well-established that the proton resonance in NMR spectra is shifted downfield on hydrogen bonding. SCF calculations have not been very successful in predicting the changes in the chemical shift and in the coupling constant accompanying hydrogen bonding. Weissman,[(1157)] employing an approximate MO wave function, has shown that the low field gradient at the deuteron in ice (compared to gas phase) results from the lengthening of the O—D bond in ice rather than from hydrogen bonding. Kollman[(618)] has computed the field gradient at the deuteron in water monomer and dimer and the results are shown in Table IV.

TABLE IV. Deuteron Coupling Constants in Water and Ice[a]

	Water monomer, r(O—H) = 0.96 Å		Water dimer, r(O—H) = 0.96 Å		Water dimer, r(O—H) = 1.01 Å	
	q	n	q	n	q	n
Kollman	0.587	0.10	0.537	0.10	0.364	0.11
Weissman	0.592	0.08	0.528	(0.08)	0.361	(0.08)
Experiment	0.479	<0.12	—	—	0.330[b]	0.1

[a] Here, q is the value of q_{zz} (largest component) in the principal axis system; the Z axis turns out to be nearly along the O—D bond; n is the asymmetry parameter $q_{\alpha\alpha} - q_{\beta\beta}/q_{\gamma\gamma}$, where $|q_{\gamma\gamma}| \geq |q_{\beta\beta}| \geq |q_{\alpha\alpha}|$.

[b] In ice I.

In an earlier section of this chapter, we briefly discussed the various contributions to the hydrogen bond energy and referred to the valence bond approach to the study of the mechanism of hydrogen bonding. Recently, Murrell and van Duijneveldt[774,1083,1084] have attempted to calculate the contributions to the hydrogen bond energy employing the perturbation method. The results clearly show that the Coulomb electrostatic energy is the dominant contribution to weak hydrogen bonds; the exchange repulsion is very small. Induction energy also makes a small contribution, while the dispersion energy is of the magnitude generally encountered in most intermolecular interactions. The delocalization (charge-transfer) energy becomes important in medium and strong hydrogen bonds. It appears that the dipole moment of the X—H bond is still a good basis for discussing the relative ability of proton donors to form hydrogen bonds. The Coulomb, exchange, and charge-transfer contributions to the hydrogen bond energy are favored by the low-*s* character of the lone-pair electrons. Orbitals with low-*s* character would be better electron donors.

Many authors have examined the charge-transfer model for hydrogen bonds. This model rationalizes the entire nonelectrostatic part in terms of an electron transfer from the proton-acceptor to the proton-donor molecule. This appears reasonable at first sight since hydrogen-bonded systems would then become a subcategory of a more general class of donor–acceptor complexes. However, the dipole enhancement upon hydrogen bound formation can only be partly explained by the charge-transfer mechanism. Similarly, enhancement of the infrared X—H stretching band intensity cannot fully be accounted for on the basis of charge transfer alone. Thus, in the water dimer,[620] the charge-transfer mechanism predicts intensification by a factor of 4, but the observed intensification is by a factor of 20. While charge transfer may play some role in hydrogen bonding, it is not essential to invoke this mechanism to explain many or most of the features of medium/weak hydrogen bonds.

Recently, Kollman and Allen[622] have broken down the total hydrogen bond stabilization energy in dimers of water and HF into different contributions within the SCFMO approximation. They have divided the SCF hydrogen bond energy into two terms, (i) electrostatic and exchange, and (ii) delocalization and charge transfer, and they use an empirical estimate of the dispersion energy contribution. This study shows that the shape and minimum of the potential curve (considering only electrostatic and exchange terms) is similar to the final SCF solution. Similar calculations on formamide linear dimer have been carried out by Dreyfus and Pullman,[276] who show that the electrostatic effect is significant ($\sim$2 kcal)

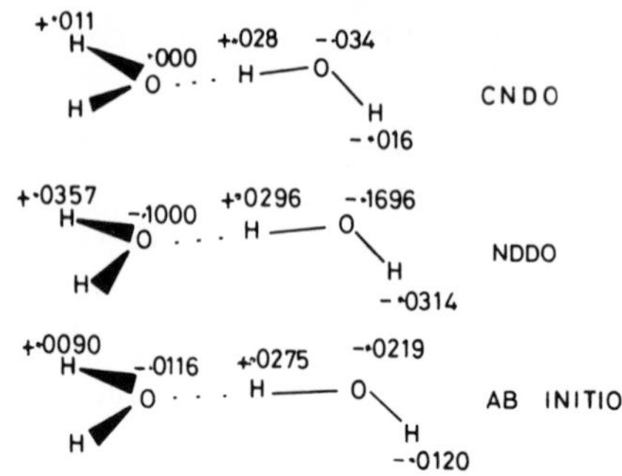

Fig. 4. Population analysis of water dimer (from Kollman and Allen[619]).

at 5 Å, but the exchange repulsion becomes nonnegligible only at an N···O distance of 3.75 Å.

Several workers have reported population analyses on monomeric, dimeric, and polymeric water. The population analyses in water dimer by the different methods are shown in Fig. 4. The important generalizations from these studies may be summarized as follows. The hydrogen in the hydrogen bond loses electrons upon hydrogen bonding. The oxygen atoms gain electrons; more electrons are gained by the oxygen atom on the proton-donor (electron-acceptor) molecule. The greatest electron loss occurs at the hydrogens immediately attached to the proton-acceptor (electron-donor) molecule. The other hydrogens attached to the electronegative atom of the proton-donor molecule gain electrons upon hydrogen bond formation. The charge shifts are as follows:

$$\overset{\delta-}{\mathrm{H}} \diagdown \overset{\delta-}{\mathrm{O}}\text{—}\overset{\delta+}{\mathrm{H}}\cdots\overset{\delta-}{\mathrm{O}} \genfrac{}{}{0pt}{}{\diagup \mathrm{H}^{\delta+}}{\diagdown \mathrm{H}^{\delta+}}$$

These charge shifts explain why the OH···OH···OH trimer is more stable than the O···HOH···O structure. In the latter, the third water molecule donates electrons to a hydrogen which already has more electrons than the monomer.

Kollman and Allen[620] have studied the charge-density difference maps for dimers of H_2O and HF. The density differences near the atoms show good agreement with the Mulliken atomic populations, and the overlap regions are more readily understandable. An important result from this study is that there is charge decrease in the H···Y region of the X—H···Y hydrogen bond on complex formation; exchange and Coulomb repulsions are probably responsible for this effect. We should note here the difference between ordinary covalent bonds and hydrogen bonds; in a covalent bond, the region of the bond gains electrons when the two atoms approach each other.

8. CONCLUSIONS

There is little doubt that we are now able to make reliable predictions regarding the energy and geometry of hydrogen-bonded aggregates of water and other molecules. Although the energy varies considerably with the basis set employed in the *ab initio* calculations, the uncertainties are not greater than in many experimental studies. If we can estimate the zero-point energy and correlation energy corrections, we shall be able to obtain better agreement between experiment and theory. The difference in zero-point energy between $(H_2O)_2$ and $2H_2O$ can be estimated from the force constants of Del Bene and Pople,[251] as well as of Morokuma and Winick,[764] to be 1.5 kcal mol^{-1}; this is about 25% of the dimerization energy. Some of the very weak dimers are likely to have much smaller zero-point energy differences. Calculations of Del Bene and Pople, as well as Kollman and Allen,[621] predict a larger dimerization energy for water than HF, contrary to experiment. This indicates that zero-point energy differences and correlation effects may possibly be important to an understanding of the relative dimerization energies.

Of the semiempirical methods used to study hydrogen bonding, the CNDO/2 is far superior to EHT and NDDO. A comparison of CNDO/2 results with those from *ab initio* calculations reveals that the CNDO/2 in most cases predicts reasonable energies of dimerization as well as geometries (Tables I and III), but the energies often depend on the monomer geometry. This is because the difference between the calculated and the experimental bond lengths in *ab initio* monomers is small compared to those of the CNDO/2 monomers. CNDO/2 and *ab initio* calculations also provide similar populations in water dimer (Fig. 4). No *ab initio* calculations seem to have been carried out on hydrogen-bonded systems where the CNDO/2 method fails. The CNDO/2 method seems to be fairly successful in predicting the proton potential functions, which is not the case with the EHT method. The EHT method is particularly bad in the case of water and its aggregates,[778] although it is somewhat useful in predicting properties of hydrogen bonds in some other systems.[775,776]

Theoretical calculations show that there can be significant deviations from the lone-pair approach. Hankins observes that the lone-pair approach ($\phi \approx 45°$) in the water dimer is due to the desire of the external hydrogens to take up a *trans* position to one another; if one fixes the external hydrogen of the proton donor to be perpendicular to the plane bisecting the HOH angle, there would be no lone-pair directionality in the dimer.

Based on the theoretical calculations discussed above, we cannot draw conclusions regarding the structure of ice or liquid water. It appears likely, however, that cyclic as well as open dimers and polymers may be present in liquid water. Each of these species may include a large variety of distorted hydrogen bonds with a distribution of energies. Although the cyclic and bifurcated structures of the water dimer are less stable than the linear dimer, they are sufficiently stable probably to contribute to water structure. Extensive potential surfaces for dimers and trimers such as those calculated by Del Bene and Pople should provide better potentials for Monte Carlo calculations.[1159]

Vibrational spectroscopic results derived from potential energy curves appear to be in qualitative agreement with experimental data. The CNDO/2 method overemphasizes the band intensity enhancements due to hydrogen bonding. Molecular orbital studies on NMR spectra of hydrogen-bonded species have been limited. More detailed theoretical studies on the spectroscopic aspects would be worthwhile.

Theoretical calculations provide useful information about the charge shifts in hydrogen bonds and permit some important generalizations about the nature of the hydrogen bond. There is, however, much scope for detailed studies on the decomposition of hydrogen bond energies in water dimers and polymers into individual contributions. Such studies will enable us to understand the mechanism of hydrogen bonding more clearly.

CHAPTER 4

The Properties of Ice

F. Franks

Unilever Research Laboratory Colworth/Welwyn
Colworth House
Sharnbrook, Bedford, England

1. INTRODUCTION

Systematic studies of ice date from the first half of the twentieth century and are closely linked with the work of Bridgman[116,120,121] and Tammann,[1051] who investigated the effects of pressure and temperature on its phase behavior. Their experiments led to the realization that there existed a number of complex structural modifications of ice, stable or metastable under a variety of conditions, and this provided the foundations for subsequent investigations of the detailed structures of these polymorphs and the internal molecular motions within the ice lattice. The structure of and molecular interactions in the solid state are generally more easily described than those in the corresponding liquid, and this has been particularly true for water. Thus, studies of the intermolecular nature of liquid water have frequently employed analogies with the known properties of ice and this will become more apparent in subsequent chapters of this volume. In a way this is hardly surprising, since many important developments in our understanding of the liquid state have been due to solid-state physicists who applied familiar techniques, such as X-ray diffraction, which had proved so successful in the elucidation of solid structures.

It is perhaps for this reason that terms such as "structure," "lattice," and "icelikeness" have found their way into discussions of liquid water, and Eisenberg and Kauzmann[301] have provided us with a timely reminder that rather different meanings may have to be attached to such terms when they are used to describe the positions, motions, and interactions of molecules in the liquid state.

Although the study of ice and its properties constitutes an important and interesting area of investigation in its own right, this chapter will concern itself mainly with those of its properties which may help us better to understand the behavior of liquid water and aqueous systems.

2. PHASE BEHAVIOR OF ICE

Most people when they talk about "ice" refer to the solid which is formed when liquid water freezes at atmospheric pressure, and most work reported in the literature is confined to this substance. However, the work of Tammann[1051] and Bridgman[116,120,121] referred to above had indicated that other forms of solid water can exist under different conditions of temperature and pressure,* and their work was intensively followed up by Whalley and his colleagues at the National Research Council of Canada,[1165,1167] who removed many of the question marks left on the phase diagram by earlier investigators, and who mapped out new areas which had become accessible with the development of high-pressure technology. Figure 1 shows the present state of the solid/liquid phase diagram and indicates the various stages in its development. In addition to the phases shown in Fig. 1 there exists a modification of normal or hexagonal ice I*h*, referred to as ice I*c* (cubic) metastable between -80 and 120°C, at which temperature it becomes stable. It cannot therefore be prepared by cooling ice I*h* but is obtained by the condensation of water vapor below -80°C. Studies of the interconversion of ices II, III, and V have shown that they will all yield ice I*c*[75] and this has provided a method for the preparation of reasonable quantities of this substance for further study. Not much is known about ice IV, which appears as a metastable form capable of coexistence with ice V. So far it has only been observed for D_2O. When water vapor is condensed onto a clean surface at liquid nitrogen temperatures, amorphous, or vitreous ice is obtained.[156,374,375] This has not received detailed study, and it is doubtful whether it can be classified as ice at all, since it does not show any evidence (spectroscopic) of hydrogen bonds. When amorphous ice is annealed at -120°C, ice I is invariably produced.

A crystalline form of ice has also been reported which is formed by

* The phase diagram developed by Bridgman was for D_2O only, but, as will be seen presently, D_2O and H_2O behave in a very similar manner.

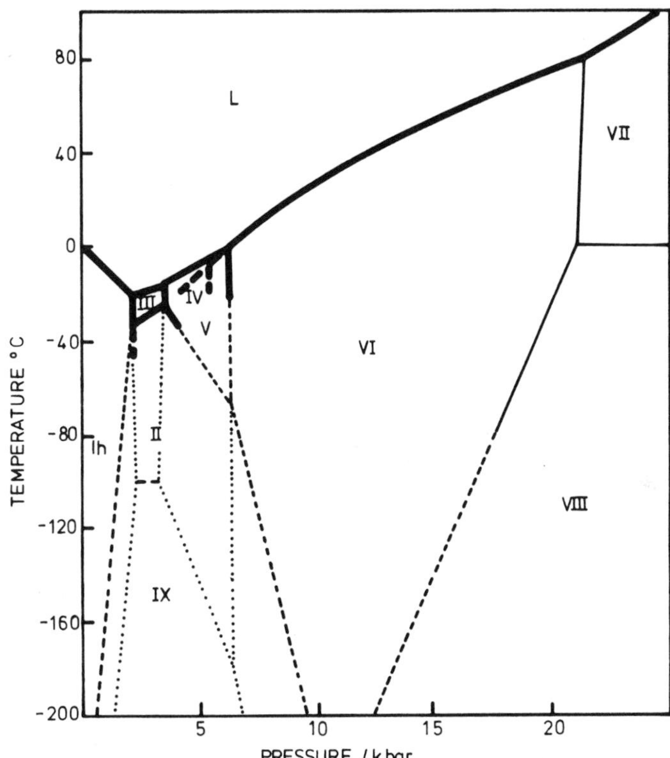

Fig. 1. Solid–liquid phase diagram of ice, after Whalley *et al.*[1167] Solid and long broken lines refer to stable and metastable regions directly measured by Bridgman (thick lines) and Whalley *et al.* (thin lines). Short broken and dotted lines represent estimated and/or extrapolated stable and metastable boundaries, respectively.

cooling when a vacuum is applied to liquid water.[215,969] It has been claimed that the solid so obtained is heavier than water, but although attempts have been made to verify its existence, they have not been successful.

Apart from pure-water solid phases, there exists a whole range of crystalline solids, known as clathrate hydrates,[559] in which the formation of water lattices is promoted by the presence of apolar, or slightly polar, molecules, such as rare gases, hydrocarbons, ethers, amines, etc., which are, however, not bonded to the crystalline water framework. Although these hydrates bear a marked structural resemblance to the ice polymorphs, they are, strictly speaking, two-component systems derived from aqueous solutions and will therefore be dealt with in Volume II of this Treatise.

TABLE I. Triple Point Data for H_2O and D_2O[a]

Phases coexisting	H_2O		D_2O	
	Pressure, kbar	Temperature, °C	Pressure, kbar	Temperature, °C
I–liquid–vapor	6.1×10^{-6}	0.01	—	—
I–liquid–III	2.07	−22.0	2.20	−18.9
I–II–III	2.13	−34.7	2.25	−31.0
II–III–V	3.44	−24.3	3.47	−21.5
III–liquid–V	3.46	−17.0	3.49	−14.5
V–liquid–VI	6.26	0.16	6.28	2.6
VI–liquid–VII	22.0	81.6	—	—
VI–VII–VIII	21	~5	—	—

[a] Taken from Eisenberg and Kauzmann.[301]

The phase diagram shows a number of invariant points which are summarized in Table I. This indicates that ices II, VIII, and IX cannot be prepared directly from liquid water but must be obtained by cooling ice III, V, VI, or VII under pressure or by decompressing one of the above polymorphs isothermally. Thus, ice II has normally been prepared by the decompression of ice V at −35°C, and ice IX by the cooling of ice III to −100°C. Solid phases corresponding to ices I–VII have also been confirmed for D_2O and in each case triple points are about 3° above those for H_2O and this was attributed by Bridgman to the lower zero-point energy of D_2O.

3. STRUCTURES OF CRYSTALLINE ICE PHASES

Several of the polymorphs have been subjected to diffraction studies and supplementary information has been derived from infrared spectra in the low-frequency region. One feature which all phases have in common is the nearest-neighbor geometry, in that each oxygen is hydrogen-bonded to four neighboring oxygen atoms, but as the pressure is raised so the $H\hat{O}H$ angle becomes progressively distorted from its ideal tetrahedral value, as it exists in ice I. Thus, the primary effect of pressure is the distortion of the lattice so that distances between nonbonded water molecules are reduced.

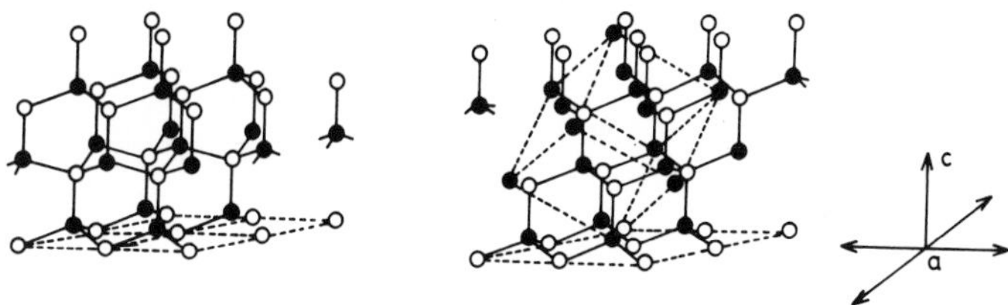

Fig. 2. Crystal structures of hexagonal (left) and cubic (right) ice I perpendicular to the *c* axis. Filled-in and open circles refer to oxygen atoms in different planes.

Figure 2 shows the crystal structure of hexagonal (stable) and cubic (metastable) ice I. It is of interest that even at low temperatures the thermal agitation, which amounts to 0.2 Å,[1195] makes the cubic diamond-type lattice unstable. The distance between oxygen atoms, identical for both polymorphs, is 2.75 Å at −80°C and 2.76 Å near the melting point. Since this distance is characteristic for the hydrogen bond, it led to the assumption that the hydrogen atoms lie along the O···O axis. Figure 3 shows the dependence of the lattice constants of ice I*h* on the temperature. If all the nearest-neighbor distances in ice were the same, then the ratio of the lattice

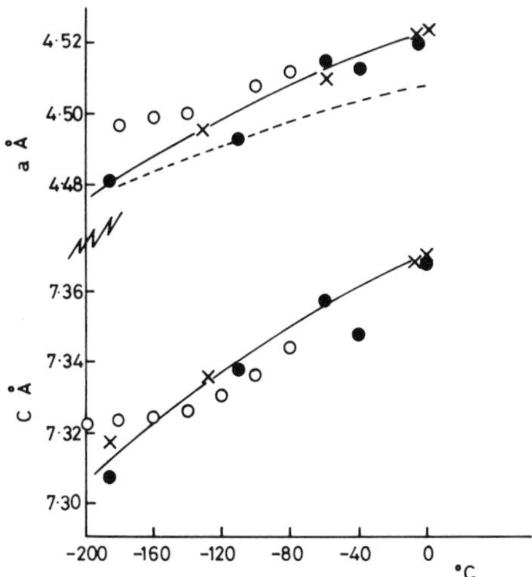

Fig. 3. Crystal lattice parameters of ice I as function of temperature; closed circles: H_2O, older data, summarized by Lonsdale;[696] open circles: H_2O, data of Brill and Tippe;[125] crosses: D_2O. The broken line represents the theoretical relationship $1.633a = c$ (see text).

parameters would be given by $c/a = 1.6330$[738]; from Fig. 3, it appears that this value may be attained at very low temperatures, although other studies have shown that the ratio c/a is independent of temperature between 0 and -180°C.[653] These, and more recent results, do, however, confirm earlier reports that the ratio c/a does deviate from its ideal value. Detailed studies,[653] covering a wide temperature range, provide an estimate of $c/a = 1.6280 \pm 0.0002$, suggesting that the O···O distances parallel to the c axis are shorter than those in other directions, and that the actual O···Ô···O angles deviate slightly from the ideal tetrahedral value. The reported existence of an anomaly in the a parameter between -100 and -130°C[653] could not be confirmed by Brill and Tippe.[125]

Although the exact position of the hydrogen atoms is not readily obtainable by X-ray methods, neutron diffraction has provided confirmatory evidence for the assumption that the hydrogen atoms lie along the O···O axis,[834,835,1180] although it has been claimed that the experimental results can be interpreted equally well on the basis of a valence angle of 104.5°, as it exists in water vapor.[182] This suggestion, together with the claim that the O—H bond in ice is longer by 0.05 Å than it is in water vapor,[834,835] implies that the hydrogen bonds in ice must be bent, i.e., the hydrogen atoms do not lie along the lines joining the oxygen atoms. Although it is not known for certain whether the HÔH angle in ice differs from the vapor value, recent NMR second moment measurements on several of the ice polymorphs can best be reconciled with calculated theoretical proton second moments by assuming a hydrogen bond angle of 104° at an O—H distance of 1.010 Å,[879] suggesting a displacement of the hydrogen atoms of 2° from the O···O line. By assuming harmonic vibrations and assigning force constants of the right order of magnitude, one can calculate that if hydrogen bond formation does indeed occur with a lengthening of the O—H distance by 0.05 Å, then the hydrogen bonding forces would also be strong enough to distort the HÔH angle by several degrees from its normal value.

Direct evidence concerning the structure of some of the high-pressure ice modifications has been provided by Kamb and his associates.[567,568,571–573] It appears that in ice II each oxygen atom is bonded to four others at 2.80 Å and has, in addition, the next nearest neighbors at 3.24 Å. Thus, the bond lengths are very similar to those in ice I, but the effect of pressure makes itself felt in the deviations from the ideal tetrahedral structure. Apparently the hydrogen bond strain energy, which must be greater than in ice I, is small enough to be offset by the extra van der Waals energy, so that the energy difference between the two species is only 10 kcal mol^{-1}.

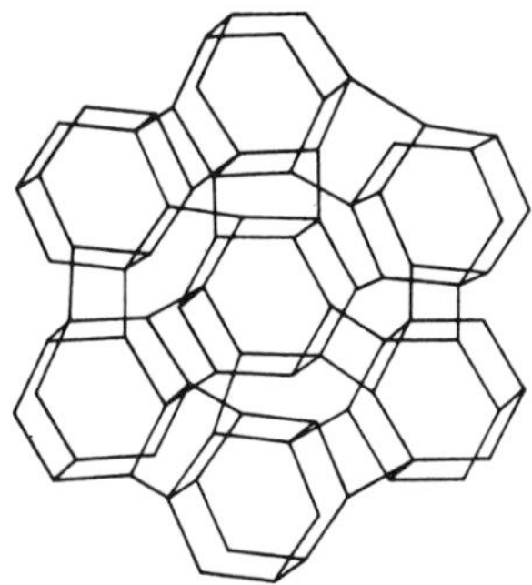

Fig. 4. Hydrogen bond topology in ice II, reproduced from Eisenberg and Kauzmann.[301] Each line represents a hydrogen bond and vertices correspond to the oxygen atom positions.

Figure 4 shows the arrangement of hydrogen bonds in ice II, and the resemblance to ice I is at once apparent, the main difference being that the hexagonal tunnels (see Fig. 2) are packed in a more economical manner.

Ice III bears signs of further distortion from the ideal tetrahedral structure and hydrogen bond lengths vary from 2.76 to 2.80 Å. The structure can be described as consisting of helical chains of hydrogen-bonded oxygen atoms held together by further hydrogen bonds. Oxygen atoms can thus either be members of the helix or act as links, bonding four separate helical segments. In ice V, hydrogen bond lengths vary between 2.76 and 2.87 Å and hydrogen bond donor angles between 84 and 135°. This rather severe pressure distortion of the ideal lattice leads to some nonbonded neighbor distances of 3.28 and 3.46 Å, compared to 4.75 Å in ice I. The high-density polymorphs VI, VII, and VIII have one structural feature in common: they consist of two discrete interpenetrating frameworks. In ice VI, bonded neighbor distances are 2.81 Å but each oxygen also has eight nonbonded neighbors (part of the other framework) at 3.51 Å. This interpenetration of two complex structures demands a wide distribution of donor angles ranging from 76° to 128°. Ices VII and VIII apparently have the same crystal structure and differ only in the position of the hydrogen atoms in the hydrogen bonds. The structure can be regarded as consisting of two interpenetrating ice I*c* frameworks. Each oxygen atom has eight near neighbors at 2.86 Å but is hydrogen-bonded to only four of these. The body-centered structure so obtained is shown in Fig. 5, which also shows what might be regarded as an even denser form of ice with a cubic close-packed structure. However, Kamb and Davis[573] have predicted that ice VII (and VIII) is the ultimate

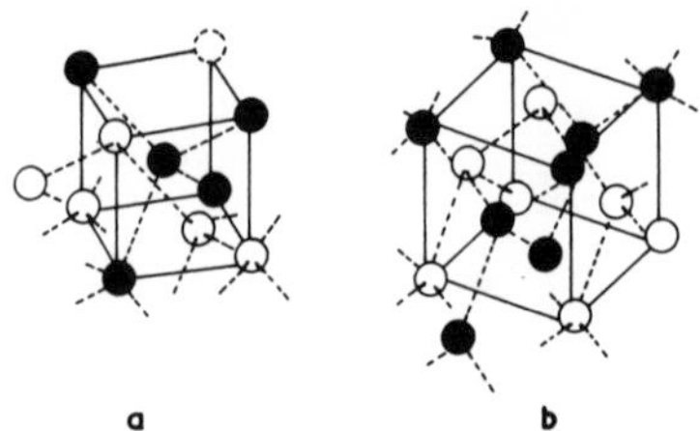

Fig. 5. (a) Body-centered-cubic structure of ice VII, showing interpenetrating hydrogen bond systems (broken lines); (b) hypothetical closest-packed ice structure. The interpenetrating hydrogen bonded frameworks are represented by filled-in and open circles (oxygen atom). Redrawn from Kamb and Davis.[573]

high-density modification which will still have the recognizable properties of ice; even at pressures of 220 kbar, the free energy of the close-packed cubic structure is higher by 2 kcal mol^{-1} than that of ice VII. Recent X-ray diffraction measurements at 230 kbar[524] have in fact confirmed the calculations of Kamb and Davis,[573] and no noticeable change in the crystal structure can be found.

The structural investigations thus show that the application of pressure on ordinary ice does not result in any breaking of hydrogen bonds, but rather in marked distortions of donor angles from the ideal tetrahedral value, bringing into closer proximity water molecules which are not bonded and which are subject to net repulsion interactions. Simultaneously, the ideal hydrogen bond lengths of 2.76 Å are increased significantly, thus resulting in weaker hydrogen bonds.

Figure 6 summarizes, in the form of a histogram, the observed O···O distances of the various ice polymorphs, reduced to atmospheric pressure. It is seen that, starting with the simple ice I lattice, the application of pressure results in the intermediate (II, III, V) complex structures with considerable distributions of O···O distances, before eventually giving rise once again to a simple tetrahedral structure (ice VII) with a few well-defined lengths. The actual hydrogen bond length gradually increases from 2.76 Å in ice I to 2.95 Å (2.86 Å at 25 kbar) in ice VII and at the same time the nearest nonbonded neighbor distance decreases from 4.5 Å in ice I to 2.95 Å in ice VII. Thus, if it were not for the lengthening of the hydrogen bond as the pressure is increased, ice VII would have twice the density of ice I.

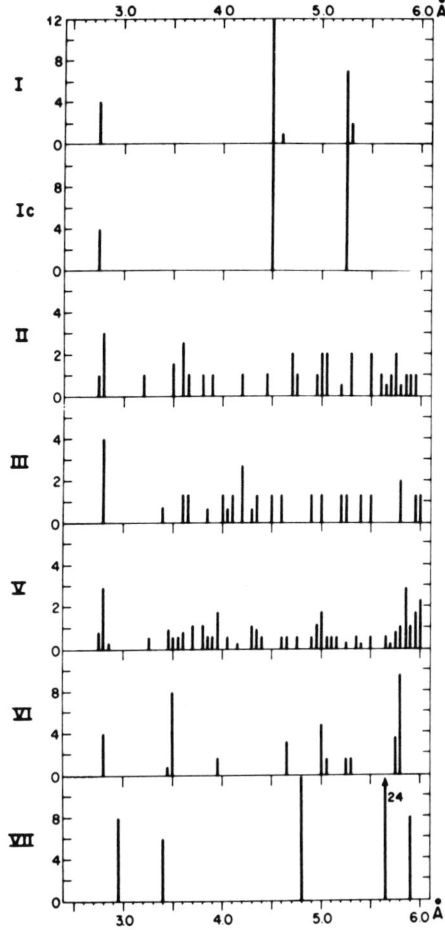

Fig. 6. Histogram showing the distribution of O···O distances in ice polymorphs, reduced to atmospheric pressure. Reproduced from Kamb, p. 507 in *Structural Chemistry and Molecular Biology*, A. Rich and N. Davidson (eds.), W. H. Freeman and Company (1968).

4. THERMODYNAMIC PROPERTIES OF ICE POLYMORPHS

The thermodynamic functions associated with the polymorphic transitions can be estimated by means of the Clapeyron equation

$$(\partial P/\partial T)_{\text{coexist}} = \Delta S/\Delta V$$

TABLE II. Thermodynamic Functions of Polymorphic Transitions in Ice

Transition	Temperature, °K	Pressure, kbar	ΔV, cm³ mol⁻¹	ΔS, e.u.	ΔH, cal mol⁻¹
I*c* → I*h*	130–210	—	—	—	38
I → II	188	2.13	−3.92	−0.76	−180
I → III	188	2.13	−3.53	0.16	40
II → III	188	2.13	0.39	0.92	220
II → V	?69	3.44	−0.72	1.16	288
III → V	269	3.44	−0.98	−0.06	−16
V → VI	273	6.26	−0.70	−0.01	−4
VI → VII	191	22	−1.05	~0	~0
VI → VIII	~268	~21	?	~−1.01	−282
VII → VIII	~268	~21	0	~−0.93	−260
III → IX	~173	~2	?	?	?

This was first done up to ice VII by Bridgman, who combined experimentally determined densities of the polymorphs with the slopes of the coexistence lines in the phase diagram. A more complete picture has been provided by Whalley and his colleagues,[130,1166] extending up to ice IX, and the data are collected in Table II. It is seen that transformations involving ice II and ice VIII are accompanied by entropy changes of ~1 e.u., whereas other transition entropies are much smaller. This provides clues as to the location of protons along the OHO bonds of the various ice structures.

If the existence of hydrogen-bonded water molecules in ice is accepted, then an idealized lattice can be constructed in which each molecule has four nearest neighbors; this is shown diagrammatically in two dimensions in Fig. 7(a). Such a crystal should exhibit polarity in the direction of the arrow. However, the neutron diffraction results indicate two half-atoms of hydrogen along each O···O axis, each half-atom being situated at a distance of 1 Å from the nearest oxygen atom.[1166] This result is best explained in terms of Pauling's representation of ice[826] involving a statistical distribution of hydrogen atoms, such that only one hydrogen atom can be found situated along each O···O axis, and that each oxygen atom is therefore surrounded by four hydrogen atoms, two of which are at a distance of 0.97 Å and the other two at a distance of 1.79 Å. This type of irregular lattice is shown in Fig. 7(b). Electron diffraction studies on ice I*c* have shown a similar disordered lattice.[526] On the basis of a statistical distribution of hydrogen atoms along O···O axes, Pauling calculated the zero-point entropy of ice as $S_0 = R \ln \frac{3}{2} = 0.814$ e.u., which agreed well with the experimental value.[376]

The various aspects of proton disorder and proton movements in ice lattices have been subjected to intensive studies by a variety of physical techniques and the results obtained from such studies are discussed in Section 7.4.

From the entropies associated with transitions involving ices II, VIII, and IX it became apparent that hydrogen atoms in these phases were probably ordered and this was later confirmed for ice II by spectroscopic investigations, although the hydrogen positions in ice VIII are not yet known in detail. In this context it is worth mentioning that no proton ordering seems to occur when ice I is cooled. There have, in the past, been suggestions that at low temperatures a piezoelectric effect can be observed in ice I.[258] This could not arise from the oxygen lattice, which is centrosymmetric down to the lowest temperatures. If there were a hydrogen-ordering transition at some low temperature, this could explain the observed piezoelectricity. There is no doubt that the piezoelectric effect can be induced in ice as a result of an inhomogeneous distribution of impurities, but the balance of evidence seems to be against the existence of a real effect.[228]

If such an ordering transition does indeed take place, then this must be reflected in the heat capacity of ice. Giauque and Stout[376] carried out high-precision specific heat determinations down to 10°K and did report

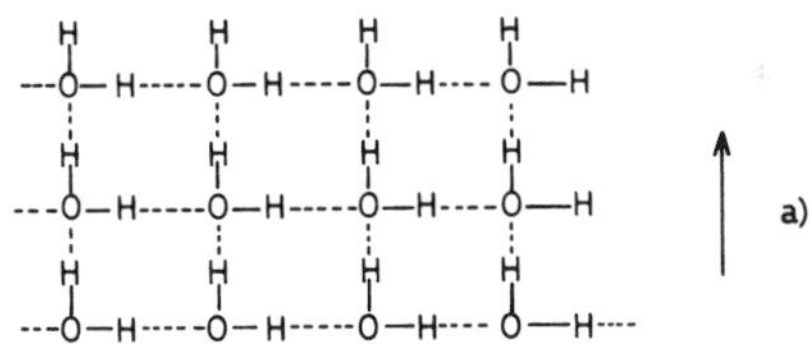

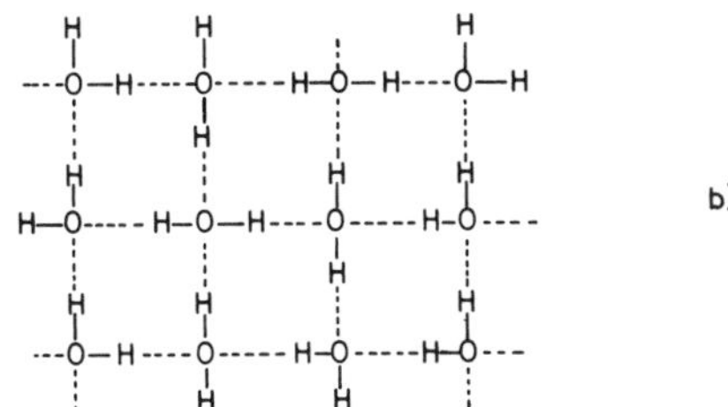

Fig. 7. (a) Hydrogen atom arrangement in ordered lattice; (b) hydrogen atom (H_2O molecule) disorder in ice I crystal, giving rise to residual entropy of 0.8 e.u.

a small anomaly between 85 and 100°K.* A recent investigation by Pick[838] using single ice crystals in a Calvet heat flux calorimeter has substantiated the existence of such an anomaly near 120°K. The experimental transition entropy thus determined amounts to about 1–3% of the zero-point entropy, indicating that any ordered zones formed in ice must be confined to a small part of the lattice. Finally, in this context, it is interesting to note that ice appears to reach thermal equilibrium very slowly. This was first reported by Giauque and Stout,[376] but aging effects have since been observed by a number of workers not only at high temperatures, where a large number of lattice faults would probably exist, but even at low temperatures (200°K).

It is now accepted that the heat capacity of ice arises from two types of lattice vibrations—hindered translation (hydrogen bond stretching) and hindered rotation (hydrogen bond twisting or libration)—but the possibility of low-temperature ordering processes has, in the past, led to difficulties in the interpretation of the available heat capacity data. Normally, the translational (calorimetric) component of C_v is represented by a Debye function θ_D, and the vibrational and rotational contributions (from infrared and Raman spectra) by Planck–Einstein functions. Giguère[381] claimed that, in the case of ice, $C_{v,\,trans}$ could not be represented by a Debye curve below 100°K. Hydrogen peroxide, on the other hand, with hydrogen bonds of the same length and the same strength, behaved normally, with $\theta_D = 234°$ up to 50°K. He believed therefore that an as-yet-undetected transition might occur in ice below 100°K. Pitzer and Polissar[851] suggested that such a transition occurs in the neighborhood of 60°K, i.e., that at very low temperatures the hypothetical, perfectly ordered ice crystal exists. Inside this temperature range, C_v increases in terms of a Debye curve with $\theta_D = 310°$; above 70°K a rapid rise in C_v is due to a cooperative disordering process. The problem could not at that time be resolved without further C_v data down to lower temperatures on ice specimens of high purity, because the presence of impurities, such as HF, were known to give lattice imperfections the concentration of which is independent of temperature.[416] Such C_v determinations in the temperature range 2–27°K were in fact reported[339] shortly after the publication of Giguère's detailed discussion. The results, which are consistent with those of Giauque and Stout,[376] indicate that the temperature dependence of θ_D can be accounted for without specifying anomalous behavior. A comparison between the thermal measurements

* Whereas the centigrade scale of temperature is employed in the remainder of this chapter, temperatures in this section are quoted in degrees Kelvin, as being more appropriate to a discussion of thermodynamic properties.

and C_v, as obtained from the elastic properties of ice, extrapolated to the lowest temperatures, shows very good agreement and it can now be concluded that the very low-frequency vibrational modes do not contribute appreciably to the measured C_v. Also, the fundamental frequencies of the H_2O molecules are so high that below 200°K they hardly contribute toward C_v. Flubacher and co-workers[339] found that the $\theta_D(T)$ curves could be accounted for on the basis of $3N$ degrees of freedom; θ_D goes through a minimum at $T \simeq \theta_0/15$, and for $T < \theta/30$, C_v can be expressed by the simple power series $C_v = aT^3 + bT^5 + cT^7 + \cdots$ characteristic of simple crystals. The decrease in θ_D as $T > \theta_0/3$ has also been observed for other substances and is ascribed to the onset of anharmonicity. Furthermore, C_v in the temperature range $80 < T < 200$°K can be expressed in terms of an Einstein function for $3N$ vibrations of frequency 620 cm^{-1}.

These findings are in agreement with calculations based on C_v of D_2O and H_2O, which show that librational contributions become significant at temperatures above 80°K.[95] Thus, the low-temperature lattice frequency spectrum of ice cannot be considered to be as abnormal as would appear at first glance.*

Recent calorimetric studies[1036] have confirmed the existence of a glass transition in the region of 135°K. As has been mentioned in Section 2, amorphous ice (glassy water) can be obtained by the low-temperature condensation of water vapor. The crystallization is accompanied by a ΔH of -400 kcal mol^{-1} and a corresponding $\Delta C_v = 8.4$ cal (mol deg)$^{-1}$, which is almost as large as the heat capacity of fusion of ice at its normal melting point. It appears that the crystalline form produced at the glass transition is ice I*c* and this had been previously noted from X-ray[89] and electron diffraction[972] experiments. There has in the past been a certain amount of dispute on whether the transition is indeed a glass transition, since the amorphous form of ice has never yet been obtained by the normal method, i.e., the supercooling of the liquid. However, Sugisaki *et al.*[1035] have observed the well-known glass transition for methanol in a sample prepared by vapor-cooling in a manner analogous to that used for water. They therefore conclude that the 135°K transition represents a true glass point.

The calorimetric studies show that the transition ice I*c* → I*h* is sluggish and irreversible and takes place over the temperature range 160–210°K.

* In fact, doubt has been expressed concerning the apparently "normal" Debye behavior of H_2O_2, NH_3, and HCN, and it has been suggested that, as a result of inseparable contributions from translational and librational modes, θ_D should possibly be referred to more than $3N$ degrees of freedom.[339] This would result in $\theta_D(T)$ curves resembling that of ice.

The average enthalpy of transition, -38 cal mol^{-1}, agrees well with earlier data[717] and is a measure of the difference between the lattice energies of the two polymorphs.

The kinetics of the transition have been investigated by X-ray diffraction[271] and Mössbauer spectroscopy[801] and the results interpreted in terms of an activated process. The time dependence of the heat capacity and enthalpy[1036] can be similarly analyzed and suggests that the transition takes place in two stages, associated with activation energies of 5 and 11 kcal mol^{-1} at low and high temperatures, respectively. This is in agreement with the analysis of the X-ray diffraction data.

The volumetric properties of ice I, i.e., the coefficients of cubic expansion α and adiabatic compressibility β_s, are available from measurements on bulk ice and single crystals and from the X-ray diffraction data. With regard to α, agreement between results based on bulk measurements and X-ray crystal parameters is reasonable at relatively high temperatures. Thus, at 260°K and 1 bar, $\alpha \simeq 150 - 160 \times 10^{-6}$ deg^{-1} and it is established that α decreases with a decrease in temperature. However, single-crystal measurements indicate[227] that α becomes negative at 60°K and passes through a minimum near 40°K and that the expansion is isotropic. On the other hand, α, as derived from crystal lattice parameters,[653] does not show a change of sign in the neighborhood of 75°K; the data are not of a high enough precision[125] to determine whether α shows any anisotropy, i.e., differences in α when measured parallel or perpendicular to the crystal c axis. Within the range 260 to 90°K, β_s, as determined from the elastic constants, changes from 160 to 50×10^{-6} deg^{-1} at 1 bar. The specific adiabatic compressibility $\varkappa_s = (\partial V/\partial P)_s$ can be expressed by

$$\varkappa_s(T) = 11.94(1 + 1.653 \times 10^{-3}T + 3.12 \times 10^{-6}T^2)\ \mathrm{cm^2\ (dyn\ g)^{-1}}$$

(T in degrees centigrade) to within $\pm 15\%$ over the temperature range 273–130°K.[229]

It is of interest to consider the role of thermal and volumetric properties in determining the nature of the phase diagram. Eisenberg and Kauzmann[301] have discussed this in some length. The phase diagram (Fig. 1) shows that those transitions which do not involve hydrogen-ordered polymorphs are associated with very small entropy changes and therefore the slopes of the $P(T)$ curves are also small and their magnitudes depend on the sign and magnitude of ΔV (see Table II). For transitions which involve hydrogen ordering, a delicate balance must exist among ΔS, ΔV, and ΔE (the internal energy). It has been established that, as the transition temperature is lowered along the coexistence line, $T\,\Delta S$ decreases. Therefore, for hydrogen ordering to occur, there must be a compensating decrease in ΔE and/or $P\,\Delta V$. However, as P increases and hydrogen bonds are being progressively distorted, ΔE increases, probably due to repulsive interactions between nonbonded neighbor molecules. Thus, for a more stable, low-temperature polymorph to exist, the transition must be accompanied by a correspondingly large decrease in $P\,\Delta V$. Of necessity, ΔV for solid–solid transitions is small (see Table II), but for phase changes taking place at high pressures, $P\,\Delta V$ can assume magnitudes of the same order as ΔE. Thus, for the transition ice VI → ice VII at 22 kbar, $\Delta E = -P\,\Delta V$, since $\Delta S \simeq 0$. For a hydrogen-ordering transition, e.g., ice I → ice II, for which ΔS is negative, and the pressure comparatively low (2 kbar), ΔV is correspondingly large, so that $P\,\Delta V$ (200 cal mol^{-1}) can still compensate $-T\,\Delta S + \Delta E$.

5. MECHANICAL PROPERTIES

The mechanical properties of ice under different conditions of pressure and temperature are of the greatest importance in geological and glaciological studies, but they will only be briefly reviewed here, as they do not contribute significantly to our understanding of the behavior of liquid water. Progress up to 1958 has been reviewed by Glen.[(396)] Since then, valuable results have been achieved from studies on natural glacial and synthetic single crystals,[(157,781)] and the effects of low concentrations of impurities on stress/strain relationships have been studied.

A typical creep curve for a glacial single crystal is shown in Fig. 8[(503,504)]; its shape resembles that observed for germanium. The temperature and stress (θ) dependence of the static creep rate s' can be described by $ds'/dt = K\theta^m \exp(-E/RT)$ over the temperature range −40 to −15°C. Figure 9 shows a typical set of stress/shear strain curves at −15°C for various strain rates. They show that ice differs from LiF, Ge, and other solids which show similar yield drops, in that the initial slope of the $\theta(s)$ curve depends on ds/dt and on the temperature. A similar behavior is observed when a tensile stress is applied. The maximum stress is very sensitive to dislocations in the lattice. Thus, the presence of 0.5 ppm of HF in ice reduces the maximum obtainable stress by 50%,[(564)] but no corresponding effect has been observed with NH_3.

The five elastic moduli of ice have recently been redetermined by Dantl,[(229)] based on measurements of longitudinal and transverse sound velocities in single crystals in the temperature range 0 to −140°C. The

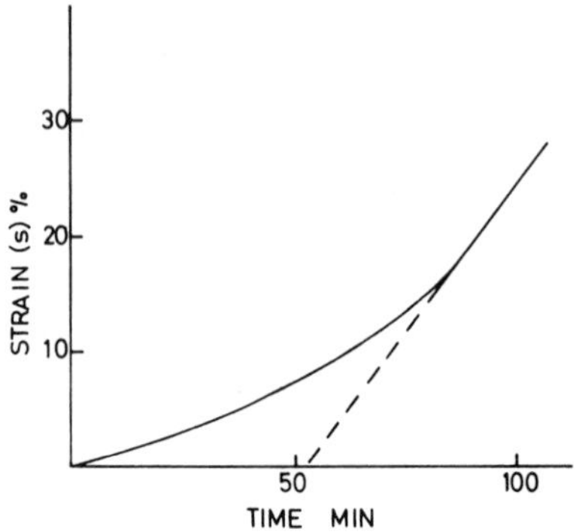

Fig. 8. Typical creep curve for ice I single crystal, after Higashi *et al.*[504] The deformation is characterized by a low initial slope which, at larger strains, increases and reaches a constant value.

results, which differ by up to 5.6% from those of previous authors (which also differ among themselves), are interpreted by the assumption of an aging process whereby the orientation of H_2O dipoles accompanying the growing of an ice crystal is followed by a slow randomization of orientations. This hypothesis is readily reconciled with experimental observations such

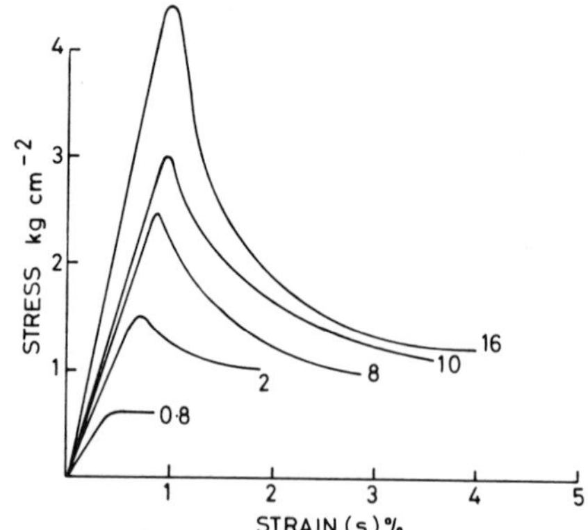

Fig. 9. Stress/strain curves for ice I at −15°C, after Higashi *et al.*[503] Curves relate to different strain rates ($ds/dt \times 10^{-6}$ min^{-1}) as indicated. The initial linear slope is a function of ds/dt and of temperature, as is also the magnitude of the yield drop; there is no indication of recovery.

as the dependence of the density of ice on conditions of crystal growth,[230, 269] freezing potentials,[1185] and claims of a piezoelectric effect (see Section 4). It also accounts for values of the elastic moduli very close to the melting point which are up to 5% higher than those reported at lower temperatures.[127]

In line with the apparent anomalies in the low-temperature heat capacity of ice, already referred to, the elastic constants of single ice crystals reveal a similar type of behavior.[488] Thus, for heating and cooling below certain critical rates, low results are obtained for the elastic constants, which eventually reach limiting values. By treating the results in terms of a rate process, an activation energy of 3 kcal mol^{-1} for the time dependence of the elastic constants has been obtained. The phenomenon is considered to arise from an ordering process, although, as has been discussed above, the nature of such a process is far from clear and is still the subject of some controversy.

6. LATTICE DYNAMICS

This section is mainly devoted to a discussion of the vibrations which take place in the ice lattice and which are reflected in the infrared and Raman spectra; the neutron inelastic scattering spectrum of ice is dealt with in Chapter 9. The subject was reviewed by Ockman[802] in 1958 and since then many of the significant advances have been reported by Whalley[77,78,79, 731,1057] and Hornig[440,532] and their respective colleagues. The recent work has been reviewed by Whalley.[1164] The compound which might be expected to show a similar behavior to ice is hydrogen peroxide. Both substances have the same crystal structure, consisting of three-dimensionally hydrogen-bonded frameworks, but hydrogen peroxide, unlike ice, does not possess a residual entropy, nor does it exhibit the phenomenon of the maximum density in the liquid state. An analysis of the molecular spectra of solid H_2O and H_2O_2 shows up a striking contrast in the low-frequency region.[381] Whereas the Raman spectrum of hydrogen peroxide exhibits eight distinct peaks below 300 cm^{-1}, the spectrum of ice consists of a broad band in the region 150–300 cm^{-1}. The three-dimensional ice lattice of 4-coordinated hydrogen bonds possesses an almost perfect symmetry, whereas in hydrogen peroxide neighboring molecules are held by short hydrogen bonds, forming infinite helices which are held together by much weaker bonds.

The complexity of the ice spectra arises from the intramolecular and intermolecular coupling of vibrations and from the large number of hydrogen bond stretching and libration modes. The assignment of observed spectral peaks to individual modes has been greatly facilitated by comparisons of spectra of protonated and deuterated ice, so that a fairly complete analysis is now possible. As a first approximation, three spectral frequency regions can be distinguished: broad bands observed at frequencies below 1200 cm^{-1} are associated with intermolecular modes classified as hydrogen bond stretching (ν_s) and libration (ν_l). In the range 1200–4000 cm^{-1}, the modified intramolecular OH stretching and HÔH bending modes are observed, which for the isolated water molecule occur at ν_1(symmetric stretch) = 3657 cm^{-1}, ν_3(asymmetric stretch) = 3756 cm^{-1}, and ν_2(HÔH deformation) = 1595 cm^{-1}. In addition, this region contains overtones of ν_s and ν_l and their combinations with the intramolecular modes. Finally, in the region above 4000 cm^{-1} a variety of weak bands appear as overtones and combinations of the fundamental and lattice modes. Although the positions of oxygen and hydrogen atoms in the ice lattices are known in a fair degree of detail, there is no information about the normal crystal vibrational modes and therefore the many observed absorption and scattering peaks cannot be directly assigned to specific modes. Progress in the elucidation of the spectra has therefore relied mainly on comparisons of observed spectral features with the known fundamental and overtone bands of the water molecule in the dilute vapor.

The problem of vibrational coupling in water has been ingeniously overcome by studying vibrations of the HDO, rather than the H_2O, molecule. Thus, the uncoupled OH stretching vibration can be studied in a dilute solution of H_2O in D_2O, and similarly, dilute solutions of D_2O in H_2O yield the uncoupled OD spectrum. The application of this technique to the elucidation of the Raman spectra of liquid water is described in detail in Chapter 5, and it has also been extended to spectroscopic studies of lattice vibrations in hydrates.[342,961] The effects of coupling on the observed spectral features of ice I are shown in detail in Fig. 10.

Apart from helping to overcome the complexities due to vibrational coupling, isotopic substitution studies have also provided additional data on the basis of which assignments of various bands could be made with a higher degree of certainty. Since ν is proportional to $m^{-1/2}$, where m is the mass of the vibrator, isotope shifts in ν_s are proportional to $(20/18)^{1/2}$, whereas the corresponding shifts for rotational vibrations ν_l are proportional to $2^{1/2}$.

Considering first the intramolecular modes, the uncoupled OH stretching absorption is observed at 3277 cm^{-1} with a bandwidth of approximately 50 cm^{-1}, reflecting the hydrogen disorder of ice I. In water vapor this band appears at 3707 cm^{-1} and the 10% reduction in frequency occurs as a

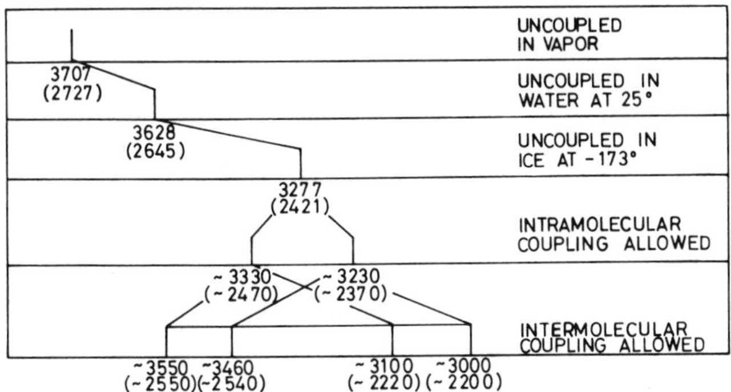

Fig. 10. The effects of intramolecular and intermolecular coupling on the O—H (O—D) valence stretching band ν_1 in ice. Adapted with changes from Whalley.[1164]

result of hydrogen bond formation. Coupling in pure H_2O ice gives rise to a broad band near 3220 cm^{-1}.

The bending mode ν_2 which in the vapor occurs at 1595 cm^{-1} gives rise to a weak band centered around 1650 cm^{-1}. This upward shift of frequency for a bending vibration is again in conformation with other substances known to form hydrogen bonds in the condensed state. The exact origin of this band is still in some doubt and it was at one time suggested that it could be assigned to an overtone of some lattice mode. Isotopic substitution experiments have led to conflicting views regarding the possibility of coupling of bending vibrations.[77,1203] In addition to the normal bending mode there is an additional intense band centered on 2270 cm^{-1}. From the observed temperature dependence of the frequency at maximum intensity this band has been assigned to a combination and/or overtone of the type $3\nu_l$, $\nu_2 + \nu_l$.

Whereas the modes described so far have involved motions of H atoms in a symmetric lattice of O atoms, the low-frequency spectrum, <1000 cm^{-1}, arises from displacements of O atoms, but to a first approximation we can regard such motions as involving whole H_2O molecules. The two types of motion, i.e., restricted translation (hydrogen bond stretching) and restricted rotation (libration) can be distinguished by the associated isotopic shifts $\Delta\nu_s$ and $\Delta\nu_l$, as explained above.

The librational band ν_l is broad and intense, centered around 840 cm^{-1} but containing structural features which have enabled Bertie and Whalley[77] to attempt its analysis. There are a possible $3N$ vibrations per crystal to be considered. Of these, three refer to the normal rotations of each molecule and the remainder are due to coupling with other molecules in the lattice. If the hydrogen bonds are symmetric and only interactions between nearest neighbors are considered, the three ν_l vibrations of the uncoupled molecule have the same frequency. But the lattice spectrum arises from strong coupling, so that a broad band, extending from 1050 to 400 cm^{-1}, is in fact observed. A detailed analysis would of course be required for the exact evaluation of the zero-point energy and a description of intermolecular forces in the ice crystal. Several attempts have been made to analyze the normal modes of small numbers of water molecules. However, these calculations have been of limited value because the exact origin of the absorption is not known and intermolecular coupling, now believed to be an important factor of the spectrum, has been neglected.

What has been said above for the librational modes applies equally well to the absorption in the region 360–50 cm^{-1} which characterizes the hindered translational vibrations in ice. The maximum absorption occurs at 229 cm^{-1} and Bertie and Whalley[(79)] have identified a less intense pake at 164 cm^{-1} and several shoulders. They assign the 164 and 50 cm^{-1} maxima to acoustic modes and the 229 cm^{-1} maximum and the shoulder at 190 cm^{-1} to optical modes. The recent comparisons of infrared, Raman, and neutron inelastic scattering spectra[(870)] have put the previous tentative assignments on a somewhat firmer basis.

The extension of infrared and Raman studies to the high-pressure ice polymorphs has provided valuable information about differences in hydrogen bond strengths, lattice distortions, and positions of hydrogen atoms in the lattices. Whalley and his colleagues[(78,731,1057)] have carried out comprehensive spectroscopic studies and were able to make several predictions concerning the nature of the various polymorphs, later to be confirmed by X-ray and neutron diffraction studies. Table III present a summary of the main observed infrared absorptions for H_2O and D_2O ice polymorphs. The studies indicate that from a spectroscopic point of view there is no substantial difference between ice I*h* and I*c* and also that ν_2 is the same for all phases. However, ν_1 increases with increasing pressure, an observation which led earlier workers to the conclusion that compression resulted in hydrogen bond rupture. Whalley, on the other hand, has postulated that application of pressure merely interferes with the tetrahedral coordination of the oxygen atoms and hence might lead to distorted (bent) hydrogen bonds. This is of course in agreement with subsequent crystallographic evidence. From the fine structure which becomes apparent in the OD stretching band ν_1 of HDO upon the application of pressure, Bertie and Whalley predicted a number of discrete hydrogen bond lengths in ices II and III, and they also suggested that the hydrogen atoms in ice II occupied fixed

TABLE III. Summary of Observed Infrared Absorption Frequencies (cm^{-1}) in Ice[a]

	ν_1	ν_2	ν_l	ν_s [d]
Ice I				
O—H in HDO	3277	1490	~822	229
O—D in HDO	2421[b]	?	~515	—
O—H in H_2O	3220[c]	1650	~840[b]	229[c]
O—D in D_2O	2425[c]	1210	~640[c]	222[c]
Ice II				
O—H in HDO	3373, 3357, 3323	—	770	—
O—D in HDO	2493, 2481, 2460	—	464	—
O—H in H_2O	3225[c]	1690	800, 642[b]	~151 R
O—D in D_2O	2380[c]	1220	593,[c] 507	~146 R
Ice III				
O—H in HDO	3318[b]	—	786	—
O—D in HDO	2461,[b] 2450	—	473	—
O—H in H_2O	3250[b]	—	812, 734	—
O—D in D_2O	2450[c]	1240	~600[c]	166 R
Ice V				
O—H in HDO	3350	—	~780	—
O—D in HDO	2461[b]	—	~445	—
O—H in H_2O	3250[b]	1680	~730	—
O—D in D_2O	2390[c]	1225	~490	159 R
Ice VI (Raman spectrum only)				
H_2O	3204	—	—	—
D_2O	2370	—	—	—
Ice VII (Raman spectrum only)				
H_2O	3348, 3440, 3470	—	—	—
D_2O	2462, 2563, 2546	—	—	—

[a] Adopted from Ref. 301.
[b] Shoulders on principal peak.
[c] Several closely spaced maxima observed.
[d] R refers to Raman spectra where no published infrared data are available.

positions, giving rise to in ordered crystal. In the ν_l region of the spectrum application of pressure leads to a small decrease in the frequency of maximum absorption, again implying the onset of hydrogen bond distortion and weakening.

7. MOLECULAR AND IONIC TRANSPORT IN ICE

The mechanisms whereby molecules, atoms, ions, or crystal faults can migrate through the solid have presented scientists with some of the most challenging problems relating to the behavior of ice. In modern times the experimental techniques which have been brought to bear on this problem have included dielectric measurements, self-diffusion, ionic conductivity, and NMR and elastic relaxation. Nevertheless, despite a considerable literature on the subject, most of the unsolved problems in our understanding of ice are concerned with aspects of molecular motion and the interpretation of measured relaxation phenomena.

7.1. Dielectric Properties

Both the static permittivity ε_0 of ice and the dependence of ε on frequency ω have been subjects of intensive study. Experimental values of ε_0 are required for determinations of the dipole moment μ of the water molecule in ice and $\varepsilon(\omega)$ provides information about the rotation of dipoles within the lattice and hence about the dipole reorientation time τ_ε. Measurements of ε_0 of polycrystalline ice covering the temperature range -60 to 0°C were carried out by Auty and Cole,[37] and Humbel *et al.*[542] studied ε_0 of single crystals over the same range of temperature. They found that ε_0 increases with decreasing temperature and that ε_0 measured parallel to the crystal *c* axis is about 10% higher than ε_0 measured perpendicular to the *c* axis. At 0°C, ε_0 of polycrystalline ice is about 10% higher than ε_0 of liquid water, an unexpected result in view of the volume contraction on fusion. Davidson, Whalley, and their colleagues made a comprehensive study of ε_0 and $\varepsilon(\omega)$ of the various ice polymorphs.[1165,1175] The outstanding results are that ε_0 increases with pressure for each polymorph, apart from ice II and ice VIII, which exhibit zero pressure dependence of ε_0. For ice I, $(\partial \ln \varepsilon_0/\partial P)_T = 14 \times 10^{-6}$ bar^{-1} at -23°C for the range $0 < P < 2$ kbar.[178] The ε_0 values for the high-pressure phases are summarized in Table IV. $\varepsilon_0(P)$ is a measure of the increase in density (i.e., closer packing of dipoles) and also shows that dipole mobility persists up to very high

TABLE IV. Dielectric Parameters of Ice

Ice phase	Temp., °C	Pressure, kbar	ε_0	ε_∞	τ_ε, μsec	$\Delta E_\varepsilon^\ddagger$, kcal mol^{-1}	$\Delta V_\varepsilon^\ddagger$, cm^3 mol^{-1}
I	−23.4	6×10^{-6}	97.5	3.1	168	14	3
	−23.4	1.6	101	—	213	—	—
	−30	6×10^{-6}	99	3.2	—	—	—
II	−30	2.3	3.66	3.66	—	—	—
III	−30	2.3	117	4.1	2.75	11.6	4.5
	−100	2.3	—	3.96	—	—	—
V	−30	5	144	4.6	7.2	11.5	4.8
VI	−30	8	193	5.1	~6	11.0	4.4
	22	21	—	—	48	13.6	—
VII	22	21.4	—	—	0.23	11.6	2.5
IX	−100	2.3	3.74	3.74	—	—	—

pressures. The low ε_0 for ice II and ice VIII are of interest in showing that here the dipoles are "frozen in" and cannot reorient upon the application of an external field. This result is, of course, related to the fact that hydrogen ordering exists in these two polymorphs.

The interpretation of ε_0 in terms of dipole moments present certain problems and these are discussed in Chapter 7, which should be referred to for details of the terminology employed in the following paragraphs. The Kirkwood theory, although strictly speaking not applicable to a substance like ice, has proved reasonably successful for the derivation of dipole moments from observed dielectric data.

The Kirkwood correlation parameter g, where

$$g = 1 + \sum_i N_i \langle \cos \gamma_i \rangle$$

and $\langle \cos \gamma_i \rangle$ is the average cosine of angles formed by dipoles of the ith molecular shell surrounding a given molecule, is a useful measure of the influence of correlated molecular orientations on the experimental bulk dielectric behavior. The estimate of g is subject to some degree of uncertainty, presuming a knowledge of the average moment m of a given molecule when surrounded by its neighbors and the vector sum m^* of the dipole moments of a given molecule and those of its neighbors. The most recent calculations, based on a multipole moment model for H_2O and taking into account three shells of neighboring molecules (a total of 84 molecules) lead to $m = 2.6$ D for ice I[(214)] and hence $g \simeq 3$. Similar values, $2.4 < g < 3.4$, have also been reported for the other polymorphs which exhibit hydrogen disorder and this is taken as evidence that, to a first approximation, the fourfold coordination persists.

It has been shown that ε_0 measurements provide a useful tool for the study of solid–solid transitions. Figure 11 shows the $\varepsilon_0(T)$ profile for ice III. It is seen that a gradual transition occurs between -65 and $-108°C$, below which temperature $\varepsilon_0 = \varepsilon_\infty$ (i.e, $g = 0$), characteristic of an ordered ice structure. The process exhibits a hysteresis loop which is quite reproducible.[1167] Similar results have been obtained for the transition ice VII → ice VIII at 0°C and 30.5 kbar.[1165]

A study of $\varepsilon(\omega)$ provides information about τ_ε, the dielectric relaxation time, which is an estimate of the time decay of the macroscopic polarization of ice when the electric field is removed. This in turn is related to τ_r, the

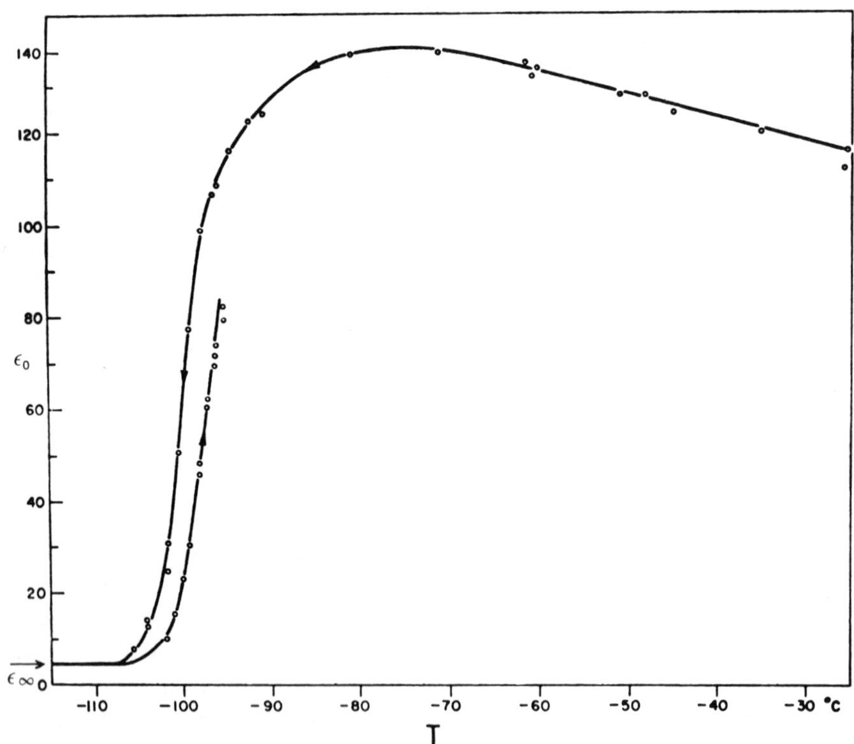

Fig. 11. The static dielectric constant ε_0 of ice III at 2.3 kbar as function of temperature (cooling and heating experiments are indicated by arrows); from Whalley *et al.*[1167] The transition to ice IX is shown by the steep drop in ε_0. Below $-108.5°C$, $\varepsilon_0 = \varepsilon_\infty$ $(g = 0)$, probably indicating the ordered structure of ice IX.

(microscopic) reorientation time, which can be compared with results derived from NMR relaxation experiments (see Chapters 6 and 7). According to the transition state theory, dipole rotation can be represented by Arrhenius-type equations and energies, entropies, and volumes of activation, evaluated from the temperature and pressure dependence of τ_ε.

As might be expected, τ_ε increases as the temperature decreases and most estimates(37,198) give $\Delta E_\varepsilon^\ddagger$ as 14 kcal mol^{-1} for the range $-3.5 > T > -55$°C and $\Delta E_\varepsilon^\ddagger = 10$ kcal mol^{-1} for $-55 > T > -73$°C. However, very recent measurements(198,1118) have illustrated the importance of crystal preparation and other experimental conditions in dielectric measurements on ice. In what may well be the most carefully controlled series of experiments carried out so far, extending over the frequency range 8×10^{-3} to 10^5 Hz, von Hippel and his colleagues have found evidence for four relaxation processes in the dielectric spectrum. In addition to the well-authenticated Debye relaxation, they were able to resolve two high-frequency relaxations and one low-frequency relaxation in the 10 Hz region. The relaxation pattern is markedly influenced by aging, so that after ten months of storage, single-crystal samples gave rise to six clearly separated dispersion regions. The $\tau_\varepsilon(T)$ relationships, too, are complex: the two high-frequency relaxations can be described by single energies of activation, but the low-frequency process and the Debye relaxation, when plotted against temperature, exhibit pronounced curvatures in opposite directions. Although at high temperatures (-20°C) reported values for ε_0 and τ_ε agree well with other recent results,(198) this is not the case at -78°C, where a discrepancy of an order of magnitude exists in the reported τ_ε data. It therefore appears that the whole question of accurate values for ε_0, ε_∞, and $\tau_\varepsilon(T)$ has been reopened, and the possible existence of several relaxation processes may well require a reassessment of the postulated mechanisms for molecular reorientation in ice (see Section 7.4).

High-pressure measurements have shown that τ_ε increases with rising pressure, with $\Delta V_\varepsilon^\ddagger = 3$ cm^3 mol^{-1}. However, dipole reorientation is much faster (by a factor of 100) in the high-pressure phases (see Table IV), but hare again for any given phase, $(\partial \ln \tau_\varepsilon/\partial P)$ is positive.* Figure 12 shows the pressure dependence of τ_ε for some of the ice polymorphs. The similarity between ices III, V, and VI is striking and may reflect the strained hydrogen bonds of these high-pressure phases compared to the "ideal" hydrogen bonds in ice I.

* Actually, the $\varepsilon(\omega)$ data for the high-pressure polymorphs cannot be interpreted in terms of a single relaxation process, but rather as a distribution of relaxation processes.

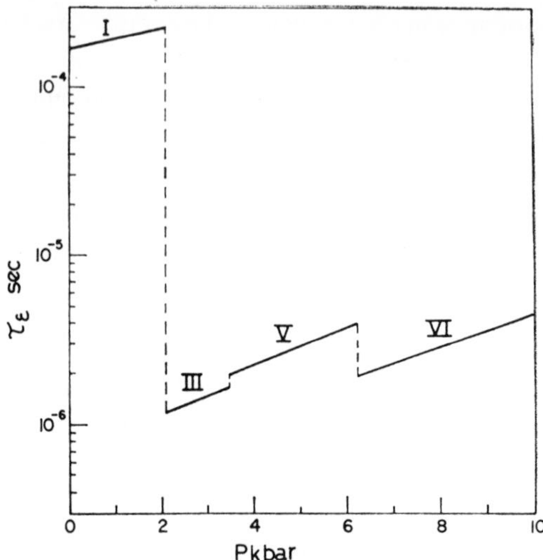

Fig. 12. Dielectric relaxation times at −23.4°C of ices I, III, V, and VI as function of pressure; from Wilson *et al.*[1175] Note the similar relaxation behavior of ices III, V, and VI, probably related to their structural similarities (see Fig. 6).

Table IV includes the high-frequency permittivity of ice. Of this quantity, 1.7 units are due to electron polarization, and the remainder would normally be assigned to atomic displacements. In the case of ice, however, the large atomic component seems to arise mainly from the intermolecular modes ν_l and ν_s referred to in Section 6. It has been suggested[1165] that changes in ε_∞ occur at some phase transitions, e.g., a reduction of 6% at the ice III → ice IX transition and that this may arise from the disorder/order processes taking place.

7.2. Self-Diffusion

Another property which might help in the solution of problems relating to the mechanism of molecular transport in ice is the self-diffusion coefficient D^*. Measurements have been performed with different isotopic species[92,253,254,549,642,889] showing that ^{2}H, ^{3}H, and ^{18}O have the same self-diffusion coefficient, $D^* \simeq 2 \times 10^{-11}$ cm^2 sec^{-1}, at −7°C. Isotopic self-diffusion measurements are not easy to perform and the summary of recent data shown in Fig. 13 illustrates the magnitude of the scatter. However, the collected results do indicate that the energy of activation associated with the self-diffusion process, $\Delta E^{\ddagger}_{D*} \simeq 14$ kcal mol^{-1}, is in good agreement with $\Delta E_{\varepsilon}{}^{\ddagger}$ for dielectric relaxation. Diffusion perpendicular to the c axis is 10% faster, an anisotropy which is comparable with that found for dipole orientation. There are not as yet any published data on the effect of pressure on D^*.

Finally, it appears(92,254) that D^* is unaffected by the doping of ice with HF or NH_4F. This is of relevance to the elucidation of the mechanism of the diffusion process in ice.

7.3. Electrical Conductivity

The fact that ice possesses a bulk dc conductivity suggests that a protonic mechanism contributes to the total transport process. The observed conductivity is a function of the concentration of ionic species and their respective mobilities. The conductivity of ice is one of the most controversial subjects and the many published data show relatively poor agreement. Uncertainties arise mainly from electrode polarization defects and from surface conductivity, which, at higher temperatures, may introduce

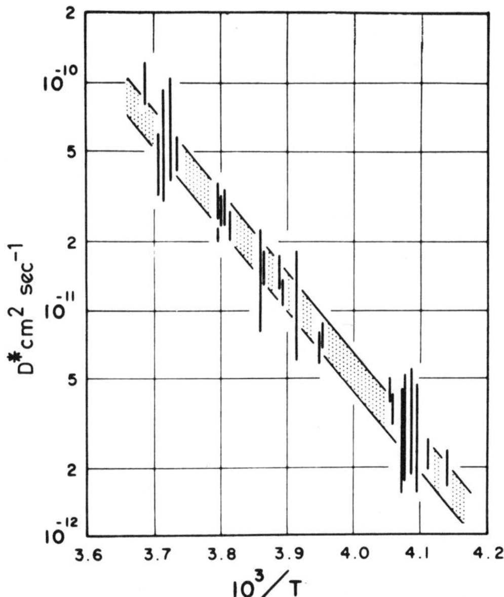

Fig. 13. Arrhenius plot of experimental self-diffusion coefficient data for ^2H, ^3H, and ^{18}O in ice, summarized by Onsager and Runnels(809) The shaded area is taken as an indication of $D^*_{H_2O}$, and gives $\Delta E_D{}^{\ddagger} = 15.0 \pm 0.5$ kcal mol^{-1}.

very large errors unless proper precautions are taken. Results covering the temperatures down to $-46°C$ were reported by Jaccard,[551] whereas Eigen and his colleagues[298] confined their measurements to temperatures above $-19°C$. Other published results include those of Bradley,[106] Gränicher,[417] and Heinmets and Blum.[487] Recent redeterminations by Bullemer *et al.*[149] incorporate extensive precautions to avoid errors due to polarization and surface conductivity, and have led to a value of $\sigma = 1.1 \pm 0.5 \times 10^{-10}$ $ohm^{-1}\ cm^{-1}$ for the bulk conductivity at $-10°C$. This is an order of magnitude lower than previous results had indicated.[106,298,487] Bullemer *et al.* also quote an energy of activation $\Delta E_\sigma^\ddagger = 8.0 \pm 0.5$ kcal mol^{-1} for the *bulk* conductivity which is again substantially lower than previously reported values (11 kcal mol^{-1}). The point is made that at temperatures above $-30°C$ surface currents exceed bulk currents and are subject to a $\Delta E_\sigma^\ddagger$ of 20–30 kcal mol^{-1}. More detailed reports on the relative importance of surface effects have also been provided by Maidique *et al.*,[722] who highlight the considerable discrepancies which exist in published $\Delta E_\sigma^\ddagger$ data, ranging from 0 to 33 kcal mol^{-1}.

The equilibrium properties and kinetics of the reaction

$$2H_2O \rightleftharpoons H_3{}^+O + OH^-$$

have been intensively studied by Eigen and his colleagues, and several reviews are available.[295] Their data are summarized in Table V together with supplementary results obtained from measurements of the pressure dependence of the conductivity. More recent data are also included and it is seen that the latest estimates of conductivity, mobility, and dissociation rate and their temperature dependences are lower than the corresponding results obtained earlier by Eigen. Maidique *et al.* have speculated that the saturation currents at high fields reported by Eigen may well have been artifacts arising from internal surface conductivity in polycrystalline samples. They find the "true" volume current to be much lower and ohmic in character. Instead of postulating a saturation current, they consider a saturation charge and from single-crystal experiments suggest an upper limit of 3×10^{11} charge carriers/cm^3, possessing a mobility of about 5×10^{-3} cm^2 (V sec)$^{-1}$.

Bullemer[150] and his associates have extended their investigations to measurements of the Hall effect, which provides an estimate of the fraction of charge carriers which are actually in motion. Although the interpretation of data is complicated by the fact that one cannot readily assume that the charge carriers (protons) are independently free to move, the results never-

TABLE V. Ionic Equilibrium and Transport Properties of Ice at −10°C[a]

$$2H_2O \underset{k_{-1}}{\overset{k_1}{\rightleftharpoons}} H_3{}^+O + OH'$$

	H_2O	D_2O
Low-field-strength bulk dc conductivity σ, $ohm^{-1}\,cm^{-1}$	1.0×10^{-9}, 1×10^{-10} [b] 6×10^{-11} [c]	3.6×10^{-10}
$\Delta E_\sigma{}^\ddagger$, kcal mol^{-1}	11, 8[b]	13
$\Delta V_\sigma{}^\ddagger$, $cm^3\,mol^{-1}$	−6[d]	
k_1, sec^{-1}	3.2×10^{-9}, 1.1×10^{-9} [b]	2.7×10^{-11}
$\Delta E^\ddagger_{dissocn}$, kcal mol^{-1}	22.5, 17.6[b]	25
k_{-1}, mol^{-1} liter sec^{-1}	8.6×10^{12}	1.3×10^{12}
Equilibrium constant	3.8×10^{-22}	0.2×10^{-22}
$\Delta H^\circ_{dissocn}$, kcal mol^{-1}	22.5	24
$\Delta V^\circ_{dissocn}$, $cm^3\,mol^{-1}$	−9[c]	
Proton mobility, $cm^2\,V^{-1}\,sec^{-1}$	7.5×10^{-2}, 2.4×10^{-3} [c] 5×10^{-3} [e]	1.2×10^{-2}
Hall mobility, $cm^2\,V^{-1}\,sec^{-1}$	1.4[c]	—
Proton concentration, mol $liter^{-1}$	1.4×10^{-10}	0.32×10^{-10}

[a] Data taken from Eigen,(295) except where noted.
[b] Bullemer *et al.*(149)
[c] Cole and Wörz.(198)
[d] Chan *et al.*(178)
[e] Maidique *et al.*(722)

theless indicate that only 0.2% of the available protons are free to migrate. The high mobility can be realistically accounted for only on the basis of correlated proton motions in ice.

Experiments at lower temperatures with proton-injecting electrodes have shown that the proton mobility increases markedly as the temperature falls,(312) with distinct changes in the current/temperature profiles at −60°C and below −80 to −100°C. Thus, at −60°C the mobility is 5 $cm^2\,V^{-1}\,sec^{-1}$. This may seem rather high, but at the low temperatures covered by these experiments the drift mobility of long-lived charge carriers, not subject to recombination, was measured and thus should be of the same order of magnitude as the Hall mobility, and this is indeed the case. At even lower temperatures the mobility falls drastically ($10^{-5}\,cm^2\,V^{-1}\,sec^{-1}$ at −130°C) and it decreases further at even lower temperatures. This rather complex temperature dependence of the proton mobility suggests that several mechanisms contribute to the transport of charge in ice.

7.4. Orientational and Ionic Defects in Ice; Theories of Transport Mechanisms

The experimental results discussed in Sections 7.1–7.3 indicate that dipole reorientation, mass transport, and charge transport all take place in ice and the question arises as to the nature of the mechanisms by which such transport processes might occur. The problem of diffusion mechanisms in ice was studied by Bjerrum[(87)] and his explanations for the observed phenomena have become generally accepted, although various modifications have from time to time been proposed.

Essentially, Bjerrum suggested that observed transport behavior could be accounted for on the basis of low concentrations of lattice defects. These were postulated to arise from the rotation of molecules. If the central molecule in Fig. 14 rotates through 120° as shown, then two defects are formed: one of them will consist of an O···O link without a hydrogen atom (*L* defect) and the other will be a doubly occupied hydrogen bond (*D* defect). Subsequent molecular reorientations then separate the two defects, which will migrate through the crystal independently, one lattice position per molecular rotation. The concentration of orientational defects is though to be of the order of 10^{-7} mol defects per mol ice at -10°C and the energy of formation has been estimated at 16 kcal per mol pair of defects.

A rather different origin of *D* and *L* defects has recently been proposed by von Hippel.[(1117)] It is suggested that a pair defect is formed as a result of near-infrared phonon excitation causing a proton to shift intramolecularly to one of the two unoccupied (lone-pair electron) sites of the H_2O molecule (see Fig. 15). The modes involved in such an intramolecular rearrangement might be a combination of ν_2 and ν_3. The defect pair having been formed,

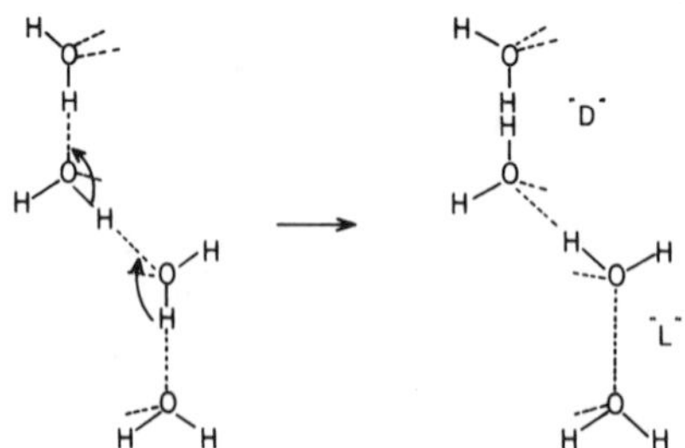

Fig. 14. Bjerrum-type orientational defects arising from molecular rotations in ice. Each rotation creates a pair of defects, one *L* and one *D* defect, which are then able to migrate independently.

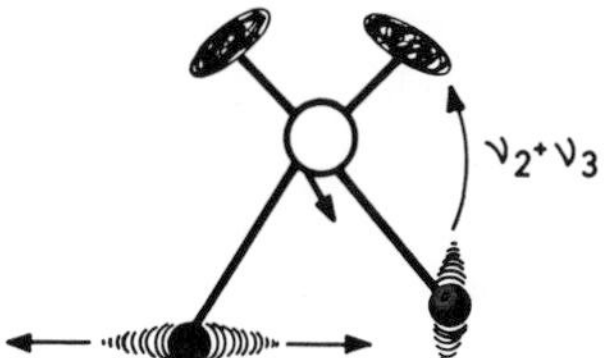

Fig. 15. Intramolecular proton migration, leading to a pair of orientational defects, according to von Hippel.[1117]

the two defects will drift apart with a slight bias toward the direction of the field, i.e., toward the electrodes. Von Hippel shows that this proposed mechanism is compatible with the accepted τ_ε and the observed deuterium isotope effect on τ_ε[37] At the time of writing it is too early to assess whether this very novel mechanism is likely to become generally accepted and will in time supersede the Bjerrum model.

Whatever the nature of the process leading to a defect pair, clearly its existence must result in crystal dislocations in its neighborhood, e.g., the "ideal" O···O distance of 2.75 Å could not readily accommodate two hydrogen atoms in a linear O···H···H···O bond. Various alternative models have been proposed[194,195,281] whereby the mutual repulsion between the hydrogen atoms in a *D* defect could be reduced, without changing the position of the oxygen atoms. An alternative approach[300] led to the suggestion that an elastic lattice relaxation could take place whereby the O···O distance increased by 0.5 Å but the hydrogen bond collinearity would remain intact.

Although there is no direct evidence for the existence of Bjerrum-type orientational defects, nevertheless, many investigators consider that the rotational reorientations occurring in ice, as determined by dielectric and NMR spin–lattice relaxation (T_1) are best accounted for in terms of a modified Bjerrum model; experimental energies and volumes of activation for molecular rotation are in good agreement with values calculated on such a basis. It has also been shown[809] that the model of migration by orientational defects is compatible with the experimentally determined ratio 1.5 for the dielectric to the elastic relaxation time. Although energies and volumes of activation of the high-pressure polymorphs differ markedly from those for ice I (see Table IV), nevertheless, there is good reason to believe that the relaxation mechanisms in the high-pressure phases resemble those in ordinary ice. Possibly, the strained hydrogen bonds result in lower energies of defect formation and this in turn would be reflected in $\Delta E_\varepsilon^\ddagger$.

It is now established that the processes of dielectric relaxation, NMR spin–lattice relaxation, elastic relaxation, and self-diffusion are all subject to the same energy of activation, i.e., 14 kcal mol^{-1} at -10°C, but whereas the first three processes could be accounted for in terms of the migration of orientational defects (as a result of molecular rotations), this is not really possible in the case of self-diffusion, where mass transport is involved. It was, however, this similarity in activation energies which caused Siegle and Weithase[(978)] to suggest that T_1 was determined by molecular jump diffusion, rather than by Bjerrum defects. A further difference between self-diffusion and other processes mentioned above lies in the effect of doping on relaxation behavior. Whereas τ_ε and T_1 are very sensitive to low concentrations of HF,[(623)] this is not the case with D^*. It is therefore unlikely that D^* and T_1 are governed by the same process.

Gränicher,[(416)] in the first critical analysis of transport phenomena in ice, came to the conclusion that dielectric relaxation and self-diffusion could not be subject to the same mechanism. It has since been suggested that, in addition to purely orientational defects, there might exist vacant lattice sites and/or interstitial molecules[(439)]* and that diffusion is controlled by the occurrence of composite defects, as shown diagrammatically in Fig. 16. Such postulated mechanisms would not affect the ratio $\tau_\varepsilon/\tau_{\text{elastic}}$, computed from a consideration of Bjerrum defects, but do enable D^* to be calculated. Thus, Onsager and Runnels[(809)] conclude that the mechanism of interstitial migration best accounts for the observed NMR linewidth and T_1 data. Such a mechanism is also compatible with the experimental self-diffusion coefficients. NMR linewidths and second moment measurements on partly deuterated ice (HDO) as function of temperature have led Tait to favor the interstitial model.[(1042)] Other NMR results on D_2O ice[(554,1128)] tend to support these findings, which are, however, in conflict with the frequency of molecular motion deduced from T_1 measurements on H_2O ice.[(54)] The two sets of results can be reconciled if it is accepted that T_1 is diffusion-controlled and that diffusion rates in D_2O or HDO are lower than in H_2O. Although it has been concluded (see Section 7.2) that the various isotopic modification have the same D^*, tracer diffusion measurements have only been performed on ^2H_2O, $^1H_2{}^{18}O$, and $^1H^3HO$, all of which have the same mass. In liquid water, on the other hand, it has been shown that D^* of H_2O–D_2O mixtures decreases with increasing D_2O concentration.[(259)] In order to fully resolve the problem of interpreting T_1 in terms of diffusional motions, further measurements are required.

* Kopp, cited in Onsager and Runnels.[(809)]

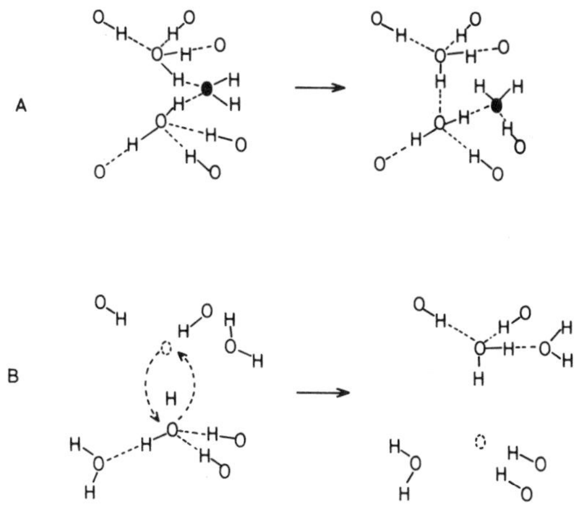

Fig. 16. Hypothetical composite lattice defects in ice. (A) association of a D defect with an interstitial molecule (denoted by a filled-in circle); (B): association of a D defect with a vacant lattice site (dotted circle). The method of transport via the vacant site is indicated by the arrow.

It now becomes apparent that the fact that the energies of activation associated with the various transport processes have identical values (within experimental error) may be coincidental rather than an indication of a common relaxation process. It must, however, again be stressed that no direct evidence exists for interstitial molecules or for any interactions between defects giving rise to aggregates or to coupling between vacant sites and interstitial defects.

Whereas the reorientation and self-diffusion behavior of ice can be reasonably well accounted for on the basis of orientational defects, the transport of charge requires the existence and propagation of ionic defects. Bjerrum postulated that hydrogen atoms could move to alternative positions along the O···O bonds and thus create H_3^+O and OH^- ionic defects.[88] This double-well model was later elaborated by Gränicher[416] and Jaccard[552] and the concept of proton tunneling in ice was invoked by Eigen and de Mayer[296] to interpret their conductivity data. From estimates of the mean residence time of a hydrogen atom in a given position (10^{-13} sec) and the concentration of ionic sites (3×10^{-12} molecule^{-1}) it follows that at $-10°C$, to which these estimates refer, the number of ionic defects experienced by a given H_2O molecule is insignificant compared to the number of orientational defects in which the same molecule is involved. This is no longer true at low temperatures (see Section 7.3), where conductivity becomes the predominating transport mechanism.

The whole issue of ionic defects and proton migration has been put in question by the recent conductivity data of Bullemer *et al.*[149] and particularly those of Maidique *et al.*[722] (see Section 7.3). The latter, in fact, doubt whether there is any justification for invoking proton tunneling or, indeed, the existence of protonic species in the bulk volume of ice. Historically, the theory of a unique mechanism for proton migration in ice developed as a result of the very high measured mobilities. However, the recent, very careful experiments on highly purified single crystals led to values for the proton mobility much lower than those reported by Eigen (see Table V) and approaching the corresponding values for liquid water ($\sim 2 \times 10^{-3}$ cm^2 V^{-1} sec^{-1} at 0°C). Accordingly, von Hippel[1117] suggests that *all* transport through the bulk volume is by orientational defects, but that these can be converted into ionic defects at the electrodes by injection or ejection. Thus, at the anode the processes occurring may be represented by $L \rightarrow H_3{}^+O$ and $OH' \rightarrow D$, whereas at the cathode, proton ejection gives rise to $H_3{}^+O \rightarrow L$ and $D \rightarrow OH'$. In addition to such defect conversions, ionic defects might also be created by the interaction of proton (or electron) exchange between electrodes and ice bonding imperfections at the electrode/ice interface. An interesting consequence of processes involving defect conversion would be the creation of low-dielectric-constant regions near the electrodes.

Although once again it is too early to judge whether von Hippel's criticisms of the accepted theories and his alternative proposals will meet general agreement, recent mobility data, arrived at by conductivity and Hall effect measurements, do suggest that some reassessment of the original Bjerrum and Eigen models is now required.

8. SUMMARY

The foregoing sections indicate that, as regards the "static" structural properties and phase behavior of ice, X-ray diffraction, thermodynamic, and spectroscopic studies have provided a firm foundation, and only minor details remain to be filled in to complete the general picture. Most of the outstanding problems relate to the dynamic aspects, both with regard to high-frequency motions and the long-term aging behavior which is observed in ice. One particular area of interest concerns the surface properties of ice, which are responsible for such phenomena as freezing potentials, adhesion (regelation), crystallization from the vapor phase, and pressure melting. Other important fields where technological developments might be coupled to an improved knowledge of the dynamic properties of ice include cryobiology, glaciology, and meteorology.

Several of the key investigations are already in progress and the future outlook for an even more complete understanding of ice in its various manifestations çan be deemed to be hopeful.

CHAPTER 5

Raman and Infrared Spectral Investigations of Water Structure

G. E. Walrafen

Bell Telephone Laboratories Inc.
Murray Hill, New Jersey

1. INTRODUCTION

The models that have been proposed to represent the structure of liquid water are too numerous to be described here, but they are discussed in Chapter 14. Two general classes of models of interest for present purposes are usually designated by the terms continuum[1132] and mixture.[356] Continuum models treat water in terms of a continuous distribution of interactions that are presumed to be spectroscopically indistinguishable, whereas mixture models generally relate distinct spectral features to structures differing in the extent of hydrogen bonding. Some of the earlier spectroscopic investigations[1132] appeared to favor continuum models, but recent laser-Raman investigations[1138] as well as results from nonlinear optical techniques such as stimulated Raman scattering and hyper-Raman or inelastic harmonic light scattering strongly favor mixture models, e.g., the consecutive hydrogen-bond disruption model.

In this chapter the Raman spectral evidence giving rise to the consecutive hydrogen-bond breakage model for water will be presented in the order of its development, namely the intermolecular studies will be described first, followed by studies of the intramolecular vibrational regions. Results from infrared absorption, neutron inelastic scattering, stimulated Raman scattering, and hyper-Raman or inelastic harmonic light scattering will be included where appropriate. Further, because the stimulated Raman scattering and hyper-Raman techniques are relatively new, they will be described briefly in a section also containing a description of the fixed-beam laser-Raman technique. Finally, tests of the consecutive hydrogen-bond breakage model involving thermodynamic and other physical properties of water will be presented.

The Raman spectral investigations conducted thus far have yielded two principal types of information. The intermolecular Raman studies appear to give evidence almost exclusively related to the tetrahedral 4-bonded configurations[1134,1135,1137,1139]; the intramolecular Raman studies for conditions between the extremes of supercooled and supercritical water give simultaneous evidence primarily for hydrogen-bonded and nonydrogen-bonded O—H···O and O—D···O units.[1137–1142] (The hydrogen-bonded O—H···O and O—D···O units of the partially hydrogen-bonded structures, 3-, 2-, and 1-bonded, have not yet been individually resolved in the intramolecular valence regions.) Intermolecular Raman evidence referring specifically to the partially hydrogen-bonded configurations, 3-, 2-, and 1-bonded, is lacking. Nevertheless, the infrared absorption[273,274] and neutron inelastic scattering[933] data from the intermolecular vibrational regions appear to give simultaneous indications suggestive of the presence of partially hydrogen-bonded, as well as for the 4-bonded, species. Further, the fact that all of the intermolecular Raman intensities are negligibly small near 100°C, where the hydrogen-bonded intramolecular Raman component intensities are large, indicates that a consecutive or stepwise breakage or disruption of hydrogen-bonds may be involved.[1138,1139]

Results from nonlinear optical techniques also provide evidence in support of conclusions obtained from Raman studies. The hyper-Raman or inelastic harmonic light scattering data,* in conjunction with stimulated Raman scattering,[885] infrared absorption, and neutron inelastic scattering results provide evidence indicating that tetrahedral 4-bonded units having an average five-molecule intermolecular point group symmetry of C_{2V} occur in water at low temperatures.[1137,1139] Further, the mutually exclusive stimulation of OH stretching components observed in stimulated Raman scattering from H_2O and HDO[199] strongly supports the decomposition of spontaneous Raman spectral contours according to hydrogen-bonded and non-hydrogen-bonded intramolecular stretching components. Again evidence for 4-bonded interactions, or primarily for nonhydrogen-bonded and hydrogen-bonded interactions, results.

Finally, strong support for the consecutive hydrogen-bond breakage model of water structure is obtained from the quantitative agreements of

* P. D. Maker, private communication.

the model with a wide range of data, in particular the thermodynamic data. Thus, good agreement with the dielectric constant and the heat capacity in the range of 0–100°C is obtained when the spectroscopic ΔH° value of 2.5_5 kcal mol^{-1} OHO is employed with mole fractions of the 4-, 3-, and 2-bonded species consistent with the Raman[1139] and near-infrared[705] results.

Spectroscopic evidence leading to the present model of water structure is now described.

2. NEW EXPERIMENTAL TECHNIQUES

2.1. Stimulated Raman Scattering

Stimulated Raman scattering was discovered in 1962.[1183] In addition to light at 6943 Å from a ruby laser, light at ~7670 Å was observed when Q-switching was accomplished with a Kerr cell containing nitrobenzene.[1183] (The light at 7670 Å corresponds to a Stokes shift of 1365 cm^{-1}, and agrees with shifts reported for the symmetric stretching of the NO_2 group from spontaneous Raman scattering.[111]) The new phenomenon was correctly interpreted by Eckhardt *et al.*[285] later in 1962, and has since been discussed rigorously by many workers, e.g., by Shen and Bloembergen[970] and Bloembergen.[94]

When intense laser light is passed, for example, through a liquid, a significant fraction is displaced to lower frequencies and generally relates to spontaneous Raman lines that are intense and sharp. The displaced radiation can give the appearance of occurring at a threshold, is highly directional and generally has a linewidth smaller than the corresponding spontaneous Raman line. The process may be regarded as a two-photon scattering between a laser quantum $h\nu_L$ and Stokes quanta $nh\nu_S$ in which the material system is excited by an amount $h(\nu_L - \nu_S)$, and $(n + 1)h\nu_S$ Stokes quanta are emitted. Alternatively, the process may be viewed as a parametric coupling between a light wave at the Stokes frequency ν_S and an optical phonon or vibrational wave having a frequency ν_V. The coupling is produced by the incident laser field at $\nu_L = \nu_S + \nu_V$. The effect essentially depends upon the coefficient of the product of the electric field strengths, $E_S E_L$, but the coefficient contains a derivative of the linear polarizability α. Hence, the selection rules for stimulated and spontaneous Raman scattering are the same.

Stimulated Raman scattering from water has been accomplished with high-intensity 6943 Å ruby excitation using pulses ranging from ~0.5

nsec[885] to ~30 nsec,[824] and more recently from water and aqueous solutions using pulses of 1 psec duration at 5300 Å obtained by doubling the frequency from a mode-locked Nd^{3+}-glass laser (1.06 μm) with KH_2PO_4.[199] Sufficiently high power densities, ~25 GW cm^{-2}, are produced by optical focusing of the 5300 Å radiation to allow for photographic detection of the scattering resulting from the picosecond pulse excitation, and an advantage of the short pulses is that spectral broadening due to self-focusing is reduced. Self-focusing of a laser beam is decreased for liquids having molecular reorientation times longer than the pulse duration.[966] In addition, an important and unexpected advantage of stimulated Raman scattering recently made evident from studies of H_2O, of HDO in H_2O and D_2O, and of $NaClO_4$ in H_2O is that mutually exclusive stimulation of vibrations, namely of the OH stretching vibrations, occurs.[199] The mutually exclusive stimulation of stretching components from H_2O and HDO provides further support for the present model of water structure.

2.2. Hyper-Raman or Inelastic Harmonic Light Scattering

The hyper-Raman effect, or inelastic harmonic light scattering, was discovered in 1965,[1060] and was subsequently employed in careful examinations of the intermolecular librational regions of H_2O and D_2O, and of the OH stretching region of H_2O.* Unlike the spontaneous and stimulated Raman effects, however, the hyper-Raman scattering depends directly upon the quadratic polarizability or hyperpolarizability—hence the name hyper-Raman.

The linear, quadratic, and cubic polarizabilities α, β, and γ occur in the expansion of the polarization P in terms of the electric field strength E as follows:

$$P = \alpha E + \beta E^2/2 + \gamma E^3/6 + \cdots$$

In the laser field E_L, an elastic harmonic component will be observed at $2\nu_L$, and inelastic components will be observed at $2\nu_L \pm \nu_V$, where $2\nu_L - \nu_V$ refers to the Stokes shift, and $2\nu_L + \nu_V$ refers to the anti-Stokes shift. The effect depends upon derivatives of β in the inelastic case, and the selection rules are different from those of Raman scattering. Hyper-Raman selection rules for various point groups have been reported,[224] and a rigorous theory of elastic harmonic light scattering has been developed recently by Maker.[724] This involves two spherical tensors, $\bar{\boldsymbol{\beta}}^{[1]}$ and $\bar{\boldsymbol{\beta}}^{[3]}$, and in the in-

* P. D. Maker, private communication.

elastic case these give rise to dipolar or to octapolar selection rules, respectively. (The two tensors are subsequently abbreviated to β_1 and β_3.) The β_3 hyper-Raman spectrum can in principle be obtained directly by use of circularly polarized laser light. Thus far, however, β_3 spectra have been obtained by calculation from results obtained by using linearly polarized laser light in the perpendicular and parallel orientations. The depolarization ratio, $\varrho_l = I_\perp/I_\parallel$, is 2/3 for a completely depolarized hyper-Raman line, and $2/3 > \varrho_l > 1/9$ for a polarized line. Further,

$$\varrho_l = (I_{\beta_1} + 2I_{\beta_3})/(9I_{\beta_1} + 3I_{\beta_3})$$

If perpendicular and parallel hyper-Raman spectra are available, the β_1 and β_3 spectra may be obtained from $I_\parallel - (3/2)I_\perp$ and $I_\perp - (1/9)I_\parallel$.

The C_{2V} point group is of interest in work dealing with intermolecular vibrations of water, and the A_2 vibrational species of that group, for example, is completely depolarized in the-hyper-Raman spectrum, $\varrho_l = 2/3$. From the preceding relations it is evident that $I_{\beta_1} = 0$, or that the A_2 intermolecular libration is inactive in the dipolar hyper-Raman spectrum. (The A_2 species is also infrared-inactive.) For liquid H_2O and D_2O, low β_1 librational intensity is observed in frequency regions of maximum β_3 intensity,* and this observation allows for the direct assignment of the A_2 librations as indicated in following discussions. The assignment of the A_2 libration of liquid H_2O and D_2O from hyper-Raman data, in agreement with assignments from Raman and infrared data, supports the presence of 4-bonded tetrahedral configurations in water, and constitutes an important success of the hyper-Raman method.

2.3. Fixed-Beam Laser-Raman Method

Before the advent of laser excitation, many accurate measurements of Raman intensities from solutions using conventional 4358 Å mercury sources had been accomplished—for example, by Young *et al.*[(1192)]—by means of photoelectric Raman techniques. In comparison, many, if not most, of the early laser-excited photoelectric Raman intensity measurements were of low accuracy, despite several obvious advantages of the laser sources, e.g., the high intensity and polarization. The conventional mercury-excitation method involved a light field of large volume and of great stability, and Raman samples could be positioned to a high degree of reproducibility

* P. D. Maker, private communication.

in the field. The intense, concentrated laser beam, on the other hand, necessitates a much more rigorous geometric reproducibility to achieve results comparable to the conventional mercury methods. The problem was not widely recognized, however, because of the small number of workers engaged in accurate Raman intensity measurements of solutions, and because many workers were content with high-quality spectra from a single sample, as opposed to quantitative intensity measurements involving a series of solutions or a wide range of temperatures. A partial solution to the problem is provided by the fixed-beam laser-Raman technique.

In the fixed-beam laser-Raman technique the position of the laser beam is maintained rigorously fixed with respect to the entrance slit of the spectrometer by use of pinholes. Deviations of the laser beam produced by changing the Raman cell or sample are observed by the failure of the laser-beam to pass through the exit pinhole. The deviations are removed by redirecting the laser beam until it again passes through the exit pinhole. Redirection is accomplished with a high-reflectance mirror placed at a 45° angle to the laser beam and moved accurately by means of a gimbal. The apparatus employed in conjunction with the Cary model-81 Raman spectrometer is shown in Fig. 1, and a similar apparatus was used recently in conjunction with the Jarrell-Ash double monochromator.[(1142)]

In Fig. 1 the beam from a high-power argon-ion laser, ~1 W at 4880 Å,[(645)] is shown entering the external optics compartment of the Cary instrument at *A*. The laser is placed horizontally about 1 m from *A*, and the beam passes through a condensing lens, a half-wave plate, and a diaphragm (all not shown) before entering the optics compartment. In the optics compartment the horizontal beam is reflected vertically downward by the front-surface dielectric mirror *B*. (The dielectric mirrors *B* and *D* employed in directing the laser beam have maximum reflectivity for 45° incidence at 4880 Å.) The gimbal at *B* is next adjusted until the laser beam passes through the *entrance* and *exit* pinholes *E* and *F* and strikes the front-surface dielectric mirror *G* (maximum reflectivity for 90° incidence at 4880 Å).

The *entrance* and *exit* pinholes *E* and *F* define the position of that portion of the laser beam giving rise to the Raman effect, for either single- or double-pass operation. Single- or double-pass fixed-beam operation is obtained by use of gimbal *G*. For a single pass the gimbal is moved until the returning laser beam fails to return through pinhole *F*. For a double pass the beam reenters at *F* and retraces the path from *E* to *A*. (For very accurate double-pass work the reflection of the returning beam on the uncoated side of the low-reflectance laser-cavity mirror can be made to coincide with the emitted laser beam by carefully adjusting gimbal *G*.)

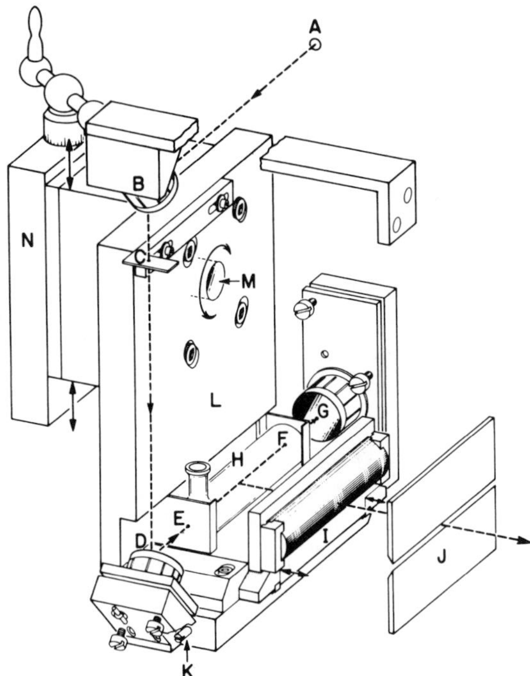

Fig. 1. Schematic illustration of the fixed-beam laser-Raman apparatus employed with the Cary model-81 Raman spectrophotometer.

The sample cell *H* is 10 cm long and 2 cm in diameter. It is positioned reproducibly by the mountings at *E* and *F*. The Raman radiation from the sample contained in cell *H* passes through the cylindrical condensing lens *I* and into the spectrometer through the slit *J* as shown by the arrow. The cylindrical lens and the slit are also 10 cm in length. The lens is brought to the focused position by rotating the shaft *K*. Rotation at *K* moves two gears attached to the shaft, which in turn move two geared racks. The motion is made even and reproducible by means of two lubricated guides, one of which is shown below *E*.

The beam passing through *E* and *F* is positioned parallel to the jaws of the slit *J* by small rotations of the precision-ground plate *L* with respect to the pinion *M*. The rotation is accomplished by a fine screw, not shown. Finally, the beam between *E* and *F* is moved into the optimum horizontal plane, the plane that passes through the slit jaws, by vertical motion of the precision slide-table *N*.

When the intensity of a Raman line has been maximized for a very small slit width by adjusting *K*, *L*, and *N* and *B*, *D*, and *G*, the apparatus is properly positioned for a given Raman cell and sample. (The positions of *K*, *L*, and *N* then remain fixed for other cells and samples.) For various samples (preferably in the same cell) slight changes in the position of the mirror *D* reestablish the correct beam position, as indicated visually at *F*, and by the maximum Raman intensity measured photoelectrically. In single-pass operation, recommended for most quantitative work, the fixed-beam technique leads to an intensity reproducibility of 1–5%.

Semiquantitative values of the depolarization ratio ϱ_l' can also be obtained with linearly polarized laser light *l* and the fixed-beam apparatus of Fig. 1 in the single-pass configuration. The intensity ratio I_h/I_v is the uncorrected depolarization ratio ϱ_l', where *h* and *v* refer to the horizontal and vertical orientations of the electric vector between *E* and *F* produced by rotating the half-wave plate. The symbols $I_\parallel$ and $I_\perp$, however, are accepted instead of I_h and I_v, and the maximum Raman intensity is obtained in the $\perp$ (or *v*) position. Additional optical refinements and calibrations are required to obtain accurate depolarization ratios ϱ_l. The accurate values are important when complete polarization, $\varrho_l = 0$, totally symmetric vibration, or complete depolarization, $\varrho_l = 0.75$, antisymmetric vibration, is approached.

3. INTERMOLECULAR VIBRATIONS OF H_2O AND D_2O

3.1. The Restricted Translational Region

The restricted translational motions of H_2O and D_2O molecules involved in directed hydrogen bond interactions are equivalent to the bending and stretching motions of the O—H···O and O—D···O units in both liquids. The hydrogen bond bending and stretching motions give rise to broad, weak bands below 300 cm^{-1}.

Raman scattering from liquid H_2O and D_2O yields a broad, weak hydrogen bond bending band centered near 60 cm^{-1}. The integrated Stokes Raman intensity of the 60 cm^{-1} band from water has been observed to decrease with temperature rise.[(1135)] Infrared absorption coefficients from liquid water at ambient temperatures also provide evidence suggestive of a broad, weak component in the approximate vicinity of 60 cm^{-1} (see subsequent discussion). In addition a component near 60 cm^{-1} has been reported from neutron inelastic scattering studies,[(933)] and the intensity of the up-scattered or anti-Stokes component in the frequency spectrum has also been observed to decrease with increasing temperature.[655] The intensity decreases observed in Raman and neutron inelastic scattering indicate that a structural breakdown occurs at elevated temperatures. Raman spectra giving evidence of the 60- and 170-cm^{-1} bands from water at temperatures of ~ -5 and $\sim$67°C are shown in Fig. 2.

In Fig. 2 the total intensities, i.e., the photoelectric tracings (neither smoothed nor averaged) are shown above, and the base lines are shown in part by the curved dashed lines. The base line intensity arises from several sources: from the reflected exciting line, from the combined Rayleigh scattered exciting line and Rayleigh wing, and from stray light in the spectrometer, which constitutes a large fraction of the base line intensity from water even in superior double monochromators. Differences between the total and base line intensities are shown below. The differences are considered to constitute the structural low-frequency Raman intensities within the uncertainties of the base line estimates. Broad components centered near 60 and 170 cm^{-1} are obvious from the intensity differences, and the integrated Raman in-

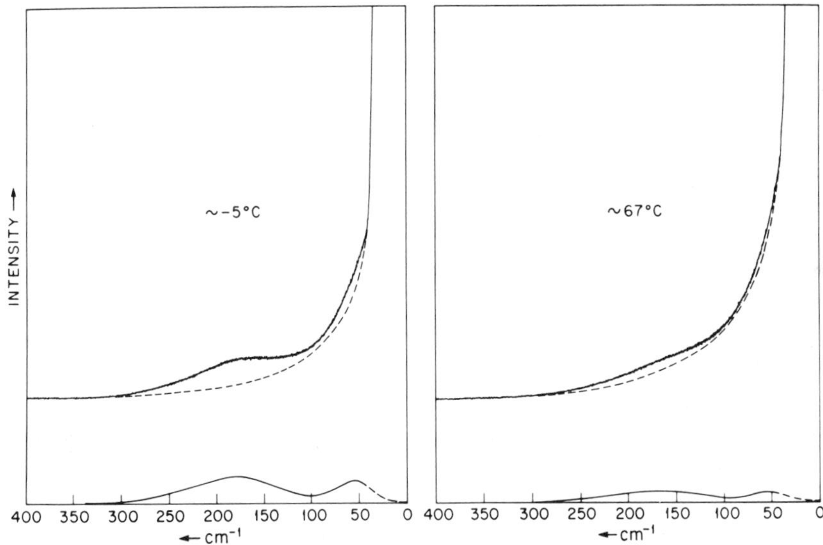

Fig. 2. Photoelectric Raman tracings from water at ~ -5 and $\sim$67°C obtained in the low-frequency region, $\Delta\bar{\nu} < 400$ cm^{-1}, with unpolarized 4358 Å mercury excitation. The baselines are indicated by curved dashed lines in the upper sections. The lower sections refer to the spectra transferred to horizontal baselines, i.e., to the differences between the photoelectric tracings and the curved baselines. The Raman intensities are quantitatively comparable at the two temperatures. The spectra were obtained with a Cary model-81 Raman spectrophotometer, using a slit width of 10 cm^{-1} and a time constant of $\sim$0.1 sec.

tensities of the 60- and 70-cm^{-1} components are seen to decrease with increase of temperature.

The total intensities, i.e., the direct photoelectric tracings of Fig. 2, also give evidence of features near 60 and 170 cm^{-1}. The peak at 170 cm^{-1} is obvious, but close inspection is required to observe the feature at 60 cm^{-1}. In the spectrum of Fig. 2 corresponding to ~ -5°C, for example, the total intensity is observed to fall rapidly and roughly linearly from 35 to ~40 cm^{-1}. Beyond 40 cm^{-1}, however, a significant decrease in slope occurs, and at ~60 cm^{-1} the curve is concave downward. The intensity beyond 60 cm^{-1} falls again until the low-frequency tail of the 170-cm^{-1} component causes the slope to decrease once more. The downward concavity evident from the spectrum corresponding to ~ -5°C constitutes evidence for a broad, weak Raman band from water near 60 cm^{-1}. The existence of the 60-cm^{-1} Raman band has also been verified by investigators employing baffle techniques,[(91)] or logarithmic recording.*

The Raman band evident in Fig. 2 near 170 cm^{-1} is produced by the stretching motion of O—H···O units in liquid H_2O, and the same motion of O—D···O units gives rise to a band near 170 cm^{-1} from liquid D_2O.[(1134)] The decrease in the integrated intensity of the 170-cm^{-1} band evident from Fig. 2 with increasing temperature has been studied quantitatively for temperatures from -6.0 to 94.7°C;[(1135)] the quantitative intensity variations have been verified by Blatz by means of the baffle technique.† The decrease in the Raman intensity with temperature has been interpreted in terms of a structural breakdown, i.e., a breakdown of O—H···O units.[(1139)] A broad infrared absorption has also been observed for liquid H_2O and D_2O near 170 cm^{-1},[(273)] and for liquid H_2O has been examined at a series of temperatures. Further, neutron inelastic scattering spectra from liquid H_2O and D_2O indicate a component near 170 cm^{-1}, and the neutron component intensity has also been observed to decrease with increasing temperature.[(655,933,934)] In addition, evidence strongly suggestive of low hyper-Raman intensity near 170 cm^{-1} has also been obtained from the intensity increase observed for the β_3 component in the wings of the elastic scattering from water.‡ The 170-cm^{-1} Raman band from water is shown with great clarity in Fig. 3, lower spectrum.

Two interference filters were employed in series to obtain the Raman spectrum shown in the lower section of Fig. 3. The upper spectrum of the

* H. J. Bernstein, private communication.

† L. A. Blatz, private communication.

‡ P. D. Maker, private communication.

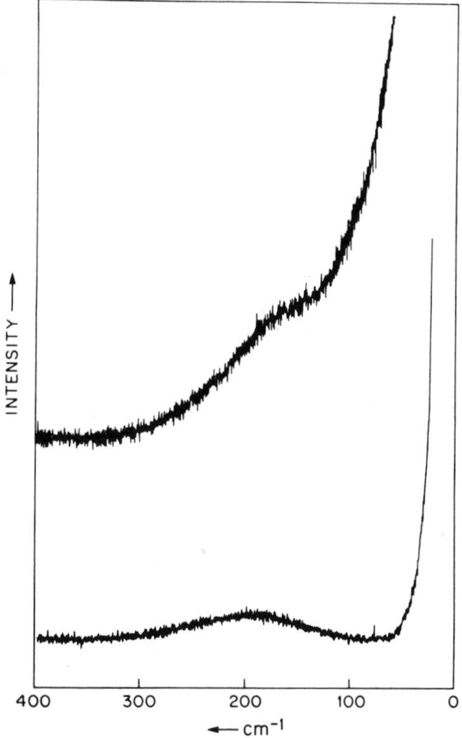

Fig. 3. Low-frequency Raman tracings from water at 25°C recorded with two interference filters in series (below) and without using interference filters (above). The spectra were obtained with unpolarized 4358 Å mercury excitation using the Cary model-81 Raman spectrophotometer. The interference filters were placed between the sample compartment and the "image slicer," and tilted with respect to the optical axis. A 10 cm^{-1} slit width was employed with a time constant of ~0.1 sec.

figure was obtained without the use of interference filters, for comparison. The pair of interference filters allowed ~40% of the radiation to pass into the Cary model-81 double monochromator for Raman shifts of 170 cm^{-1} or above, but only 10% of the radiation was admitted at 4358 Å. The spectral improvement produced by the reduction of intensity at 4358 Å is obvious—the *apparent* base line intensity at about 75 cm^{-1} is equivalent to the base line intensity between 300 and 400 cm^{-1} (lower section).

The results shown in Fig. 3 indicate that the stray-light level was reduced further in the Cary model-81 double monochromator, in which the stray light is normally very low. (Comparisons between a Cary model-81 Raman instrument and a Jarrell-Ash double monochromator indicate that the stray-light level of the Cary instrument is *at least* as low as that of the Jarrell-Ash, for which a value of 10^{-10} was obtained at 4890 Å in measurements with the 4880 Å line from an argon-ion laser.*) Further, the improvement evident from Fig. 3 indicates that stray light, rather than real intensity from water, gives rise to a considerable part of the intensity at low frequencies. The spectra of Fig. 3 also suggest that quasielastic scattering, i.e., intensity from the Rayleigh wing, is not very important above roughly 75 cm^{-1}. In addition, it should be noted that neutron inelastic scattering studies indicate that quasielastic scattering occurs primarily below 30 cm^{-1} for water at ambient temperatures.†

An argon-ion laser Raman spectrum from water in the low-frequency region is also shown in Fig. 4. The base line was estimated as before (the curved dashed line) and the differences between the total and base line intensities are shown below. Analog computer analysis was accomplished using Gaussian components, and the components are shown by dashed lines in the region of overlap. The components are centered near 60 and 160 cm^{-1}, and the half-widths are $\sim$40 and $\sim$130 cm^{-1}. The half-width value of $\sim$130 cm^{-1} for the hydrogen bond stretching component at 160 cm^{-1} is in satisfactory agreement with the previous mercury-excited value of 110 cm^{-1} for the component centered near 166 cm^{-1} at 25°C.[(1135)]

The far-infrared spectra from liquid H_2O and D_2O also give evidence of broad bands centered near 170 ± 5 cm^{-1} and 167 ± 5 cm^{-1}, respectively.[(273)] Accurate determinations of the absorption coefficients are difficult, however, because the thicknesses of thin films must be measured accurately and maintained constant during the measurements. The most accurate and extensive values of the absorption coefficients of use for present purposes are those of Bagdade[(42)] and of Draegert *et al.*[(273)] (See also the accurate low-frequency absorption coefficients reported by Chamberlain *et al.*[(175)]) The absorption coefficients are shown in Fig. 5 for ambient temperature. The dashed line in the figure is considered to represent both sets of data (Bagdade and Draegert *et al.*), and was employed with the two baselines *A* and *B* in analog computer analyses employing Gaussian components.

* G. E. Walrafen, unpublished work.

† G. J. Safford, private communication.

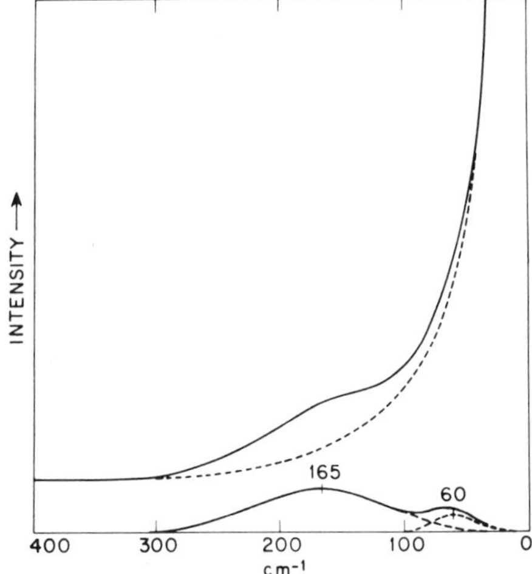

Fig. 4. Argon-ion-laser-excited (4880 Å) Raman spectrum from water obtained at 25°C with a high-pressure cell and the Cary model-81 Raman spectrophotometer.[1139] The differences between the total intensity and the baseline intensity (upper solid line and lower dashed line) were obtained and are shown below (solid line). The Gaussian components (below) are indicated by dashed lines in the region of overlap, $\Delta\bar{\nu} < 100\ \text{cm}^{-1}$.

The *B* baseline of Fig. 5 was employed previously by Draegert.[272] The position of the baseline is determined by the absorption coefficient at $\sim$350 cm^{-1}, and at low frequencies, e.g., 20 cm^{-1} or less. The area above the baseline is assumed to comprise the absorption from intermolecular vibrations such as hydrogen bond bending and stretching, and the area below is assumed to arise from an extremely broad, non specific absorption, and, to a smaller extent, from the neighboring broad librational contour.

Both baselines *A* and *B* were employed in Gaussian analog computer decompositions of the infrared spectrum of Fig. 5. Analog decomposition employing baseline *A* yielded two Gaussian components centered near 65 and 200 cm^{-1}, with half-widths of $\sim$85 and $\sim$175 cm^{-1}. However, the spectrum could not be fitted above 270 cm^{-1}, and the half-width values are large in comparison to the corresponding Raman values. Analog decomposition employing baseline *B* yielded an excellent fit to 350 cm^{-1}, and components

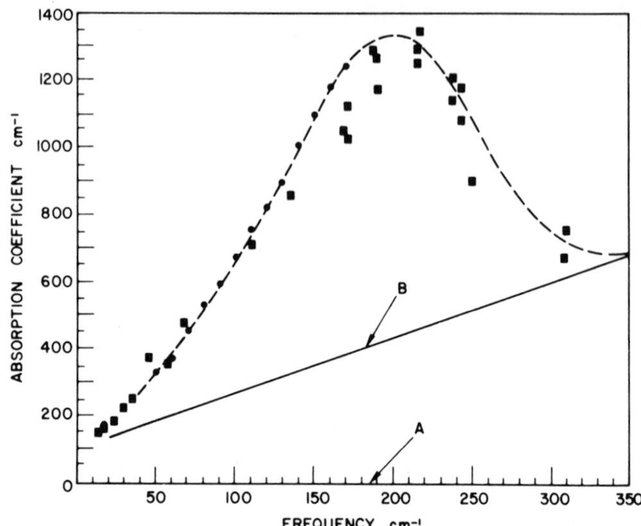

Fig. 5. Far-infrared absorption coefficients in units of cm^{-1} (Np cm^{-1}) plotted versus frequency in cm^{-1}. Data from Bagdade[42] (■); data from Draegert *et al.*[273] (●). Note the two baselines employed, *A* and *B*.

centered near 72 and 188 cm^{-1} having half-widths of ~80 and ~140 cm^{-1}, respectively, were obtained. The component frequencies and half-widths are in satisfactory agreement with the corresponding Raman values (the half-width of ~80 cm^{-1} seems somewhat large), and it should be noted that the presence of two broad components is indicated at roughly the same frequencies for widely different baselines. Certainly, a weak, broad infrared component centered in the general vicinity of ~70 cm^{-1} is suggested by the analog analyses.

Infrared as well as Raman activity from the hydrogen bond bending vibration was predicted previously from treatment of a fully hydrogen-bonded, five-molecule tetrahedral structure in terms of approximate C_{2V} point group symmetry.[1134,1135,1137,1139] In that treatment the broad 60-cm^{-1} band was considered to include the $\nu_3 A_1$ and $\nu_4 A_1$ hydrogen bond bending motions. A similar approximation could also be applied to ice, for which the hydrogen bond bending motions should be infrared-active, and in this regard a broad, weak infrared absorption band has recently been reported for ice I*h* in the vicinity of 65 cm^{-1}.[76] The presence of an infrared absorption in ice I*h* at 65 cm^{-1} strengthens the analog computer indications of a weak, broad band in the liquid at ~70 cm^{-1}.

The area below baseline *B* of Fig. 5 is thought to arise in part from a broad, nonspecific absorption. The (total) absorption coefficients from water have been observed to increase,[42,272] whereas the area above the baseline has been found either to remain constant or to decrease slightly with temperature rise.[272] In the Raman spectrum the corresponding integrated intensity has been observed to decrease markedly with increasing temperature,[1134,1135]* but this effect is thought to be specific to the 4-bonded species,[1139] and a less rapid decrease would be expected from methods sensitive to all of the hydrogen bond bending and stretching interactions. Thus, the temperature dependence observed for the infrared absorption above baseline *B* is reasonable, and supports the baseline employed. Little is known, however, about mechanisms giving rise to the nonspecific absorption. The effect may be related to the increase in the concentration of nonhydrogen-bonded OH units, and to the corresponding increase in rotational freedom with increase of temperature. Investigations of the far-infrared absorption coefficients over a wider range of temperatures are indicated.

3.2. The Librational Region

The librational or restricted rotational motions of H_2O and D_2O molecules in the liquid arise from restraints produced by the hydrogen bonds. Three librational components analogous to the rotations of isolated H_2O and D_2O molecules result. The librational components of the liquid are broad and the components overlap. The breadth and overlap give rise to broad contours, although clear indications of the component substructure remain, particularly in the Raman, hyper-Raman, and neutron inelastic scattering spectra. The librational Raman bands from liquid H_2O, for example, occur within the region 200–1100 cm^{-1}. In the case of D_2O, however, overlap of the Raman hydrogen bond stretching and librational contours occurs in the region 150–300 cm^{-1} because the librational region involves the range from about 150 to 800 cm^{-1}.

Quantitative Raman spectra obtained from water at two temperatures with unpolarized 4358 Å mercury excitation are shown in Fig. 6 for the region 200–1100 cm^{-1}. A decrease in the total Raman contour intensity is apparent with increase of temperature. The Raman contour shapes, however, do not appear to change markedly from 10 to 90°C.

An argon-ion-laser Raman spectrum in the region below 1800 cm^{-1} from liquid H_2O is also shown in Fig. 7. The librational contour, shown

* Also L. A. Blatz, private communication.

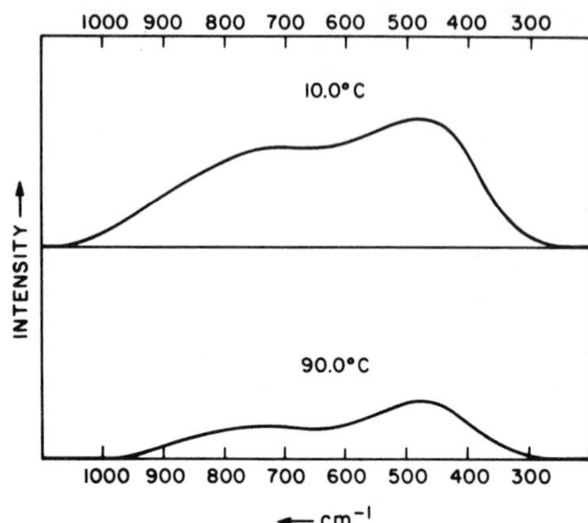

Fig. 6. Raman spectra obtained with unpolarized 4358 Å mercury excitation in the intermolecular librational region from liquid H_2O at 10 and 90°C. The spectra were obtained with the Cary model-81 Raman spectrophotometer using the double-slit technique and method *A* (see Ref. 1137). The original photoelectric tracings were transferred to horizontal base lines; the intensities shown are quantitatively comparable.

above, was transferred to a horizontal baseline (below) by estimating the baseline (upper dashed curve) as described previously. The transferred contour was then decomposed into Gaussian components by the analog computer technique, and the three components required to fit the contour are shown by dashed lines in the figure. The librational components are centered near 425, 550, and 740 cm^{-1} and their half-widths range from about 200 to 250 cm^{-1}. The breadth and overlap of the components mentioned previously are evident from the figure. (The band indicated at about 1650 cm^{-1} arises from the intramolecular bending motion.)

Analog decomposition was also accomplished for argon-ion-laser Raman spectra from liquid D_2O. Overlapping between the hydrogen bond stretching and librational contours was observed, but the analog analyses and the Raman spectra yielded components or contours centered at ~60, ~175, 350, and 500 cm^{-1}. The contour centered at 500 cm^{-1}, however, definitely refers to the envelope of at least two components, but further analysis leading to accurate component frequency values was not possible. The component at 350 cm^{-1} was also severely overlapped by the component at ~175 cm^{-1}, and by the components giving rise to the envelope centered near 500 cm^{-1}.

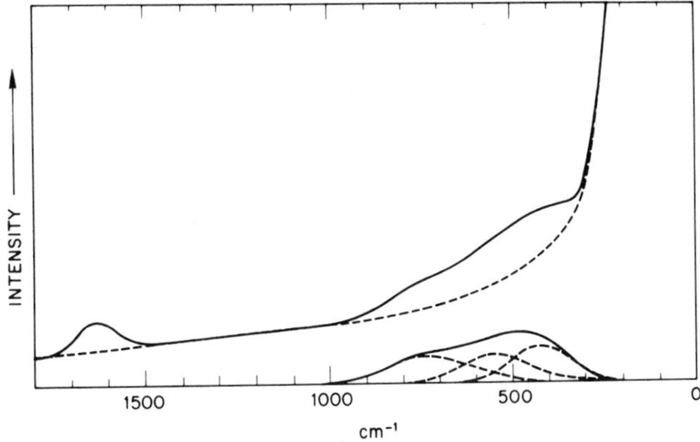

Fig. 7. Argon-ion-laser-excited (4880 Å) Raman spectrum obtained in the 200–1800 cm^{-1} region from liquid H_2O at 25°C. The spectrum was obtained by passing the laser beam through a Raman cell positioned directly in front of the slit of the Cary model-81 spectrophotometer (cf. Fig. 1 of Ref. 1139). The double-slit technique was employed and it produced the high background below 300 cm^{-1}. Differences between the total and baseline intensities were transferred to a horizontal baseline (below) and decomposed into Gaussian components (dashed lines).

Infrared transmittance spectra from liquid H_2O and D_2O have also been obtained in the librational region by Draegert *et al.*[273] The transmittance spectra are shown in Fig. 8. Maximum absorption occurs for liquid H_2O near 685 cm^{-1}, and at about 505 cm^{-1} for D_2O. The infrared librational contours are broad, and overlapping with the hydrogen bond stretching region is indicated, especially for liquid D_2O.

Neutron inelastic scattering from water has been investigated by many workers, e.g., see Refs. 365, 655, 933, 1025. A broad librational peak from

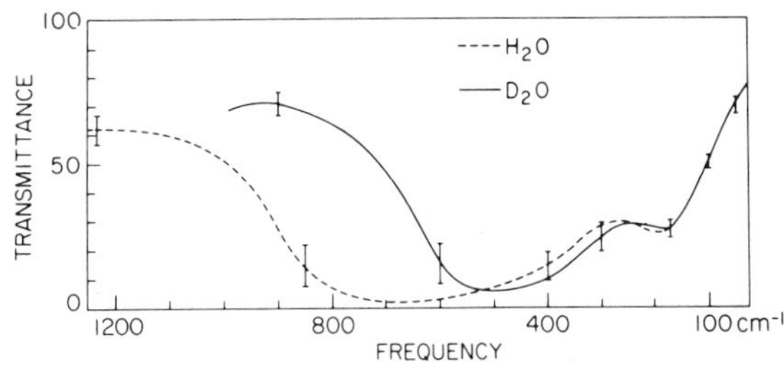

Fig. 8. Infrared transmittance spectra from liquid H_2O (broken curve) and D_2O (solid curve) obtained by Draegert *et al.*[273]

liquid H_2O centered near 494 cm^{-1} was observed in the early work,[655] and evidence for a shoulder at about 740 cm^{-1} (or above) was also uncovered, but clear evidence for three definite librational components was not obtained. The neutron inelastic scattering spectra recently reported by Safford *et al.*[933] for temperatures of 1 and 25°C, however, indicate definite components. Maxima were observed at 454 $\pm$ 30 and 590 $\pm$ 45 cm^{-1}, and a third, unresolved component was also indicated. The frequency of 860 $\pm$ 70 cm^{-1} reported for the unresolved component, however, falls in a region of very low resolution, and that value should be regarded solely as an indication of the presence, rather than the position, of the high-frequency component.* A more reliable frequency value for the third component is probably that obtained from the work of Stiller and Danner[1025] 766 cm^{-1}. Neutron inelastic scattering spectra from liquid D_2O have also been reported.[655] The spectra yield peaks at ~50, ~160, and ~465 cm^{-1}. (For a more detailed analysis of neutron inelastic scattering, see Chapter 9.)

Stimulated Raman spectra from water have been obtained recently by Rahn *et al.*[885] The work is of special interest because stimulated librational scattering from water was reported—a librational intensity maximum was observed at about 550 cm^{-1}. The stimulated librational spectrum reported is unique in that all other known spectra indicate either similar (590 $\pm$ 45 cm^{-1}), as well as additional maxima (neutron inelastic scattering[933]), maxima at other frequencies (infrared absorption[273]), or no obvious features at 550 cm^{-1} (Raman spectra[1137]), with the possible exception of the intensity maximum in the β_3 hyper-Raman spectrum from water, which nevertheless occurs slightly below, at 500 cm^{-1}.† (All spectra, however, contain a component, unresolved in many cases, near roughly 550 cm^{-1}).

A further interesting feature of the stimulated scattering is that the spontaneous Raman librational component cross sections from water are very low relative to the spontaneous intramolecular OH stretching component cross sections.[1133,1134] Thus, the simultaneous stimulation of the librational, ν_L, and intramolecular OH stretching, νOH, regions was unexpected. The simultaneous stimulation may involve coupling between components of ν_{OH} and ν_L. Peaks in the stimulated OH stretching region have been observed near 3425 cm^{-1}[885] and between 3400 and 3450 cm^{-1}.[199] The difference, 3425 cm^{-1} $-$ 550 cm^{-1} = 2875 cm^{-1}, and (very approximate) Gaussian analog decomposition of the stimulated OH stretching contour[885] (this work) indicates that the reported marked low-frequency asymmetry

* G. J. Safford, private communication.

† P. D. Maker, private communication.

TABLE I. Frequencies $\Delta\bar{\nu}$ (cm^{-1}), Half-Widths $\Delta\bar{\nu}_{1/2}$ (cm^{-1}), and Integrated Intensities I_G (in %) for Gaussian Components Obtained from Analog Computer Analysis of Unpolarized, β_1 (Dipolar), and β_3 (Octapolar) Inelastic Harmonic Light Scattering or Hyper-Raman Spectra[a]

Unpolarized			β_1			β_3		
$\Delta\bar{\nu}$	$\Delta\bar{\nu}_{1/2}$	I_G	$\Delta\bar{\nu}$	$\Delta\bar{\nu}_{1/2}$	I_G	$\Delta\bar{\nu}$	$\Delta\bar{\nu}_{1/2}$	I_G
H_2O								
470	200	13		Absent		470	260	49
550	320	49	535	320	41	530	210	19
775	260	38	755	265	59	715	320	32
D_2O								
330	170	26		Absent		305	150	31
425	260	36	415	225	37	400	200	60
565	210	38	570	200	63	535	245	9

[a] From P. D. Maker, private communication.

is centered near 2900 cm^{-1}. Further, the librational component at 550 cm^{-1} is thought to refer to the in-plane motion,[141,1134,1137,1139] and coupling with the OH stretching motion would provide a mechanism for librational stimulation when the spontaneous cross section is low. Simultaneous stimulation of νOH and ν_L would thus also give rise to stimulation at the difference frequency $\nu_{OH} - \nu_L$.

Librational hyper-Raman spectra, unpolarized, β_1, and β_3, have been obtained recently from liquid H_2O and D_2O by Maker.* The librational hyper-Raman spectra provide unequivocal experimental evidence for three broad, definite librational components from both liquids, and significantly for the operation of selection rules.[1139] The unpolarized, β_1, and β_3 hyper-Raman spectra were analyzed by analog computer techniques using Gaussian component shapes. The results of the analyses are shown in Table I. The agreements between half-widths and, particularly, between frequencies are good, and they reflect the high quality of the hyper-Raman spectra. The range of hyper-Raman component half-widths for liquid H_2O, 200–320 cm^{-1}, is in satisfactory agreement with the Raman component half-widths, 200–250 cm^{-1}, and the agreement between the hyper-Raman and Raman compo-

* P. D. Maker, private communication.

TABLE II. Selected Values of Intermolecular Vibrational Frequencies Obtained from Liquid H_2O and D_2O

Raman scattering		Infrared absorption	Inelastic harmonic light scattering (Hyper-Raman)		Neutron inelastic scattering	Stimulated Raman scattering
cm^{-1}	Polarization		β_1	β_3		
H_2O						
~60	—	~70	—	—	60	—
170	*P*	170	—	—	175	—
425–450	*dp*	—	Absent	470	454	—
550	*dp*	685	535	530	590	550
720–740	*dp*		755	715	766	—
D_2O						
~60	—	—	—	—	~50	—
~175	—	165	—	—	~160	—
350	—	—	Absent	305	—	—
500	—	505	415	400	~465	—
			570	535		

nent frequencies is quite good, 470, 530–550, and 715–775 cm^{-1} compared with 425–450, 550, and 720–740 cm^{-1}, respectively. (A similar complete comparison is not yet possible for liquid D_2O). Further, the absence of components in the β_1 spectra, at 470 cm^{-1} for H_2O, and at 305–330 cm^{-1} for D_2O, was clearly indicated by the computer analyses, although evidence for the operation of selection rules is also obvious from direct visual examinations of the β_1 and β_3 hyper-Raman spectra from H_2O as well as from D_2O.

Selected values for all of the intermolecular vibrational frequencies from liquid H_2O and D_2O obtained from Raman scattering, infrared absorption, β_1 and β_3 hyper-Raman, neutron inelastic, and stimulated Raman scattering are compared in Table II, in which laser-Raman depolarization data are included for liquid H_2O.[1137]

3.3. The Five-Molecule, Fully Hydrogen-Bonded C_{2V} Structural Model

Examination of Table II indicates that the intermolecular librational components are completely depolarized in the Raman spectrum from water (cf. Ref. 1137). Further, the librational component from liquid H_2O at

TABLE III. Selection Rules for Raman, Infrared, and β_1 and β_3 Hyper-Raman Spectra Corresponding to the C_{2V} Point Group[a]

Raman scattering		Infrared absorption	Inelastic harmonic light scattering (hyper-Raman)		
	Polarization[b]		β_1	β_3	Polarization[c]
A_1	p	A_1	A_1	A_1	p
A_2	dp	Inactive	Inactive	A_2	dp
B_1	dp	B_1	B_1	B_1	p
B_2	dp	B_2	B_2	B_2	p

[a] Depolarization ratios are included for the Raman and hyper-Raman spectra. The depolarization ratio ϱ_l refers to the ratio for linearly polarized excitation. The order of entries does not refer to the order in Tables I and II.

[b] $\varrho_l = 3/4$, dp. $\varrho_l < 3/4$, p.

[c] $\varrho_l = 2/3$, dp. $2/3 > \varrho_l > 1/9$, p.

470 cm^{-1} and the corresponding component from liquid D_2O at 305–330 cm^{-1} are absent in the β_1 (dipolar) hyper-Raman spectra (Tables I and II). In addition, the infrared librational spectra from liquid H_2O and D_2O give evidence of maximum absorption at frequencies much higher than 470 cm^{-1} and 305–330 cm^{-1}, in agreement with the observed hyper-Raman inactivity. Accordingly, the spectroscopic data are all consistent with the requirements of the C_{2V} point group.[224,1134,1135,1137,1139] Point groups of lower symmetry, e.g., C_2, C_S, are not consistent with the data, however, and an examination of many point groups having symmetries higher and lower than C_{2V} reveals that the other groups violate the data of Tables I and II, and/or are geometrically unrelated to water. The selection rules and depolarization ratios for the C_{2V} point group are listed in Table III.

The C_{2V} symmetry demanded by examination of the data from Tables I and II was not unexpected. The early X-ray diffraction data of Morgan and Warren[761] (cf. also Brady and Romanow[107]) indicated that there is a tendency for water to engage in tetrahedral coordination, and ice I*h* unquestionably involves tetrahedral coordination. Further, the C_{2V} point group is a subgroup of T_d and arises naturally for the C_{2V} water molecule in a tetrahedral field of next-nearest oxygen neighbors, as shown in Fig. 9.

In Fig. 9 a five-molecule, fully hydrogen-bonded structure taken from the ice I*h* lattice is illustrated. The overall symmetry of the structure shown, i.e., the symmetry when the conformations of the six outer protons are

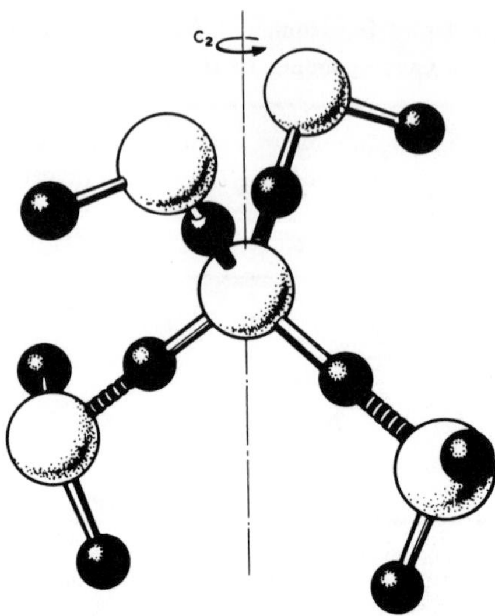

Fig. 9. Five-molecule, fully hydrogen-bonded structure having intermolecular C_{2V} symmetry. Small spheres, H atoms; large spheres, O atoms. Disks refer to hydrogen bonds. The O—O distances are equal, and the oxygen atoms form a regular tetrahedron. The hydrogen bonds are equal in length, and in liquid H_2O are about 1.8–1.9 Å. The OH bonds (rods) are also equal in length, and for the liquid are about 1.0 Å.

considered, is C_2. However, in terms of the central oxygen atom, of the tetrahedral nearest-neighbor oxygen atoms, and of the four intervening protons, the symmetry is C_{2V}. The symmetry when the outer proton conformations are excluded will be referred to as intermolecular C_{2V} symmetry.

The intermolecular C_{2V} model of Fig. 9 is useful in describing and designating the intermolecular vibrations of the fully hydrogen-bonded structures that constitute most of the structures present in liquid H_2O and D_2O at very low temperatures. In the C_{2V} model nine intermolecular vibrations must be considered—the intramolecular vibrations are excluded. Of the nine intermolecular vibrations, six refer to restricted translations and three to restricted rotations.

Two of the six restricted translations are equivalent to the hydrogen bond bending motions. They are the totally symmetric deformations belonging to the A_1 vibrational species, designated $\nu_3 A_1$ and $\nu_4 A_1$.[1134] The remaining four restricted translations, the hydrogen bond stretching motions, involve the A_1, B_1, and B_2 vibrational species. They are designated $\nu_1 A_1$, $\nu_2 A_1$, $\nu_7 B_1$, and $\nu_9 B_2$.[1134] The hydrogen bond bending vibrations $\nu_3 A_1$ and $\nu_4 A_1$ are unresolved and occur near 60 cm^{-1}; the hydrogen bond stretching vibrations $\nu_1 A_1$, $\nu_2 A_1$, $\nu_7 B_1$, and $\nu_9 B_2$ are also unresolved and give rise to the broad band near 170 cm^{-1} for liquid H_2O and D_2O.

The restricted rotational or librational components, unlike the restricted translations, are resolved in various spectra, and show isotope shifts as indicated in Table II. The A_2 libration, designated $\nu_5 A_2$,[1134] involves the restricted rotation that is symmetric only with respect to the C_2 axis (Fig. 9). It is forbiden in all dipolar spectra, namely in infrared and β_1 hyper-Raman spectra (Table III). The B_2 band, designated $\nu_8 B_2$,[1134] is assigned to the libration that is symmetric with respect to the plane formed by the atoms of the central H_2O molecule of Fig. 9. That plane is arbitrarily designated B_2. The remaining libration, designated $\nu_6 B_1$,[1134] is only symmetric with respect to the B_1 plane (perpendicular to the B_2 plane and passing through the C_2 axis). Because none of the librations is totally symmetric with respect to all symmetry elements (C_2, B_1, B_2), each is completely depolarized in the Raman spectrum for C_{2V} symmetry.

For liquid H_2O and D_2O the $\nu_5 A_2$ librations are centered near 425–470 cm^{-1} and 305–350 cm^{-1}, the $\nu_6 B_1$ librations are centered near 715–766 cm^{-1} and 535–570 cm^{-1}, and the $\nu_8 B_2$ librations are centered near 530–590 cm^{-1} and 400–415 cm^{-1}; Table II and Refs. 1134, 1135, 1137, 1139.

The structure shown in Fig. 9 is consistent with the data of Table II, and it is useful in making vibrational assignments according to C_{2V} symmetry, but it necessarily refers to a fixed or average configuration. Distortions from C_{2V} symmetry involving angles and distances are present in those molecules of the real liquid that are involved in the fully hydrogen-bonded structures, and the various distortions give rise to the breadth observed for the individual intermolecular components from liquid H_2O and D_2O. (Distortions are also present in the partially hydrogen-bonded structures.) The data of Tables I and II, however, refer to the real liquid and they can provide information that limits the types of distortions expected for the fully hydrogen-bonded case, irrespective of models. Of various possible distortions, one proposed recently from X-ray diffraction studies of liquid H_2O and D_2O involves unequal O—O distances[784] (see Chapter 8).

The X-ray radial distribution function is closely reproduced by a model having, on the average, and for a time scale long with respect to the Raman time scale, at least two distinctly unequal O—O distances. (The intermolecular Raman vibrational periods are about 10^{-12} sec.) However, distortions involving two unequal distances can refer to point groups below C_{2V}, e.g., C_S, and distortions involving various combinations of unequal distances can lead to no symmetry. Accordingly, all vibrational species would be active in the β_1 (dipolar) hyper-Raman spectrum, in disagreement with the spectroscopic results. Further, infrared activity would be indicated for all librational components, and one or more of the Raman librational components would be polarized. The interpretation of the radial distribution function is therefore not consistent with the spectroscopic data summarized in Tables II and III, which indicate that, on the average, and for times of roughly 10^{-12} sec, the O—O distances are equal and the intermolecular symmetry is C_{2V}, *for that fraction of the* H_2O *and* D_2O *molecules involved only in fully hydrogen-bonded interactions*, i.e., excluding the partially hydrogen-bonded structures.

The five-molecule, fully hydrogen-bonded C_{2V} structural model applies only to the 4-bonded configurations and is thus applicable at low temperatures, where the 4-bonded concentration must be high. Partially hydrogen-bonded structures, 3-, 2-, and 1-bonded, become increasingly important in liquid H_2O and D_2O at elevated temperatures. At present, however, the techniques that appear to give intermolecular evidence relating to the partially hydrogen-bonded species—infrared absorption and neutron inelastic scattering spectroscopy—have only been employed to temperatures of 50 and 75°C, respectively. Accurate absorption coefficients in the intermolecular hydrogen bond bending and librational regions as well as neutron inelastic scattering spectra are required at elevated temperatures. Such data will provide vibrational frequencies of value in normal coordinate analyses involving partially hydrogen-bonded structures.[141]

4. INTRAMOLECULAR VIBRATIONAL SPECTRA FROM HDO IN H_2O AND D_2O

4.1. Raman Spectra

When H_2O and D_2O are mixed in equimolar amounts, HDO is formed and the solution contains HDO, H_2O, and D_2O in the approximate molar ratio of 2 : 1 : 1.[81] Accordingly, Raman spectra from a 50 mole % solution of H_2O and D_2O will contain contributions from H_2O and D_2O in the intramolecular valence region, 2000–4000 cm^{-1}, but the principal contributions may come from the OH and OD stretching vibrations of HDO. A Raman spectrum obtained from a 50 mole % solution of H_2O and D_2O is shown in the intramolecular valence region in Fig. 10.

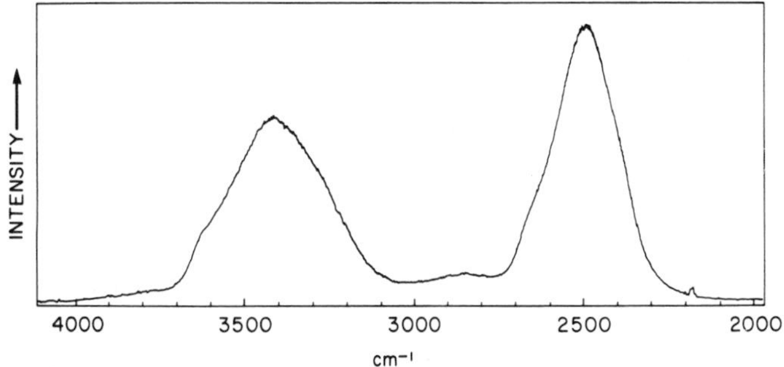

Fig. 10. Argon-ion-laser Raman spectrum (4880 Å) from a 50 mole % solution of H_2O and D_2O at 25°C obtained photoelectrically with the Jarrell-Ash double monochromator using the fixed-beam laser-Raman technique (cf. Ref. 1142). A slit width of $10\ cm^{-1}$ was employed with a time constant of 1 sec. The spectrum was neither smoothed nor averaged.

The principal Raman intensity maxima evident from Fig. 10 occur at $3415 \pm 5\ cm^{-1}$ and at $2495 \pm 5\ cm^{-1}$, and they refer to the OH and OD stretching vibrations of HDO, and of H_2O and D_2O. A minor intensity maximum is also evident from Fig. 10 at about $2860 \pm 10\ cm^{-1}$. This weak maximum is thought to arise from the overtone of the fundamental intramolecular bending vibration of HDO near $1450\ cm^{-1}$.[1134]

In addition to the two principal intensity maxima, prominent high-frequency shoulders are evident, in the OH and OD stretching regions near $3625 \pm 10\ cm^{-1}$ and $2650 \pm 10\ cm^{-1}$. The shoulders are thought to arise from the nonhydrogen-bonded OH and OD stretching vibrations,[1137–1142] whereas the intensity maxima arise from the corresponding hydrogen-bonded stretching vibrations.[1137–1142] Other minor features in addition to the intensity maxima and high-frequency shoulders are also of importance.

Careful examination of the spectrum near $3300\ cm^{-1}$ and $2400\ cm^{-1}$ reveals weak, broad, low-frequency shoulders. The regions at $3275\ cm^{-1}$ and $2375\ cm^{-1}$ are concave downward, whereas the opposite positions above the peaks at about $3525\ cm^{-1}$ and $2625\ cm^{-1}$ give evidence of upward concavity. The broad, weak, low-frequency shoulders are thought to arise from coupling between hydrogen-bonded OH and OD stretching vibrations from HDO as well as from H_2O and D_2O.[1141] The HDO concentration is maximal in a 50 mole % solution, and the coupling interactions between HDO molecules are also maximal. Coupling components from H_2O and D_2O also occur at about $3250\ cm^{-1}$ and $2370\ cm^{-1}$, and they can also make contributions to the spectrum.

The intermolecular coupling of intramolecular OH and OD stretching vibrations of HDO can be greatly reduced, and the spectra simplified, by lowering the HDO concentration. (The weak intramolecular coupling between OH and OD stretching vibrations within the HDO molecule, however, is not greatly affected.) If a very small amount of H_2O, for example, is added to pure D_2O, the formation of HDO will be very nearly complete. The corresponding Raman spectrum will give evidence of OH and OD stretching vibrations at the approximate positions of the peaks in Fig. 10, but the OH stretching peak will be very weak and will refer almost exclusively to HDO. The OD stretching peak will be very intense and will refer almost exclusively to D_2O. A Raman spectrum obtained from a 1 mole % solution of H_2O in D_2O as solvent is shown in the OH stretching region in Fig. 11.

The photoelectric tracing of the Raman spectrum from the 1 mole % solution of H_2O in D_2O is shown in the upper part of Fig. 11. The differences between the observed intensity and the baseline intensity (dashed line) were determined and transferred to a horizontal baseline (lower part of Fig. 11). The resulting spectrum was then decomposed by the analog computer technique using Gaussian components. Two Gaussian components were found to give an adequate fit, and the positions and half-widths of the components are shown in the figure. The Gaussian components are centered near 3435 cm^{-1} and 3628 cm^{-1}, and the corresponding half-widths are roughly 260 cm^{-1} and 90 cm^{-1}. The component from HDO near 3435 cm^{-1} refers to the hydrogen-bonded OH stretching vibrations, and the component near 3628 cm^{-1} refers to the nonhydrogen-bonded OH stretching vibrations.

A Raman spectrum from a 1 mole % solution of D_2O in H_2O as solvent is shown in Fig. 12 (upper part). The spectrum was transferred to a horizontal baseline (lower part) as described in Fig. 11. Analog computer decomposition yielded two Gaussian components centered near 2530 cm^{-1} and 2650 cm^{-1}; the half-widths are approximately 145 cm^{-1} and 80 cm^{-1}. The half-width of 80 cm^{-1} for the nonhydrogen-bonded OD stretching component at 2650 cm^{-1} is nearly equal to that of the corresponding nonhydrogen-bonded OH stretching component from Fig. 11, but the half-widths of the hydrogen-bonded OD and OH stretching components are larger and unequal, 145 cm^{-1} versus 260 cm^{-1}. Small half-widths, 100 ± 10 cm^{-1}, that are approximately equal for the OD and OH stretching components have been found to characterize the nonhydrogen-bonded components from 10 mole % solutions, whereas large, unequal half-widths, 160 ± 10 and 250 ± 10 cm^{-1}. characterize the hydrogen-bonded OD and OH stretching components.[1142]

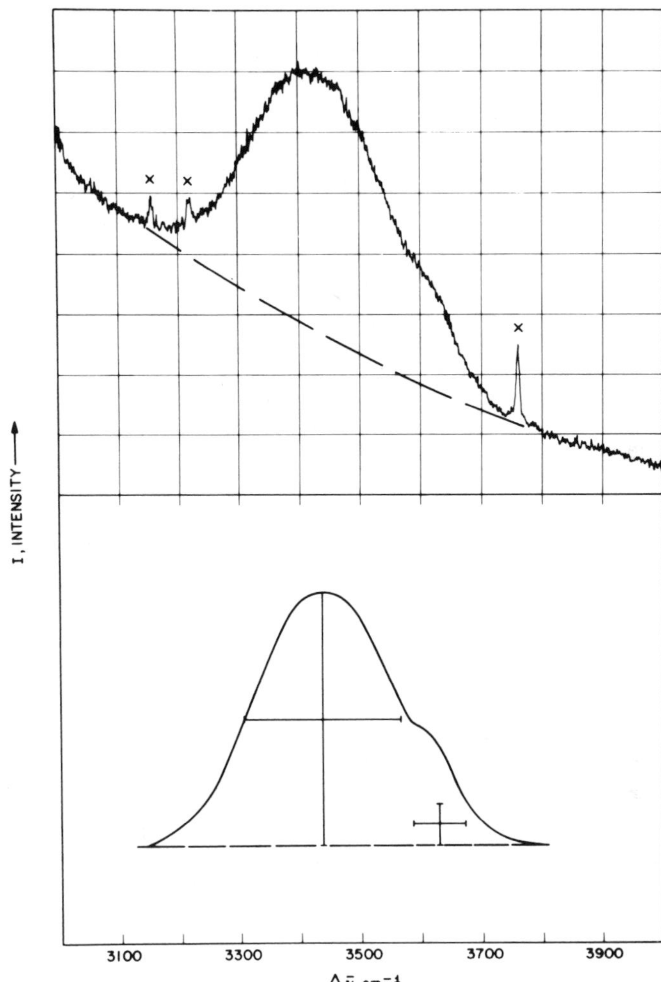

Fig. 11. Argon-ion-laser Raman spectrum (4880 Å) from a 1 mole % solution of H_2O in D_2O at 25°C obtained photoelectrically with the Cary model-81 Raman spectrometer (cf. Ref. 1140). A slit width of 10 cm^{-1} and a time constant of 0.1 sec were employed. Spectral lines from argon are indicated by X's. The baseline (above) is indicated by the dashed line. The spectrum was transferred to a horizontal baseline (below) and decomposed into two Gaussian components. The positions and half-widths of the components are indicated by vertical and horizontal bars.

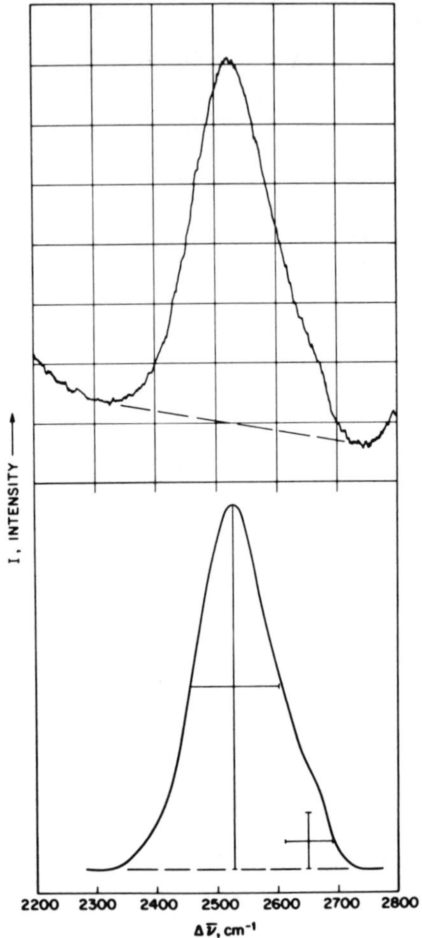

Fig. 12. Argon-ion-laser Raman spectrum (4880 Å) from a 1 mole % solution of D_2O in H_2O at 25°C obtained photoelectrically with the Cary model-81 spectrometer under conditions described in Ref. 1140. A slit width of 10 cm^{-1} and a time constant of 0.1 sec were employed. The baseline (above) is indicated by the dashed line. The spectrum was transferred to a horizontal baseline (below) and decomposed into two Gaussian components. The positions and half-widths of the components are indicated by vertical and horizontal bars.

Low signal-to-noise ratios are involved in the spectra of Figs. 11 and 12, because of the low HDO concentration. In addition, large base-line slopes resulted from the intense neighboring bands of D_2O and H_2O. The Raman intensities of the coupling components, however, are not large even in 50 mole % solutions of H_2O in D_2O (Fig. 10). Accordingly, reliable information, can be obtained without resorting to extreme dilutions. The coupling components are weak, and the formation of HDO is sufficiently complete even at 12 mole %.[1141] Further, the signal-to-noise ratios can be increased significantly and the base-line slopes reduced by use of moderate HDO concentrations. Raman spectra from a 5 mole % solution of H_2O in D_2O and from a 6 mole % solution of D_2O in H_2O are shown in Fig. 13, and the shapes are seen not to differ greatly from those of the 1 mole % solutions.

The ratios of signal to root-mean-square noise, S/N_{rms}, from the Raman spectra of Fig. 13 are about 165 for (a) and about 110 for (b). The high S/N_{rms} values, and the ability to reproduce the spectral shapes, demonstrated by repeated scannings, indicated that derivative spectra of reasonable accuracy could be obtained. The derivative spectra are shown in Fig. 14.

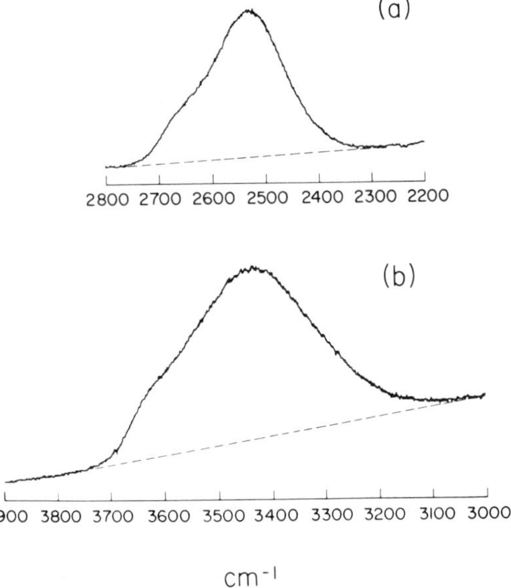

Fig. 13. Argon-ion-laser Raman spectra (4800 Å) from (a) 6 mole % D_2O in H_2O and (b) 5 mole % H_2O in D_2O, obtained at 25°C with a complete Jarrell-Ash photoelectric Raman instrument. A slit width of 10 cm^{-1} and a time constant of 2 sec were employed.

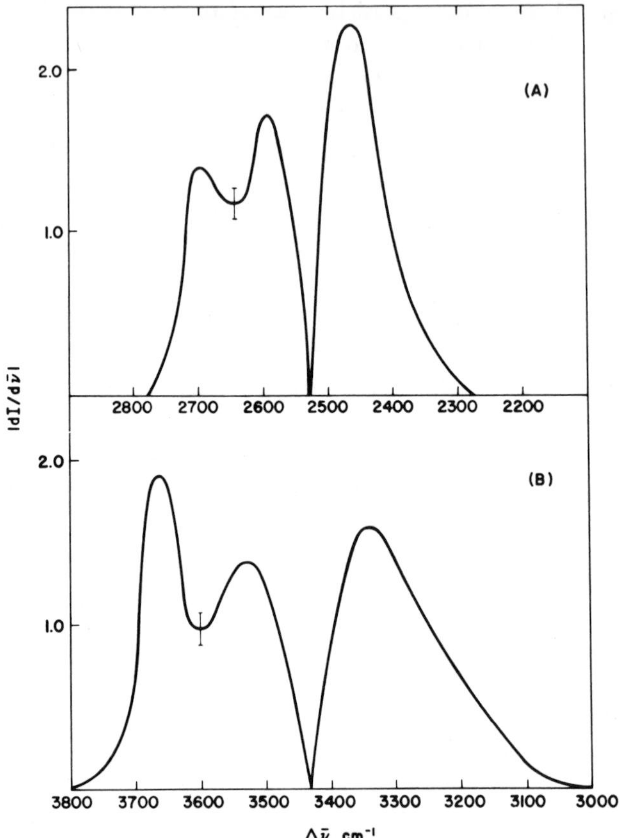

Fig. 14. Absolute values of the derivative $|dI/d\bar{\nu}|$ plotted versus the Raman shift $\Delta\bar{\nu}$ in cm^{-1} for (A) the OD stretching contour and (B) the OH stretching contour. The derivative values were obtained from the spectra of Fig. 13, and the values of $|dI/d\bar{\nu}|$ were corrected for baseline slope. Error bars are shown at 2645 and 3600 cm^{-1}.

Absolute derivative values $|dI/d\bar{\nu}|$ are shown as functions of $\Delta\bar{\nu}$ for the 5 and 6 mole % solutions. Values of $|dI/d\bar{\nu}|$ uncorrected for base-line slope were obtained directly from the photoelectric tracings. The base-line slopes were then determined and the values of $|dI/d\bar{\nu}|$ were corrected accordingly. (The slope corrections are small.) The corrected values of $|dI/d\bar{\nu}|$ are shown in the figure. In the OD case (A), $|dI/d\bar{\nu}| = 0$ at 2530 $\pm$ 10 cm^{-1}, and $|dI/d\nu|$ is minimal at 2645 $\pm$ 10 cm^{-1}. In the OH case (B), $|dI/d\bar{\nu}| = 0$ at 3435 $\pm$ 10 cm^{-1} and $|dI/d\bar{\nu}|$ is minimal at 3600 cm^{-1}.

The derivative values of $2530 \pm 10\ cm^{-1}$ and $3435 \pm 10\ cm^{-1}$ are thought to refer to the central positions of the hydrogen-bonded OD and OH stretching components, and they compare favorably with the component frequency values of $3435\ cm^{-1}$ and $2530\ cm^{-1}$ obtained from Gaussian analog analysis of the spectra of Figs. 11 and 12. Similarly the results, from the derivative method of $2645 \pm 10\ cm^{-1}$ and $3600 \pm 10\ cm^{-1}$ compare favorably with the Gaussian component frequency values of $2650\ cm^{-1}$ and $3628\ cm^{-1}$ corresponding to the centers of the nonhydrogen-bonded OD and OH stretching components. The agreements between the frequencies obtained by the derivative method and by the analog computer technique using Gaussian components provide support for the analog method, and for the use of Gaussian shapes. (Additional support for the use of Gaussian components will be presented in subsequent sections.)

In previous work, argon-ion-laser (4880 Å) as well as quantitative mercury-excited (4358 Å) photoelectric Raman spectra were obtained from a solution of 11 mole % (6.2 *M*) D_2O in H_2O.[1138] The quantitative mercury-excited Raman spectra were obtained at temperatures from 16 to 97°C, and an isosbestic frequency was observed in the OD stretching region of HDO at $2570 \pm 5\ cm^{-1}$. The quantitative mercury-excited Raman spectra are shown in Fig. 15 for temperatures from 32.2 to 93.0°C.

An isosbestic frequency in the case of Raman spectra refers to the frequency of a spectral contour at which, for constant intensity of excitation, the Raman intensity is independent of sample temperature.[1137–1142]

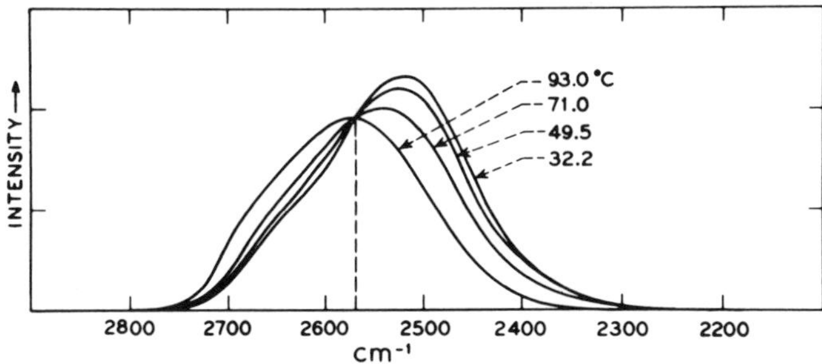

Fig. 15. Quantitative Raman spectra from a solution of 11 mole % (6.2*M*) D_2O in H_2O obtained in the range 32.2–93.0°C with unpolarized 4358 Å mercury excitation using the Cary model-81 spectrophotometer (cf. Ref. 1138). A slit width of $15\ cm^{-1}$ and a time constant of 0.1 sec were employed. The original photoelectric tracings were transferred to horizontal baselines after unwanted mercury lines were removed. The isosbestic frequency of $2570 \pm 5\ cm^{-1}$ is indicated by the dashed vertical line.

Further, for HDO the isosbestic frequency observed for the OD stretching contour separates the nonhydrogen-bonded and hydrogen-bonded OD stretching component centers, and it indicates that an equilibrium exists between the two broad classes of OD stretching interactions. The isosbestic frequency also implies that the component shapes, positions, and half-widths are approximately invariant in the temperature range involved, and that the integrated component intensities, have opposite temperature dependences.[962,1137–1142]

Argon-ion-laser Raman spectra (4880 Å) have also been obtained recently by Lindner[686] at temperatures and pressures from 20°C and 100 bar (density, 1.0 g cm^{-3}) to 400°C and 3900 bar (density, 0.9 g cm^{-3}). The spectra were obtained from a 6.2 M or 11 mole % solution of D_2O in H_2O for comparison with the previous work.[1138] The preliminary Raman spectra are shown in Fig. 16 (see also Chapter 13).

Lindner's results are of importance because of the large temperature range involved, and because the density was maintained constant to 10%. At nearly constant density, the density of cohesive interactions is not expected to change significantly, and the evidence for hydrogen-bonded and non-hydrogen-bonded components provided by Fig. 16 indicates that the hydrogen bond breaking process does not involve large changes in O—O distances.

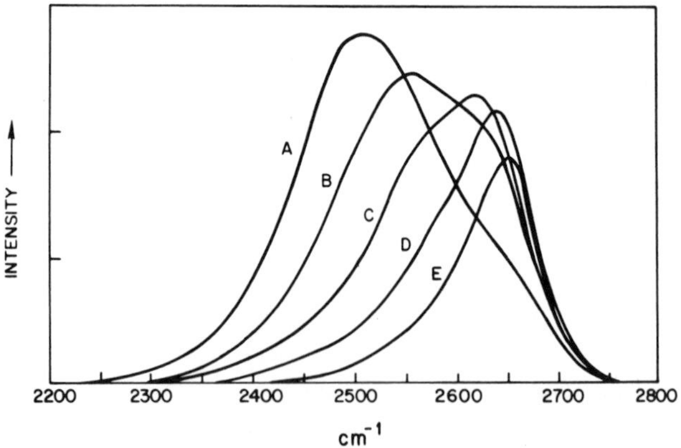

Fig. 16. High-temperature, high-pressure argon-ion laser Raman spectra (4880 Å) obtained by Lindner[686] from a solution of 6.2M or 11 mole % D_2O in H_2O. Spectrum A: 20°C, 100 bars; spectrum B: 100°C, 1000 bars; spectrum C: 200°C, 2800 bars; spectrum D: 300°C, 4700 bars; spectrum E: 400°C, 3900 bars. Spectra A–D: density, 1.0 g cm^{-3}; spectrum E: density, 0.9 g cm^{-3}. The spectra were obtained with a three-prism Steinheil spectrograph and a slit width of 18 cm^{-1}.

Evidence for structured OD stretching components is apparent from Fig. 16 for all temperatures and pressures involved. In spectrum *A*, corresponding to 20°C, maximum intensity occurs near 2507 cm^{-1}, and a high-frequency shoulder is evident near 2650 cm^{-1}. In spectrum *E*, corresponding to 400°C, maximum intensity occurs near 2652 cm^{-1}, i.e., at the position of the high-frequency shoulder in spectrum *A*, and a low-frequency tail is also apparent although the center occurs at a significantly higher frequency than that of the spectrum *A* intensity maximum. The intermediate spectra, *B*, *C*, and *D*, corresponding to 100, 200, and 300°C, respectively also give clear evidence of contour structure. Two broad, major OD stretching components are indicated for each of the intermediate spectra, and the centers are separated by ~110–160 cm^{-1} in the range 100–300°C.

The Raman spectra of Fig. 16, unlike the quantitative spectra of Fig. 15, however, do not give indications of an isosbestic frequency. A temperature range of 380°C is involved in the Fig. 16 spectra, compared to only 61°C for Fig. 15, and an increase in the frequency of the major hydrogen-bonded component from about 2510 to 2620 cm^{-1} is indicated from 20 to 400°C. Under such conditions an isosbestic region, rather than a single frequency, would be expected from quantitative Raman spectra corrected for temperature.

Some aspects of the Fig. 16 spectra, however, indicate that the intensities are not quantitatively intercomparable. The *A* and *B* spectra of Fig. 16, corresponding to 20 and 100°C, cross at 2550 cm^{-1}, whereas the quantitative spectra of Fig. 15 that refer essentially to the temperature range of the *A* and *B* spectra cross at 2570 ± 5 cm^{-1}. The discrepancy indicates that the intensity of the *B* spectrum is somewhat too large relative to spectrum *A*. Such minor discrepancies are not uncommon in laser-Raman work, however, as indicated in the earlier discussion of the fixed-beam technique.

The *A* and *C* spectra, corresponding to temperatures of 20 and 200°C, on the other hand, cross near 2570 cm^{-1}, in agreement with the isosbestic frequency of 2570 ± 5 cm^{-1}. The agreement suggests that a relatively narrow frequency range may characterize the crossings to temperatures as high as 200°C. The *D* and *E* spectra, however, unquestionably cross the *A* spectrum at frequencies well above 2570 cm^{-1}, as expected from the indicated increase in the hydrogen-bonded component frequency. Corrections of the Stokes Raman spectra for ground-state depopulation would raise the intensities at high temperatures relative to those at low temperatures, and would thus lower the crossing frequencies somewhat, although an isosbestic range is still indicated from subsequent analyses.

The shapes of the *A* and *B* spectra of Fig. 16 agree closely with the argon-ion-laser and mercury-excited Raman spectra obtained at 25 and 93.0°C, respectively. The *A* spectrum of Fig. 16 compares very favorably with the corresponding laser-Raman spectra obtained previously, (cf. Fig. 1 of Ref. 1138) and the *B* spectrum compares favorably with the mercury-excited Raman spectrum of Fig. 15 corresponding to 93.0°C. The contour shapes obtained by Lindner are thus of sufficient accuracy for Gaussian analog decompositions, and such decompositions were performed for all of the spectra.

Component frequencies from three-Gaussian analog decompositions of the Fig. 16 spectra are shown in Fig. 17 (b), where component frequencies obtained in previous two-Gaussian decompositions are included.[(1138)] Frequencies corresponding to the intensity maxima from the Fig. 16 spectra are also included in Fig. 17 (a) with corresponding frequencies from previous work.[(1138)] These will be discussed first.

Frequencies corresponding to intensity maxima of OD stretching contours are shown from 16 to 400°C. Below 100°C the frequencies are seen to increase nearly linearly with temperature rise, but the rate of increase decreases with increase of temperature from 100 to 400°C. The behavior from 200 to 400°C suggests that a limiting value near 2650 cm^{-1} is approached, and the limiting value falls well below the dilute gas value of 2722 cm^{-1}.[(686)] The limit of 2650 cm^{-1} defines the frequency of the nonhydrogen-bonded state for a density near 1 g cm^{-3}, and is in agreement with previous values (cf. Figs. 10, 12, and 13). Between 16 and 100°C the combined results from Lindner[(686)] and Walrafen[(1138)] were fitted by linear least squares which yielded the equation $\Delta\bar{\nu} = 2497._3 + 0.6598_1 t$, where t is the temperature in °C. Much below 0°C, however, it is probable that another limiting frequency value corresponding to the intensity maximum is approached. The second limit, if observable, would probably occur above or near ~2450 cm^{-1}, the value observed from studies of isotopically substituted ice,[(1191)] and the limit would also define the frequency of the fully hydrogen-bonded state.

In the early argon-ion-laser Raman work approximate two-Gaussian analog decompositions of the OD stretching contours were employed, although evidence for a third low-frequency component was obtained.[(1138)] In the analog decompositions of the Lindner spectra the third low-frequency component was included. Four Gaussian components, however, were recently uncovered in OD stretching contours from HDO,[(1141)] but the hydrogen-bonded coupling component near 2400 cm^{-1} is broad and weak, and the nonhydrogen-bonded coupling component near 2600 cm^{-1} is of virtually negligible intensity in all but the most concentrated HDO solutions.[1141] Thus, the three-Gaussian analyses are quite adequate, and variations in the two- and three-Gaussian parameters for the hydrogen-bonded and non-hydrogen-bonded components have been found to be very small.[1142] Fre-

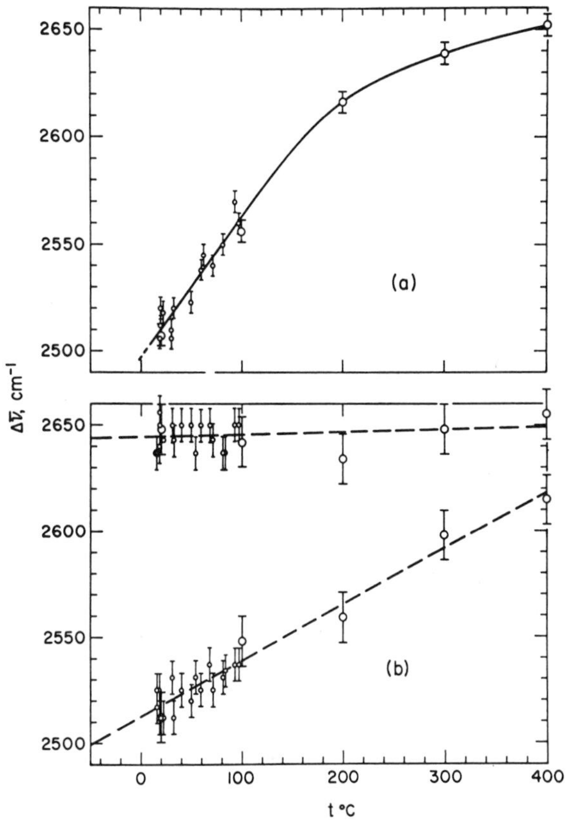

Fig. 17. Raman shifts $\Delta\bar{\nu}$ in cm^{-1} referring to (a) intensity maxima of the OD stretching contour and (b) to the corresponding nonhydrogen-bonded (upper) and hydrogen-bonded (lower) components, obtained from Gaussian analog computer decompositions of OD stretching contours from a 6.2*M* or 11 mole % solution of D_2O in H_2O, as functions of temperature. Small circles, Walrafen.[1138] Large circles, Lindner.[686] The straight solid line shown in (a) below 100°C was obtained by a linear least squares treatment of the combined data for temperatures from 16 to 100°C; the dashed lines shown in (b) for both components also refer to linear least squares treatments of the combined data from 16 to 400°C.

quencies of the two important components obtained from the three-Gaussian analyses are shown in Fig. 17 (b). The results corresponding to the low-frequency coupling component (not shown in the figure) ranged from about 2415 to 2560 cm^{-1} for temperatures from 20 to 400°C. Results from previous two-Gaussian analyses for temperatures from 16 to 97°C are included in the figure.[1138]

From Fig. 17 (b) the agreement between the previous[1138] and present results from Gaussian analog decompositions is seen to be excellent. In the previous work the least squares results were plotted[1138] but the equations were not given. The previous least squares equation from the nonhydrogen-bonded component frequencies is $\Delta\bar{\nu} = 2644._5 + 0.01127_7 t$, and for the hydrogen-bonded component frequencies is $\Delta\bar{\nu} = 2513._1 + 0.2443_6 t$, where $\Delta\bar{\nu}$ is the Raman frequency shift in cm^{-1} and t is the temperature in °C. In the case of Fig. 17 (b) all of the component results were treated by least squares. For the nonhydrogen-bonded component frequencies the least squares equation is $\Delta\bar{\nu} = 2644._1 + 0.01220_1 t$, and for the hydrogen-bonded component frequencies is $\Delta\bar{\nu} = 2512._2 + 0.2646_3 t$.

For temperatures from 16 to 400°C the nonhydrogen-bonded component frequency may be regarded as constant. According to the least squares equation, the frequency varies from 2644 cm^{-1} at 0°C to 2649 cm^{-1} at 400°C, a change of 0.2% in 400°C. The constancy of the nonhydrogen-bonded component frequency indicates that the nature of the interaction is essentially unchanged when the density of the HDO–H_2O mixture is near 1 g cm^{-3}, or when the density of cohesive interactions is nearly constant. It should also be emphasized that the range 2644–2649 cm^{-1}, like the high-temperature limit mentioned previously, is well below the dilute gas value of 2722 cm^{-1}.[686] The hydrogen-bond breaking or disruption at liquid densities thus gives rise to nonhydrogen-bonded OD groups engaged in strongly cohesive interactions that are not highly directional and are clearly not comparable to the OD groups in the low-density gas. Attempts to estimate the concentration of nonhydrogen-bonded groups in water from comparisons between vibrational frequencies of the liquid and of the dilute gas, or of dilute solutions of H_2O in organic solvents, can therefore only yield misleading results.

In the temperature range 16–400°C the hydrogen-bonded component frequency increases from 2516 cm^{-1} to 2618 cm^{-1} according to the least squares equation. The increase is unquestionably significant and is in the direction of the nonhydrogen-bonded component frequency, as expected for consecutive hydrogen bond breakage. The increase of intensity in the nonhydrogen-bonded region relative to the hydrogen-bonded region, clearly evident from Fig. 16, indicates that the hydrogen-bonded OD stretching component must refer to partially hydrogen-bonded, as well as to fully hydrogen-bonded, structures, because structures having fewer hydrogen bonds are increasingly favored at higher temperatures. Accordingly, the increase in the hydrogen-bonded OD stretching frequency must arise from the consecutive breakdown of hydrogen-bonded structures. The broad hydrogen-bonded OD stretching components obtained from analog computer analyses involving a wide temperature range can only be regarded, therefore, as the unresolved envelopes of the broad components from the 4-, 3-, 2-, and 1-bonded structures.

Finally, the component frequencies from Fig. 17(b) can be employed in conjunction with the isosbestic frequency (Fig. 15) to estimate the isosbestic region from 32 to 400°C. In the range 32.2–93.0°C the isosbestic frequency was observed at 2570 $\pm$ 5 cm^{-1}, and it was suggested previously that the value of 2570 cm^{-1} may essentially characterize the crossings to 200°C. At 400°C the nonhydrogen-bonded and hydrogen-bonded component centers occur near 2650 and 2620 cm^{-1}. Hence, an isosbestic region from 2570 cm^{-1} to roughly 2635 cm^{-1} is expected for the 32–400°C range, in fair agreement with Fig. 16.

The increase in the frequency of the hydrogen-bonded component with temperature rise was stated previously to be a consequence of the breakdown of hydrogen-bonded structures, and the presence of an isosbestic region in the Raman spectra from HDO referring to a wide temperature range must also be a consequence of the consecutive breakdown. Similar conclusions have been reached from extensive high-temperature and high-pressure near-infrared studies of water.[701,702,705]

In addition to the evidence for hydrogen-bonded and nonhydrogen-bonded components resulting from studies of the OD stretching contour from HDO at a series of temperatures, additional strong evidence for the two classes of components has recently been obtained from laser-Raman spectra of ternary solutions of $NaClO_4$ and HDO in H_2O or in D_2O.[1142] The recent argon-ion-laser (4880 Å) and quantitative mercury-excited (4358 Å) Raman spectra from such ternary solutions yielded isosbestic frequencies (upon changing the $NaClO_4$ concentration at constant temperature and HDO concentration) in the OH and OD stretching regions from HDO that agree quantitatively with (temperature) isosbestic frequencies from the binary solutions.[1142] In addition, the component parameters obtained from studies of the ternary HDO solutions were found to be in excellent agreement with the corresponding quantities obtained from studies of the binary HDO solutions. Quantitative Raman spectra obtained with un-

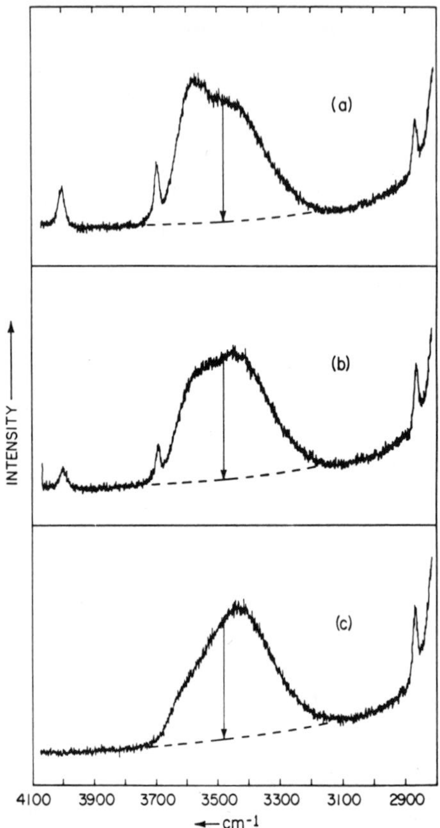

Fig. 18. Quantitative photoelectric tracings of mercury-excited (4358 Å) Raman spectra obtained at 25°C with the Cary model-81 spectrophotometer in the OH stretching region of HDO from a binary solution, 5.53*M* or 10 mole % H_2O in D_2O (c), and from ternary perchlorate solutions, 2.00*M* $NaClO_4$ (b), and 4.00*M* $NaClO_4$ (a), with 5.53*M* H_2O in D_2O. The sharp lines at 2875, 3700, and 4000 cm^{-1} are mercury lines that were not removed by the optical filter. The position of the isosbestic frequency is indicated by arrows at 3480 ± 5 cm^{-1}. A slit width of 15 cm^{-1} and a time constant of 3 sec were employed.

polarized 4358 Å mercury excitation from binary and ternary solutions having a stoichiometric concentration of 5.53 *M* or 10 mole % H_2O in D_2O are shown in the OH stretching region in Fig. 18.

The quantitative Raman spectra of Fig. 18 give evidence of a (concentration) isosbestic frequency at $3480 \pm 5\ cm^{-1}$ and of nonhydrogen-bonded (3620 ± 20 to $3570 \pm 5\ cm^{-1}$) and hydrogen-bonded (3435 ± 5 to $3450 \pm 20\ cm^{-1}$) components in the OH stretching contour from HDO for $NaClO_4$ concentrations of O–4 *M* at 25°C.[1142] The (concentration) isosbestic frequency of $3480 \pm 5\ cm^{-1}$ is in good agreement with the (temperature) isosbestic frequency of $3470 \pm 10\ cm^{-1}$ obtained recently from quantitative mercury-excited (4358 Å) Raman spectra from binary HDO–D_2O solutions.* The nonhydrogen-bonded and hydrogen-bonded OH stretching component frequencies are also in satisfactory agreement with values obtained from Figs. 10, 11, and 13.

The marked effects of $NaClO_4$ on the OH stretching contour from HDO clearly indicate that ClO_4^- is a strong structure-breaker.[1142] The effects of adding $NaClO_4$ are similar to the effects produced by increasing the temperature, of the corresponding binary solution, and the effect of adding $NaClO_4$ to produce a 1 *M* solution has been shown to be equivalent to the effect produced by a temperature rise of ~34°C.[1142] The results from additional studies of aqueous solutions of $NaClO_4$ will be discussed in a subsequent section.

4.2. Infrared Absorbance Spectra

Infrared absorbance spectra from 6.2 *M* or 11 mole % solutions of D_2O in H_2O were obtained in previous work for comparisons with Raman spectra.[1138] The infrared absorbance contours in the OD stretching region from HDO yielded some direct visual evidence for high-frequency asymmetry, but the high-frequency nonhydrogen-bonded component appears to be weak in the infrared spectrum compared to the Raman spectrum. Nevertheless, analog computer decompositions revealed broad Gaussian components centered near 2620 ± 10, 2520 ± 5, and $2413 \pm 5\ cm^{-1}$, in reasonable agreement with the corresponding three-Gaussian Raman component frequencies, 2650, 2520, and 2420 cm^{-1}.[1138]

Infrared absorbance spectra from dilute solutions have also been reported.[1191] The absorbance spectrum from 0.8 mole % D_2O in H_2O was carefully enlarged photographically (this work) and subjected to Gaussian

* T. H. Lilley and G. E. Walrafen, unpublished work.

analog decomposition. Three broad Gaussian components centered near 2630, 2500, and 2380 cm^{-1} were definitely required to fit the contour in acceptable agreement with the preceding Raman and infrared component frequency values. (The analog computer results are in contradiction to results claimed from an elaborate study of the infrared contour.[1191] However, the reported contour can readily be seen to be asymmetric with respect to the frequency of maximum absorbance, and the claimed absence of secondary absorbances[1191] is almost certainly false.) The infrared components assigned to nonhydrogen-bonded and hydrogen-bonded OD stretching vibrations thus persist over a wide range of concentrations, although the low-frequency hydrogen-bonded coupling component (2380–2413 cm^{-1}) is present at high dilutions and appears to be more intense in infrared than in Raman spectra.

Quantitative infrared absorbance spectra have also been obtained recently by Senior and Verrall[962] from a 12 mole % solution of D_2O in H_2O at temperatures from 29 to 87°C. (In the present context quantitative infrared absorbance spectra refer to accurate values of $\log_{10}(I_0/I)$ obtained *at constant film thickness* throughout a series of measurements, e.g., from 29 to 87°C.) The quantitative absorbance spectra are shown in Fig. 19. Unlike the Raman spectra, they do not provide clear evidence for asymmetry in the form of a distinct high-frequency shoulder. However, the infrared spectra

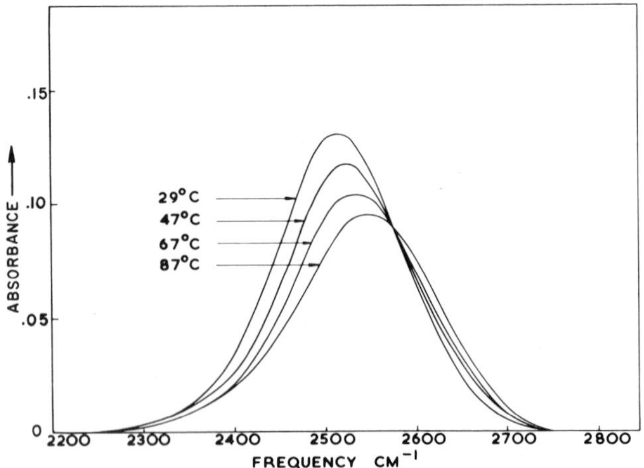

Fig. 19. Quantitative infrared absorbance spectra [$\log_{10}(I_0/I)$ at constant film thickness] obtained by Senior and Verrall[962] from a 12 mole % solution of D_2O in H_2O at temperatures from 29 to 87°C using the double-beam technique. An isosbestic frequency or frequency of constant absorbance is evident at 2575 cm^{-1}.

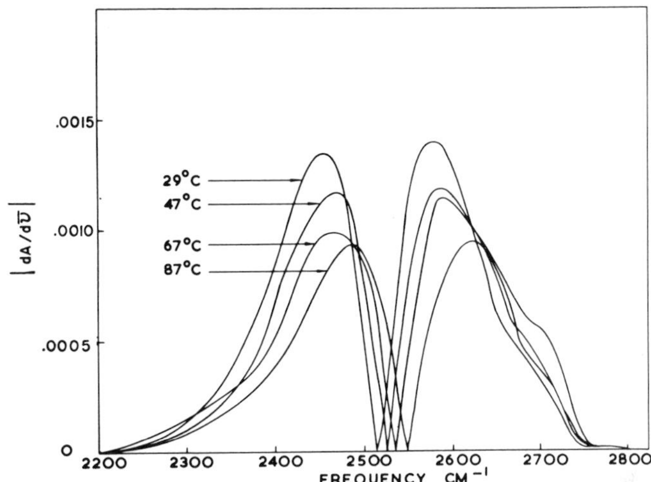

Fig. 20. Absolute values of the derivative of absorbance with respect to vibrational frequency $|dA/d\bar{\nu}|$ plotted versus $\bar{\nu}$ in cm^{-1} obtained from the infrared absorbance spectra of Fig. 19 by numerical differentiation.[962]

are not symmetric with respect to the frequencies of maximum absorbance. Low-frequency tails are present at all temperatures examined. Further, the frequencies of maximum absorbance shift from 2517 cm^{-1} at 29°C to 2553 cm^{-1} at 87°C, in good agreement with the Raman results of Fig. 17(a). The quantitative absorbance spectra also give evidence of a well-defined isosbestic frequency at 2575 cm^{-1}, in excellent agreement with the Raman value of 2570 ± 5 cm^{-1}.[1138] The presence of two (or more) broad infrared components is clearly indicated.

The absorbance spectra reported by Senior and Verrall were examined by various techniques, including Gaussian analysis by means of a digital computer.[962] However, for present purposes results obtained from numerical differentiations of the spectra[962] are of special interest. The results of the differentiations are shown in Fig. 20, in which absolute values of the derivative of absorbance with respect to vibrational frequency $|dA/d\bar{\nu}|$ are plotted versus the frequency in cm^{-1}.

The absolute derivative values shown in Fig. 20 are similar in many respects to the corresponding absolute derivative values shown in Fig. 14, although there are important differences in the high-frequency region. From the spectrum of Fig. 20 corresponding to 29°C it is evident that $|dA/d\bar{\nu}| = 0$ near 2200 and 2800 cm^{-1}, and at 2517 cm^{-1}, in reasonable agreement with the spectrum of Fig. 14 (25°C), for which $|dI/d\bar{\nu}| = 0$ near 2275 and 2775 cm^{-1}, and at 2530 $\pm$ 10 cm^{-1}. Further, from Fig. 20 (29°C), $|dA/d\bar{\nu}|$ is maximal at about 2450 and 2585 cm^{-1}, and inflections in $|dA/d\bar{\nu}|$ occur near 2650 and 2700 cm^{-1}, From Fig. 14 (25°C), $|dI/d\bar{\nu}|$ is maximal near 2475 and 2600 cm^{-1}, in reasonable agreement with Fig. 20, but a minimum and a maximum in $|dI/d\bar{\nu}|$ occur near 2645 $\pm$ 10 and 2700 cm^{-1}, respectively instead of inflections.

The presence of inflections in the infrared absorbance derivative spectra at the frequencies of pronounced minima and maxima in the Raman derivative spectra indicates that the nonhydrogen-bonded component may be weak in the infrared spectrum compared to the Raman spectrum. Further, the excellent agreements between the Raman and infrared frequencies corresponding to various features in the derivative spectra indicate that the two types of spectra possess the same component substructure, namely the same number of components, and the same component shapes, positions, and half-widths (but not relative component intensities), as indicated previously in analog computer decompositions of comparable Raman and infrared absorbance spectra using Gaussian functions.[1138]

The agreements between frequencies corresponding to various features of the infrared absorbance and Raman derivative spectra are entirely consistent with a model of water structure involving nonhydrogen-bonded and hydrogen-bonded OH (or OD) groups. However, the findings are hardly consistent with the continuum model. In the continuum model the infrared and Raman contours are considered to be composed of a very large number of sharp, overlapping components having parameters, and in particular intensities, that are individually variable.[301,1132] The OD and OH stretching contours from that model would not have the common features observed in the infrared and Raman derivative spectra that are clearly associated with broad, definite components.

Finally, some mention should be made of the high-temperature and high-pressure infrared spectra obtained by Franck and Roth[354] in the OD stretching region from HDO (see Chapter 13). Those workers reported extinction coefficients from an 8.5 mole % solution of HDO in H_2O at temperatures and pressures to 400°C and 3900 bars (density 0.9 g cm^{-3}), and they observed an absorption maximum near 2605 cm^{-1}. The infrared extinction contour reported at 400°C and 3900 bars, however, gives indications of marked low-frequency asymmetry, and Gaussian analog analyses (this work) suggest the presence of broad components centered near 2615 and 2525 cm^{-1}.

The frequencies of 2615 and 2525 cm^{-1} are close to the values expected for the nonhydrogen-bonded (2590–2610 cm^{-1} [1141]) and hydrogen-bonded

(2560 cm^{-1} from the three-Gaussian analysis of the 400°C spectrum of Fig. 16) coupling components. Further, the value of 2615 cm^{-1} is very close to the high-temperature frequency limit approached by the infrared absorption maxima,[354] and therefore the high-frequency component must relate to nonhydrogen-bonded OD groups. In addition, the frequency of 2525 cm^{-1} is too low to arise from nonhydrogen-bonded OD groups and must refer to hydrogen-bonded interactions. Also, the component at 2615 cm^{-1} is intense and relatively narrow, and the component at 2525 cm^{-1} is broad and weak, as expected for a temperature of 400°C, at which nonhydrogen-bonded HDO monomers predominate.[704,705,1139] More work is required, however, before assignments of the high-temperature and high-pressure infrared components can be made with complete certainty, but in any case the continuum interpretation originally favored by Franck and Roth[354] is clearly untenable, especially in view of the recent laser-Raman spectra provided by Lindner.[686]

4.3. Stimulated Raman Spectra

Stimulated Raman spectra from mixtures of H_2O and D_2O have been obtained recently by use of picosecond pulse excitation at 5300 Å.[199] In the OH stretching region from solutions of 50% H_2O–50% D_2O (by volume) and 60% H_2O–40% D_2O, mutually exclusive stimulation of nonhydrogen-bonded components (50% H_2O–50% D_2O) and of hydrogen-bonded components (60% H_2O–40% D_2O) was observed, and was produced predominantly by HDO. The observation of mutually exclusive stimulation according to the two classes of OH stretching components, and the close agreements evident from comparisons between the stimulated frequencies and the frequencies of the spontaneous Gaussian component centers, constitute unequivocal evidence for the reality of the broad, spontaneous components. The observation of stimulated Raman scattering from HDO, however, necessarily involves highly concentrated HDO solutions, because preferential stimulation of the intense stretching bands from H_2O and D_2O occurs in dilute HDO solutions. Figure 21 shows the stimulated (a, b), and spontaneous (c) Raman spectra from the concentrated HDO solutions.

The stimulated Raman spectrum shown in Fig. 21 (a) gives evidence of exclusive stimulation of nonhydrogen-bonded OH stretching components. Intensity maxima are evident at 3640 ± 20 and 3540 ± 20 cm^{-1}, in very good agreement with values of spontaneous nonhydrogen-bonded components from less concentrated solutions of HDO (Figs. 11 and 13, and Ref. 10), and evident from the spontaneous Raman components (c) from the 50 mole % solution of H_2O in D_2O.

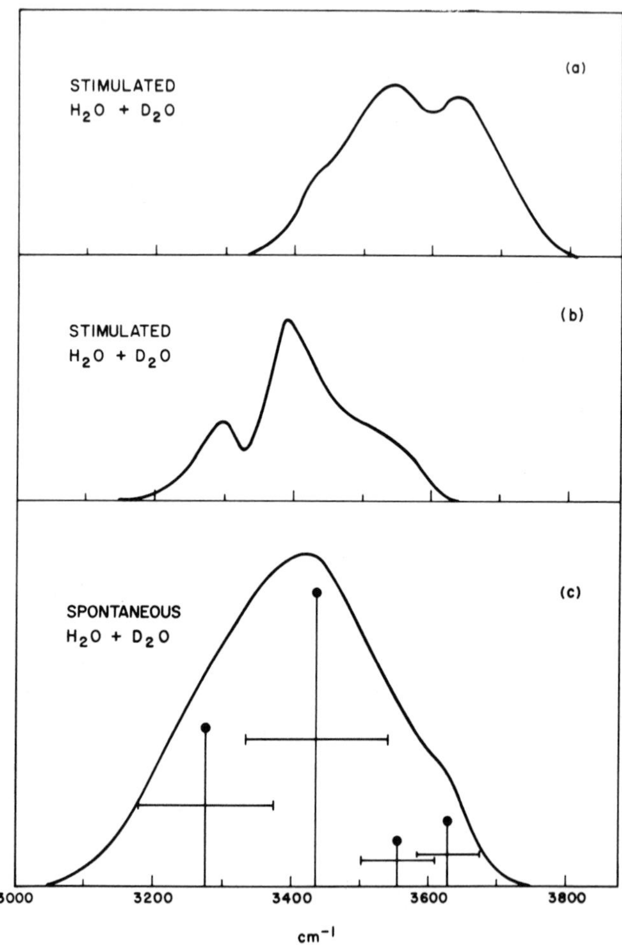

Fig. 21. Microdensitometer tracings corresponding to the stimulated Raman scattering from (a) 50% H_2O–50% D_2O (by volume) and (b) 60% H_2O–40% D_2O in the OH stretching region compared with (c) the spontaneous Raman spectrum from 50 mole % H_2O in D_2O, transferred to a horizontal baseline from Fig. 10. The spontaneous Raman spectrum was decomposed into Gaussian components by the analog technique and the component positions and half-widths are indicated in (c).

The stimulated Raman spectrum shown in Fig. 21 (b), on the other hand, gives evidence of exclusive stimulation of hydrogen-bonded OH stratching components when the stoichiometric D_2O concentration, and thus the HDO concentration, is lowered slightly. The intensity maxima evident from spectrum (b) at 3390 $\pm$ 20 and 3300 $\pm$ 20 cm^{-1} are also in acceptable agreement with values of spontaneous hydrogen-bonded components (Figs. 11 and 13, and Ref. 10) and with hydrogen-bonded components shown in spectrum (c), which adequately approximates the spontaneous Raman spectrum from 60% H_2O–40% D_2O.

The agreements between the frequencies of the stimulated intensity maxima and the frequencies of the spontaneous Gaussian component centers strongly support the analog computer decomposition and the use of Gaussian shapes. The agreements also provide unquestionable evidence for the reality of the broad, spontaneous components. The shapes of the stimulated contours, however, aside from 10% uncertainties in contour heights[(199)] and uncertainties of $\leq$ 20 cm^{-1} in frequency,[(199)] would not be expected to compare favorably with the envelopes of the two classes of corresponding spontaneous components. Stimulated Raman contours are sharpened by gain narrowing.[(94,970)] They are also broadened by frequency-sweeping when self-focusing occurs,[(199,885)] and self-focusing is possible with picosecond pulses if processes involving subpicosecond time scales are present.[(199,966)] In addition, modulation between stimulated Raman components can occur when the component separation is small.[(199)] Further, it should also be recognized that the photographic image is the sum of the exposures produced by the stimulated radiation throughout its temporal development. Thus, the primary information to be gained from stimulated Raman scattering that is relevant for present purposes involves the frequencies of highest stimulated intensity, and that information is in agreement with the spontaneous Raman component data.

4.4. Near-Infrared Spectra

The near-infrared spectral region that has been investigated intensively for solutions of HDO in H_2O and D_2O[(701,702,704,705,1186)] involves wavelengths from about 0.9 μm (11,111 cm^{-1}) to 2.1 μm (4762 cm^{-1}), and refers to various overtones and combinations of the fundamental vibrations ν(OH), ν(OD), and δ(HDO) (deformation) at roughly 2.9 μm, 4.0 μm, and 6.9 μm, respectively. The region from about 1.4 to 1.7 μm has been investigated by Worley and Klotz,[(1186)] who obtained near-infrared absorbance spectra from 6 *M* solutions of HDO in D_2O at temperatures from 7 to 61°C. The region from 0.9 to 2.1 μm has been investigated by Luck *et al.*[701,702,704,705] for various concentrations of HDO in H_2O or D_2O, over a wide range of temperatures and pressures.

The quantitative near-infrared absorbance spectra obtained by Worley and Klotz refer to the overtones of the nonhydrogen-bonded and hydrogen-bonded OH stretching components, 2ν(OH), from about 1.4 to 1.6 μm, and to combination vibrations, 1.6–1.7 μm and above. The region of 1.6 μm and above was not resolved, as expected, and only a gradual increase in wavelength was observed with decreasing temperature. The region from about 1.4 to 1.6 μm, however, was resolved into contributions from the nonhydrogen-bonded and hydrogen-bonded OH stretching overtone components, separated by an isosbestic wavelength at 1.468 μm, or 6812 cm^{-1}.

The isosbestic frequency value of 6812 cm^{-1}, when compared with twice the corresponding value of ~3473 cm^{-1} obtained in the fundamental infrared region by Hartman,[469] indicates a reasonable average anharmonicity for the 2ν(OH) region of 134 cm^{-1}, or about 2%, and supports the assignment of the region to 2ν(OH). Similarly, the nonhydrogen-bonded peak at 1.416 μm, below the isosbestic wavelength of 1.468 μm, and the hydrogen-bonded peaks above, at about 1.525 and 1.556 μm, yield anharmonicities of about 2.5%, 4.4%, and 0.8%, respectively, when compared with nonhydrogen-bonded (3620 cm^{-1}), hydrogen-bonded (3430 cm^{-1}), and hydrogen-bonded coupling (3240 cm^{-1}) fundamental OH stretching components obtained from Gaussian analog computer analyses of Raman contours.[1141] The detailed assignments are thus also supported by reasonable anharmonicities.

The near-infrared region from 0.9 to 2.1 μm has been investigated exhaustively by Luck *et al.*[701,702,704,705] In the work of Luck and Ditter,[705] near-infrared extinction coefficients from a solution of 50 ml H_2O/liter D_2O are shown from 0 to 340°C in the region from 1.3 to 1.8 μm, and the data from 10 to 60°C were found to be in excellent agreement with the data of Worley and Klotz. An isosbestic wavelength was obtained near 1.472 μm, in excellent agreement with the value of 1.468 μm,[1186] but from 136 to 340°C a sizeable shift occurred that yielded an isosbestic region from 1.472 to 1.425 μm for the complete temperature range of 0–340°C.

The isosbestic region from 1.472 to 1.425 μm corresponds to a shift of 224 cm^{-1}, and it indicates that a series of partially hydrogen-bonded components gives rise to the broad contour centered near 1.525 μm. The position of the nonhydrogen-bonded component, however, varies only from about 1.41 μm (or less) to 1.40 μm between 0 and 340°C, or by about 50 cm^{-1} or less, and therefore is too nearly constant to explain the observed isosbestic region. Thus, the shift in the isosbestic wavelength must be produced predominantly by the broad hydrogen-bonded OH stretching overtone region.

A shift from 2570 to roughly 2635 cm^{-1} was inferred previously from analysis of the Raman data obtained by Lindner,[686] and such a shift would give rise to an OD overtone shift of $\sim$130 cm^{-1}, neglecting anharmonicity. Further, in the case of the OH stretching vibrations, a shift of $130\sqrt{2}$ or 185 cm^{-1} might be expected, in fair agreement with the shift of 224 cm^{-1} observed by Luck and Ditter. Again, the isosbestic region was attributed to partially hydrogen-bonded structures.

5. INTRAMOLECULAR VIBRATIONAL SPECTRA FROM H_2O AND D_2O

5.1. Spontaneous and Stimulated Raman Spectra

Photoelectric Raman spectra from liquid H_2O and D_2O give obvious visual evidence for the presence of at least three broad components in the OH and OD stretching regions.[1133,1134,1137,1142] For liquid H_2O an intensity maximum occurs at 3410 $\pm$ 10 cm^{-1}, an intense shoulder occurs near 3225 $\pm$ 20 cm^{-1}, and a weak shoulder is evident near 3625 $\pm$ 20 cm^{-1}; for liquid D_2O the corresponding features occur near 2510 $\pm$ 10 cm^{-1}, 2390 $\pm$ 20 cm^{-1}, and 2675 $\pm$ 20 cm^{-1}.[1142] Four major Gaussian components at 3250, 3425, 3530, and 3625 cm^{-1}, however, were required to fit the OH stretching region from H_2O obtained in the early mercury-excited Raman studies,[1137] and a low-frequency residual also resulted. Further, a definite broad, weak component near 3060 cm^{-1}, corresponding approximately to the residual in the mercury-excited work, in addition to four major Gaussian components at 3225, 3415, 3565, and 3640 cm^{-1} have now been obtained from analog computer analysis of argon-ion-laser excited (4880 Å) Raman spectra from H_2O, and computer analysis of laser-Raman spectra in the OD stretching region from liquid D_2O has also yielded four Gaussian components near 2380, 2500, 2590, and 2660 cm^{-1} (a significant low-frequency residual or fifth component was not observed for D_2O.)[1142] Thus, four (or five) Gaussian components give rise to the broad OH and OD stretching contours from liquid H_2O and D_2O, although only three features are visually apparent from the spectra.

The integrated intensities of the four major Gaussian components from liquid H_2O were examined in the mercury-excited Raman work for temperatures from 10.0 to 90.0°C.[1137] Intensities of the two major high-frequency components occurring above 3460 cm^{-1} (the isosbestic frequency) were found to increase with temperature rise, whereas intensities of the two strong major components having frequencies below 3460 cm^{-1} were found to decrease. The pairs of high- and low-frequency components, accordingly, were assigned to nonhydrogen-bonded and to hydrogen-bonded interactions, respectively.

The four-component substructure of the OH and OD stretching contours from H_2O and D_2O (and from HDO) is thought to arise from the presence of two major classes of interactions, nonhydrogen-bonded and hydrogen-bonded, and from the fact that vibrational coupling between molecules is involved in each of the major classes. A coupling somewhat analogous to that present in water has been studied for aqueous solutions of selenious acid.(1136)

In the case of H_2SeO_3 in water, the Se=O vibrations of the H_2SeO_3 molecules appear to be coupled through hydrogen bonds, and the couplings give rise to a weak, additional high-frequency stretching component (shoulder) that is highly depolarized. The weak component is thought to refer to the out-of-phase coupling, and the intense, principal Se=O component refers to the in-phase coupling. The situation in water is similar, but the phase relationships are not developed, and the weak additional components occur at low rather than at high frequencies. Two nonhydrogen-bonded and two hydrogen-bonded components arise from the coupling interactions, and the low-frequency component of each class is generally the weaker.

For H_2O and D_2O (and HDO), the components centered near 3240–3260 and 3530–3560 cm^{-1} for the OH stretching region, and near 2370–2415 and 2590–2615 cm^{-1} for the OD stretching region, have been designated as the coupling components.(1141) This designation, however, is somewhat oversimplified, for convenience. The coupling components from pure H_2O and D_2O are intense relative to those corresponding to the maximum HDO concentration attainable, and the coupling component intensities appear to decrease with the HDO concentration.(1141) Nevertheless, each member of each of the two major classes is probably influenced by coupling interactions at high concentrations of H_2O, D_2O, or HDO.

Evidence for the four-component substructure has been greatly strengthened by the results of two recent investigations.(199,1142) One investigation involved Raman spectra from aqueous solutions of $NaClO_4$,(1142) and the other involved stimulated Raman spectra from solutions of H_2O and of $NaClO_4$ in H_2O.(199) Important results from the two investigations are shown in Fig. 22.

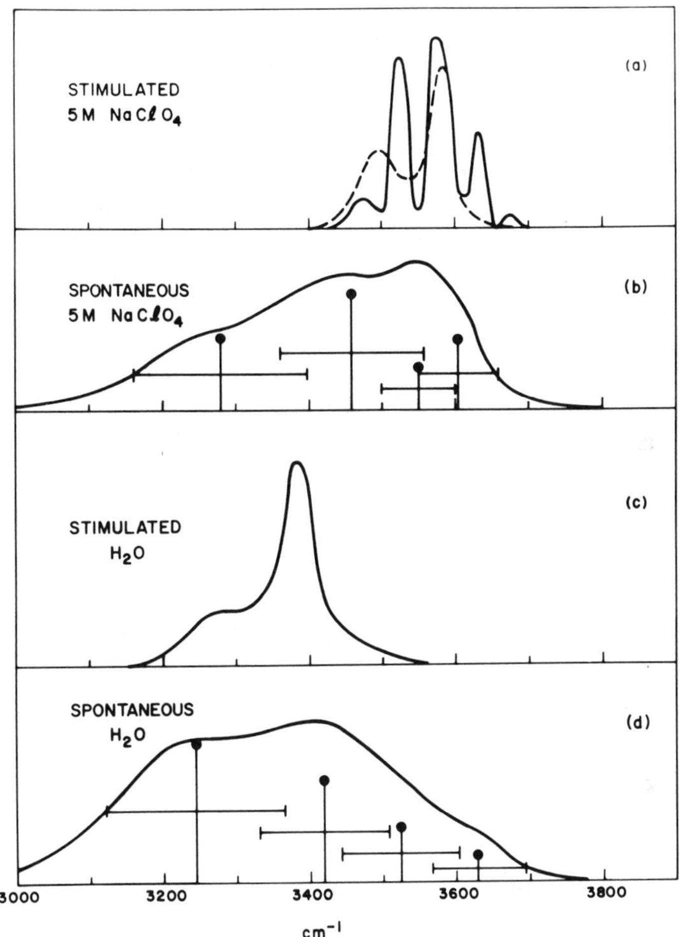

Fig. 22. Spontaneous and stimulated Raman spectra from pure H_2O (c, d) and from 5*M* $NaClO_4$ in H_2O, (a, b). The spontaneous spectra (b, d) were obtained photoelectrically with argon-ion-laser (4880 Å) excitation using a Jarrell-Ash double monochromator and the fixed-beam technique.[1142] The stimulated spectra (a, c) were obtained photographically by use of picosecond pulse excitation at 5300 Å; the results shown refer to microdensitometer tracings. Spectra (b) and (d) were obtained by transferring the original photoelectric tracings to horizontal baselines. The half-widths and peak heights of Gaussian components are indicated by horizontal and vertical bars.

In Fig. 22 (d) and (b) spontaneous Raman spectra from pure H_2O and from 5 *M* $NaClO_4$ in H_2O, respectively, obtained photoelectrically with argon-ion-laser (4880 Å) excitation are shown. The corresponding stimulated Raman spectra obtained photographically with picosecond pulse excitation (5300 Å) are included in parts (c) and (a) for comparison. The spontaneous Raman contours were decomposed into Gaussian components and the positions and half-widths are also shown.

In Fig. 22(d) the four major spontaneous Gaussian OH stretching components from pure H_2O are indicated near 3245, 3420, 3520, and 3620 cm^{-1}. (The weak fifth component centered near 3060 cm^{-1} is not indicated.) Comparison of (d) with the corresponding stimulated Raman spectrum (c) indicates that the hydrogen-bonded components were stimulated exclusively. The stimulated intensity maxima occur near 3270 ± 20 and 3380 ± 20 cm^{-1}, in acceptable agreement with the centers of the hydrogen-bonded spontaneous Gaussian components, and with the values from Fig. 21(c) of 3300 ± 20 and 3390 ± 20 cm^{-1}, which refer to exclusive stimulation of hydrogen-bonded OH stretching components from HDO.

In Fig. 22(b) four major spontaneous Gaussian OH stretching components from 5 *M* $NaClO_4$ in H_2O are indicated near 3280, 3460, 3550, and 3605 cm^{-1}. The values are not greatly different from those indicated for pure H_2O in (d), and they compare favorably with intensity maxima and shoulders evident in (b) at 3455 ± 5 and 3545 ± 5 cm^{-1}, and at 3220 ± 20 and 3615 ± 20 cm^{-1}, respectively. In addition, in agreement with the pronounced structure-breaking effect of ClO_4^- (cf. Fig. 18), the combined intensities of the two high-frequency components compared to the combined intensities of the two low-frequency components from (b) can be seen to have increased relative to the corresponding quantities from (d).

Comparison between the stimulated (a) and spontaneous (b) Raman spectra from 5 *M* $NaClO_4$ in H_2O indicates that stimulation has occurred exclusively in the nonhydrogen-bonded OH stretching region. In both of the (a) spectra (solid and dashed lines) the principal stimulated intensity occurs between 3500 and 3600 cm^{-1}, or in the nonhydrogen-bonded region. Further, little or no intensity occurs below 3460 cm^{-1}, in the hydrogen-bonded region. Evidence for modulation or beating between the stimulated nonhydrogen-bonded components is clearly present, however, and requires examination.

Two stimulated Raman spectra were shown in Fig. 22(a) to give an indication of the variations that resulted in the spectra from the $NaClO_4$ solutions. In one of the (a) spectra (shown by the solid line), the spacings between the five peaks shown is nearly constant and equals ~ 53 cm^{-1}.

In the other spectrum, shown by the dashed line in (a), only two peaks are evident and the separation is $\sim$90 cm^{-1}. The spacing between the nonhydrogen-bonded spontaneous components from (b) at 3605 and 3550 cm^{-1} is 55 cm^{-1}, in excellent agreement with the stimulated beat spacing of $\sim$53 cm^{-1}. The temporal development of the stimulated scattering from the (dashed) spectrum, however, was insufficient to produce well-defined modulation. The highest stimulated intensity, nevertheless, was observed at $\sim$3590 cm^{-1}, or near the spontaneous Raman component at 3605 cm^{-1}.

Finally, it should be noted that the combined results of Figs. 21 and 22 provide evidence for the principal four-component substructure of the OH stretching contours of H_2O and HDO, and for each case exclusive stimulation of pairs of components corresponding to the nonhydrogen-bonded or hydrogen-bonded classes was observed.

5.2. Infrared Spectra

The component substructure evident from spontaneous and stimulated Raman contours corresponding to the OH and OD stretching regions from liquid H_2O and D_2O can also be demonstrated for infrared spectra,[(1137)] but accurate absorbance data are required, and detailed spectral features are difficult to observe. Accurate absorption coefficient data are even more difficult to obtain by absorption methods because the OH and OD stretching coefficients are very large and films of very small and uniform thicknesses are involved, and the thicknesses must be measured accurately. However, problems arising in absorption work with thin films may be circumvented by use of the attenuated total reflectance (ATR) technique, and the ATR method has recently been employed by Crawford and Frech[(218)] to obtain accurate absorption coefficients in the OH stretching region from water. The ATR results are shown and compared with absorbance spectra in Fig. 23.

In Fig. 23(b) the quantity $\alpha\lambda_0/4\pi$ obtained by Crawford and Frech[(218)] is plotted versus frequency in cm^{-1}, and semiquantitative values of $\log_{10}(I_0/I)$ or absorbance, obtained in this work, are shown in (a) for comparison. The two quantities $\alpha\lambda_0/4\pi$ and $\log_{10}(I_0/I)$ are not, of course, strictly comparable because the absorption coefficient α is multiplied by $\lambda_0/4\pi$ in (b), but the differences in shape are not great when the range of λ_0 is relatively small.

Comparisons between the spectra shown in Fig. 23(a) and (b) indicate common features that suggest the presence of at least four components. In (a) the absorbance is maximal at about 3425 cm^{-1}, and $\alpha\lambda_0/4\pi$ is maximal in (b) near 3390 cm^{-1}. Further, low-frequency shoulders are evident in (a)

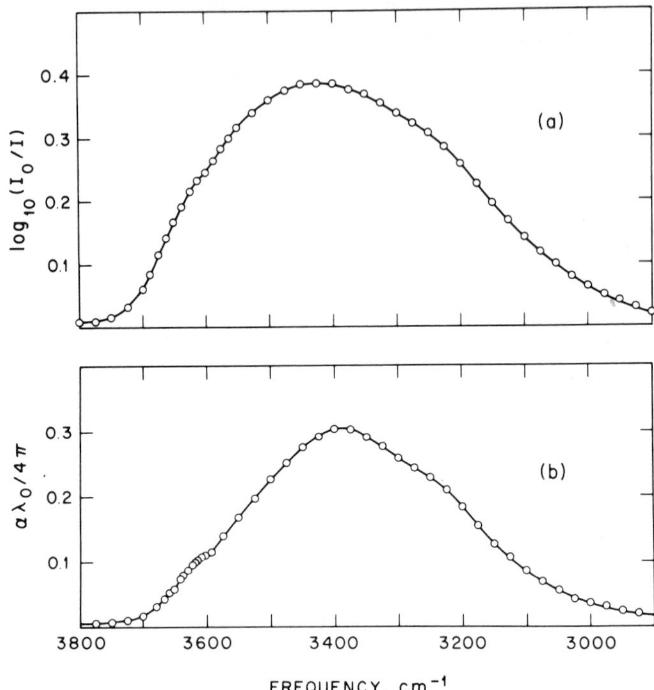

Fig. 23. Comparisons between (a) semiquantitative infrared absorbance data, $\log_{10}(I_0/I)$ obtained in this work, and (b) the quantity $\alpha\lambda_0/4\pi$ obtained by Crawford and Frech.[218]

and (b) near 3225 ± 25 cm^{-1}, and high-frequency inflections are also evident near 3600 cm^{-1} in both spectra. In addition, the region of (b) near 3300 cm^{-1} is concave upward, whereas the corresponding high-frequency region near 3500 cm^{-1} is concave downward, and a similar situation is evident from (a), but the downward concavity near 3500 cm^{-1} is greater. Thus, the (a) and (b) comparisons suggest that broad components near 3225, 3390–3425, ~3500, and ~3600 cm^{-1} are present.

The (a) and (b) contours of Fig. 23 were decomposed into Gaussian components by means of the analog computer technique. For (a) broad components centered near 3050, 3215, 3420, 3555, and 3645 cm^{-1} resulted. For (b) broad components at 3090, 3220, 3395, 3540, and 3625 cm^{-1} were obtained. The half-widths corresponding to the nonhydrogen-bonded OH stretching components from (a) and (b), 3625–3645 cm^{-1} and 3540–3555 cm^{-1}, were 90 cm^{-1} and 135–150 cm^{-1}, respectively; the half-widths corresponding to the hydrogen-bonded OH stretching components, 3395–3420

cm^{-1} and 3215–3220 cm^{-1}, were 220–300 cm^{-1} and 195–215 cm^{-1}, respectively. The half-width range for the fifth component, 3050–3090 cm^{-1}, was 190–260 cm^{-1}.

The frequencies and half-widths of the five Gaussian infrared components are in reasonable agreement with the corresponding five Gaussian quantities from argon-ion-laser Raman spectra (cf. Ref. 1142). Further, the fifth component near 3050–3090 cm^{-1}, although weak and broad, is somewhat stronger in the infrared than in the Raman spectrum.

The fifth component may arise from Fermi resonance between the overtone of the intramolecular bending vibration at 1650 cm^{-1}[(1133,1134)] and the coupling component near 3215–3220 cm^{-1}, and in regard to this it should be noted that a similar component is apparently absent in the OH stretching region of HDO, for which Fermi resonance is not expected.[(1141)]

6. RELATION OF COMPONENT PROPERTIES TO WATER STRUCTURE

In recent laser-Raman work involving ternary solutions of $NaClO_4$, HDO, and H_2O or D_2O,[(1142)] the hydrogen-bonded OD and OH stretching components from HDO were partially isolated at $NaClO_4$ concentrations of 4 M, and the shapes of the components could be examined directly at frequencies near, and considerably above, the component centers. The examinations indicated that the component shapes are Gaussian even in the high-frequency wings, and they provided strong additional evidence for the use of Gaussian component shapes in analog computer decompositions.

The Gaussian shapes of the nonhydrogen-bonded OH and OD stretching components, as well as the large half-widths, can be explained in terms of random distortions involving changes in O—O distances and O—H···O angles in the case of the hydrogen-bonded components, or random dipole–dipole interactions, for example, in the case of the nonhydrogen-bonded interactions. The various distortions affect both of the major classes of interaction, but they give rise to hydrogen-bonded component half-widths that are characteristically greater than those of the nonhydrogen-bonded components.[(1142)] For the fully hydrogen-bonded C_{2V} structures the distortions may be regarded as random deviations from the nearest-neighbor tetrahedral O—O distances and O—H···O angles. For temperatures near 25°C, at which the 4- and 3- bonded structures predominate, the liquid may be regarded as a hydrogen-bonded network that is locally tetrahedral on the average, but lacks long-range order, and is composed of polygons of varying numbers of sides and conformations, with significant bridging, free chain ends (nonhydrogen-bonded O—H···O or O—D···O units or 3-bonded structures), and innumerable random polyhedra (cf. Ref. 1027).

The overlapping of the broad components is a consequence of the large half-widths and thus of the distortions. The presence of frequency regions common to the two major classes of components, however, does not preclude the division according to those classes. The energy of interaction between H_2O molecules is a function of the O—O separation, of two polar angles, and of three Euler angles.* Further, the nonhydrogen-bonded and hydrogen-bonded classes of interaction may be regarded as separate volume elements in the six-dimensional space of distances and angles. In the frequency region common to components of the two major classes, the projection of the volume elements on the frequency coordinate would, when examined only as a function of that coordinate, not allow the projected contributions to be distinguished separately. Thus, the regions of common frequency in spontaneous Raman scattering or in infrared absorption would refer to the combined contributions of the two volume elements in the six-dimensional space, but when the spectral contribution from one of the volume elements is absent, e.g., as in the case of mutually exclusive stimulated Raman scattering, contributions from the remaining class remain.

With regard to component breadth and overlap, the continuum model of water should be mentioned. In early Raman work involving HDO, the OH and OD contours were (incorrectly) observed to be very nearly symmetric, and the breadth and near-symmetry of the contours were thought to provide unequivocal evidence for a range of spectroscopically inseparable interactions.[1132] However, the detailed features observed in more recent and accurate laser-Raman spectra indicate that the breadth of the *contours* was overemphasized. Further, the O—O separation was related directly to the vibrational frequency in the continuum interpretation,[1132] but that relationship is clearly untenable in terms of the six-dimensional space required to define the interaction energy. In addition, the contours according to the continuum model were considered to be envelopes of a large number of very narrow overlapping components capable of individual variation, but such variations are clearly not independent of or different from the variations observed experimentally according to the nonhydrogen-bonded and hydrogen-bonded classes. Further, the recent observation of mutually

* F. H. Stillinger, private discussion.

exclusive, stimulated Raman scattering is in excellent agreement with a mixture model of water, i.e., of a model in which distinct classes of interactions are recognized, but it is in complete disagreement with the continuum model. Accordingly, the continuum model in its traditional sense should be abandoned, and attempts to use it in the face of evidence to the contrary cannot be considered to be well-founded.[301]

7. THERMODYNAMIC TESTS OF THE CONSECUTIVE HYDROGEN-BOND DISRUPTION MODEL

In previous Raman work[1138] involving a 6.2 *M* solution of H_2O and D_2O at temperatures from 16 to 97°C, ratios of integrated component intensities I_{2645}/I_{2525} obtained from a two-Gaussian decomposition of the OD stretching contours were found to yield a van't Hoff $\Delta H°$ value of 2.5 ± 0.6 kcal mol^{-1} for the disruption of O—D···O units. The ratios of integrated intensities corresponding to the nonhydrogen-bonded (I_{2645}) and hydrogen-bonded (I_{2525}) components are shown on a logarithmic plot versus $10^3/T$ in Fig. 24. The plot is linear, within the precision of the analog data, and a least squares treatment yielded the equation $\log_{10}(I_{2645}/I_{2525}) = 0.8593_3 - 546.8_1/T$, shown by the solid line in the figure.

The results of Gaussian analog computer decompositions of the high-temperature and high-pressure laser-Raman spectra obtained by Lindner[686] from a 6.2 *M* solution of D_2O in H_2O were discussed in a preceding section. The OD stretching contours obtained in the temperature range from 20 to 400°C were decomposed into three Gaussian components, and the fre-

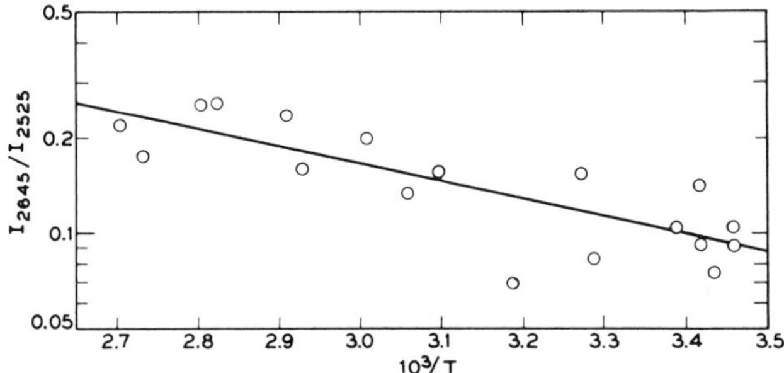

Fig. 24. The ratio of integrated component intensities I_{2645}/I_{2525} obtained from a two-Gaussian decomposition of the OD stretching contours from 6.2 *M* D_2O in H_2O in the range 16–97°C plotted on a logarithmic scale (the logarithms are not plotted) versus $10^3/T$.[1138]

quencies of the two major components were plotted as functions of temperature [cf. Fig. 17(b)]. The three-Gaussian analyses also yielded integrated component intensities of use in comparisons with the results shown in Fig. 24. However, the data of Fig. 24 involved two-Gaussian decompositions, and thus two-Gaussian decompositions of the high-temperature, high-pressure contours were accomplished as well. The ratios of the principal nonhydrogen-bonded and hydrogen-bonded component intensities were found, fortunately, to compare favorably from either two- or three-Gaussian analyses. (Residuals were involved, of course, in the two-Gaussian case.) Accordingly, the three-Gaussian intensity data corresponding to the frequency data of Fig. 17(b) were employed.

In Fig. 25 the least squares equation corresponding to the Fig. 24 data is represented by a solid line for temperatures from 16 to 97°C, and by a dashed line for the extrapolation to 400°C and above. The ratios of integrated intensities of the major nonhydrogen-bonded (I_U) and hydrogen-bonded (I_B) components (U refers to unbonded, and B to bonded OD units) obtained from the Lindner data are also shown. Within the error limits indicated, the agreement between the least squares result from Fig. 24 and the high-temperature, high-pressure analog data is highly satisfactory. The slopes corresponding to the two results can be seen to be nearly equal, and the apparent slight upward displacement evident in the high-temperature, high-pressure results is probably not significant.

The analog results from Lindner's spectra were also treated by least squares. The least squares equation which resulted is $\log_{10}(I_U/I_B) = 0.8608_5 + 513.1_1/T$. The agreement with the least squares equation from the Fig. 24 data is good. Further, the slope ($513.1_1 \times 4.576$) yields a ΔH of 2.4 kcal mol^{-1} ODO for the 20–400°C range, in good agreement with the previous ΔH° value of 2.5 ± 0.6 kcal mol^{-1} ODO. The agreement indicates that a ΔH value of 2.4–2.5 kcal mol^{-1} characterizes the processes of consecutive hydrogen-bond disruption that give rise (between 16 and 400°C, and from 1 bar to 3900 bars and above) to the partially and completely nonhydrogen-bonded species.

In previous quantitative Raman measurements the temperature dependences of the intermolecular hydrogen bond stretching and librational intensities were determined.[1135,1137] As indicated in an early section, the intermolecular Raman bands arise almost exclusively from the fully hydrogen-bonded or 4-bonded C_{2V} structures, and thus 4-bonded concentrations resulted from the intensity measurements.[1139] The temperature dependences of the intermolecular Raman intensities also yielded a ΔS° value of 8.5 cal(deg mol)$^{-1}$ OHO and a ΔH° value of 2.5_5 kcal mol^{-1} OHO,[1135,1139]

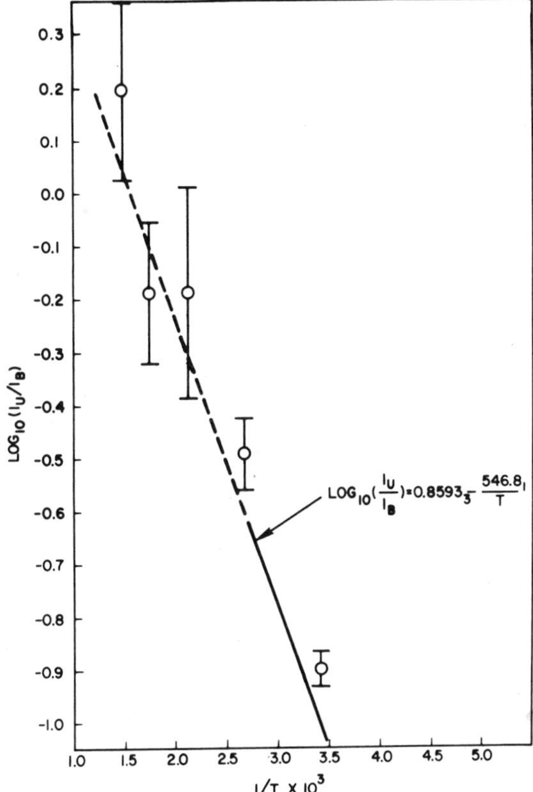

Fig. 25. Plot of $\log_{10}(I_U/I_B)$ versus $1/T \times 10^3$ from the high-temperature, high-pressure laser-Raman data of Lindner.[686] I_U and I_B refer to the integrated Gaussian component intensities of the nonhydrogen-bonded and hydrogen-bonded OD stretching components from a 6.2*M* solution of D_2O in H_2O. The solid line refers to the least squares result from Fig. 24, and it is extrapolated (dashed line) to above 400°C; the least squares equation shown *does not* come from the data of this figure. The error bars refer to several analog computer trials involving three-Gaussian fits of the contours [cf. Fig. 17(b)].

in good agreement with the ΔH° and ΔH values from the plots of Figs. 24 and 25. The 4-bonded concentrations were then employed in conjunction with the ΔH° value of 2.5_5 kcal mol^{-1} OHO, and the concept of stepwise or consecutive hydrogen-bond disruption was used to calculate values of the heat capacity $C^\circ_{p(l)}$.[1139]

For temperatures below 25°C, at which 4-bonded and 3-bonded species predominate, the (standard, 1 atm) heat capacity $C^\circ_{p(l)}$ was calculated from

$$C^\circ_{p(l)} = C^\circ_{p(3)} + [f_4(1 - f_4)/R](\Delta H^\circ/T)^2 \quad (1)$$

where f_4 refers to the mole fraction of H_2O molecules involved in 4-bonded interactions and $C^\circ_{p(3)}$ is the standard heat capacity of the 3-bonded species, taken as 9.00 cal(deg mol)$^{-1}$.[1139] A value of 18.2 cal(deg mol)$^{-1}$ resulted from eq. (1) for 0 and 25°C, in excellent agreement with the corresponding experimental values of 18.15 and 17.99 cal (deg mol)$^{-1}$ for $C^\circ_{p(l)}$.

Above 25°C the values of $C^\circ_{p(l)}$ calculated from eq. (1) were found to be lower than the experimental values. The differences were considered to relate to the presence of 2-bonded species, not treated by eq. (1).[1139] Accordingly, eq. (1) was transformed to

$$C^\circ_{p(l)} = C^\circ_{p(3,2)} + \{[f_4'(1 - f_4')/R] + [f_3'(1 - f_3')/R]\}(\Delta H^\circ/T)^2 \quad (2)$$

where $f_4' = f_4/(f_4 + f_3)$ and $f_3' = f_3/(f_3 + f_2)$. The value of $C^\circ_{p(3,2)}$ was taken as 9.00 cal (deg mol)$^{-1}$, as before for $C^\circ_{p(3)}$, and the mole fraction f_4 was obtained directly from intermolecular Raman intensity data. The mole fractions f_3 and f_2, where $1 = f_4 = f_3 + f_2$, were then determined by iteration using eq. (2) and the experimental heat capacity values. The experimental and calculated heat capacity values are shown in Fig. 26, and the values of f_3 and f_2 that resulted are listed for temperatures of 0–100°C with the f_4 values obtained from intermolecular Raman intensities. The mole fractions f_4, f_3, and f_2 are also plotted versus temperature in Fig. 27.

The mole fraction f_4 shown in Fig. 27 is proportional to the intermolecular hydrogen bond stretching intensity, and below 25°C, $1 - f_4 = f_3$. In addition, the assumption that $C^\circ_{p(3)} = 9.00$ cal. (deg mol)$^{-1}$ at 0 and 25°C was confirmed by the agreement between the calculated and experimental heat capacities. Above 25°C, however, the agreement between calculated and experimental heat capacities depends upon the assumption that

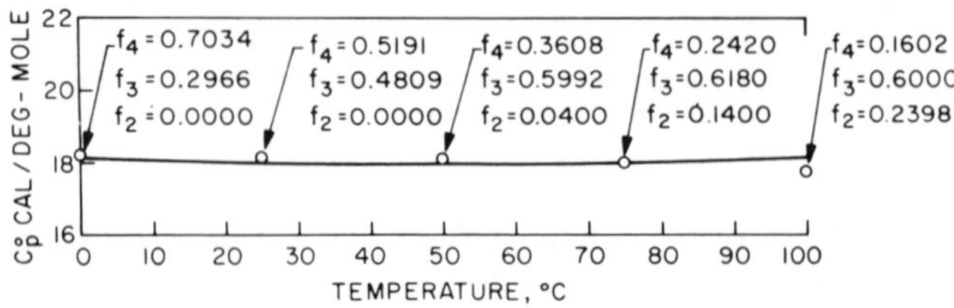

Fig. 26. Comparisons between (standard, 1 atm) heat capacities C_p° and heat capacities calculated from a ΔH° of 2.5_5 kcal mol^{-1} OHO. The 4-, 3-, and 2-bonded mole fractions are indicated for increments of 25°C.

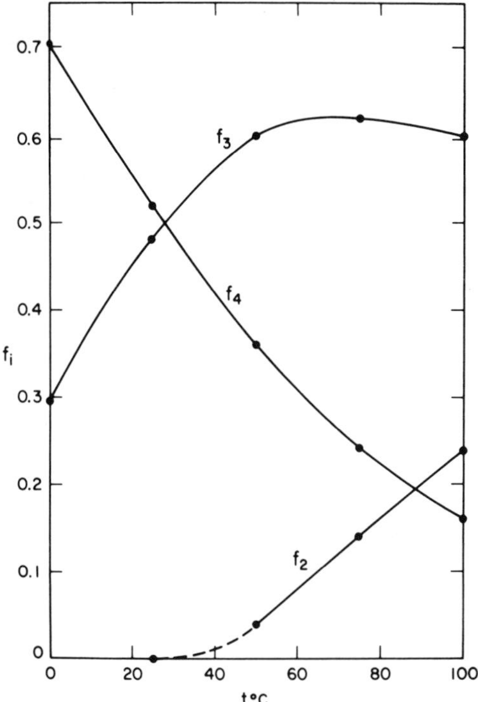

Fig. 27. Mole fractions of H_2O molecules involved in 4-, 3-, and 2-bonded interactions obtained from intermolecular Raman intensities, f_4, and calculated from the heat capacity, f_3, and f_2 (cf. Fig. 26).

$C^{\circ}_{p(3,2)} = 9.00$ cal (deg mol)$^{-1}$, and upon the validity of eq. (2) and the associated relationships between the f and f' values. Thus, an independent test of the mole fraction values seemed desirable, although the variations shown in Fig. 27 are reasonable.

The percentage of hydrogen bonds broken in water has been determined at a series of temperatures from near-infrared measurements by Luck.[701,702,704] Several near-infrared bands were examined and the resulting range of percentage values is shown in Fig. 28 (shaded area) for temperatures from 0 to 100°C. For comparison, the percentage of broken hydrogen-bonds, $\%B$, was calculated from values of f_4, f_3, and f_2 by means of the relation

$$\%B = 10^2 \sum_{i=0}^{4} (4 - i)f_i/4,$$

and calculated values of $\%B$ are indicated in Fig. 28.

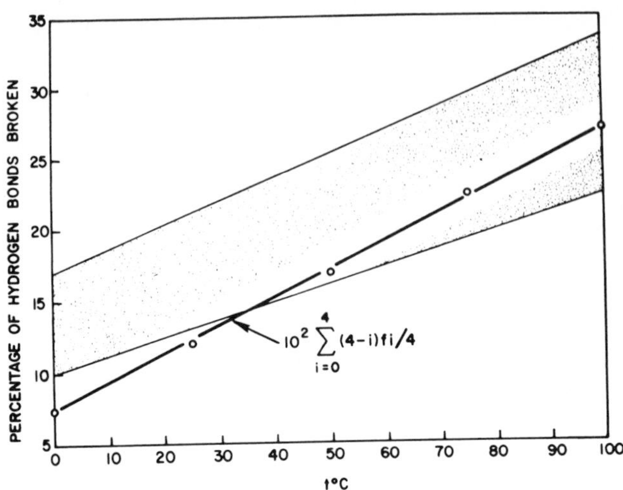

Fig. 28. Comparisons between percentages of hydrogen bonds broken in water obtained by Luck[701,702,704] (shaded area) and calculated from the 4-, 3-, and 2-bonded mole fractions f_4, f_3, and f_2 by use of the relation $\%B = 10^2 \sum_{i=0}^{4} (4 - i)f_i/4$ (cf. Fig. 27 for f values).

The agreements between the calculated values of $\%B$ and the values provided by Luck are very satisfactory in view of the experimental errors involved in obtaining the f_4 values, and the various assumptions employed. The agreements also strongly support the concept of consecutive hydrogen bond breakage, and the procedure employed in relating the Raman intensity data to the experimentally observed heat capacities. The agreements indicate further that the mole fraction values are of sufficient accuracy to be employed with reasonable confidence in other calculations, e.g., in calculations of the dielectric constant of water (see also Chapter 7).

The dielectric constant of water ε_{H_2O} was calculated in the temperature range 0–100°C from values of the mole fractions f_4, f_3, and f_2, and from the approximate Kirkwood relationship[301]

$$\varepsilon = 2\pi N^* m^2 g/kT \tag{3}$$

where N^* is the number of molecules per unit volume, m is the dipole moment, g is the correlation parameter, k is the Boltzmann constant, and T is the absolute temperature. In terms of f_4, f_3, and f_2, the dipole moment and the correlation parameter required for eq. (3) were obtained from

$$m = \sum_{i=2}^{4} m_i f_i \qquad \text{and} \qquad g = \sum_{i=2}^{4} g_i f_i$$

Further, two methods were employed to obtain the g_i and m_i values. In the first method, linear interpolations were employed between g_0 and g_4 and between m_0 and m_4 to obtain g_2 and g_3 and m_2 and m_3, respectively. A value of 2.81 was used for g_4,[443] and various values of g_0 were chosen from 1 to 1.10. Further, m_4 was taken as 2.45 D,[443] and several values of m_0 were tried, from 1.88 to 2.07 D. Values of ε_{H_2O} were then calculated from eq. (3) for a series of temperatures by means of a digital computer employing many estimated g_0 and m_0 values. The results for favorable g_0 and m_0 values (1.10 and 2.07 D) are shown in terms of $\log_{10}\varepsilon_{H_2O}$ in Fig. 29. Experimental values of $\log_{10}\varepsilon_{H_2O}$ obtained in the work of Kay *et al.*[581] are shown for comparison.

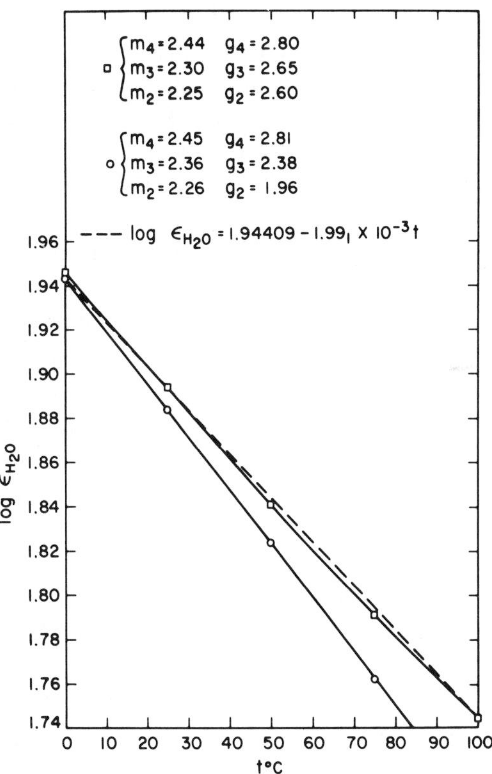

Fig. 29. Comparisons of the logarithm of the dielectric constant of water $\log_{10}\varepsilon_{H_2O}$ in the range 0–100°C from the data of Kay *et al.*[581] (broken curve) with values obtained from the mole fractions f_4, f_3, and f_2 shown in Figs. 26 and 27, and from values employed in two methods of calculation (solid curves).

From Fig. 29 the first method for determining values of g_i and m_i is seen to provide agreement between the calculated and experimental $\log_{10}\varepsilon_{H_2O}$ values at 0°C. The calculated value of ε_{H_2O} is 87.64 and the experimental value is 87.92 at 0°C. However, the calculated and experimental values of $\log_{10}\varepsilon_{H_2O}$ are seen to diverge between 0 and 75°C, although the agreement is still reasonably good. The calculated value of ε_{H_2O} is 57.78 at 75°C, and the experimental value is 62.34 according to the experimental relationship $\log_{10}\varepsilon_{H_2O} = 1.94409 - 1.99 \times 10^{-3}t$. The discrepancies between calculated and experimental values, however, were removed in the second method of determining the g_i and m_i values.

In the second method less restrictive conditions were imposed on the g_i and m_i values, namely $g_4 > g_3 > g_2$ and $m_4 > m_3 > m_2$, rather than linear interpolations. Values of ε_{H_2O} were calculated for small increments of g_i and m_i using a digital computer and the results from the linear interpolation method as a guide. The results for favorable g_i and m_i values taken from a large number of calculations are shown in Fig. 29. The agreement between the calculated and experimental values is seen to be excellent for the entire temperature range, and an interesting relationship between the two methods was revealed.

The values of m_4, m_3, and m_2 obtained from the first and second methods of obtaining the g_i and m_i values are seen from Fig. 29 to be nearly equal for favorable cases, i.e., the linear interpolation seems to apply essentially to both methods as far as the m_i values are concerned. This suggests that the m_0 values from both methods are significantly above 1.88 D. The range of the g_i values, i.e., the decrease from g_4 to g_2 is seen from Fig. 29 to be smaller for the second method. The small range was found to be characteristic of a large number of computer trials, where good fits of the experimental data were obtained at all temperatures, and in all such cases the m_i values were near those from the linear interpolation method. Thus, the g_i values from the second method are considered to be significantly better than those from the first method, and a g_0 value considerably in excess of 1 is indicated.

With regard to both methods of calculation, it should be noted that the g_i and m_i values were taken to be independent of temperature. Temperature-dependent g_i and m_i values could have been employed,(443) but they would involve additional assumptions, and the Raman data appear to provide evidence in opposition to the use of such quantities, at least from 0 to 100°C. For example, the frequency of the nonhydrogen-bonded OD stretching component from HDO is constant from 16 to 400°C, and the hydrogen-bonded component frequency does not vary greatly from 0 to 100°C. In addition, the integrated molar intensities of the nonhydrogen-bonded and hydrogen-bonded components appear to be roughly constant from 0 to 100°C.

Some features of the g_i and m_i values obtained from the second method are also of considerable interest in relation to the Raman observations. In a preceding section the intermolecular Raman bands were stated to arise almost exclusively from the 4-bonded structures. In this regard, it should be noted that the maximum dipole moment of 2.44 D refers to the 4-bonded structure. Further, the dipole moment decreases from m_4 to m_2, and intermolecular Raman intensity does not appear to come from the 3-bonded and 2-bonded structures. The previous analysis of the Lindner data also indicated that the frequencies of the nonhydrogen-bonded component and intensity maximum from the OD stretching contour of HDO at 400°C and 3900 bars occur near 2650 cm^{-1}, or well below the dilute gas value of 2722 cm^{-1}. The g_i values from the second method suggest that g_0 is considerably greater than 1. The g_0 from the second method would refer to the liquid density because the g_4, g_3, and g_2 values as well as their extrapolation refer to the liquid. A correlation parameter of unity refers to a random situation, but a value in excess of unity is indicative of some order and thus of additional interactions.

Finally, it should be emphasized that the results shown in Figs. 24–28, as well as the dielectric constant calculations, depend upon a ΔH value of about 2.5 kcal mol^{-1} for the disruptions of O—H···O and O—D···O units, and further, that values near 2.5 kcal mol^{-1} characterize a wide range of measurements. The quantitative Raman measurements have yielded values generally near this figure,[1135,1137–1139] and in addition, near-infrared[1186] and fundamental infrared measurements[962] give values of 2.3–2.4 kcal mol^{-1} ODO. In addition, ultrasonic studies of water have provided a value of 2.547 kcal mol^{-1} (for $d = 2.82$ Å),[240] and 2.5 kcal mol^{-1} has been obtained recently from the temperature dependence of neutron scattering residence times.* (See also values quoted in Ref. 660.)

In conclusion, it is evident from the foregoing results that nonhydrogen-bonded and hydrogen-bonded O—H···O units can be considered separately in water, and that a consecutive disruption of the hydrogen bonds occurs with rise of temperature. Further, it would appear that the consecutive hydrogen-bond disruption model seems to provide a satisfactory model of water structure. It is consistent with the ultrasonic and neutron scattering results, with the near- and fundamental-infrared results, with the

* G. J. Safford, private communication.

laser-Raman results, with the heat capacity and dielectric constant, with the definitive results from stimulated Raman scattering, and with all of the various detailed results described in preceding sections of this chapter.

8. CURRENT AND FUTURE WORK

In previous work[1139] preliminary high-pressure laser-Raman results were obtained in the intermolecular hydrogen bond stretching region at constant temperature, and an estimate of $\Delta\bar{V}$, the change in the partial molal volume involved in the disruption of O—H···O units, was made. The high-pressure laser-Raman results, however, were obtained only at pressures up to 1200 bars. Accordingly, work on a two-window laser-Raman cell capable of withstanding higher pressures has begun, and the study of solutions of HDO in H_2O or in D_2O to pressures of 4000 bars is anticipated, in addition to high-pressure laser-Raman studies involving the effects of xenon on water structure.

Laser-Raman spectra from binary solutions of HDO in H_2O or in D_2O have been reported,[1138,1140] and in addition, spectra from ternary solutions of $NaClO_4$ or $NaIO_4$, and HDO in H_2O or in D_2O have been obtained recently.[1142] As yet, however, Raman spectra from binary solutions of HTO in H_2O or in T_2O have not been reported, and a laser-Raman study of such solutions is anticipated.

Stimulated Raman spectra have also been obtained recently from H_2O, D_2O, H_2O–D_2O mixtures, and solutions of $NaClO_4$ in H_2O and D_2O, using picosecond pulse excitation at 5300 Å.[199] The stimulated Raman spectra provide unequivocal evidence for the presence of definite OH and OD stretching components at frequencies corresponding to the spontaneous Gaussian components, and they also give evidence for mutually exclusive stimulation of nonhydrogen-bonded and hydrogen-bonded components. The stimulated Raman scattering investigation was preliminary, however, and detailed systematic investigations are planned.

Finally, with regard to stimulated Raman scattering investigations of water, and hyper-Raman investigations as well, it should be noted that such techniques have provided information of definitive importance in the few studies that have been conducted. Further use of nonlinear optical techniques will unquestionably provide valuable new information bearing on the detailed interactions and structures present in water and aqueous solutions, and indeed, the nonlinear techniques may be vital to future progress.

CHAPTER 6

Nuclear Magnetic Resonance Studies on Water and Ice

Jay A. Glasel

Department of Biochemistry
University of Connecticut
Medical School
Farmington, Connecticut

1. THEORETICAL AND EXPERIMENTAL FOUNDATIONS OF MAGNETIC RESONANCE

Nuclear magnetic resonance spectroscopy is, in essence, a semiclassical form of experiment. Thus, the separation of spectral energy levels is in a region such that the frequencies that cause transitions are produced by macroscopic oscillators (radiofrequency transmitters) as phase-coherent trains. Irradiation is achieved by antennas which may be physically arranged so as to cause the sample to be subject to waves with accurately known polarization. The signal-receiving antennas may be arranged so that the NMR signal has a known phase relation to the irradiating waves, and the signal is amplified by standard radiofrequency techniques. The method is semiclassical in so far as the interaction time of the irradiation with the system studied is concerned. For purely classical experiments, such as measurements of viscosity, the experiment is sensitive only on a very long time scale, say greater than 1 min. In purely nonclassical methods such as UV spectroscopy, the time scale is of the order of 10^{-15} sec. In the case of NMR measurements we are dealing with times ranging from 10^{-7} sec to very long times.

These remarks do not imply in any way that the basic physical property upon which the method depends—the magnetic properties of *nuclei* in molecules—is not quantum mechanical (more correctly, quantum electrodynamic) in description. However, we are not as yet able to accurately predict the nuclear magnetic properties of *any* nucleon on the basis of an *a priori* theory.

Quantum mechanics only becomes necessary in discussions of ultrahigh-resolution NMR, where spectral energy levels are perturbed by electron-transmitted nuclear spin–spin coupling, or when the details of the effect of molecular energy levels upon nuclear magnetic behavior is needed. Even in these cases—and only the latter will be discussed in this chapter—the quantum mechanical problem is one of *molecular* properties and the inaccuracies are due to our incomplete knowledge of the solution to the many-electron-in-a-multicentered-potential problem.

1.1. Properties of Nuclei in Electric and Magnetic Fields

We will now enter into a brief discussion of nuclear properties giving rise to magnetic resonance phenomena. More complete discussions can be found, in order of increasing rigor, in the books by Emsely *et al.*,[308] Slichter,[999] and Abragam.[1]

The quantities of interest result from a description of the electric and magnetic fields at distances away from a nucleus. Nuclei possess positive charge, but the question arises as to whether or not these are describable by treating the nucleus as a charged sphere. The potential of a general nonspherical body is given by a series expansion in spherical harmonics, whose first term is the "monopole" moment, and whose higher-order terms, the "dipole," "quadrupole," "octapole," moments, etc., are usually of decreasing importance. For the nuclear potential problem it is found that as far as the electrical moments are concerned, only nuclear *electric* monopole and quadrupole moments are of importance. In the case of the magnetic field distribution, only nuclear *magnetic* dipole moments are of importance. This is not to say that all isotopes possess these. The rules of thumb in given Table I are for Z (number of protons in a nucleus) and A (number of protons plus neutrons.)

Nuclear magnetic resonance is a process whereby interactions between nuclear *magnetic* dipoles and electromagnetic radiation are observed. No such experiment can be performed on nuclei that do not possess a magnetic moment. Nuclear electric quadrupole resonance (NQR) may be observed for nuclei possessing such moments via the interaction of the quadrupole moment with the electric component of radiation. We will not discuss this phenomenon here.

TABLE I. Nuclear Properties of Stable Isotopes[a]

Z	$A - Z$	Spin number I
Even	Even	0
Even	Odd	$\frac{1}{2}, \frac{3}{2}, \frac{5}{2}, \ldots$
Odd	Even	
Odd	Odd	$1, 2, 3, 4, \ldots$

[a] All have electric monopole moments. $I \geq 1$ implies electric quadrupole moment; $I \geq \frac{1}{2}$ implies magnetic dipole moment.

A nuclear magnetic dipole moment is described by a vector whose magnitude is a fundamental nuclear constant but whose direction is a dynamic variable. In classical terms it acts as a bar magnet of given strength. For our purposes the nuclear electric quadrupole moment, if it exists for a particular nucleus, can be described by a symmetric tensor whose components are of fixed magnitude and direction with respect to the magnetic dipole moment. This means that perturbation on the latter by a magnetic field perturbs the former and vice versa.

The properties mentioned above are all codified under the formalism used to describe nuclear spin angular momentum. Thus, a nucleus may possess an internal degree of freedom known as spin, and therefore has associated with this degree of freedom a quantum number (usually designated by **I**). This may have magnitudes

$$I = (0, \tfrac{1}{2}, 1, \tfrac{3}{2}, 2, \tfrac{5}{2}, \ldots) \times \hbar$$

The magnetic moment $\boldsymbol{\mu}$ is given by

$$\boldsymbol{\mu} = \gamma \hbar \mathbf{I}$$

where $\hbar$ is Planck's constant divided by 2π and γ is the magnetogyric ratio.

The nuclear electric quadrupole moments Q are also, of course, fundamental nuclear properties. The nuclei of importance and their properties are listed in Table II.

In the absence of an external magnetic field, space is isotropic and the nuclear magnetic dipole has equal probabilities of being pointed in any direction. If an external magnetic field is applied, because of the interaction of the two dipoles—the applied north–south macroscopic dipole and the microscopic nuclear magnetic dipole—this isotropic condition is lifted and

TABLE II. Properties of Nuclei Important in NMR Studies on Water

Nucleus	Spin number I	Magnetic dipole moment μ, Bohr magnetons	Electric quadrupole moment Q, cm^2	Resonant frequency at $H_0 = 10{,}000$ G, ω_R
^1H	$\frac{1}{2}$	2.79270	0	42.577
^2H	1	0.85738	0.00277×10^{-24}	6.536
^{17}O	$\frac{5}{2}$	-1.8930	-0.0265×10^{-24}	5.772

space becomes anisotropic for the nuclear dipole. No longer are all directional probabilities equal.

1.2. Spectroscopic Parameters Describing Nuclear Magnetic Resonance

The quantum mechanical description of the nuclei's behavior is that the total wave function for a nuclear magnetic dipole is given by

$$\Psi_{\mathrm{TOT}} = a_I \psi_I + b_{I-1} \psi_{I-1} + \cdots + z_{-I} \psi_{-I}$$

where the coefficients α_i are generally time-dependent probabilities of finding the dipole in eigenstate ψ_i. The time dependence means that the possibility exists for a nucleus originally in a given state at an initial time to assume another state at a later time. It is found, in addition, that in the absence of a steady magnetic field the ψ_i are degenerate. This means that the energy levels for the system are $(2I + 1)$-fold degenerate. In the presence of the field this degeneracy is lifted, yielding $2I + 1$ energy levels whose splitting is proportional to the strength of the applied field. For a large number of nuclei in a field, such as for the protons of H_2O, the deuterons of D_2O, and the oxygens of $H_2{}^{17}O$, we may use the principle of ensemble averaging and find that at thermodynamic equilibrium a Boltzmann distribution holds. Thus, there is a time-independent excess of spins in the lower states. Using this principle, we may lay aside quantum mechanics and deal only with the direction of this excess—now a *macroscopic* magnetic moment vector—and its dynamic behavior as we disturb the equilibrium of the system. In order to achieve this, we need to cause a "flow of probability" from one energy state to another. The only way to do this is to cause an oscillating magnetic dipole field to interact with the nuclear magnetic dipole. At the steady magnetic fields practically obtainable at the present moment this involves using the magnetic component of electromagnetic radiation in the frequency range from 2 kHz to 300 MHz. This perturbing radiation is supplied to the sample in such a way that it has the effective component of magnitude H_1 rotating in a plane perpendicular to the direction of the steady field H_0. That is, H_0 is given by

$$\mathbf{H}_0 = H_{0z}\mathbf{k}$$

and

$$\mathbf{H}_1 = H_{1x} \cos \omega t \cdot \mathbf{i}$$

The solution of the macroscopic, or ensemble-averaged, equations of motion results in the following: For values of H_{1x} much less than H_0, as ω approaches the value of $E_0/\hbar(E_0$ is the energy splitting), a component of macroscopic magnetization is *induced* in the sample in the XY plane where there was none before the application of H_{1x}. In a perfect system this induced magnetism is a vector representing the summation of all the microscopic induced magnetic moments and has zero magnitude at every frequency except for $E_0/\hbar$, at which frequency in rotates in the XY plane with a finite magnitude. This rotating magnetic field vector can be made to cause an induced emf in a receiver coil which in turn can be amplified and detected, resulting in an "NMR signal." Hence, to summarize, as the irradiating frequency approaches $E_0/\hbar$, a flow of probability takes place from lower to upper energy state, tending to equalize their populations, and in addition, a magnetism is induced in the XY direction. This contrasts to the situation without the perturbing H_{1x}, where there are unequal upper- and lower-energy-state populations in a field H_0, resulting in a nuclear paramagnetism in the Z direction and zero magnetism in the XY planes, since all directions of the microscopic moments are equiprobable.

In every spectroscopic problem there arises the question of how and when systems whose energy states are raised "relax" to the lower state once again. This is because a basic principle of spectroscopy states that in a system where there are transitions between equally populated energy levels no signal results. Thus, in NMR spectroscopy the question arises: How does a nuclear system raised to an upper energy state return to a lower state with resultant transfer of energy to the rest of the world? In more succint terms: How does it reach thermodynamic equilibrium?

In addition, because of the nature of the radiation, the wavelength is large compared to the size of the sample and its phase coherence: the question arises as to how rapidly the XY component of induced magnetism returns to its equilibrium value of zero after H_{1x} is turned off. This last question does not exist in other forms of spectroscopy using noncoherent radiation.

In fact, both forms of relaxation take a unique form. In UV spectroscopy, for instance, decay from an upper to a lower state takes place via the mechanism of spontaneous emission in times of the order of 10^{-8} sec. This is not true of nuclear magnetism. It may be easily shown[(1)] that a nuclear magnetic moment is almost completely insulated from the rest of the world. If there were no mechanism other than spontaneous emission, the lifetime of an upper state would be of the order of 2000 years. A spectacular demonstration of the existence of another mechanism is the fact that the times actually observed are from 10^{-5} to 100 sec. Because the relaxation of magnetism induced in the XY plane need not involve energy transitions, in general this will take a different amount of time than the process just mentioned. Of course, every energy level transition causes loss of phase coherence and hence, in general, the relaxation of the XY magnetism will take the same amount of time as the approach to thermal equilibrium, if *no* other mechanisms are operative.

These arguments may be put in more compact mathematical format, as follows. For small deviations from equilibrium the nuclear paramagnetic susceptibility (in the Z direction) decays to its equilibrium value M_0 according to the differential equation

$$d\langle M_z\rangle/dt = -(1/T_1)\{\langle M_z\rangle - \langle M_0\rangle\}$$

which has as solution

$$\langle M_z(t)\rangle = M_0(1 - e^{-t/T_1})$$

The characteristic value $1/T_1$ is called the "relaxation rate" and T_1 is called the longitudinal (for Z direction) relaxation time. In the XY plane where the equilibrium value is zero, the differential equation is

$$d\langle M_x\rangle/dt = -(1/T_2)\langle M_x\rangle$$

with solution

$$\langle M_x(t)\rangle = e^{-t/T_2}$$

The relaxation time T_2 for the magnetism in the XY direction is called the transverse (because it is perpendicular to the direction of H_0) relaxation time.

How can T_2 be different from T_1? Consider the case (which is very real indeed) where, in a macroscopic sample, microscopic defects exist which are fixed in space. Because of these, the steady magnetic field at the nucleus is perturbed slightly from the value H_0. In short, the magnitude of the H_0 local field at the nucleus is either slightly greater or slightly less than H_0, with direction other than along H_0. This means that following the perturbation, causing magnetism in the XY direction, a nucleus experiences a steady perturbation not in the direction of H_0, due to the defects. Since these are in random directions over a macroscopic sample of nuclei, their magnetic moments tend to become randomized from their fixed perturbations following their bunching due to the H_{1x} perturbation. All this is an adiabatic process but nonetheless the XY magnetism disappears. Thus, T_2 may be very much shorter than T_1 since every T_1-type process causes a T_2-type relaxation but not vice versa.

This arguments also gives us some insight into the processes in solids and liquids. In order to have a T_1-type relaxation, that is, an energy release in going from upper to lower state, it is necessary to *stimulate* emission by having an oscillating magnetic field moment, mimicking H_{1x}, at a frequency of $E_0/\hbar$. For a sample of real matter this is brought about by the motions of electrons in molecules through space as they collide with one another. That fraction of collisions which results in motions with frequencies of around $E_0/\hbar$ may stimulate emission. That is, they couple the nuclear magnetic moments to the molecular system surrounding them. This molecular system is often loosely called the lattice and hence the T_1 process is called the spin–lattice relaxation time. In a liquid the distribution of collision frequencies is very broad; in a solid it is very narrow. This much is intuitively clear. Thus, unless for some special reason the sharp distribution peak in a given solid falls at the frequency $E_0/\hbar$, there will be very little energetic coupling of a nucleus to the lattice and so T_1 will be very long. In practice, it may take hours or even days. On the other hand, in a solid the defects are fixed and hence T_2 may be very short. For a liquid the T_1 process is much more efficient. On the other hand, if the defects are forming and breaking up, as is probable for a nonviscous liquid, then a nucleus is in an isotropic environment and so the T_2 process caused by defects is ineffective. One therefore expects T_1 to equal T_2 and this is found to be true for nonviscous liquids.

In summary, there are three basic parameters which describe an NMR signal: (1) the frequency of absorption; (2) its energy coupling to its surroundings, demonstrated by the spin–lattice relaxation time T_1; and (3) the characteristic time for loss of XY magnetism, given by the transverse time T_2.

The frequency of absorption is, of course, proportional to the energy splitting caused by the steady magnetic field. However, again we must consider the local field at the nucleus. The presence of the electronic structure of the molecule surrounding a nucleus causes the local field to differ from H_0. In general the field at the nucleus is given by a sum of terms,

$$H_{\text{local}} = H_0 - H_{\text{dia}} + H_{\text{para}}$$

where H_{dia} is the induced diamagnetic term (Lenz diamagnetism) resulting when any free electronic structure is placed in a magnetic field, and H_{para} is an additive paramagnetic term due to the presence in the total wave function of the molecule of contributions from states with unpaired electron spins, which are low-lying enough to have a finite occupancy probability at room temperature. Because of these terms the NMR absorption frequencies of a given nucleus will be different in different electronic environments. The NMR signals from the nucleus in these different surroundings are said to be "chemically shifted" from one another. Anticipating a later discussion, we can easily see the difficulty in the theoretical interpretation of chemical shifts since, not only are the observed quantities the sum of two terms, but the terms themselves immediately involve quantum chemistry and thus all the uncertainties inherent in that art.

Given the frequency position of the NMR signal, a spectroscopist's next question is, "What is the width of the line?" Neglecting any effects due to imperfections in the apparatus, we may invoke the uncertainty principle and state that the width of the line is proportional to the lifetime of the state of the system. In the case of an ensemble of nuclei this means that the shortest lifetime, T_2, governs the width of the signal. In special cases $T_2 = T_1$, but in general, $T_2 \ll T_1$. The important parameter T_1 is therefore the only one which cannot be measured by a single NMR spectrum of the kind familiar to most spectroscopists, i.e., constant irradiation of a sample while the frequency is varied, and subsequent detection of a signal. The T_1 information is usually hidden and special methods must be used for its extraction.

1.3. Experimental Arrangements

There are two basic types of NMR spectrometer. One is designed specifically to observe the *frequencies* of resonance from nuclei in various environments. These are the so-called "high-resolution" spectrometers. They are typified by the use of magnets with a high degree of field homogeneity at the location of the sample, very accurate control of H_0, both as to steadiness in time and resettability, and low H_1 power capability. There are several commercial models available. The second type is not truly a spectrometer, but is a magnetic resonance experiment set up specifically to observe only the magnitude of the component of nuclear magnetism in the XY plane or along the Z axis. Because the experiment involves pulsed rather than continuous radiofrequency fields, it is called "pulsed NMR." The pulsed rf experiment is typified by an H_1 source of high power and a medium-quality magnet. With such a unit, by changing experimental conditions slightly, both T_1 and T_2 may be measured within a few minutes of each other. By adding a small pulsed field to the steady magnetic field, T_1 may by measured by a method called "adiabatic fast passage."[(391)]

Since the linewidth of a signal is proportional to T_2 and since, as the rate of lower to upper energy level transitions is increased by increasing the intensity of H_1, it is possible in a few cases to measure T_1 and T_2 with a conventional, unmodified, high-resolution spectrometer.[(1085)] The principle is simply that when the rate of low to high transitions becomes greater than the rate of downward transitions due to spin–lattice energy exchange, the signal begins to decrease in magnitude, i.e., it saturates. This can only be used if: (1) the analytical line shape is known, (2) the available rf power is great enough to saturate the system, and (3) linewidths can be measured accurately (i.e., signal-to-noise ratio is high).

In pure liquid all these methods work well for protons. However, since basically all the experiments on pure water consist of measurements of the temperature and pressure dependences of T_1, T_2, and ω for ^1H_2O, ^2H_2O and $H_2{}^{17}O$, the opportunities are rapidly exhausted. For more complex experiments—measurements of diffusion coefficients, for instance, in electrolytic solutions—versatility in a system is not only a virtue but a necessity.

2. MAGNETIC SHIFT AND SPIN COUPLING PHENOMENA IN WATER VAPOR, WATER, AND ICE

2.1. Theory of Chemical Shift

In this section we are concerned with the *frequency* of NMR absorption of water in its various phases. The magnitude of the shift in frequency in going from a bare, isolated proton to a hydrogen atom can be easily estimated. In a field H_0 of a few thousand gauss the resonant frequency of a bare proton is $\sim 10^7$ Hz. We want to know

$$\delta H = H_{0,l} - H_0$$

that is, the difference between the external field H_0 and the local field $H_{0,l}$ due to the electron circulating outside the proton. Consider an electron orbiting about a proton in a fixed circle of radius r. The system has, in a steady magnetic field, a classical Larmor angular precession whose frequency is given by

$$\omega = (e/2m_e c)H_0 \tag{1a}$$

and thus has an angular velocity

$$V/r = \omega$$
$$V = r(e/2m_e c)H_0 \tag{1b}$$

This in turn produces a Lenz field at the proton, given by,

$$H_\mathrm{L} = i/cr, \qquad i = \text{current} = eV/r$$
$$= (e/c)V/r^2 \tag{1c}$$

so that the total field is reduced by the amount

$$\delta H = (e^2/2m_e c^2)H_0/r \tag{1d}$$

δH is about 1 part in 10^5 and for a frequency of resonance of 10^7 Hz corresponds to ~100 Hz. This frequency shift, called the chemical shift, is expressed in either Hz or parts per million. The estimate given above is based on observations on protons.

The approach to a more accurate theoretical expression is indicated by the following. In the presence of a magnetic field H_{0z} of a few thousand gauss the electronic wave function of a molecule is given by

$$\psi = \psi_0 - (e\hbar/2m_e c)H_{0z} \sum_n [\langle n \mid L_z \mid 0\rangle/(E_n - E_0)]\psi_n \tag{2}$$

where L_z is the Z component of orbital angular momentum, ψ_0 is the wave function in the absence of H_{0z}, and E is the energy of the nth electronic state ψ_n.

For an H_{0z} field of finite magnitude the velocity of the electron is given by

$$\mathbf{v} = (1/m_e)[\mathbf{p} + (e/2c)\mathbf{H} \times \mathbf{r}] \tag{3}$$

and using the relation $\mathbf{r} \times \mathbf{p} = \mathbf{L}\hbar$, the complete eqn. (1) becomes

$$\delta\mathbf{H} = -(e\hbar/2m_e c)(\mathbf{L}/r^3) - (e^2/2m_e c^2)[\mathbf{r} \times (\mathbf{H} \times \mathbf{r})] \tag{4}$$

We may calculate the quantum mechanical analog by replacing eqn. (3) by the quantum mechanical average over the wave function (2). This has three components, corresponding to the three space axes. For example, the Z component is

$$\delta H_z = -(e\hbar/2m_e c)\langle\psi \mid 2L_z/r^3 \mid \psi\rangle - (e^2/2m_e c^2)H_{0z}\langle\psi \mid (x^2 + y^2)/r^3 \mid \psi\rangle \tag{5}$$

and by combining (2) and (4),

$$\delta H_z = \Big((e^2/2m_e c^2)\langle 0 \mid (x^2 + y^2)/r^3 \mid 0\rangle - (e\hbar/2m_e c)^2 \Big\{\sum_n \langle 0 \mid L_z \mid n\rangle\langle n \mid 2L_z/r^3 \mid n\rangle + [\langle 0 \mid 2L_z/r^3 \mid n\rangle\langle n \mid L_z \mid 0\rangle/(E_n - E_0)]\Big\}\Big)H_0$$

For a circular orbit, where classically

$$\langle 0 \mid (x^2 + y^2)/r^3 \mid 0\rangle = 1/r \tag{6}$$

the first term gives the estimate shown above. Quantum mechanically, it must be derived from the ground-state wave function and is called the diamagnetic shielding term σ_D. The second term is called the paramagnetic shielding term σ_P. It is clear that the two are of opposite sign, so that

$$\delta H = - \mid \sigma_D \mid H_0 + \mid \sigma_P \mid H_0. \tag{7}$$

The calculation of δH from theory requires an accurate knowledge of ground- *and* excited-state wave functions.

One further equation must be introduced before applying the above considerations to H_2O. In the presence of a strong electric field, say of magnitude E_z along the Z axis at a hydrogen atom, the total σ is reduced by an amount

$$\Delta\sigma = -(881/216)E_z^2 a_0^4/m_e c^2 \tag{8}$$

due to distortion of the electronic cloud by the field.

It should be clear that the diamagnetic term is temperature-independent while the paramagnetic term, depending upon the population of low-lying states, is temperature-dependent.

The shielding of a proton in a water molecule with respect to a bare proton has been determined experimentally and the experimental values are (in c.g.s. units)

$$p^+(g) - H_2(g) = 26.6 \times 10^{-6} \qquad \text{(Refs. 793, 890)} \tag{9a}$$

$$H_2(g) - CH_4(g) = 4.20 \times 10^{-6} \qquad \text{(Ref. 956)} \tag{9b}$$

$$CH_4(g) - H_2O(g) = -0.559 \times 10^{-6} \qquad \text{(Refs. 512, 956)} \tag{9c}$$

resulting in

$$p^+(g) - H_2O(g) = 30.2 \times 10^{-6} \tag{9d}$$

It was found[512] that the shift $CH_4(g) - H_2O(g)$ was temperature- and pressure-independent from 130 to 190°C and from 4.5 to 12.1 bars. These experimental values yield an average σ, that is, they must be interpreted on the basis of a diamagnetic shielding tensor. The paramagnetic contributions may be found from microwave measurements of spin rotation constants[340] of HDO and are[97]

$$\begin{aligned} \sigma_{aa}^{P} &= -73.79 \times 10^{-6}, & \sigma_{bb}^{P} &= -34.37 \times 10^{-6} \\ \sigma_{u}{}^{P} &= -107.59 \times 10^{-6}, & \sigma^{P} &= -71.92 \times 10^{-6} \end{aligned} \tag{10}$$

where the principal inertial axis system[468] is used. Combining these values with the experimental value of the total shielding, one finds a diamagnetic shielding of

$$\sigma^{D} = 102 \times 10^{-6} \tag{11}$$

We see that for H_2O the paramagnetic antishielding term is very important.

Referring to eqns. (5) and (1) and the "experimental" value of $\langle 1/r \rangle$, the reciprocal average distance of *all* the electrons from a proton in H_2O may be calculated. The same value may be calculated, given a molecular wave function. The most recent and successful work of this sort for water is on the basis of minimal Gaussian-lobe wave functions.[468] These give for the proton $\langle 1/r_H \rangle$ a calculated value of 10.71 Å^{-1} versus an "experimental" value of 10.85 Å^{-1}.

Thus, there is quite good agreement between experiment and theory. It should be emphasized that this means that theoretically the *diamagnetic* contribution can be found accurately on the basis of these wave functions. No theoretical calculations have been attempted with respect to calculating the total contributions to σ. The shift corresponding to $H_2O(g) \rightarrow H_2O(l)$ is temperature-dependent. Since the gas shift is independent of temperature, this means that the dependence is due to changes in shielding of a proton in the liquid. The magnitude of the change is[512]

$$\sigma = -4.846 \times 10^{-6} \quad \text{at} \quad -15°\text{C}$$
$$= -3.661 \times 10^{-6} \quad \text{at} \quad 100°\text{C}$$

The temperature dependence of the shielding is not linear, but the equations

$$\sigma(t) = -4.656 + 11.884 \times 10^{-3}t - 2.006 \times 10^{-5}t^2 \quad (\text{rms error } \pm 0.004) \tag{12a}$$

$$\sigma(t) = -4.64 + 10.3 \times 10^{-3}t \quad (\text{rms error } \pm 0.022) \tag{12b}$$

(t is the temperature in °C) are obeyed within the accuracy given from -15

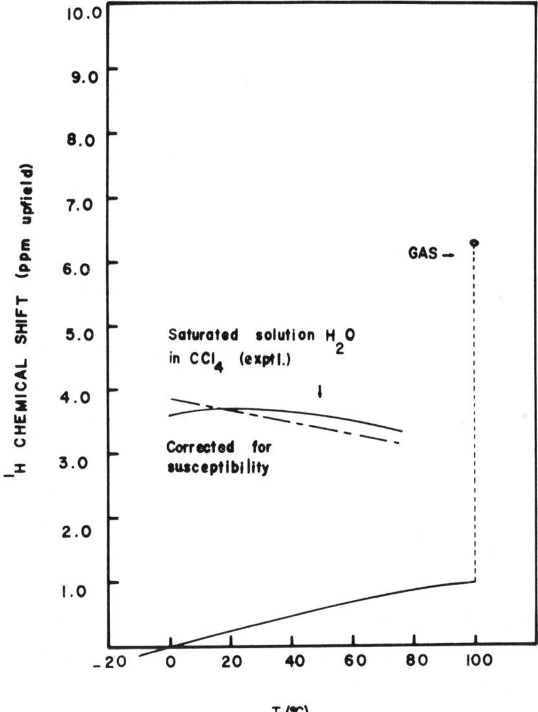

Fig. 1. Proton chemical shift versus temperature for liquid and vapor H_2O. Also shown is proton chemical shift versus temperature for a saturated solution of H_2O in carbon tetrachloride. Solid lines are corrected for susceptibility.

TABLE III. ^{1}H Chemical Shifts and Shielding Constants in Liquid H_2O as a Function of Temperature (referred to water at 0°C)

t, °C	δ, ppm	δ(corr), ppm	σ, ppm
−15	0.170	0.190	−4.846
−10	0.115	0.126	−4.782
−5	0.055	0.058	−4.714
0	0.000	0.000	−4.656
5	−0.058	−0.060	−4.596
10	−0.117	−0.121	−4.535
15	−0.175	−0.179	−4.477
20	−0.231	−0.235	−4.421
25	−0.235	−0.288	−4.368
30	−0.340	−0.341	−4.315
35	−0.394	−0.394	−4.262
40	−0.446	−0.443	−4.213
45	−0.498	−0.492	−4.164
50	−0.549	−0.541	−4.115
55	−0.600	−0.590	−4.066
60	−0.650	−0.636	−4.020
65	−0.700	−0.683	−3.973
70	−0.750	−0.727	−3.929
80	−0.850	−0.820	−3.836
90	−0.948	−0.909	−3.747
100	−1.045	−0.995	−3.661

to 100°C. Figure 1 illustrates the observed changes in chemical shift, and the values are listed in Table III.

Another experimental value of interest is the chemical shift incurred in transferring a molecule of water from the pure liquid to a nonpolar organic solvent. This has been determined experimentally for water dissolved in carbon tetrachloride,[(512)] as shown in Fig. 1.

Since these measurements are for saturated solutions, the observed shifts may still reflect clusters of molecules, and hence the true shift for a single molecule is not known. Changing cluster size may be responsible for the temperature changes observed. For the general case, in discussing shifts in pure liquids, the total observed shielding is given by

$$\sigma_{\mathrm{TOT}} = \sigma_1{}^{\mathrm{D}} + \sigma_2{}^{\mathrm{D}} + \sigma_3{}^{\mathrm{D}} + \sigma_4{}^{\mathrm{D}} + \sigma_1{}^{\mathrm{P}} + \sigma_2{}^{\mathrm{P}} + \sigma_3{}^{\mathrm{P}} + \sigma_4{}^{\mathrm{P}} \tag{13}$$

where σ_1 includes the effect of van der Waals forces on diamagnetic (D) and paramagnetic (P) terms, σ_2 includes the effect of the shielding of a non-polar solute by a nonpolar solvent,[(104,145)] σ_3 includes the effects of the polarization of *solvent* molecules by polar *solute*, and σ_4 includes the effect of repulsive overlap on shielding.

2.2. Measurement of Deuteron and ^{17}O Chemical Shifts in Water

Figure 2 shows the observed values of the chemical shift of ^{17}O in $H_2^{17}O$ as a function of temperature. Included is the shift in transferring to the vapor phase, where one finds

$$\delta(t) = -38.0 + 47.3 \times 10^{-3}t \tag{14}$$

The linearity (shown in Fig. 2) in the shift with respect to temperature changes is in serious dispute, since two independent studies find the dependence very nonlinear in the 4–75°C range.(336,707) Whichever is correct, the magnitudes of the changes are interesting for purposes of later discussion. Based on the crude estimate of the diamagnetic (Lamb) term given by eqn.

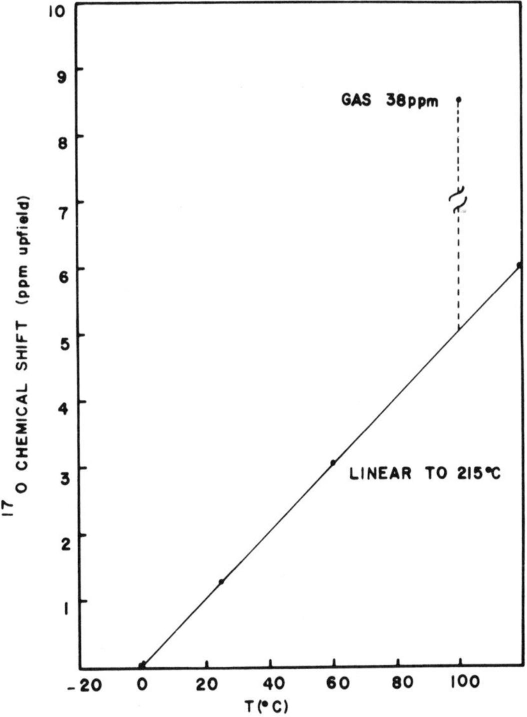

Fig. 2. Oxygen-17 chemical shift versus temperature for liquid and vapor $H_2^{17}O$.

(1b), which gives $\sigma^D \approx 1.8 \times 10^{-5}$ for ^1H and $\sigma^D \approx 3.95 \times 10^{-4}$ for ^{17}O, the fractional change for ^1H over the whole temperature range is $\sim 5.5 \times 10^{-2}$, while for ^{17}O it is only $\sim 1.3 \times 10^{-2}$. Hence, the changes relative to this base are greater for ^1H than for ^{17}O.

Neither ^1H nor ^{17}O shifts have been successfully measured in ice, where lines are very broad. The two features of these experiments which one hopes to explain are (1) the upfield shift of both ^1H and ^{17}O in going from liquid to vapor and (2) the upfield shift of both with increasing temperature. The direction of both is that of increased *total* shielding.

Since there is not even a completely acceptable semiempirical theory of quantitative chemical shifts for any nucleus, one must be prepared to accept some qualitative rationalizations of the phenomena on the basis of the formulas given above. The problem arises, of course, from our inability to form the sums over excited states needed in the evaluation of the paramagnetic term,

$$(e^2N/3m_e^2c^2)\langle 1/r^3\rangle \sum_n [\langle 0 \mid L^2 \mid n\rangle/(E_n - E_0)]$$

2.3. Structural Interpretations of Chemical Shift Measurements

For protons we have seen that, experimentally, both shielding terms are important in contributing to the observed chemical shift. It is easy to see that formation of a hydrogen bond between two water molecules—whether it be described as electrostatic or covalent (or as a mixture) in origin—will lead to a loss of electron density at the proton involved in the bond. Consequently, $\langle 1/r\rangle$ and $\langle 1/r^3\rangle$ will be reduced, leading to a deshielding of the proton. This effect is largest for maximum H bonding and decreases as bonds are broken. This idea has been made more quantitative on two levels. The first, more superficial, one results from the assumption that the proton chemical shift varies linearly with temperature, and therefore a "two-state" model may be set up with free and hydrogen-bonded protons so that

$$\delta_{\text{obs}} = X_{\text{bonded}}\,\delta_{\text{bonded}} + (1 - X_{\text{bonded}})\delta_{\text{free}} \tag{15}$$

Since the δ_{free} must only differ from the vapor phase by 0.5–0.7 ppm, as suggested by experiments on proton shifts from water in various solvents,[(512)] one finds $X_{\text{bonded}} \approx 0.80$ at 0°C and $\sim$0.65 at 100°C. This result agrees with some other estimates on the basis of two-state theory,[(360)] but not with certain other calculations.

A procedure which begs the question a little less is to try to calculate the temperature dependence of δ from a more basic standpoint. This has been attempted[471] by an electrostatic approach[732] based on the formula given above for a proton in an electrostatic field. The formula includes both paramagnetic and diamagnetic contributions. The electrostatic field is assumed to be directed along the O—H···O′ bond axis and to be due to the O′ oxygen. The field strength would clearly be dependent upon the H bond length and this may be accounted for by assuming a potential energy function for the H bond, e.g., a Morse function. Since the H bond is so weak, it is to be expected that higher vibrational states with different H bond distances would be excited at room temperature, and these must be taken into account. The final results show that, depending upon the potential function and the dissociation energy and fundamental H bond frequency associated with it, the H bond shift should be from 6.6 to 3.9 ppm downfield from the vapor and that the temperature dependence should be linear with a slope of from 2 to 8×10^{-3} ppm deg^{-1}. These values are well within the range of those observed for water. However, we have ignored the possibility that an increase in temperature might be accompanied by a breaking of H bonds. The temperature dependence enters because higher vibrational states with larger vibrational amplitudes become more populated at higher temperatures. Thus, we may explain the experimental results by means of two models which have completely different physical aspects.

The ^{17}O shift data are perhaps more interesting, but have not yet been treated in the literature. Consider first a lone oxygen atom: the diamagnetic contribution to the chemical shifts for atoms with $Z > 1$ has been calculated from Hartree–Fock wave functions.[262] It rises monotonically with Z approximately as $Z^{3/2}$, ranging in the first two rows of the periodic table from 0.18 ppm for H to 4.64 ppm for F. However, the percentage contribution of the *outer* electrons to σ^D decreases rapidly with increasing Z. Changes in the one valence electron of hydrogen, due to chemical modifications, effect σ^D very profoundly. For oxygen with two valence electrons it is expected that, as with fluorine, the effect of chemical modifications on σ^D will be much less. In fact, for a wide range of compounds the chemical shifts vary by over $\sim$700 ppm. This is taken to be evidence, in common with all higher-Z elements, that the paramagnetic contribution σ^P is of prime importance in determining chemical shifts involving these elements. The calculation of σ^P is limited to only the outer shell electrons, since the inner core has spherical symmetry (zero angular momentum) and therefore contributes nothing to the paramagnetic term.

The problem of the ground state of an oxygen atom has been treated[688] using Slater wave functions, and it is found that the hybrid orbitals from the mixing of the $2s$ and $2p$ levels show maxima in four directions which are in two orthogonal planes through the nucleus. In one plane the orbitals are at an angle of 103° from each other and in the other the angle is 133°30′. The latter, containing the highest charge density, are the "lone-pairs."

Thus, the oxygen nucleus finds itself in almost a tetrahedral environment in a bare oxygen atom. In a free water molecule it is known from spectroscopic data that the angle between hydrogens is 104°30′, and hence the formation of the molecule does not affect the fundamental electrical symmetry about the oxygen nucleus. In such a state we may expect the paramagnetic term to be small and the chemical shift to be dominated by the diamagnetic term, in *contrast* to the usual case for high-Z elements. In fact, $H_2^{17}O$ has its resonance very far upfield, in common with other molecules where oxygen retains similar electrical symmetry.(186,260) In contrast, carbonyl compounds have a paramagnetic shift (downfield) of 200–600 ppm. It is found, moreover, that the ^{17}O resonance undergoes an *upfield* shift of some 11 ppm when H_2O is diluted radically with acetone at room temperature.(900)

These facts imply that in liquid water at 0°C an oxygen nucleus is in an asymmetric environment (ensemble-averaged) which grows more symmetric as the temperature is raised. According to the Bernal and Fowler rule, where an oxygen atom in water can have at most two valence-bonded protons and two hydrogen-bonded ones at the same time, one would predict the general trend of the observed chemical shifts on the basis of increasing importance of paramagnetic contributions as the system is distorted from spherical symmetry at low temperatures. In turn, this would imply the dominant existence of small clusters at low temperatures.

The linearity of the shift versus temperature data may be used, as before, in a two-state calculation with $\delta = -38 + 47.3 \times 10^{-3}t$ [eqn. (15)]

$$\delta = X_{\text{bonded}}\,\delta_{\text{bonded}} + (1 - X)_{\text{bonded}}\,\delta_{\text{free}}$$

and with X_{bonded} and $(1 - X_{\text{bonded}})$ the same as for previous calculations for proton shifts,

$$\delta_{\text{bonded}} = -43 \text{ ppm} \qquad \text{and} \qquad \delta_{\text{free}} = -15.8 \text{ ppm}$$

The apparently large shift from water vapor to unbonded water in liquid water (δ_{free}) can be explained in the same way as the much smaller value for protons in H_2O(~0.9 ppm), on the basis of dispersion shifts which are proportional to the mean square of the fluctuating induced-dipole–induced-dipole field, $\sigma_d = \phi\langle E_d{}^2\rangle$ at the nucleus, ϕ being a constant characteristic of the nucleus in question. A quantitative calculation of the dispersion interaction is not possible at present. However, experiments have been performed on the shift in covalent fluorine compounds in various nonpolar solvents[536] which suggest that $\phi_F = 20\phi_H$. This in turn suggests that $\phi_{^{17}O}$ is characterized by a value at least an order of magnitude greater than that for ^1H, thus explaining the large difference mentioned above between shifts of free and nonhydrogen-bonded water. The obviously desirable experiments involving $H_2^{17}O$ in inert organic solvents have not yet been performed.

2.4. Measurements of Spin–Spin Coupling Constants

The scalar coupling constants in HDO ($J_{^1H\text{-}^2H}$) and $H_2{}^{17}O$ ($J_{^1H\text{-}^{17}O}$) cannot be observed directly in the liquid state, due to rapid exchange (see later section). In the gas phase, however, the appropriate splittings may be seen and the values are[520]

$$J_{^1H\text{-}^2H} = 1.1 \text{ Hz}, \quad \text{implying that} \quad J_{^1H\text{-}^1H} = 7.18 \text{ Hz} \tag{16}$$

and[338]

$$J_{^1H\text{-}^{17}O} = 79 \pm 2 \text{ Hz}$$

The latter value is not very different from that obtained from very dilute labeled water in acetone (where exchange effects are absent), where $J_{^1H\text{-}^{17}O} = 73.5 \pm 2.1$ Hz.[900] According to the valence-bond theory of spin–spin coupling constants,[575] the coupling constant $J_{^1H\text{-}^{17}O}$ should be proportional to the product $\varphi_{^{17}O}(0)\varphi_{^1H}(0)$ of the electron densities at each nucleus. The results stated above for "free" water and acetone solutions indicate that, in accord with one of the basic assumptions of the previous section, the changes between free and bound water are due to changes in outer electron structure. This is not a very unusual conclusion.

A value of $J_{^1H\text{-}^{17}O} = 92 \pm 15$ Hz, derived indirectly from exchange studies,[739] is doubtless too high and the reasons for this will be discussed later.

3. MEASUREMENT AND INTERPRETATION OF MAGNETIC RELAXATION TIMES IN WATER AND ICE

In contrast to chemical shift studies on water, where differences in behavior of ^1H and ^2H are negligible, the nuclear relaxation behavior of the two isotopes is profoundly different. There are three species of water,

^1H_2O, ^2H_2O, and $H_2{}^{17}O$, which are of importance in these studies. Nuclear magnetic relaxation phenomena are of interest to chemists in so far as they provide information on the dynamic properties of molecules in liquids and solids. These properties are (1) rotational diffusion, (2) translational diffusion, and (3) intermolecular exchange.

It is beyond the scope of this discussion to enter into a detailed theory of nuclear relaxation. However, some aspects of the theory pertaining to the above three properties will be discussed for relaxation phenomena associated with the various isotopically differing water molecules. Given the phenomenon of relaxation, the basic problem is to relate this macroscopic observation to microscopic (molecular) motion. The discussion of properties 1 and 2 will be cast largely within the framework of measurements of T_1 since in these cases this is the physically significant quantity, except in ice. The phenomenon 3 is dependent upon measurements of T_2 and will be so discussed.

3.1. Dipole-Induced Relaxation

The dominant cause of exchange of nuclear resonance energy with the molecular thermal sink in the case of ^1H_2O is *magnetic* dipole coupling between the two protons. Random motions of protons in the vicinity of a given proton produce a time-dependent magnetic field. A certain fraction of the power spectrum of the Fourier transform of these motions corresponds to the resonant frequency and is therefore effective in causing induced downward transitions. The fundamental formula relating the macroscopic relaxation time and the microscopic motion is

$$1/T_1 = (9/8)(\gamma_H{}^4\hbar^2/r^6)[J_2{}^1(\omega) + J_2{}^2(2\omega)] \tag{17}$$

where $J_l{}^m(m\omega)$ are the Fourier transforms of the autocorrelation functions $G_l{}^m(t)$,

$$J_l{}^m(m\omega) = \int_{-\infty}^{\infty} e^{-im\omega t} G_l{}^m(t)\, dt \tag{18}$$

For ^1H_2O the $G_l{}^m(t)$ take two forms. First, there is an intramolecular contribution due to the random angular motion of the other proton on a molecule, and second, there is an intermolecular contribution from random translational motion of neighboring molecules. It is a basic experimental and theoretical fact that for ^1H_2O these two contributions are of the same order of magnitude.[392]

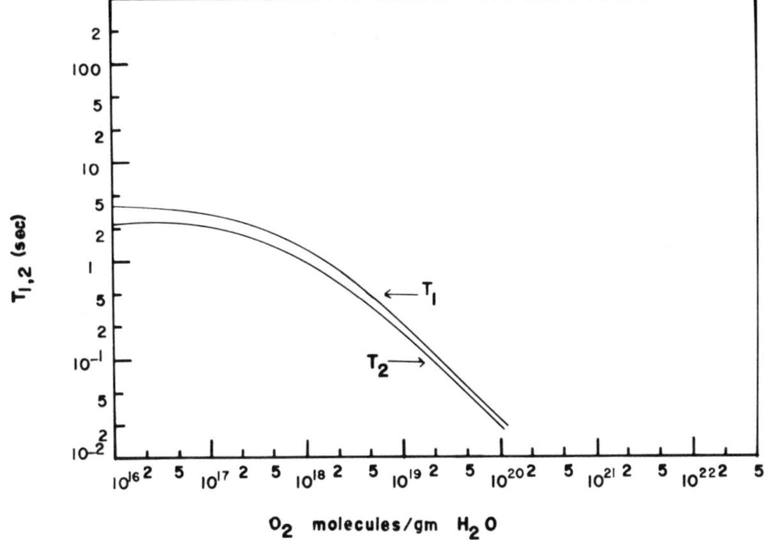

Fig. 3. Proton spin–lattice (T_1) and spin–spin (T_2) relaxation times versus oxygen concentration in liquid H_2O.

Special note should be made of the fact that in all nuclear relaxation measurements of protons in aqueous systems great care must be taken to eliminate dissolved oxygen from the sample. Oxygen, being paramagnetic, has the very important property of lowering the relaxation time of the protons. In water containing dissolved air this amounts to a decrease of 40% from the time for degassed water at room temperature.[(475)] This effect is shown in Fig. 3.

The variations of the relaxation rates with oxygen concentration follow the relationships

$$1/T_1 = 0.284 + 4.55 \times 10^{-19} N_{\mathrm{O}} \tag{19a}$$

$$1/T_2 = 0.455 + 5.25 \times 10^{-19} N_{\mathrm{O}} \tag{19b}$$

where N_{O} is the number of oxygen molecules cm^{-3}. The effect appears to be partly due to a weak oxygen–water complex.

3.2. Measurement and Interpretation of Proton Relaxation in Water

Krynicki[(638)] has made the most careful studies of pure ^1H_2O relaxation times up to the time of writing. The measured times are tabulated in Table IV and shown in Fig. 4. The measured rates, as a function of either

TABLE IV. Proton Spin–Lattice Relaxation Times and Rates for Air-Free Water as a Function of Temperature[a]

t, °C	T_1, sec	$1/T_1$, sec^{-1}
0	1.73	0.578
5	2.07	0.483
10	2.39	0.418
15	2.76	0.362
20	3.15	0.317
25	3.57	0.280
30	4.03	0.248
35	4.50	0.222
40	5.00	0.200
45	5.50	0.182
50	6.03	0.166
55	6.60	0.152
60	7.20	0.139
65	7.81	0.128
70	8.46	0.118
75	9.11	0.110
80	9.80	0.102
85	10.50	0.095
90	11.25	0.089
95	12.00	0.083
100	12.75	0.078

[a] From K. Krynicki.

T or $1/T$, are not notably linear. Krynicki has attempted a breakdown of the observed values into intermolecular and intramolecular contributions. The technique is to use different models for the intermolecular contribution and to subtract the results from the total observed relaxation rates to find the intramolecular contribution, which may then be compared with theoretical results for internal consistency. Three models have been used:

(a) Classical Stokes–Einstein translational diffusion, where the characteristic time τ_c for diffusion is $\tau_c = \langle r^2 \rangle / 12D$, where D is the macroscopic diffusion coefficient and r^2 is the mean square diffusion distance. $1/T_1$ involves the density of water and the macroscopic diffusion as variable parameters.

(b) Classical continuous diffusion, taking into account the fact that the proton relaxation in water involves relative motion between nuclei *not* situated at the center of mass of the molecule–spin rotation interaction,[539] and *not* distributed continuously about a given origin, but according to a

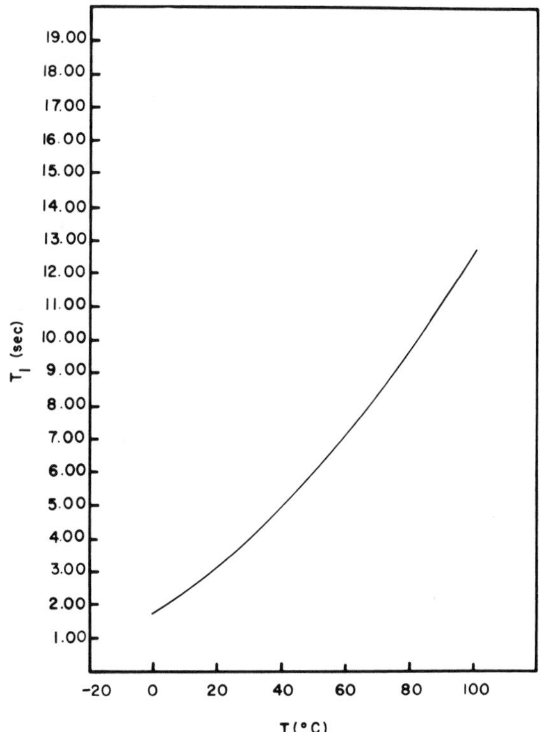

Fig. 4. Observed proton spin–lattice (T_1) relaxation time versus temperature for oxygen-free H_2O.

radial distribution function $G(r) = 4\pi\varrho(r)r^2$ (i.e., a nonuniform distribution of classical diffusion spheres). $1/T_1$ involves a radial distribution function and a macroscopic diffusion coefficient.

(c) Nonclassical jump diffusion, where a molecule librates in place for a time τ_0 and then diffuses continuously for a time τ_1 before returning to the librational state. This calculation involves, as parameters, the macroscopic diffusion coefficient and the Debye temperature for a crystalline lattice (obtainable from a neutron diffraction measurement).

The results of calculations on the basis of these models are shown in Table V at a single temperature and are compared to an experimental value. Figure 5 shows Arrhenius plots of $1/T_1$ on the basis of models (a) and (b) combined.

The thrust of all these remarks is to demonstrate that the calculations are model-independent and that the intermolecular contribution to dipolar

TABLE V. Calculated Values of $(1/T_1)_{inter}$ for Protons in H_2O at 20°C for Various Models

Model[a]	$1/T_1$, sec^{-1}
(a)	0.122
(b)	0.139
(c) $\tau_1 \gg \tau_0$	0.132
$\tau_1 \ll \tau_0$	0.127
$\tau_1 \approx \tau_0$	0.126
Experimental[(406)]	0.11

[a] $\tau_0 \approx 4 \times 10^{-12}$ sec.

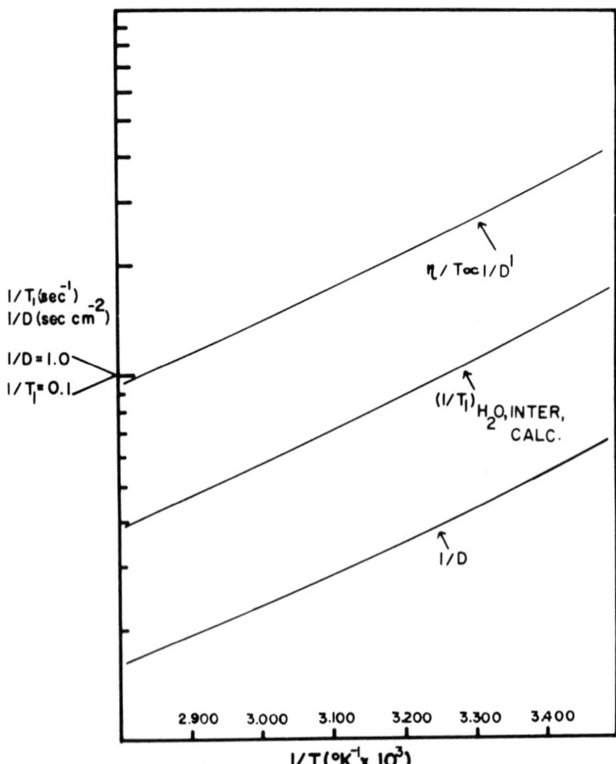

Fig. 5. Bulk viscosity, calculated intermolecular contribution to proton spin–lattice relaxation rate, and reciprocal bulk self-diffusion coefficient versus reciprocal absolute temperature.

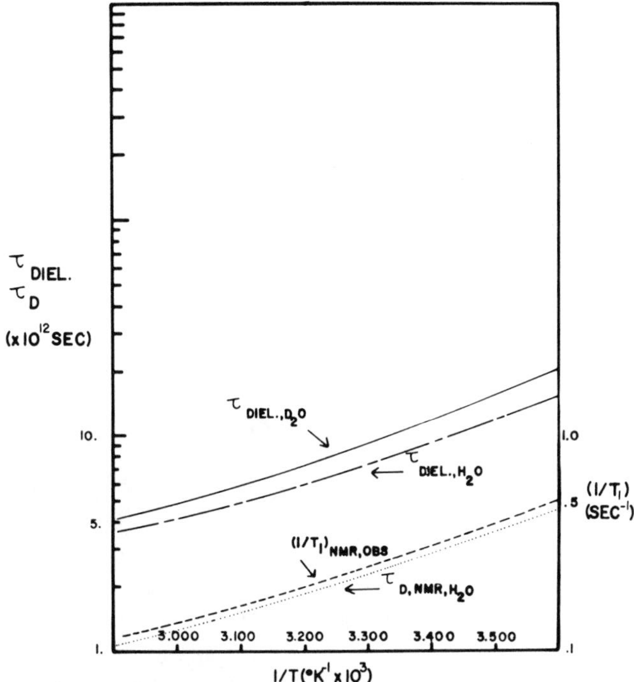

Fig. 6. Dielectric relaxation times for H_2O and D_2O, observed proton spin–lattice relaxation rates, and calculated NMR intramolecular reorientation times versus reciprocal absolute temperature. Observed ratio between dielectric relaxation time and calculated NMR reorientation time is 3.7 and is temperature-independent.

relaxation is inappropriate as a tool for probing the "fine structure" of molecular translational diffusion.

The mean apparent activation energy for proton spin–lattice relaxation in H_2O is 3.77 kcal mol^{-1} above 40°C, although Arrhenius behavior is not obeyed over the entire temperature range.

The intramolecular contribution, derived on the basis of the subtraction procedure just described, has essentially the same temperature dependence as the measured T_1 and as that of the dielectric relaxation time as shown in Fig. 6.

Proton T_1 in H_2O have been measured as a function of pressure from 0 to 9500 bars.[(62)] Figure 7 shows the more recent results of Hertz and Radle[(496)] in the form of a three-dimensional model, expressing T_1 as function of pressure and temperature. This shows the same anomaly as does the

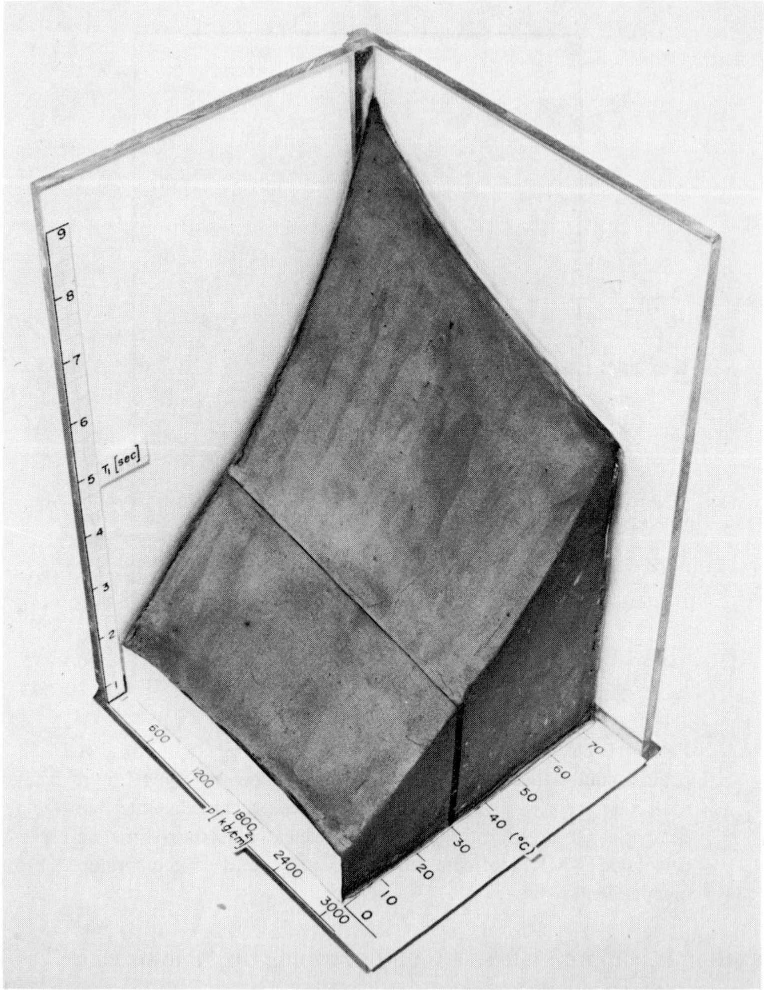

Fig. 7. Three-dimensional relationship among proton spin–lattice relaxation time, pressure, and temperature for oxygen-free H_2O.

pressure dependence of the macroscopic viscosity of water.[530] That is, with respect to viscosity, water behaves as a "normal" liquid only at temperatures above 25°C. It is found that over the whole range, in accord with the simple Debye law, the relaxation rate is proportional to the viscosity. The relaxation rate is 11% lower at $P_{\max} = 1400$ bars at 0°C than at $P = 1$ bar at the same temperature. Hertz and Radle have also made preliminary measurements of D in the same pressure and temperature ranges and found that it, too, goes through a maximum at $P_{\max} = 1500$ bars at 0°C, but the per cent increase is much smaller than for the maximum in $1/T_1$. Since $(1/T_1)_{\text{inter}}$ is proportional to $1/D$, the results may be interpreted to mean that $(T_1)_{\text{intra}}$ undergoes a minimum at about $P = 1400$ bars at 0°C.

The reason for the intensitivity of these NMR measurements to the details of microscopic motion is not difficult to find. The relaxation times or rates are proportional to the Fourier transform of the relevant autocorrelation functions. That is, they are proportional to an *integral* of the correlation functions, and not to the functions themselves. Since the details of molecular motion are tied up in the form of the correlation functions,[412,639] it is not surprising that the relaxation times are relatively insensitive measures.

3.3. Measurement and Interpretation of Proton Relaxation in Ice

Proton spin–lattice relaxation times in carefully purified ice have been measured[54] between 0 and -60°C. The results are shown in Fig. 8 in the form of an Arrhenius plot. In contrast to the liquid, the process is characterized by a single apparent activation energy of 14.1 ± 0.1 kcal mol^{-1}. Also in contrast to the liquid, T_1 is frequency-dependent, with $T_1 \propto \omega_0{}^2$.

The interpretation of T_1 measurements in solids is difficult.[302] This is because the macroscopic quantity T_1 is related to τ, the reorientation time for magnetic field fluctuations in the solid, by the equation

$$T_1 = CH_{0,l}^2\tau \tag{20}$$

where $H_{0,l}$ is the static local magnetic field strength and involves evaluation of dipolar lattice sums. This latter may be computed for hexagonal ice with some difficulty. The above results may be interpreted roughly on the basis of a model where the proton undergoes a random walk between the six possible arrangements of a proton about a central oxygen atom, with a simple mean jump time τ_j. This leads to the relationship $T_1 = 2.14 \times 10^5\tau_j$ and to the results shown in Table VI.

There is clear discrepancy between the two sets of data. The activation energy derived from dielectric relaxation, 13.2 ± 0.05 kcal mol^{-1}, also differs from the NMR result outside the limits of error. Recently, by means of a relaxation technique called the "rotating-frame" method, τ has been evaluated in a more straightforward way.[93] These experiments are based on the use of a very high rf field (Fig. 9). The rotating frame relaxation rate observed, $1/T_{1\varrho}$, is then given by

$$1/T_{1\varrho} = \tfrac{3}{2}C'[\tau/(1 + 4\omega_1{}^2\tau^2)] \tag{21}$$

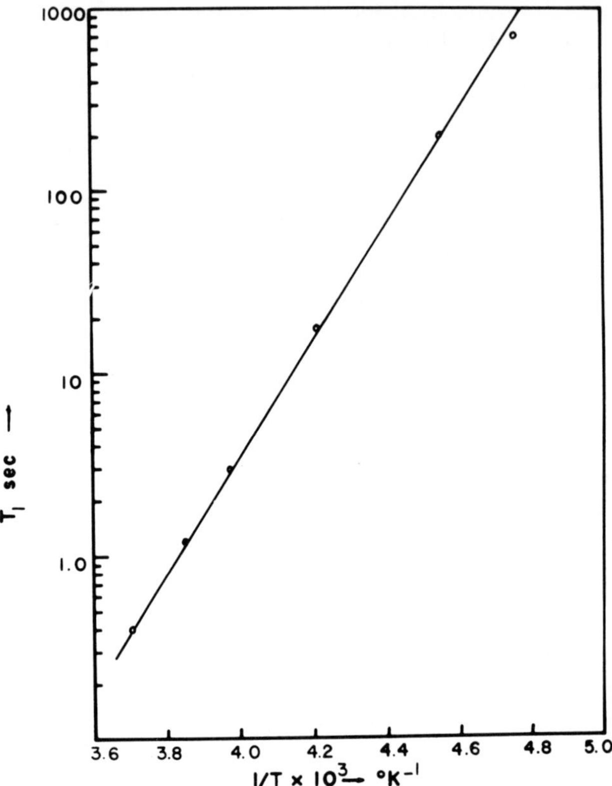

Fig. 8. Proton spin–lattice relaxation time versus reciprocal absolute temperature for single-crystal hexagonal ice.

where $\omega_1 = \gamma H_1$ and H_1 is the magnitude of the magnetic vector of the rf field. Thus, a knowledge of ω_1 permits τ to be evaluated without knowledge of G', from a plot of $T_{1\varrho}$ versus $\omega_1{}^2$. The value found is $\tau(-0.8°\text{C}) = 1.2 \times 10^{-6}$ sec with an activation energy of 14.3 kcal mol^{-1}. The agreement with the

TABLE VI. Comparison of the Time between Rotational Jumps of H_2O in Ice from NMR and Dielectric Relaxation Measurements[1016]

t, °C	τ_D, sec	τ_{NMR}, sec
-3	5×10^{-5}	2.5×10^{-6}
-10	1×10^{-4}	5.5×10^{-6}
-36	1.8×10^{-3}	1.1×10^{-4}

previous indirect measurements is very good. These values agree with those for the diffusion times in ice.(809) In short, the results indicate that in pure hexagonal ice NMR relaxation is dominated by a hopping motion of molecular vacancies (diffusion of Schottky defects) and not by migration of Bjerrum defects, as reflected by the dielectric phenomenon.

In liquid water the dielectric and NMR relaxation phenomena both have the same activation energy *and* reorientation times, because on any model, structural diffusion is at least 10^5 times faster than these defect diffusions.

Information on the positions and motions of the protons in ice may also be obtained from studies of the line *shape*. These may be performed in two ways: (1) by following the time-dependent decay of the NMR signal

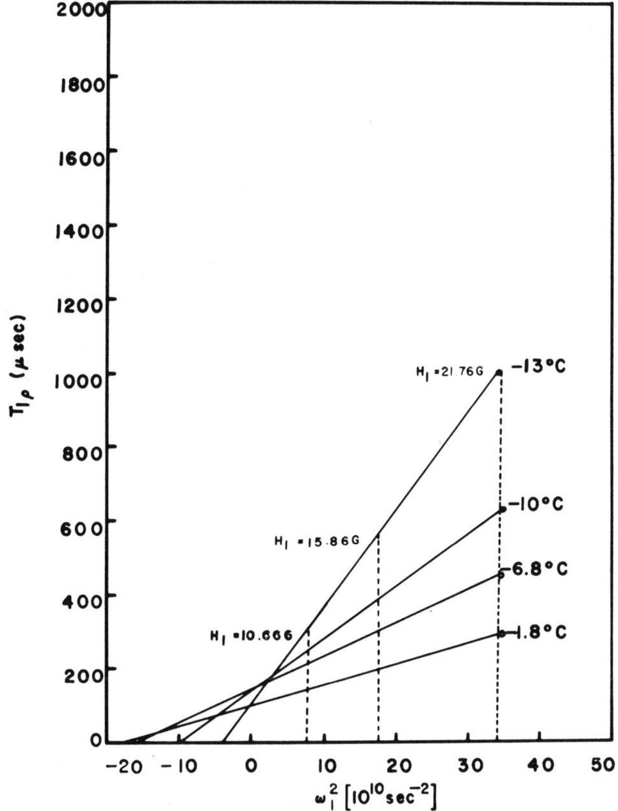

Fig. 9. Rotating-frame spin–lattice relaxation time versus temperature and frequency ($\omega_1 = \gamma H_1$) for single-crystal hexagonal ice.

TABLE VII. Experimental and Theoretical Second Moments in Ice (in gauss²)

Theoretical M_z	Polycrystalline	Single crystal		
		$H_{\hat{c}}$	$H_{\hat{y}}$	$H_{\hat{x}}$
D_2O lattice parameters rigid	30.4	38.5	33.3	33.0
vibr. corr.	29.2	36.8	31.9	31.6
1.01 Å OH, tet angle rigid	30.8	39.4	34.0	33.8
vibr. corr.	29.6	37.7	32.6	32.4
1.00 Å OH tet angle rigid	31.8	40.7	35.2	34.9
vibr. corr.	30.6	38.9	33.6	33.4
Experimental	32.1 ± 2	37.8 ± 2.5	33.15 ± 2	

following application of a 90° pulse (free induction decay), and (2) by considering the NMR line shape.

The value desired is the second moment M_2 of the NMR signal, which may be related to the relative positions of the protons in the ice lattice via the van Vleck formula,[1093] as corrected for vibrational and librational moments.[545]

The experiments using wide-line techniques do not appear to have been successful, due to the difficulties associated with saturating the NMR signal and with impurities in the ice.[643]

The values obtained from free induction decay experiments, along with theoretical values for polycrystalline and hexagonal ice, are given in Table VII. From changes in the shape of the decay with temperature it appears that at -25°C a molecular rotation or proton translation takes place every 3–5 × 10^{-15} sec, whereas at -5°C the time is 1–5 × 10^{-6} sec. Furthermore, the data fit a model based upon normal libration superimposed upon a rigid lattice, within experimental error. Theoretical studies show that a 25% increase in the second moment would take place if high-frequency quantum mechanical tunneling were present.[8] Therefore, absence of tunneling can be inferred.

3.4. Quadrupole-Induced Relaxation

In the case of labeled water molecules containing ^{2}H and ^{17}O, the relaxation of these nuclei is governed completely by the intramolecular process at ordinary temperatures.[392] Thus,

$$(1/T_1)_i = K_i(e^2qQ/\hbar)_i^2[J_2^1(\omega)_{Q,i} + J_2^2(2\omega)_{Q,i}] \tag{22}$$

where the index refers to the ith nucleus and where coupling between molecule and nucleus is characterized by an energy $e^2gQ/\hbar$ involving the electric field gradient at the nucleus, q, and the nuclear electric quadrupole moment Q. The only way that the problem becomes tractable is to assume the following:

(a) $J_l^m(\omega) = J_l^m(2\omega)$, where

$$J_l^m(m\omega)_i = m\tau_{r,i}/(1 + m^2\omega^2\tau_{r,i}^2)$$

and $\tau_{r,i}$ is the intramolecular reorientation time of species i;

(b) $J_l^m(m\omega)_{^{17}\mathrm{O}} = J_l^m(m\omega)_{^2\mathrm{H}}$;

(c) $J_l^m(m\omega)_Q = J_l^m(m\omega)_{\mathrm{DIPOLE\text{-}DIPOLE}}$.

The implications of the second and third assumptions have not been explored because of the difficulty of finding the recorrelation function for an asymmetric oscillator even in the frictional limit.[1014]

From a knowledge of $1/T_1$ for protonic water, and upon the assumption that the contribution to it from the *inter* term is small, one may compute τ_r for a molecule in the liquid from the gas value for the H—H distance. With this value and the aid of assumptions (a), (b), and (c), the liquid values of coupling constants may be *estimated.*

3.5. Measurement and Interpretation of Deuteron and ^{17}O Relaxation in Water

Some discussion has developed based on experiments arising from this line of reasoning. Work on D_2O[1177] shows a break in the curve of $(1/T_1)_{^2\mathrm{H}}$ versus $1/T$ not found for protonic water (Fig. 10). This indicates that the deuteron coupling constant is temperature-dependent. Such dependence is what one expects for a hydrogen-bonded system where the O—H—O distances change as a function of temperature. In hydrogen-bonded *crystals* a successful empirical relationship between $e^2qQ/\hbar$ and O—O distance has been developed.[806,1004]

$$(e^2qQ/\hbar)_{^2\mathrm{H}} = 315 - \frac{3\times 194.7}{R_{\mathrm{O-O}}^3} \tag{23}$$

where $R_{\mathrm{O-O}}$ is in Å.

X-ray diffraction measurements[485] suggest that liquid water consists of a distribution of O—O distances due to lattice and interstitial molecules (see Chapter 8). The relative amounts, and their bond distances, change with temperature, and hence it is clear that several combinations could lead to the deuteron relaxation results.

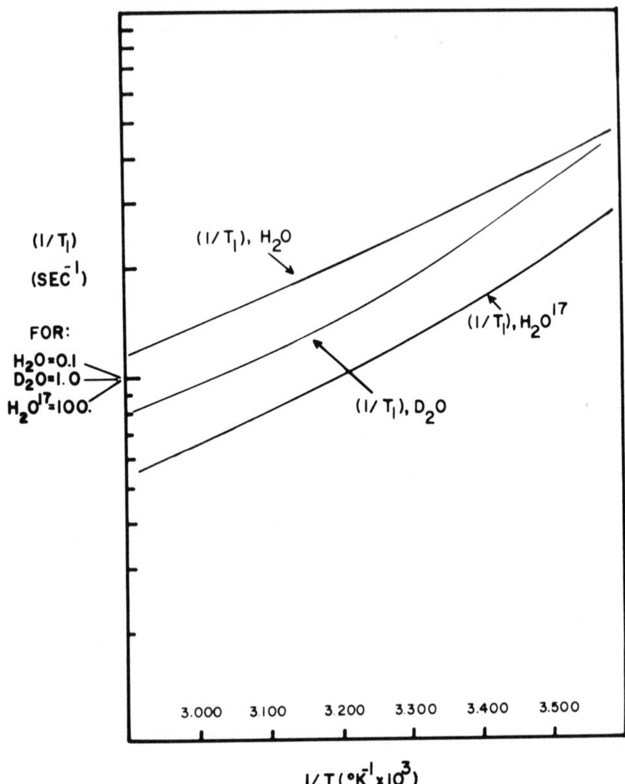

Fig. 10. Observed spin–lattice relaxation rates versus reciprocal absolute temperature for protons in H_2O, deuterons in D_2O, and oxygen-17 in $H_2{}^{17}O$.

The value estimated for the deuteron coupling constant in the liquid at 25°C, 220 kHz,[868] is about 2/3 the gas value of 315 kHz.[1074] These may be compared to the solid state value of 215 kHz.[1129]

A reasonable theoretical interpretation of these results may be obtained using a linear, 13-electron $O_1 \cdots D \cdots O_2$ system with molecular orbitals constructed with sp^3 hybrids.[806,1157] The valence-bond wave-function treatment,[212] where the hydrogen bond is treated on the basis of three configurations (see also Pauling[828]) and emphasis is placed on covalent contributions to hydrogen bond stabilization, does not explain the variation of deuteron coupling constant with bond length in crystals, and presumably therefore not in water either. Hence, the valence-bond treatment, which is the basis of much discussion in the literature concerning the cooperative nature of hydrogen bonds,[361] does not satisfy at least this

TABLE VIII. Deuterium Coupling Constants for Gas-Phase D_2O

Model	e^2qQ/h, kHz	Asymmetry parameter, η
SCF-MO-LCAO[1157]	385	0.08
SCF-Gaussian lobe[468] contracted	347.0	0.107
SCF-Gaussian lobe[791] uncontracted	343.9	0.133
Experimental[468]	315 ± 7	0.115 ± 0.061

experimental test. The free-molecule coupling constants have been calculated on several models as shown in Table VIII.

The value for solid D_2O on the above model is calculated to be 219 kHz (Table VIII). The conclusion to be drawn from these measurements is that the lowering of the coupling constant from the gas phase is simply another reflection of the associated nature of water. No conclusion as to percentage of hydrogen bonds may logically be drawn, since there is doubtlessly a distribution of bond lengths in the liquid.

Oxygen-17 quadrupole coupling constants are of more potential value in yielding information on the state of liquid water. This is because the coupling constant for oxygen should be very sensitive to changes in *p*-electron density, so that hydrogen bonding involving the lone-pair electrons should be reflected in large changes in the coupling constant. A comparison of the experimental gas-phase value and the various theoretical values is shown in Table IX.

The most successful theoretical values (compared with deuteron calculations[791]) are those involving Slater-type orbitals with contracted and uncontracted Gaussian functions. Unfortunately, these are of little use in obtaining physical insight into the factors contributing to the final value.

TABLE IX. Oxygen-17 Quadrupole Coupling Constants and Asymmetry Parameters in $H_2{}^{17}O$ (principal axis system)

	e^2Qq/h, MHz			η
	xx	*yy*	*zz*	
Harrison[468]	11.37	−9.97	−1.40	0.76
Neumann and Moscowitz[791]	10.44	−9.47	−0.975	0.813
Experimental (gas) Stevenson and Townes[1023]	6.96	−9.88	2.92	1.84

TABLE X. Calculated Values of ^{17}O Quadrupole Coupling Constants (MHz) in D_2O Ice ($t = -10°C$)

Parameter	Ref. 373	Ref. 1008
$(e^2qQ/h)_{xx}$	$\mp 1.80 \pm 0.05$	—
$(e^2qQ/h)_{yy}$	0.01 ± 0.05	—
$(e^2qQ/h)_{zz}$	1.79 ± 0.05	6.66 ± 0.10
	1.01 ± 0.04	0.935 ± 0.01

The solid-state value appears somewhat uncertain. The results from the only laboratory working on the problem are given in Table X. Both sets of results were calculated from the same experimental data. Thus, at present it can be stated with certainty only that the coupling constant is lower in ice than in the gas.

Using the same method as for D_2O, the coupling constant in the liquid may be estimated. Table XI shows the derived values at 25°C.[390,1127]

The value seems to be the same for H_2O and D_2O. No work has been done to correlate the coupling constants for ^{17}O to hydrogen bond formation, as discussed above for the deuteron constants. If the constant is indeed temperature-dependent, then the effect is rather small over the range from 0 to 80°C (change less than 10%), and hence either (1) it is insensitive to the changes in liquid structure, or (2) there is little change in the structure on a time scale of the order of the molecular reorientation time. At present there seems to be little chance of distinguishing between these possibilities.

The activation energies for the relaxation rates in all cases show non-Arrhenius behavior. They vary between 5 and 3.5 kcal mol^{-1} in the range 0–100°C. The values for D_2O relaxation[1177] are a little higher than the other measurements. However, the variation with temperature in all cases follows very closely that for macroscopic viscosity, self-diffusion coefficient, and dielectric relaxation.

TABLE XI. Estimate of ^{17}O Quadrupole Coupling Constants in Water

Species	Ref. 1008	Ref. 390	Ref. 514
$D_2{}^{17}O$	7.7 ± 1	—	—
$H_2{}^{17}O$	7.7 ± 1	8.4 ± 1	7.9 ± 0.3

3.6. Summary of Relaxation Experiments on Liquid Water

The proton, deuteron, and ^{17}O measurements are consistent in so far as they yield reasonable parameters when calculated from a rotational reorientation time of $2.5 \pm 0.2 \times 10^{-12}$ sec (25°C). The value for the dielectric relaxation time 8×10^{-12} sec[864] can be made to conform to the value by application of various necessary correcting factors. In fact, however, one of the unsolved problems is that the macroscopic dielectric measurements cannot distinguish between a single microscopic relaxation time and a distribution,[947] while the NMR measurements which theoretically could do so cannot be performed in the frequency range required (10^6 MHz). In short, the nature of the relaxing species is unknown. There is evidence that the τ for water decreases in "inert" organic solvents such as CCl_4,[513] which in turn indicates that the relaxing species in pure water may not be individual molecules. However, until techniques are much more sensitive than at present, these matrix isolation experiments cannot be extrapolated to the low concentrations necessary for quantitative conclusions. It has begun to appear that NMR relaxation measurements will be very useful in determining *differences* between bulk water and water in the hydration spheres of dissolved ions and macromolecules in water, but not very helpful in determining the absolute structure of either form.[310,311,393,394]

3.7. Measurement of Self-Diffusion Coefficient

The self-diffusion coefficient in water may be measured by a modification of the usual relaxation experiment in which a steady or pulsed magnetic field gradient is introduced in the magnet system. The results of measurements from 0°C to the critical point are shown in Fig. 11.[474]

As mentioned above, the *inter* term in proton relaxation is rather model-insensitive. It is this term which includes the self-diffusion coefficient. Figure 11 also shows the theoretical curve obtained using the classical Stokes–Einstein equation,

$$D = kT/6\pi\eta a \tag{24}$$

where a is the molecular radius, k is Boltzmann's constant, and η is the macroscopic viscosity. The agreement is quite good. In fact, Egelstaff[288] has pointed out that the self-diffusion coefficients for almost *all* liquids (including condensed rare gases) are of the same order, i.e., $\sim 10^{-5}$ cm^2 sec^{-1}. Hence, not much structural information could be expected either from classical tracer or NMR diffusion measurements. The activation

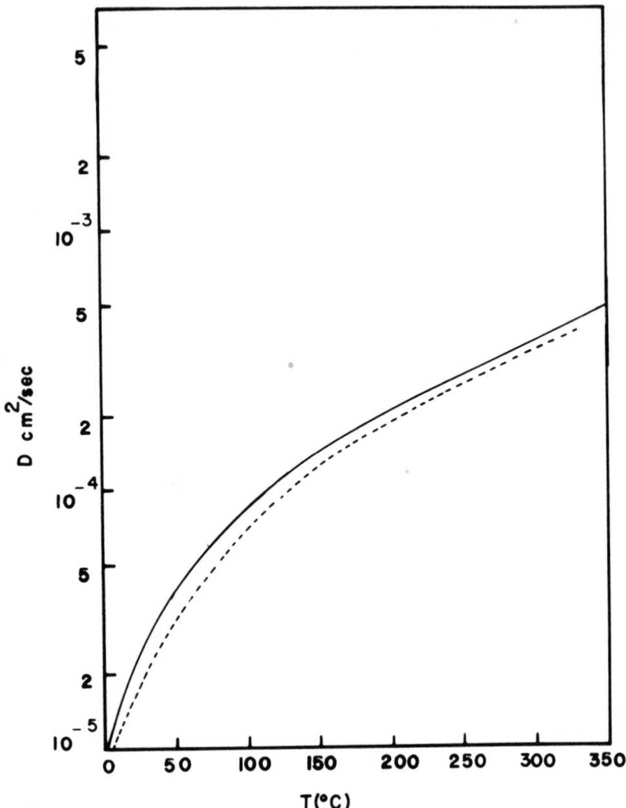

Fig. 11. NMR self-diffusion coefficients for protons in water versus temperature. Solid line is observed, dashed line is calculated on basis of the classical model.

energy for D is again non-Arrhenius in form below 170°C and has the same range of values as has η. Above 170°C the behavior is "normal," with $E_A = 1.6$ cal mol^{-1}.

3.8. Measurement of Kinetic Rate Constants in Water

One of the oldest and most fundamental problems in the physical chemistry of water is the evaluation of the rate constants characterizing the following processes[296]:

$$H_2O + H_3O^+ \xrightarrow{k_1} H_3O^+ + H_2O \tag{25}$$

$$H_2O + OH^- \xrightarrow{k_2} OH^- + H_2O \tag{26}$$

It has proved possible to arrive at an *estimate* for the constants k_1 and k_2 by NMR spectroscopic methods. The measurements make use of the following phenomenon: when observing T_1 and T_2 for both protons and ^{17}O in water containing some ^{17}O, it is found that T_1 is pH-independent while T_2 is not. The form of the curve is shown in Fig. 12.[739,880]

The qualitative explanation stems from the fact that the proton and ^{17}O are spin–spin coupled. When exchange is slow (at $p\text{H} = 7$) the shape of both the proton and ^{17}O signals, the widths of which are proportional to $1/T_2$, are partly collapsed multiplets which are obtained when transfer is faster than molecular reorientation processes in acid or basic solutions. Thus, in the latter case, $T_1 = T_2$, but at $p\text{H} = 7$, $T_2 \ll T_1$. The measured effect on proton relaxation is of course proportional to the ^{17}O concentration. Unfortunately, the accurate determination of k_1 and k_2 requires an

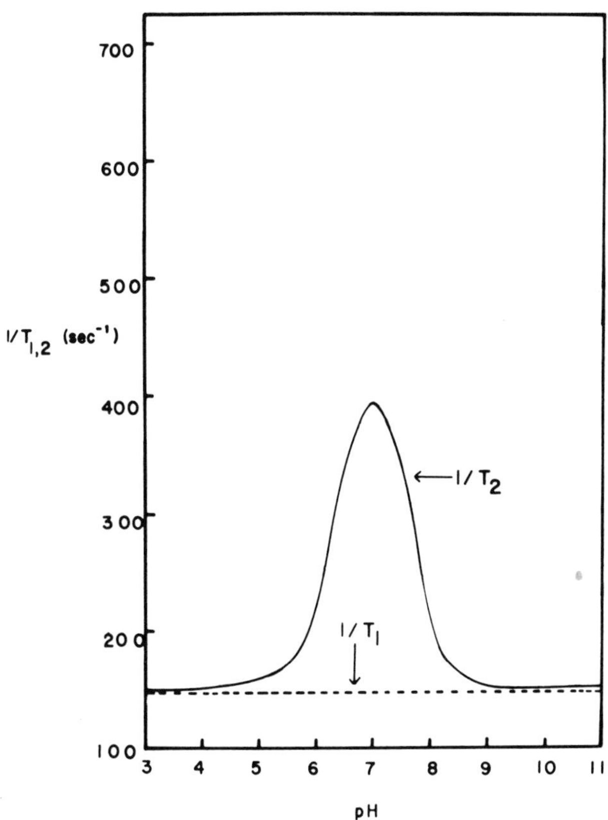

Fig. 12. Dependence of ^{17}O relaxation times upon pH for $H_2{}^{17}O$ at 29.0 ± 0.2°C.

accurate value for the spin–spin coupling constant $J_{^1H\text{-}^{17}O}$. However, even in the absence of transfer—such as in enriched water diluted by dry acetone—the spin multiplet is partly collapsed due to the effect of quadrupole relaxation of the ^{17}O on the ^1H. (See Kintzinger and Lehn[606] for the analogous problem of protons coupled to ^{14}N.) The correction for this effect is difficult, and hence the ambiguity in the derived rate constants. The experimental values can be fitted with appropriate rate constants for any value of the coupling constant greater than 46 Hz. The value used in previous work[739,397] is 92 ± 15 Hz. Using that latter value, it is found that, at 26°C,

$$k_1 = 8.2 \times 10^9 \quad \text{mol}^{-1} \text{ liter sec}^{-1} \tag{25'}$$

$$k_2 = 4.6 \times 10^9 \quad \text{mol}^{-1} \text{ liter sec}^{-1} \tag{26'}$$

with

$$1/\tau = \tfrac{2}{3}k_1[\text{H}^+] + \tfrac{1}{2}k_2[\text{OH}^-] \tag{27}$$

and activation energies

$$k_1: \qquad E_{A1} = 2.6 \quad \text{kcal mol}^{-1}$$

$$k_2: \qquad E_{A2} = 2.7 \quad \text{kcal mol}^{-1}$$

The changes in k due to changes in J are linear in J,

$$\Delta k \quad (\text{mol}^{-1} \text{ liter sec}^{-1}) = 0.04 \times 10^9 \, \Delta J \quad (\text{Hz}) \tag{28}$$

so that, for instance, the difference between $J = 80$ and 90 Hz is about 7%. In any case the k_1' derived are of the magnitude expected from a classical calculation based on the self-diffusion coefficient and anomalous mobilities of the hydroxyl and hydronium ions.[739]

4. THE CONTRIBUTIONS OF NMR METHODS TO THE STRUCTURE PROBLEM IN LIQUID WATER

4.1. Results from Chemical Shift Measurements

The interpretation of chemical shift data is so complicated for protons that the measurements seem to indicate only that on a time scale of 10^{-8} sec all microstructure effects are averaged out, and that there are no sudden changes in properties (on this time scale). Doubtless, the experimental results can be fitted on the basis of a number of models. The proton shift in the solid state has not yet been obtained. The ^{17}O shift data, which are of potentially more use, are much more inaccurate. It is not even clear why there is disagreement between laboratories on the linearity of the temperature dependence. Accurate ^{17}O shift measurements for the solid state do not exist.

Until an accurate theoretical treatment of the chemical shift phenomenon is available, it seems useless to apply the data to what is a very subtle structural manifestation. It is much easier to find differences than absolutes.

4.2. Results from Relaxation Measurements

The quantities of most theoretical interest in the structural problem are the rotational and translational recorrelation *functions*. The NMR relaxation times depend upon integrals of these functions, and hence can be expected to be relatively insensitive to the functions themselves. This is borne out in practice because there is no property yet measured which varies greatly from a theoretical value obtained on the basis of a simple classical calculation. This also holds true for the temperature dependence. It appears to be a problem for theoreticians to explain why water behaves microdynamically like a classical fluid. Certainly, relaxation experiments show that a water molecule in the liquid does not undergo quantized rotation for times longer than 10^{-13} sec. However, this is another way of saying that water is highly associated.

The quadrupole coupling constants are of potential value in evaluating an ensemble-averaged electronic structure for the liquid for comparison with models.

While rotating-frame NMR relaxation experiments have not been performed with liquid water, it appears that there is good agreement between dielectric and NMR recorrelation times and their activation energies. Therefore, in contrast to ice, the mechanism of molecular reorientation is the same. Dielectric relaxation depends upon the recorrelation function $G_1{}^1(t)$, while dipole–dipole and quadrupolar NMR relaxation depend upon $G_2{}^m(t)$. That is, in the former, dependence is on first-order spherical harmonics, while in the latter it is on second-order ones. It is useful to know that the time dependence of the two is the same, since there is no *a priori* reason why they should be.

The major conclusions from NMR work on pure water systems are concerned with: (1) the inadequacy of the valence-bond approach to predict quadrupole coupling constants correctly; (2) the relative importance of paramagnetic and diamagnetic terms in the chemical shifts of ^1H and ^{17}O;

(3) the experimental derivation of some basic kinetic constants involving proton and hydroxyl ion mobility in water; and (4) the almost classical motional behavior of the water molecule in liquid water.

The fact that there are large experimental differences between water in its pure state and the water found in solvated systems means that the NMR tool will be very useful in building upon the empirical and theoretical knowledge described in this chapter for the purpose of describing the microenvironments of water molecules in different surroundings.

CHAPTER 7

Liquid Water: Dielectric Properties

J. B. Hasted
Birkbeck College
University of London
London, England

1. INTRODUCTION

Dielectric investigations are directly relevant to questions of molecular structure. During the 1930's many measurements of the dielectric constants of vapors and dilute solutions were made and the electric dipole moments of polar molecules deduced therefrom. Interpretations of the electric dipole moment in terms of bond length, bond angle, and electronegativity gave way to interpretations in terms of molecular wave functions. The electric moment is one of the most important molecular constants and its reproduction is a valuable test of the accuracy of a wave function. The molecular polarizability, or dipole induced by unit electric field, is another important quantity obtainable from dielectric measurements.

At the same time dielectric theory proved somawhat inadequate to interpret the properties of condensed media, and its improvement came only as a result of intense analytical effort. A method of relating the static field dielectric constant to the structural correlation of polar molecules in a liquid was worked out by Kirkwood[(607)] and Fröhlich[(371)]; its application is of value in studies of water, and to some extent it is a "model-sensitive" theory. The frequency dependence of the complex dielectric constant is also of great interest, in that information can be obtained about the energies of the processes of molecular orientation. The study of dielectric relaxation became popular in the 1950's but the theoretical background to the subject is only approximate.

In this chapter these topics are considered in the order indicated in the foregoing paragraph. The electrical properties of the free molecule will be considered first.

2. THE ELECTRICAL PROPERTIES OF THE FREE WATER MOLECULE

The permanent electric dipole moment of the water molecule H_2O has been deduced from measurements of the temperature variation of the dielectric constant of the vapor, as well as from other experiments. A review of the measurements[(711)] leads to a value $\mu = 1.84 \pm 0.02 \times 10^{-18}$ esu cm (that is, 1.84 D, the symbol D denoting the Debye). One of the most accurate results is that due to Sänger and Steiger[(943)]; the treatment has been reexamined by Moelwyn-Hughes, who considers the most probable value of μ to be 1.83_4 D.

The dipole moment vector bisects the H—O—H angle. It points from the negative oxygen atom to the positive region between the hydrogen atoms. The OH bond lengths are 0.95718 Å, and H—O—H bond angle 104.474°; these free-molecule values are not maintained in the liquid. A number of point-charge models of the H_2O dipole have been constructed using these values; but it is even more important to attempt to reproduce the permanent dipole moment by a wave function which completely describes the electron density.

The mean quadrupole moment Q is known from experiment[(387)] to be $-5.6 \pm 1.0 \times 10^{-26}$ esu cm^2, but the individual values Q_{xx}, Q_{yy}, and Q_{zz} are at present unknown experimentally, although they can be calculated from wave functions. The individual quadrupole moments are defined by

$$Q_{ab} = \int r_a r_b \varrho(\mathbf{r})\, d\tau \tag{1}$$

where a and b represent any two of coordinates x, y, z; $\varrho(\mathbf{r})$ is the total charge density; r_x, e.g., is the x component of the vector $\mathbf{r}$; and $d\tau$ an element of volume. The calculated values of the individual moments are $Q_{xx} = -6.56 \times 10^{-26}$, $Q_{yy} = -5.18 \times 10^{-26}$, and $Q_{zz} = -5.73 \times 10^{-26}$ esu cm^2. These values are referred to the molecular centre of mass as origin. From them one calculates the mean quadrupole moment $Q = -5.8 \times 10^{-26}$ esu cm^2. The individual moments Q_{xy}, Q_{yz}, and Q_{zx} are zero and

$$Q = \tfrac{1}{3}(Q_{xx} + Q_{yy} + Q_{zz}) \tag{2}$$

The only octopole moments which are nonvanishing are calculated to be R_{xxz}, R_{xzx}, $R_{zxx} = -1.08 \times 10^{-34}$; R_{yyz}, R_{yzy}, $R_{zyy} = -0.50 \times 10^{-34}$; $R_{zxz} = -2.75 \times 10^{-34}$ esu cm^3. The definition of the octopole moments is

$$R_{\alpha\beta\gamma} = \int r_\alpha r_\beta r_\gamma \varrho(\mathbf{r}) \, d\tau \tag{3}$$

where α, β, and γ can all refer to coordinates x, y, and z.

The negative signs of the higher moments show that the contributions to them from the electrons are greater than the contributions from the nuclei. The near equality of the quadrupole moments shows that the charge distribution is not far from spherical. The electric charge is certainly not confined to the nuclear plane.

The molecular polarizability α is available experimentally, as a mean value, from dielectric and optical refraction measurements:

$$\bar{\alpha} = \tfrac{1}{3}(\alpha_{xx} + \alpha_{yy} + \alpha_{zz}) = 1.444 \times 10^{-24} \quad \text{cm}^3 \tag{4}$$

The components of the tensor are not known, although a method does exist for deriving them from the Kerr constant together with unavailable data on the depolarization of Rayleigh scattering.[(105)] Measurements of the Kerr[(816)] constant indicate that the anisotropy of polarizability is small.

A further electric constant of the molecule is the change of dipole moment during molecular vibration.[(210)] It is possible to deduce this change from the line strengths for vibrational absorption, and although these are not presently available in absolute form, electron spectroscopic investigation is in progress.* However, the assumption must be made that the total dipole moment of the molecule is equal to the vector sum of the bond moments, and owing to possible contributions from the oxygen lone-pair electrons, this may not be valid.

Point-charge models for reproducing the dipole moment of H_2O have been fairly numerous.[(471)] Verwey[(1109)] took account of the screening of the proton field by surrounding electrons, and proposed a planar model, illustrated in Fig. 1(a). Planar models are less realistic than nonplanar ones, and an attempt to represent the effect of the lone-pair electrons was made by Pople,[(854,855)] whose model is shown in Fig. 1(b). The bending of $2p$ bonds which is predicted by wave-function calculations was shown by Burnelle and Coulson[(152)] to give rise to dipole moments perpendicular to those directed along the bonds; they are in the nuclear plane, and for a

* D. Bostock and J. B. Hasted, unpublished data (1970).

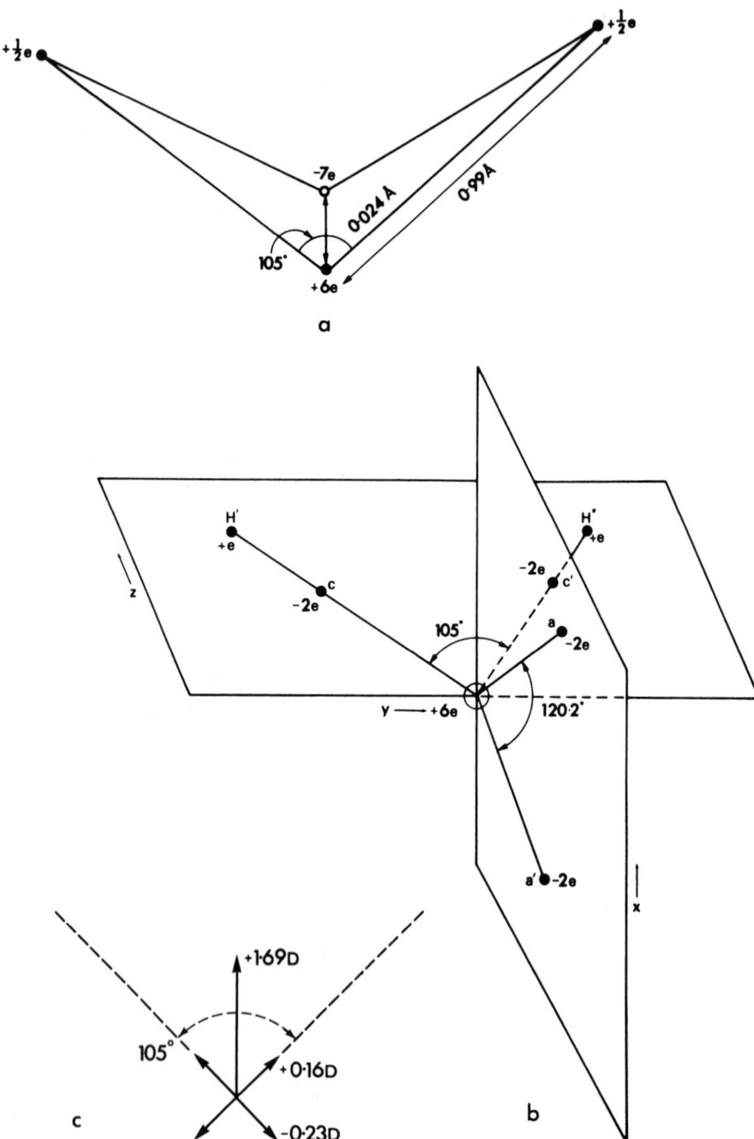

Fig. 1. Proposed electrical models for the H_2O molecule: (a) according to Verwey;[1109] (b) according to Pople;[854,855] (c) according to Burnelle and Coulson.[152]

bond angle of 105° they take the values given in Fig. 1(c). A multipole expansion model was put forward by Coulson and Eisenberg,[214] including a point dipole, a point quadrupole, and a point octopole, all at the oxygen nucleus.

Wave functions for the water molecule were calculated by Duncan and Pople,[280] then by Ellison and Schull,[307] and during the 1960's many calculations have appeared. Use of the Hellmann–Feymann theorem[40] enables tests to be carried out on the wave functions, which must produce a charge density in the "binding region" (inside the H—O—H angle) in excess of the charge density that would arise from the unbonded atoms.

In this way it is found that the lone-pair oxygen electrons must approximate to *sp* hybrids, that the orbitals of the oxygen that overlap the hydrogens must be almost pure 2*p*, and that the angle between them must be much smaller than the H—O—H angle. The lone-pair electron lobes project above and below the nuclear plane of the molecule, and contribute to the dipole moment to an extent which is not exactly known. An electron density contour map[41] in the nuclear plane is shown in Fig. 2.

The accuracy with which the different wave functions reproduce the experimental dipole moment (1.83 D) can be gauged from Table I, where the calculated values are listed.

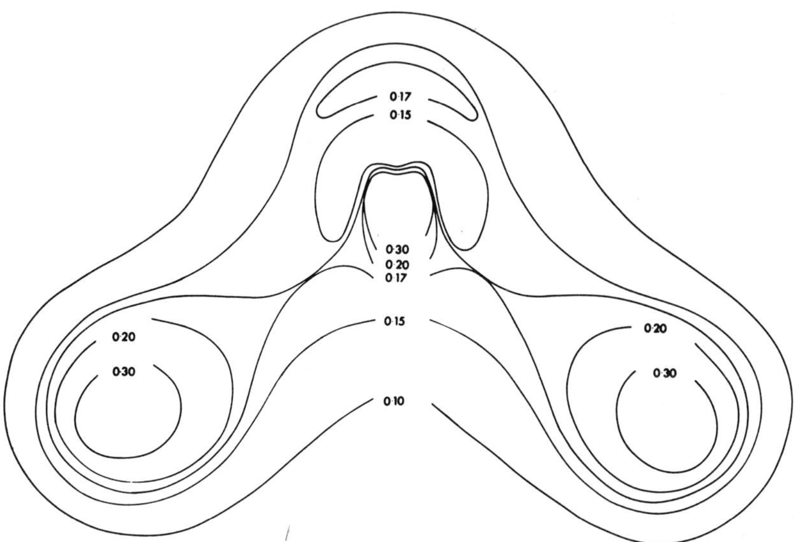

Fig. 2. Electron density contours for the H_2O molecule in the plane of the nuclei.

TABLE I. Calculations of H_2O Dipole Moment (1.83 D) from Wave Functions

Author	Date	μ, D
Ellison and Shull[(307)]	1955	1.52
Pitzer and Merrifield[(849)]	1969	1.434
McWeeny and Ohno[(719)]	1960	1.76
Moccia[(758)]	1964	2.085
Moskowitz and Harrison[(767)]	1965	1.99
Whitten, Allen, and Fink[b]	1966	2.50
Pitzer[a]	1966	1.916

[a] R. M. Pitzer, private communication (1966) quoted by Eisenberg and Kauzman.[(301)]
[b] J. L. Whitten, L. C. Allen, and W. H. Fink, private communication (1966) quoted by Eisenberg and Kauzman.[(301)]

3. THE MEASURED STATIC DIELECTRIC CONSTANT

Over 30 determinations of the static field dielectric constant ε_s of water have been reported in the literature, and between 0 and 100°C the data are very accurate. No significant frequency dependence is observed below 100 MHz.

The accepted modern measurements of the static dielectric constant of water are those due to Malmberg and Maryott,[(725)] who measured ε_s at 5° intervals from 0 to 100°C at atmospheric pressure. Measurements were made at frequencies between 3 and 96 kHz and were found to be 0.3% lower than those previously obtained by other workers between 0.5 and 81 MHz. A low-frequency ac bridge was used, of the equal ratio arm capacitance-conductance type, with a Wagner ground. Several cells of different designs, both two- and three-terminal, were connected to it, and a critical investigation of high-frequency methods was made. Malmberg and Maryott claimed a maximum uncertainty of $\pm$ 0.05 units. The best fit to their data is using the equation

$$\varepsilon_s = 87.740 - 0.4008T + 9.398 \times 10^{-4}T^2 - 1.410 \times 10^{-6}T^3 \qquad (5)$$

where T is the temperature in °C. The temperature coefficient $d(\ln \varepsilon_s)/dT$ derived from this equation is almost constant at $-4.55(\pm 0.03) \times 10^{-3}$ deg^{-1}

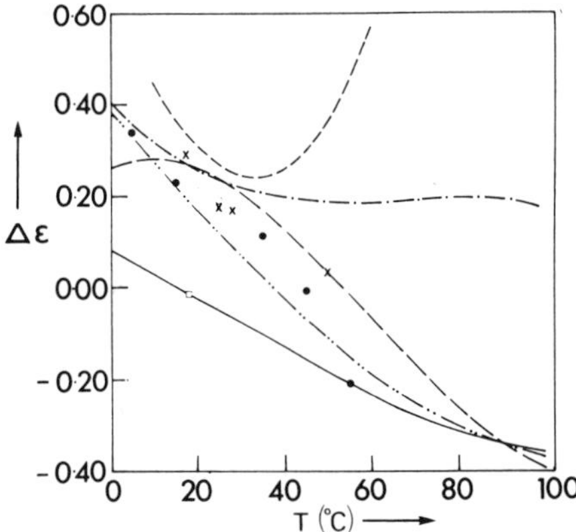

Fig. 3. Differences between measurements of static dielectric constant of water as a function of temperature; $\Delta\varepsilon = \varepsilon_s - \varepsilon_s$(Maryott and Malmberg).[725] (□) Mean of 17 values selected from literature by Lattey *et al.*;[658] (×) Albright (earlier work); (—·—) Wyman and Ingalls.;[1187] (——) Lattey *et al.*;[658] (○) Jones and Davies; (●) Albright and Gosting;[6] (- - -) Drake *et al.*;[275] (— —) Wyman (earlier work); (—··—) Akerlof and Oshry.[4]

from to 0 to 100°C. Figure 3 shows a graphic comparison of the differences between previous data and those of Malmberg and Maryott.

Recent measurements by Rusche and Good[930] are in substantial agreement with the latter. It is clear from mathematical analysis of these two sets of measurements that the "kink" supposed[277] to be present around 15°C in the temperature dependence of ε_s is not confirmed.

Earlier measurements by Wyman and Ingalls[1187] show almost the same temperature dependence but are larger by about 0.25. The measurements of Vidulich *et al.* between 0 and 40°C are also larger by 0.17. Wyman and Ingalls fit their data to the equation

$$\varepsilon_s = 78.54[1 - 0.00460(T - 25) + 0.0000088(T - 25)^2] \tag{6}$$

Measurements have been made by Akerlof and Oshry[4] from 100°C right up to the critical point. The closed vessel pressure is maintained, that is, the liquid is in equilibrium with its vapor. The authors used the method of Wyman at frequencies between 430 and 840 MHz. The static dielectric constant falls continuously with rising temperature to 9.74 at 370°C. The best fit to the data is using the equation

$$\varepsilon_s = 5321T_K^{-1} + 233.76 - 0.9297T_K + 0.1417 \times 10^{-2}T_K{}^2 - 0.8292 \times 10^{-6}T_K{}^3 \tag{7}$$

where T_K is the temperature in °K (see also Chapter 13).

The static dielectric constant of D_2O was reported by Malmberg (1958) to be fitted between 0 and 100°C by the equation

$$\varepsilon_s = 87.482 - 0.40509T + 9.638 \times 10^{-4}T^2 - 1.333 \times 10^{-6}T^3 \tag{7a}$$

The variation of the static dielectric constant of H_2O in intense electric fields was first investigated by Malsch,[726,727] who found that in a field $E = 250$ kV cm^{-1} the dielectric constant is lower than in weak fields by an amount $\Delta\varepsilon_s = 0.007\varepsilon_s$. Although this "saturation effect" is qualitatively correct according to theories of Debye,[245] Böttcher,[105] van Vleck,[1092] and Booth,[102] criticisms have been made of the measurements, and further investigations are in progress in the laboratory of Piekara.[840,841]

A variety of complications are to be expected in intense electric fields. The dependence of ε_s upon the field is given by Booth as[102]

$$\varepsilon_s = n^2 + (\varepsilon_{s0} - n^2)(3/u)[(1/\tanh u) - (1/u)] \tag{8}$$

where

$$u = \beta E \tag{9}$$

with

$$\beta = \tfrac{5}{2}(n^2 + 2)\mu/kT = 4.18 \times 10^{-4} \qquad \text{at} \quad 25°\text{C} \tag{10}$$

and from Malsch's measurements the value of β is deduced to be 3.9×10^{-4} at 25°C. An inverse, or positive saturation effect has been observed in polar liquids,[839,840,841] and was interpreted qualitatively by Piekara and Kielich[842] as arising from the molecular interactions in the liquid. There is also electrostriction in strong electric fields which increases the density and therefore the dielectric constant. The inverse saturation effect is balanced in water by a strong saturation predicted by Schellman[951] on the basis of Kirkwood correlation between dipoles. The balance in water of strong saturation and inverse saturation may well be such that the actual observed saturation is weak, and in approximate coincidental agreement with the older theories. Further experimental data are awaited with interest.

4. THEORY OF THE STATIC DIELECTRIC CONSTANT

The dielectric constant of a polar liquid arises from the orientation of molecular permanent electric dipoles μ and the electrical distortion of the molecules. Both effects are expressible in terms of a quantity P, the polarization, or dipole moment per unit volume of a continuous material:

$$P = (N_0/V)\bar{\mu}_0 \tag{11}$$

where $\bar{\mu}_0$ is the average dipole moment per molecule, N_0 is Avogadro's number, and V is the molar volume. For an electric field E_m acting on the molecule the dielectric constant ε is given in the electrostatic system of units by

$$4\pi P = (\varepsilon - 1)E_m \tag{12}$$

The electrical distortion of a molecule is expressible in terms of a polarizability α such that the induced dipole moment is equal to αE_m. This induced moment gives rise to the optical refractive index n in the infrared region and beyond.

The distribution of orientations of permanent dipoles in a steady, uniform electric field is not random, but is a Boltzmann distribution in which the thermal fluctuations govern the populations in different energy states, that is, different orientations. The permanent dipole orientation polarizability which results from this distribution is

$$\alpha_{\text{or}} = \mu_0^2/3kT \tag{13}$$

where k is Boltzmann's constant.

The contribution to the dielectric constant which arises from permanent dipole orientation is then

$$\varepsilon - n^2 = 4\pi N_0\mu_0^2/3VkT \tag{14}$$

Application of this formula to liquid water at 20°C with $\mu_0 = 1.85$ D yields a static dielectric constant of 13 (six times too small) but the temperature variation is approximately of the correct form. This large discrepancy arises because the "local" electric field E_m acting on the molecular dipole is quite different from the applied field E, and has incorrectly been assumed to be identical with it. Debye attempted to approximate E_m by the expression due to Mossotti[(768)] and to Lorentz[(698)]:

$$E_m = E + (4\pi P/3) \tag{15}$$

But this introduces feedback into the algebra, so that infinite dielectric constants are postulated, and universal ferroelectricity of polar media should be displayed above a critical temperature. This is known as the "Mossotti catastrophe."

To rectify the situation, Onsager[808] made a calculation of the local field based upon a model with a point dipole at the centre of a cavity of radius a and polarizability α, inside a continuous material of dielectric constant ε. The effect of this assumption is that the effective dipole moment is enhanced by an amount

$$\Delta\mu_0 = [(\varepsilon - 1)/(2\varepsilon + 1)]\mu_0 \tag{16}$$

This is equivalent to the application of an inner reaction field

$$E_r = 2[(\varepsilon - 1)/(2\varepsilon + 1)]\mu_0/a^3 \tag{17}$$

parallel to the dipole; there is also a cavity field

$$E_c = [3\varepsilon/(2\varepsilon + 1)]E \tag{18}$$

which is parallel to E. The vector resolution of these two fields makes up E_m:

$$\mathbf{E}_m = \mathbf{E}_r + \mathbf{E}_c \tag{19}$$

The total moment in the cavity is therefore

$$\mu_c = \mu_0 + E_m[n^2 - 1)/(n^2 + 2)]a^3 \tag{20}$$

which leads eventually to the Onsager expression:

$$(\varepsilon - n^2)(2\varepsilon + n^2)/\varepsilon(n^2 + 2)^2 = 4\pi N_0\mu_0{}^2/9VkT \tag{21}$$

Application of this equation with $n^2 = 1.85$ yields for liquid water at 20°C, $\varepsilon = 27$, which is still much lower than the experimental static value 80.4. Most of this discrepancy arises because no account has been taken of the structural correlation which can exist between permanent dipoles. If a model can be set up which describes this correlation, then the static dielectric constant can be calculated. The theory derived by Kirkwood[607] for polar liquids and extended to general applicability by Fröhlich[371] is highly sensitive to this correlation.

The theory is strictly applicable only to isotropic substances composed of coupled permanent dipoles which are not polarizable. In sufficiently high-frequency alternating electric fields the permanent dipoles are unable to reorient in the rapidly changing local field, and cannot contribute to the dielectric constant; the only remaining dielectric constant ε_∞ must arise from the atomic and electronic polarization. In water and ice this high- ("infinite"-) frequency dielectric constant is only a few per cent of the static value, and Kirkwood found it convenient simply to add the polarizable molecule contribution to the permanent dipole contribution, to obtain the static dielectric constant. However, this is an inadequate approximation, in that the polarizable molecule contribution is not necessarily the same in the presence of field-oriented dipoles as it is in the absence of field orientation.

The important quantities in the Kirkwood theory are: (1) $\boldsymbol{\mu}$, the average dipole moment of an H_2O molecule surrounded by its neighbors. This moment exceeds the moment $\boldsymbol{\mu}_0$ of the isolated molecule by the polarizable molecule contribution, equal to the product of field at the molecule $\mathbf{E}_m$ and polarizability α, so that $\boldsymbol{\mu} = \alpha\mathbf{E}_m + \boldsymbol{\mu}_0$. (2) $\boldsymbol{\mu}^*$, the vector sum of the dipole moments of all the neighbors of a given molecule, which is held in fixed orientation. $\boldsymbol{\mu}^*$ is a measure of the correlation between the orientation of dipoles. To place a limit upon the number of neighbors considered in the calculation of $\boldsymbol{\mu}^*$ is artificial. Except in ferroelectrics, the orientation is random and the vector sum zero at infinite distance. The only limit is that imposed by the accuracy which is required in the calculation. A starting-point is the general result [see eqns. (14), (18)]

$$\varepsilon - 1 = (4\pi/3V)[3\varepsilon/(2\varepsilon + 1)]\overline{\mathbf{M}^2}/kT \tag{22}$$

where

$$\overline{\mathbf{M}^2} = \int \mathbf{M}^2(X) \exp[-U(X)/kT]\, dX \Big/ \int \exp[-U(X)/kT]\, dX$$

is the average value of $\mathbf{M}^2(X)$ in the absence of a macroscopic field. A macroscopic spherical volume is assumed to consist of a number of charges each of which can be described in terms of the displacement from its lowest energy level position; X describes the set of all displacement vectors, and $\mathbf{M}(X)$ is the dipole moment of the spherical region. The projection of $\mathbf{M}(X)$ in the direction of an applied macroscopic electric field $\mathbf{E}$ is $\mathbf{M}(X) \cos\theta$, and the average value of $\mathbf{M}(X) \cos\theta$ is

$$\mathbf{M}_E = \int \mathbf{M}(X)(\cos\theta) \exp[-U(X, E)/kT]\, dX \Big/ \int \exp[-U(X, E)/kT]\, dX \tag{23}$$

where $U(X, E)$ is the energy of the system with configuration X and applied field E. It follows from electrostatics that

$$U(X, E) = U(X) - [3\varepsilon/(2\varepsilon + 1)]\mathbf{M}(X)E\cos\theta \tag{24}$$

where $U(X)$ refers to the energy in the absence of applied field. Expanding and using only the first term of $\exp[-U(X, E)/kT]$, and taking

$$\int \mathbf{M}(X)(\cos\theta)\exp[-U(X)/kT]\,dX = 0 \tag{25}$$

and $\overline{\cos^2\theta} = \frac{1}{3}$, leads to eqn. (22).

For further development the spherical region is divided into N units, each of which makes the same contribution to M,

$$\mathbf{M}(X) = \sum_{j=1}^{N} \boldsymbol{\mu}(x_j) \tag{26}$$

where $\boldsymbol{\mu}(x_j)$ is the dipole moment of the jth unit. Some substitutions lead to

$$\overline{\mathbf{M}^2} = \sum_{j=1}^{N} \int \boldsymbol{\mu}(x_j)\boldsymbol{\mu}^*(x_j)p(x_j)dx_j \tag{27}$$

where $\boldsymbol{\mu}^*(x_j)$ represents the average moment of the whole sphere for a fixed set x_j of displacements of the jth unit leading to a moment $\boldsymbol{\mu}(x_j)$; $p(x_j)$ is the probability of finding the jth unit with this particular set of displacements. Further substitutions lead to

$$\varepsilon - 1 = [3\varepsilon/(2\varepsilon + 1)](4\pi N_0/3V)(\overline{\boldsymbol{\mu}\boldsymbol{\mu}^*}/kT) \tag{28}$$

$\boldsymbol{\mu}^*$ is the average dipole moment of a sphere embedded in its own medium, with one unit held in a given orientation and possessing a dipole moment $\boldsymbol{\mu} \cdot \overline{\boldsymbol{\mu}\boldsymbol{\mu}^*}$ is the average value of $\boldsymbol{\mu}\boldsymbol{\mu}^*$, taking into account all possible configurations and weighting them according to the probability of finding the unit in this configuration.

The electronic polarization is introduced in terms of the refractive index n which arises from the induced dipole moment αE_m using eqn. (12). This leads to the Kirkwood equation

$$\varepsilon - n^2 = [3\varepsilon/(2\varepsilon + n^2)](4\pi N_0/3V)(\overline{\boldsymbol{\mu}\boldsymbol{\mu}^*}/kT) \tag{29}$$

$\overline{\boldsymbol{\mu}\boldsymbol{\mu}^*}$ is often expressed as $g\boldsymbol{\mu}^2$, where the constant g is known as the correlation factor.

An equation of this type was derived independently by Fröhlich[371] by a very general method, and in Appendix B1 of his "Theory of Dielectrics"[371] he improves the treatment by including in it the Onsager reaction field calculated with the aid of the static dielectric constant of the continuous medium in which the Kirkwood correlation sphere is imbedded. This leads to his equation (B1.41):

$$\varepsilon_s - \varepsilon_\infty = [(\varepsilon_\infty + 2)/3]^2[3\varepsilon/(2\varepsilon + \varepsilon_\infty)](4\pi N_0 g \mu_0^2/3VkT) \qquad (30)$$

which we shall call the Fröhlich equation; it is superior to eqn. (29), but is difficult to apply for the following reasons. The quantity ε_∞ refers to the dielectric constant at a frequency sufficiently high for the intermolecular contributions to have no time to contribute, but not sufficiently high for intramolecular contributions (induced polarization and vibrations of molecule) to be damped out. The choice of appropriate frequency for ε_∞ is not simple. This issue has been discussed for the case of water by Hill(506,507) and von Hippel(1115); it will be considered further below.

Note that in eqn. (30) the Kirkwood "surrounded dipole" μ is not used; the free molecule dipole μ_0 is preferred; however, the correlation calculation is carried out as before.

A different treatment, including the distortion polarization, was given by Harris and Alder(465,467); their calculations have been successfully criticized by Fröhlich,(370) O'Dwyer,(804) and Cole.(197)

5. THE STATIC DIELECTRIC CONSTANT CALCULATED ON VARIOUS MODELS

A number of calculations of the static dielectric constant of water have been made using the Kirkwood–Fröhlich theory. The parameter g must be calculated; $\mu^* = g\mu$ is the average dipole moment of a macroscopic sphere of the material immersed in a medium of the same dielectric constant, the central molecule being kept in the fixed orientation μ.

In the first calculation, due to Oster and Kirkwood,(818) it was assumed that the directions of only the N_1 nearest neighbors were correlated with the direction of the central molecule; this yields

$$g = 1 + N_1\langle\cos\gamma_1\rangle \qquad (31)$$

where $\langle\cos\gamma_1\rangle$ is the mean of the cosines of the angles γ_1 between the dipole moments of the neighbors and of the central molecule. The number of nearest neighbors was calculated from the area under the first peak of the X-ray diffraction radial distribution function due to Morgan and Warren.(761) The calculated value of g varied from 2.63 at 0°C to 2.82 at 83°C, a result which was questioned by the authors themselves; one would expect increased thermal agitation to distort or break down the structure and so decrease the

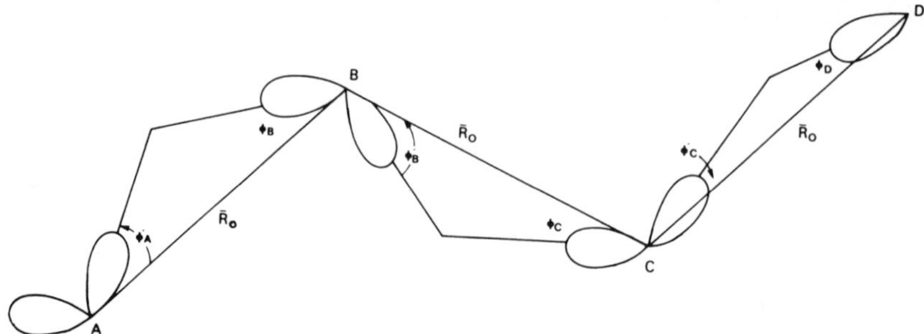

Fig. 4. Bending of hydrogen bonds proposed by Pople.[854,855]

angular correlation. The calculated static dielectric constant calculated by Oster and Kirkwood decreases too slowly with increasing temperature, as can be seen from the comparison made in Fig. 6a.

Pople[854,855] applied the Kirkwood theory to his own hydrogen bond bending model. In this theory the four nearest neighbors in the tetrahedral ice I tridymite structure are assumed to be distributed in a Gaussian manner centered on an appropriate distance. The water molecule is taken to be exactly tetrahedral; that is, the orbitals are tetrahedral, but bending of the bonds is possible, both the OH bond and the lone-pair to which it is hydrogen-bonded. The bending is at an angle ϕ to the line of the O—O internuclear vector, in the manner of Fig. 4. The energy of the system is then increased by a quantity

$$\Delta U = k_\phi(1 - \cos\phi) \tag{32}$$

where k_ϕ is the hydrogen bond bending force constant. A value for this force constant must be assumed in order to solve the statistical mechanics over a range of temperature. The radial distribution function of the second and third neighbors is derived by statistical mechanics and analytical geometry, and fitted to the Morgan and Warren[761] X-ray diffraction data over a range of temperature, for an empirically chosen k_ϕ. The best fit is obtained using a hydrogen bond bending force constant of 3.78×10^{-13} erg rad^{-2}, with 11 second neighbors and 22 third neighbors; the average nearest-neighbor distance $\bar{R}_0$ is 2.80 Å at 1.5°C and 2.95 Å at 83°C. The average bond distortion at 0°C is given as $\text{arc}\cos(\overline{\cos\phi}) = 26°$.

The application of Kirkwood theory to this model yields

$$g = 1 + \sum_{i=1}^{\infty} N_i 3^{(1-i)} \cos^{2i}\alpha[1 - (kT/k_\phi)]^2. \tag{33}$$

where i refers to the number of the coordination shell of hydrogen-bonded neighbors, and 2α is the H—O—H angle. It is found that $g \simeq 2.60$ at 0°C, decreasing to 2.46 by 83°C. Pople calculated μ taking into account the field due to the dipoles of the four nearest neighbors, and obtained $\mu = 2.15$ D at 0°C, decreasing to 2.08 at 83°C. This value is probably too small, since other electric fields are neglected; it leads, using eqn. (29), to a dielectric constant which is 20% too low. Coulson and Eisenberg[(214)] have revised μ upwards to 2.45 D, a value which had also been obtained by Verwey.[(1109)] The recalculation of ε_s then agrees fairly well with experiment over the range 0–83°C, as will be seen from Fig. 6a (P′). Harris and Alder[(465,467)] extended the Pople theory to include polarization resulting from distortion by the electric field; Fig. 6a (H) includes their results.

At about the same time as Pople made his calculations, Haggis *et al.*[(443)] calculated the static dielectric constant taking no account of bending but making a statistical analysis of hydrogen bond breaking, as follows. The broken ice I tridymite liquid was considered to be a statistical assembly of molecules bonded to 0, 1, 2, 3, 4 neighbors, there being n_i per cent of the ith type and m_i bonds between the ith and jth types. It was assumed that the probabilities of bond formation and bond breakage were independent of the numbers of other bonds possessed by the two molecules engaged. The rates of formation and destruction of each type of bond were equated under equilibrium, yielding ten equations of the type

$$m_{ij} f_1(T) \exp(-\Delta H/RT) = n_{i-1} n_{j-1} f_2(T, p) \tag{34}$$

where ΔH is the hydrogen bond energy and p (in %) the proportion of bonds broken. Sum rules were written,

$$\sum_{i=0}^{i=4} n_i = 100 \tag{35}$$

$$\sum_{i=1}^{i=4} \sum_{j=1}^{j=4} m_{ij} = 2(100 - p) \tag{36}$$

and five equations of the type

$$in_i = m_{ii} + \sum_{j=1}^{j=4} m_{ij} \tag{37}$$

Under these assumptions there are unique, calculable bond state populations n_i for a given value of p. The latent heat of vaporization L is made up of the unknown energy W necessary to overcome the van der Waals forces and the heat required to break the hydrogen bonds, so that

$$p = 1 - [(L - W)/2\,\Delta H] \tag{38}$$

W was taken to be temperature-invariant, so that the temperature variation of p and all n_i followed from the known temperature variation of L.

The calculation of g for a tridymitelike ice I system was made on the basis of the proton probability distribution shown in Fig. 5(a), which yields

$$g = 0.862(4 + \{4 + [2 + (24/9)] \cos 71°\} \cos 71°) \cos 54.5° = 2.91 \tag{39}$$

This value g_4 was taken as appropriate to those molecules bonded to four neighbors and surrounded by a medium in which $p = 0$. For a medium in which $p = 1$, only the nearest neighbors contribute to g_4 (Fig. 5b), so that

$$g_4(p = 1) = 0.862(4 + 2 \cos 71°) \cos 54.5° = 2.34 \tag{40}$$

For a molecule unbonded to any neighbor, $g_4 = 1$. Intermediate values of $g_i(p)$ were linearly interpolated, and are shown in Table II.

Variation of μ with temperature is to be expected, quite apart from the configurational effects leading to μ. Verwey's calculated value[1109] of the dipole moment enhanced by the polarization due to its neighbors, $\mu = 2.45$ D, is appropriate to ice I, and was found to give a very good value $\varepsilon_s = 92.7$ for the static dielectric constant at 0°C when used with the Kirkwood equation. Since four neighbors are responsible for the polarization, this value is appropriate for four-bonded molecules in water at 0°C. But the value of μ_4 becomes less satisfactory as the temperature is raised, since the influence of the neighboring permanent dipoles on the induced dipole is reduced by the increase in the nearest-neighbor distance. A correction was made by a method described in the original paper. The dipole moment for an unbonded molecule was taken as its free-space value $\mu_0 = 1.85$ D, and linear interpolation was made for $\mu_{3,2,1}$ as in Table II. This table includes the calculated values of p, n_i, μ_i, g_i, and ε_s.

Only one choice was made entirely empirically in this calculation, namely the van der Waals energy W. This was chosen so that ε_s for water at 0°C was in agreement with the experimental value, just as it is for ice at 0°C. The unexpectedly small difference between these two static dielectric constants could mean that the degree of configurational interaction (presumably by hydrogen bonding) in liquid water at 0°C is very large. This is reflected in the low value of p, the proportion of broken bonds.

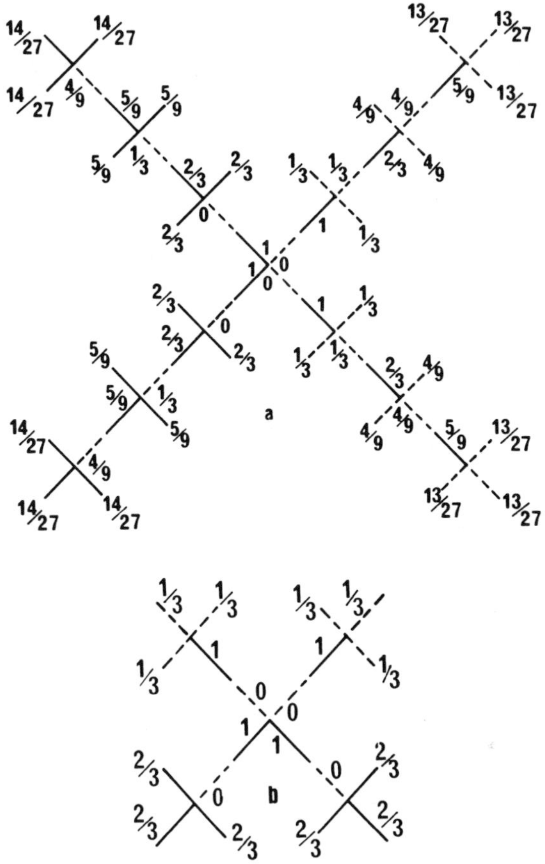

Fig. 5. Schematic representations of proton distributoin probabilities around a fixed molecule, proposed by Haggis *et al.*[443] (a) A fully four-bonded system such as ice. (b) A four-bonded molecule surrounded by no bonds other than its own, as is the case at temperatures close to 370°C.

In Fig. 6a the temperature variation of the static dielectric constant is compared with calculations on the bond-bending and bond-breaking models. It is apparent that the simple bond-breaking model is capable of interpreting the static dielectric constant over the entire temperature range up to the critical point. The bond-bending model is reasonably adequate over a smaller temperature range, but it appears that at higher temperatures, particularly above 100°C, the calculated dielectric constant would be systematically high.

TABLE II. Parameters in Calculation of Static Dielectric Constant on Bond-Breaking Model

T, °C	L, kcal mol^{-1}	p, %	n_4	n_3	n_2	n_1	n_0	μ_4, D	μ_3, D	μ_2, D	μ_1, D	g_4	g_3	g_2	g_1	ε_s (calc.)	ε_s (exp.)
0 (ice)	12.2	0	100	0	0	0	0	2.45	—	—	—	2.91	—	—	—	92.7	92
0 (water)	10.7	9.0	72.4	20.0	6.0	1.5	0.1	2.45	2.30	2.18	2.03	2.81	2.36	1.90	1.45	89.0	88.2
25	10.5	11.3	67.0	23.2	7.6	2.0	0.2	2.44	2.29	2.17	2.03	2.80	2.35	1.90	1.45	78.3	78.2
60	10.1	15.8	58.5	25.8	11.0	3.8	0.9	2.42	2.27	2.17	2.02	2.79	2.34	1.89	1.45	65.8	66.6
100	9.7	20.2	49.8	28.3	15.0	6.0	1.5	2.39	2.25	2.16	2.02	2.76	2.28	1.86	1.43	53.7	55.4
200	8.4	34.1	29.8	29.1	24.4	14.1	4.9	2.29	2.21	2.10	2.00	2.69	2.24	1.83	1.42	31.6	34.6
200	6.0	61.3	5	14.7	26.3	33.0	21.0	2.18	2.10	2.03	1.96	2.55	2.17	1.77	1.39	15.2	17.7
370	1.9	100	0	0	0	0	100	—	—	—	—	—	—	—	—	7.4	9.7

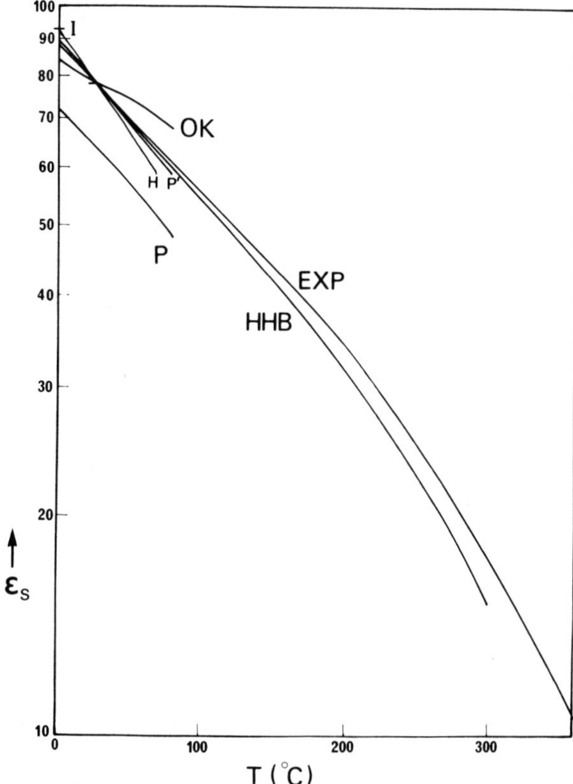

Fig. 6a. Experimental and calculated static dielectric constant of water as a function of temperature: (*I*) Ice at 0°C; (EXP) experimental (Akerlof and Oshry[4]); (OK) calculations of Oster and Kirkwood;[818] (P) calculations of Pople;[854,855] (P′) calculations of Pople improved by Coulson and Eisenberg;[214] (HHB) calculations of Haggis *et al.*;[443] (H) calculations of Harris and Alder.[465,467]

It is not difficult to adduce arguments against pure breaking and pure bending models, since both processes are likely to occur. Therefore, each model, considered on its own, is an inadequate approximation. It would be a complicated task to attempt a combination of the two. At temperatures above 70°C it does appear that the pure bending theory is inadequate for dielectric interpretation, but at low temperatures the static dielectric constant cannot really be adduced as evidence for breaking rather than bending. Yet it is a parameter highly sensitive to the amount of structural correlation between the dipoles. Whether bond bending or bond breaking is assumed, the dielectric evidence favors a small proportion of unbonded states at low temperatures. This is not the conclusion supported by theories of thermodynamic properties on the bond-breaking model due to Nemethy and Scheraga[788,790] and others.[146,147] These theories require considerable bond breakage, including a proportion of monomer molecules; the latter are also assumed in the treatment of the radial distribution function due to Narten *et al.*[784] A graphic comparison of bond state populations proposed by different authors is given in Fig. 6b.

A more serious criticism of the old bond-breaking theory in its simple form is that there is evidence from other types of measurements that bond breakage is a much more complex, even a cooperative, phenomenon. The disruption of structure does not imply that a unique amount of energy is necessary to break each bond. If there is rearrangement to a denser packing, then an appreciable part of the energy might be returned by an increase in the van der Waals energy. Buijs and Choppin[146,147] estimated the equilibrium energy requirement for breaking the first hydrogen bond as ~1.1 kcal mol^{-1} and for the second as ~1.6 kcal mol^{-1}. These are much smaller

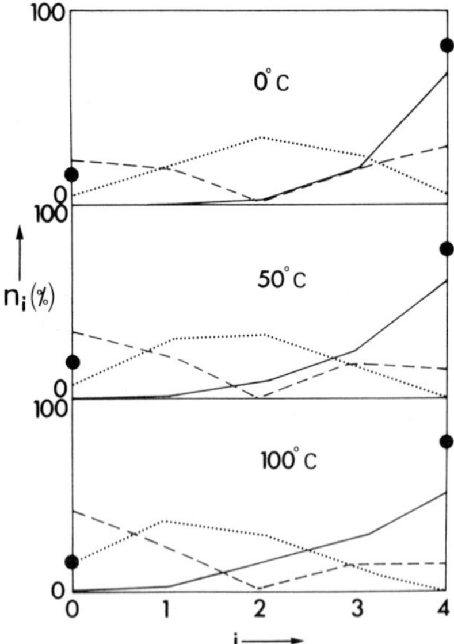

Fig. 6b. Bond state populations at $T = 0$, 50, 100°C proposed by different authors: (broken line) Nemethy and Scheraga; (dotted line) Buijs and Choppin,[146,147] Vand and Senior;[1082] (full line) Haggis *et al.*;[443] (closed circles) X-ray data of Narten *et al.*[784]

than the $\sim$5 kcal mol^{-1} activation energy for the relaxation process, originally estimated from the sublimation energy of ice to be the bond energy.

6. INTERPRETATION OF STATIC DIELECTRIC CONSTANT WITH INDUCED-DIPOLE CONTRIBUTION

In the dipole correlation model which results in eqn. (29), the square of the refractive index appears as a term which contributes the proportion of the static dielectric constant arising from induced, as opposed to permanent dipoles. But in the form proposed by Fröhlich (equations B1.41, Appendix B1, "Theory of Dielectrics"),[371] the static dielectric constant is related to the free-molecule dipole moment $\boldsymbol{\mu}_0$ and to the square of the refractive index by eqn. (30):

$$(\varepsilon_s - n^2)(2\varepsilon_s + n^2)/\varepsilon_s(n^2 + 2)^2 = (4\pi N/9kTV)g\boldsymbol{\mu}_0{}^2 \tag{41}$$

The dipole correlation interpretations discussed in the previous section have taken n^2 to be small and temperature-invariant; for example, Haggis *et al.*[443] added only a contribution due to lattice vibrations, to give a temperature-invariant value 3.2, which is close to the observed value of the high-frequency dielectric constant ε_∞ for ice; but they did not use the Fröhlich form [eqn. (30)] for spherical molecules, preferring the Kirkwood form [eqn. (29)]. Moreover, they interpreted the high-frequency dielectric constant ε_∞ for liquid water, which is appreciably higher, as arising from permanent dipole processes, relaxing in the submillimeter or infrared regions. In this interpretation it is not correct to use ε_∞ in eqn. (30).

Hill[506,507] claimed that if infrared and submillimeter processes in liquid water are interpreted as arising from induced-dipole rather than permanent dipole processes, then it is permissible to replace n^2 by ε_∞, which is in the region of 4–5 and is temperature-dependent. In the Fröhlich equation ε_s is much more sensitive to this quantity than it is in eqn. (29). Hill replaced n^2 by ε_∞:

$$(\varepsilon_s - \varepsilon_\infty)(2\varepsilon_s + \varepsilon_\infty)/\varepsilon_s(\varepsilon_\infty + 2)^2 = (4\pi N/9kTV)g\boldsymbol{\mu}^2 \tag{42}$$

and succeeded in reproducing the temperature variation of ε_s with a correlation factor $g = 1$, using a free-molecule dipole $\mu = 1.85$ D, and taking ε_∞ values empirically as shown in Table III.

It is seen that a systematic fall of ε_∞ with rising temperature is necessary to reproduce the temperature dependence of ε_s. Neither the relatively

TABLE III. Values of ε_∞ Required for Calculation of ε_s with $g = 1$

T, °C	ε_∞ (calc.)
0	4.35
10	4.32
20	4.28
30	4.23
40	4.17
50	4.11
60	4.05

small variation of molar volume with temperature, nor the kT term, is sufficient in itself to achieve this effect. The left-hand denominator term $(\varepsilon_\infty + 2)^2$ of eqn. (42) is the dominating factor, and accounts for the achievement of the large ε_s with a free-molecule dipole moment $\mu = 1.85$ D and with a correlation factor $g = 1$.

The importance of this result lies in the strong effect upon the inferred value of g, and so upon the structural model for the liquid. It is an extreme view to suggest that $g = 1$, indicating no correlation whatever between the permanent dipoles. Before the appearance of data in the submillimeter region, the ε_∞ values inferred from microwave data were insufficiently accurate, and could without great loss of plausibility be reconciled with values in Table III. Now, it is difficult to make this reconciliation; the temperature variation of ε_∞ is known to be different (see Fig. 16). The detailed situation is described in Section 12. But it is also necessary that the value of $n^2 + \varepsilon_{\text{vib}}$ employed in the calculations of Haggis *et al.* be revised, along with the estimates of bond state populations.

7. MICROWAVE AND SUBMILLIMETER DIELECTRIC CONSTANTS

The dielectric constant of pure water is, as far as is known, invariant with frequency from dc up to beyond 10^8 Hz. It is to be expected that at sufficiently high frequencies the real part of the dielectric constant will fall to a value of the order of the square of the optical refractive index, while a broad absorption band contributes to the imaginary part. The theory of the frequency variation of the complex dielectric constant is discussed in Section 10.

TABLE IV. Approximate Values of Relaxation Wavelength λ_s Estimated from Early High-Frequency Measurements (20°C)[a]

Date	Author[201]	Method	λ_s, cm
1937	Esau and Bäz	Free wave	1.85
1937	Knerr	Wire wave	1.6–1.8
1939	Hackel and Wien	Thermometric	1.6
1939	Slevogt	Wire wave	1.8
1939	Kebbel	Free wave	1.85
1940	Divilkowski and Masch	Thermometric	1.56
1943	Connor and Smyth	Wire wave	2.0
1946	Abadie and Girard	Wire wave	1.5

[a] Accepted modern value, 1.80 cm.

During the late 1930's some measurements were made of the dielectric constant of water in the MHz region and some idea of the relaxation process was obtained. Some measurements of historical interest are tabulated in Table IV.[58,201] The rapid development of microwave valves and guided wave techniques during the second world war made possible the detailed investigation of the GHz, or microwave, region. Definitive measurements were made by Collie *et al.*[201] and independently by Saxton and Lane.[945,946] These and subsequent measurements are detailed in Tables Va–Vh and VI. It will be seen from these tables that eight different microwave frequencies

TABLE Va. Microwave Dielectric Data and Relaxation Parameters Deduced from Constraint Fitting Program

T, °C	ε_s	ε_∞	τ, $\times 10^{-11}$ sec	α
0	88.3	4.46	1.79	0.014
10	84.1	4.10	1.26	0.014
20	80.4	4.23	0.93	0.013
30	76.8	4.20	0.72	0.012
40	73.2	4.16	0.58	0.009
50	70.0	4.13	0.48	0.013
60	66.6	4.21	0.39	0.011
75	62.1	4.49	0.32	—

TABLE Vb. Data of Grant and Shack[419,421] for $\nu = 34.88$ GHz

T, °C	ε'	ε''
0	11.2	21.7
10	14.3	26.5
20	19.2	30.3
30	24.6	32.8
40	30.4	33.7
50	35.4	33.1
60	39.7	31.6

have been studied, most of them by more than one author or team. In general, the agreement between different authors is better than 1 or 2%.

It is of interest to make some brief critical assessment[471] of the various microwave measurement techniques employed on liquid water. They can be listed as follows:

1. Free-wave techniques. (a) Attenuation by a variable thickness of liquid measures the absorption coefficient.[650,1190] It will be recalled that the refractive index n and the absorption coefficient $\varkappa$ are related to the complex dielectric constant $\varepsilon = \varepsilon' - j\varepsilon''$ as follows:

$$\varepsilon' = n^2 - \varkappa^2, \qquad \varepsilon'' = 2n\varkappa$$

(b) Measurements of reflection coefficient and Brewster angle. The deduction of ε' requires that the absorption coefficient be known; ε' is not very

TABLE Vc. Data of Hasted and Sabeh[472]

T, °C	$\nu = 23.68$ GHz		$\nu = 9.141$ GHz		$\nu = 3.254$ GHz	
	ε'	ε''	ε'	ε''	ε'	ε''
0	14.5	27.5	46.0	41.0	80.0	26.5
10	22.0	33.0	56.0	37.0	79.3	19.4
20	31.0	35.0	63.0	31.5	77.8	13.9
30	38.7	35.5	65.8	26.0	75.8	10.2
40	43.5	34.0	66.3	20.5	73.0	8.4
50	48.0	31.0	65.5	16.5	69.7	6.5
60	50.5	26.5	63.5	13.5	66.0	5.0

TABLE Vd. Data of Saxton and Lane[945,946]

T, °C	$\nu = 24.19$ GHz		$\nu = 9.346$ GHz	
	ε'	ε''	ε'	ε''
0	14.89	26.32	42.63	41.27
10	21.29	31.61	54.89	38.06
20	29.64	35.18	62.26	32.56
30	37.76	35.78	65.34	26.37
40	44.60	34.22	65.90	21.21
50	48.75	31.10	64.93	17.56

sensitive to the reflection coefficient, which is described by the equation

$$R = (n^2 + \varkappa^2 + 1 - 2n)/(n^2 + \varkappa^2 + 1 + 2n)$$

(c) Free-wave interferometer[895]: this yields both ε' and ε''.

2. Guided wave transmission techniques. (a) Attenuation by a variable thickness of liquid measures the absorption coefficient.[201] (b) Attenuation by variable thicknesses of liquid in tubes of different diameters, close to waveguide cutoff, measures both ε' and ε''.[201] (c) Guided wave interferometer.[142,143]

3. Methods of equivalent circuits, using guided waves. (a) Measurement of phase and magnitude of standing wave in front of liquid backed by a short circuit.[220,225,663,913] (b) In standing wave measurements an

TABLE Ve. Data of Collie et al.[201]

T, °C	$\nu = 23.62$ GHz		$\nu = 9.346$ GHz		$\nu = 3.000$ GHz	
	ε'	ε''	ε'	ε''	ε'	ε''
0	16.22	28.26	44.82	41.64	79.66	24.74
10	22.33	32.30	53.85	37.57	78.07	17.46
20	30.88	35.75	61.41	31.83	77.42	13.05
30	38.43	36.05	63.31	25.47	76.78	9.80
40	43.24	33.6	65.58	21.16	72.57	7.54
50	48.26	30.64	65.32	17.12	68.44	5.80
60	49.79	27.30	63.09	13.82	65.37	4.55
75	51.71	22.34	60.70	10.51	60.48	3.30

TABLE Vf. Data of Grant et al.[420] and Chamberlain *et al.*[175]

T, °C	Grant *et al.*[420]				Chamberlain *et al.*,[175]	
	$\nu = 1.744$ GHz		$\nu = 0.5769$ GHz		$\nu = 890$ GHz	
	ε'	ε''	ε'	ε''	ε'	ε''
0	85.3	16.5	88.1	5.85	—	—
10	82.4	11.0	84.0	3.85	—	—
20	79.2	7.9	80.3	2.75	4.30	2.28
30	76.1	5.9	76.6	2.05	4.30	2.78
40	72.9	4.4	73.3	1.55	4.27	3.20
50	69.7	3.6	69.9	1.25	4.30	3.51
60	66.7	2.9	66.8	0.96	4.40	3.80
75	—	—	—	—	4.58	4.29

additional reflecting element may be introduced, or an impedance transformer. (c) The liquid may also be backed by an open circuit, or (d) by a variable reactance.[85,447,894,1037]*

4. Perturbation, or cavity resonator measurements. (a) H_{01} circular cavity with disk specimen.[415,531,556,831,1168] (b) E_{010} circular cavity with capillary tube specimen and tuning plunger.[139,181] (d) H_{01} circular cavity with capillary tube specimen.[200]

TABLE Vg. Data of Cook[207]

T, °C	$\nu = 3.00$ GHz		$\nu = 4.63$ GHz		$\nu = 9.39$ GHz		$\nu = 23.77$ GHz	
	ε'	ε''	ε'	ε''	ε'	ε''	ε'	ε''
0	80.2	24.0	71.4	33.4	43.4	41.4	14.5	26.5
10	79.4	17.5	74.3	25.2	54.4	37.1	22.0	32.5
20	77.7	13.0	74.0	18.8	61.5	31.6	31.0	35.7
30	75.3	9.90	73.1	14.6	64.8	25.6	38.5	35.6
40	72.6	7.55	70.7	12.0	65.5	20.9	43.0	34.0
50	69.7	5.95	68.5	9.4	64.6	17.0	46.3	30.8
60	66.4	4.60	66.3	7.4	—	—	—	—

* Also H. Lueg, Institute of Physics, Sarre University, unpublished.

TABLE Vh. Data of Buchanan[142] and Sandus and Lubitz[942]

T, °C	Buchanan[142]				Sandus and Lubitz,[942]	
	ν = 23.81 GHz		ν = 9.375 GHz		ν = 9.368 GHz	
	ε'	ε''	ε'	ε''	ε'	ε''
0	15.5	28.0	44.0	41.4	—	—
10	22.8	32.7	54.0	37.8	—	—
20	30.5	35.0	62.0	32.0	62.8	31.5
30	37.5	35.5	65.0	26.4	65.7	25.7
40	43.0	33.7	65.2	21.2	66.2	20.9
50	48.0	30.5	64.5	17.0	65.6	16.9
60	—	—	63.0	—	—	—

Schematic diagrams of these methods are shown in Figs. 7–10, respectively. Methods classified under 3 and 4 are capable of yielding both ε' and ε''.

In general, the cavity resonator techniques are superior at longer wavelengths ($\geq$2 cm). Free-wave techniques are only competitive at wavelengths less than 1 cm. Since the accuracy of all guided wave techniques is proportional to the accuracy of machining of moveable parts, it decreases proportionately to wavelength decrease. For high-loss liquids, methods depending upon the measurement of standing wave ratios are less accurate than those depending only upon phase change and attenuation, or on the sharp resonance of a cavity. These are capable of better than 1%, but the cavity is only at its best for low-loss liquids.

All techniques require that the test specimen be maintained with an accurately formed geometric shape, that the temperature be uniform, and that the bounding surfaces have good dielectric properties and introduce the minimum of impurities. Fused silica satisfies these requirements best. A free surface or water-air interface is under some circumstances permissible, but in guided wave measurements a method must be found of minimizing the effect of the meniscus. The precision fabrication of planar windows, cylindrical bottles, and capillary tubes in fused silica should be given adequate attention. It is important to ensure that the dielectric properties of bulk water and not those of interfacial water are being measured. It is not demonstrated, but can be plausibly argued, that the properties are different.

TABLE VI[a]

T = 0°C			T = 5°C			T = 25°C			T = 35°C			T = 50°C		
ν, GHz	ε'	ε''	ν, GHz	ε'	ε''	ν, GHz	ε'	ε''	ν, GHz	ε'	ε''	ν, GHz	ε'	ε''
5.305	66.7	35.6	1.15	85.1	8.56	2.61	76.9	9.77	5.341	71.7	14.57	7.7075	67.2	14.49
6.565	59.5	38.8	1.80	83.7	13.18	3.75	75.6	13.74	7.738	68.4	20.26	12.473	61.5	21.41
7.688	53.6	40.3	2.786	80.8	19.49	5.30	72.9	18.58	12.465	60.3	28.1	17.463	56.25	27.15
12.56	33.9	38.6	3.723	76.0	24.58	7.95	67.6	25.74	17.45	51.2	32.6	26.78	44.2	32.0
15.15	28.2	36.05	5.345	69.8	31.75	12.50	56.6	33.3	26.84	36.7	34.6	36.30	34.3	32.6
17.40	23.9	33.5	7.762	58.6	37.8	17.43	45.8	36.1	36.53	26.3	31.8			
26.83	14.95	25.2	12.82	38.85	39.2	26.70	31.0	34.6						
36.55	11.1	19.9	17.35	28.1	35.8	37.81	21.05	29.6						
			26.49	17.7	28.0									
			37.94	12.5	21.3									

[a] These recent unpublished measurements on H_2O due to Dr. R. Pottel and his collaborators at the University of Göttingen were received after the completion of the computer analysis of the data in Tables Va–Vh.

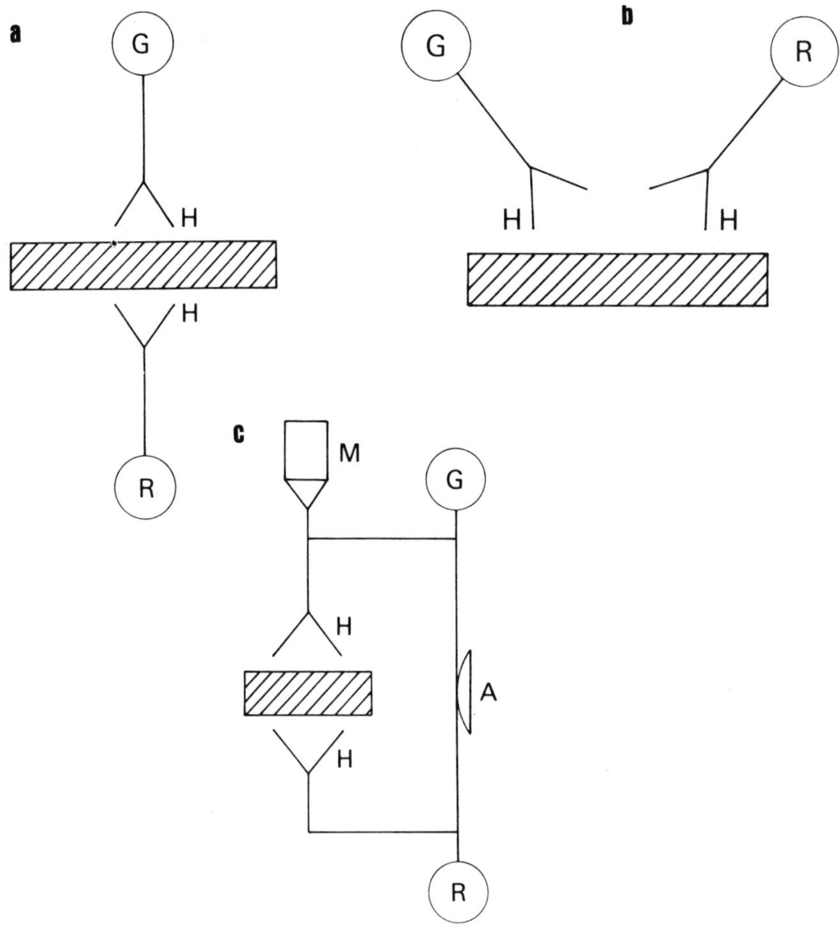

Fig. 7. Schematic representation of microwave circuit for measurement of dielectric properties of a lossy liquid such as water using free-wave techniques 1(a)–(c) as described in the text. The key for Figs. 7–10 is as follows: *L*, load; *RA*, variable reactance; *M*, micrometer adjustment; *H*, horn; *S*, standing wave indicator; ○, circular cross section; *R*, receiver (detector); *G*, generator (oscillator); *A*, attenuator (variable); *T*, hybrid *T*-junction; *P*, phase changer; H_{01}, rectangular waveguide, H_{01} mode; ⊙*L*, concentric line; *IT*, impedance transformer; H_{011}, E_{010}, resonator modes.

Most of the published measurements were made in the early 1950's before many of the electronic and commercial advances in microwave engineering had been made. If the need were felt at present, it would be possible to utilize these advances to obtain perhaps 0.1% accuracy over the relevant range of temperature and frequency. In particular, there is now a much

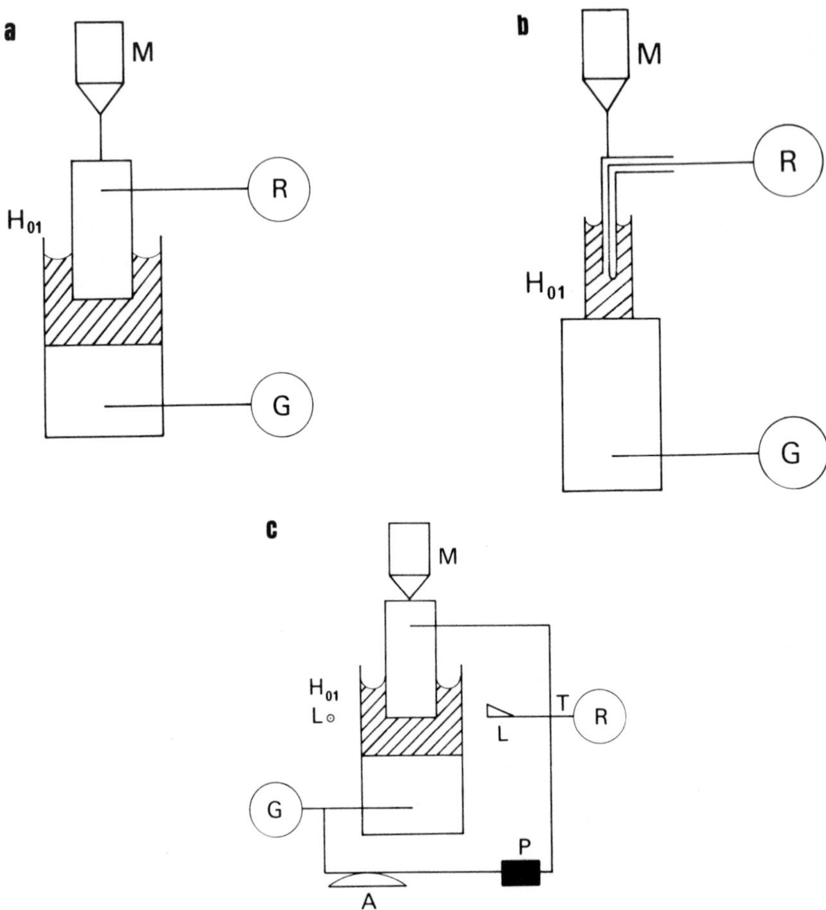

Fig. 8. Schematic diagrams of microwave circuits for measurement of dielectric properties of a lossy liquid such as water using guided wave transmission techniques 2(a)–(c) as described in text. Key given in legend to Fig. 7.

wider range of frequencies "available" than was the case at that time. The oscillators are now tunable over a wide range, and a larger number of wavelength bands of guided wave components are available. Moreover, there are now a number of extremely stable maser frequencies which could be utilized for resonant cavity measurements. There are other frequency-stabilization systems which are extremely accurate, and frequency measurement by counting is highly developed. Electroforming and interferometric monitoring of metal-cutting machinery is now well advanced. Sensitivity of detection (i.e., low noise level of solid-state detectors) is much improved,

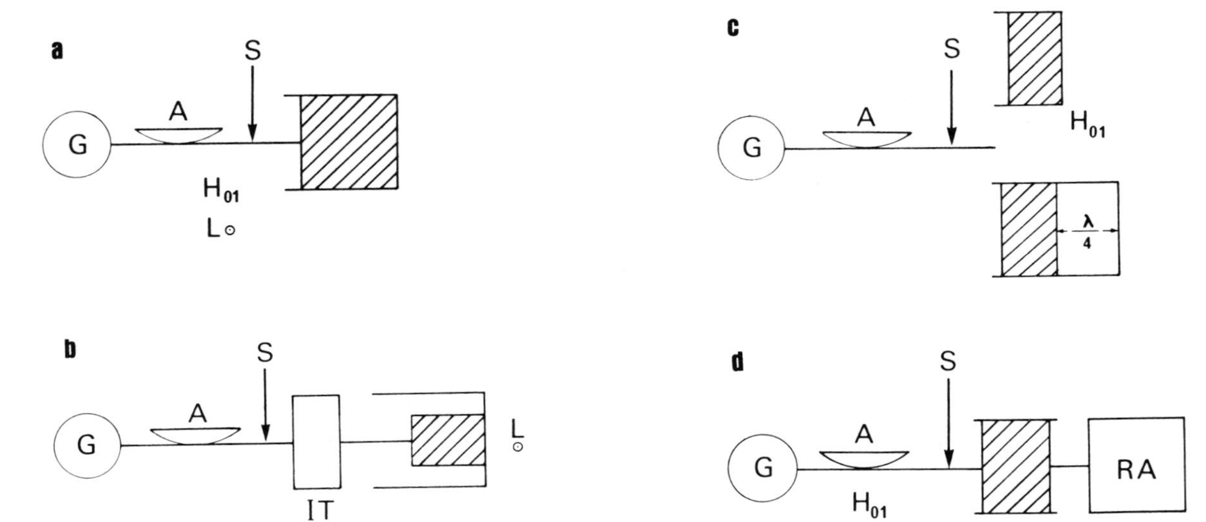

Fig. 9. Schematic diagrams of microwave circuits for measurement of dielectric properties of a lossy liquid such as water using equivalent circuits and guided waves as described in methods 3(a)–(d) of text. Key given in legend to Fig. 7.

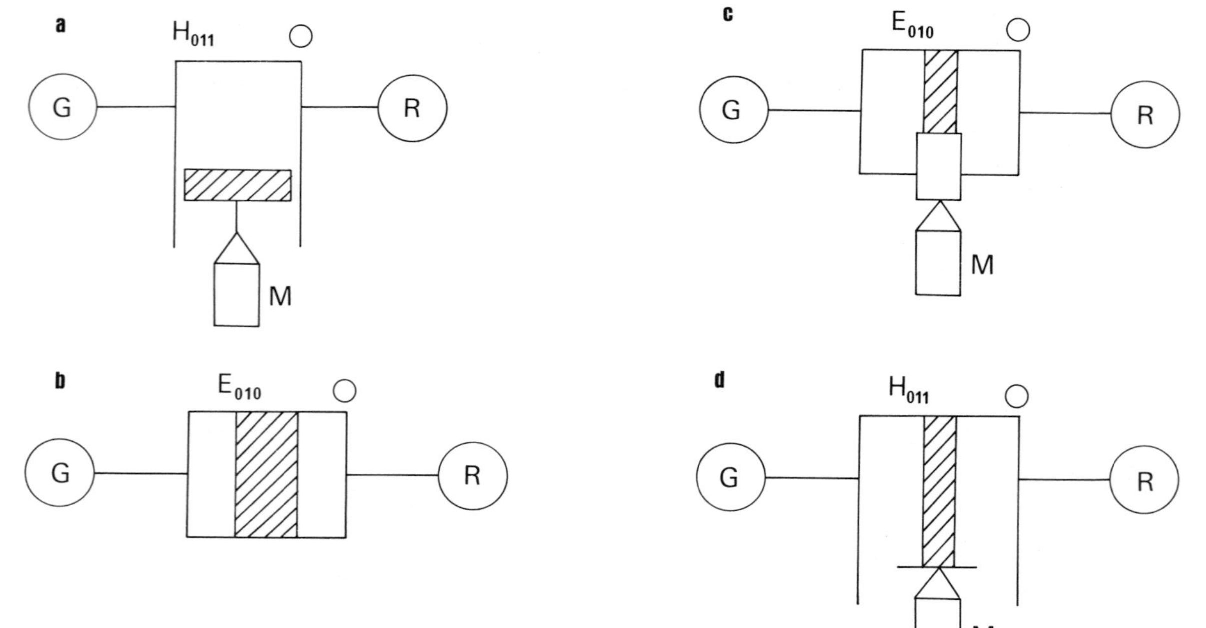

Fig. 10. Schematic diagrams of microwave circuits for measurement of dielectric properties of a lossy liquid such as water using perturbation, or cavity resonator measurements as described in methods 4(a)–(d) of text. Key given in legend to Fig. 7.

so that a greater range of attenuation can be measured, and "padding" attenuators and the large insertion loss "cutoff" attenuator can readily be employed.

The question must be asked, would the achievement of 0.1% accuracy over the entire frequency range of the Debye relaxation be worthwhile? In the opinion of the author, the project, although expensive, would be valuable from the point of view of detecting spread of relaxation times and departure from the Debye[245] equations (51) and (52), for example, due to lag in the reaction field [eqns. (56) and (57)]. There are also technological reasons why improved accuracy might be valuable. The technique of "time-domain spectroscopy"[329,1034] (application of single electrical pulse and recording and Fourier-transforming the response) will also be of value.

In the meantime the optimum treatment of the existing data requires the use of mathematical techniques of nonlinear regression analysis by which, with the aid of a computer program, the least square percentage errors on both the real and the imaginary parts of the data can be simultaneously minimized by the best choice of parameters in the Debye, or similar, equations (discussed in Section 10). The graphic Cole–Cole plot[196] is an incomplete description of these equations in that the distribution of points on the semicircle is not given; it is insufficient for a best fit that the points should merely lie on a semicircle, as they do in Fig. 20.

The question must be asked, whether a single relaxation time or a small spread of relaxation times is the more appropriate. The first attempt made by Collie *et al.*[201] to fit the data to the Debye equations (51) and (52) made use of a single relaxation time, but Grant *et al.*[420] subsequently claimed that a small spread α of relaxation times gave a better fit, and a specialized fitting procedure for spread determination was worked out.[419,421] The present analysis supports this claim.

It is found that, for liquids believed to possess a distribution of relaxation times, the characteristic ε'' (ε') plot is a circular arc cutting the ε' axis at ε_s and ε_∞, but with center below the ε' axis, on a line making an angle $-\alpha$ with the point ε_∞, 0. This arc corresponds to the equation

$$\hat{\varepsilon} = \varepsilon_\infty + \{(\varepsilon_s - \varepsilon_\infty)/[1 + (j\omega\tau)^{(1-\alpha)}]\} \tag{43}$$

which rationalizes to

$$\varepsilon' = \varepsilon_\infty + \frac{(\varepsilon_s - \varepsilon_\infty)[1 + (\omega\tau)^{1-\alpha}\sin(\alpha\pi/2)]}{1+2(\omega\tau)^{1-\alpha}\sin(\alpha\pi/2)+(\omega\tau)^{2(1-\alpha)}} \tag{43a}$$

$$\varepsilon'' = \frac{(\varepsilon_s - \varepsilon_\infty)(\omega\tau)^{1-\alpha}\cos(\alpha\pi/2)}{1+2(\omega\tau)^{1-\alpha}\sin(\alpha\pi/2)+(\omega\tau)^{2(1-\alpha)}} \tag{43b}$$

α is taken as a spread parameter of relaxation times. There is no fundamental theoretical basis for these equations, but there is much evidence for their empirical success; for $\alpha = 0$ they of course reduce to the Debye equations.

The claim of Grant *et al.* was that $\alpha = 0.015$ over a range of temperature, but Grant and Shack[419,421] recently gave best values of $\alpha = 0, 0.023, 0.015, 0.007, 0.01, 0.012, 0.015$, respectively at $T = 0, 10, 20, 30, 40, 50, 60$°C. Regression analysis programs have now been written by Mason *et al.*[735] both for the fitting of eqns. (51) and (52) with zero spread parameter α, and for the fitting of eqn. (43) including the best choice of α. In both programs the static dielectric constant measurement is weighted ten times more heavily than the microwave measurements. The fitting programs choose the values of ε_∞, τ, and α which minimize the square of the experimental errors. These program have been applied to all the microwave data listed in Table Vb–Vh, weighted equally; the values of ε_∞, τ, and α calculated from these programs are detailed in Table Va. It is found that there is a small temperature-independent spread of relaxation times for liquid water; its interpretation will be discussed in Sections 11, 12. The data for D_2O are essentially similar, although they have not yet been analyzed by computer; the relaxation wavelengths are given by Collie *et al.*[201] It is not possible at present to assign errors to ε_∞, α, and τ, but we have available the ranges of displacement of each parameter required to double the rms errors; these are sufficiently small to imply that the values are a substantial improvement in accuracy upon previous calculations. Extension of the program to provide statistical tests for confidence limits is in hand.

There are several new interesting features which emerge from this regression analysis program. First, the slight spread of relaxation times is approximately temperature-independent. Second, the values of ε_∞ are rather different from those obtained in previous, approximate, calculations; their temperature dependence is illustrated in Fig. 16. These values are due almost entirely to the inclusion for the first time of the 890 GHz data of Chamberlain *et al.*, which are of much greater importance than other data for ε_∞ determinations, because of their very high frequency.

At the low temperatures of 0 and 10°C, at which no submillimeter data are available, the values of ε_∞ in Table Va are erratic. The tabulated values of ε_∞ at low temperatures are quite close to those shown in Table III as being those necessary to interpret the static dielectric constant using the Fröhlich equation (42) with $g = 1$ and n^2 replaced by ε_∞, as was proposed by Hill. But at higher temperatures a discrepancy appears.

It is possible to show from the regression analysis program that certain results are "out of line" with all those at different frequencies. It is difficult not to be arbitrary in excluding these data from the final program, but it was decided to do so in two cases.[888,931] There is also one other case of willful exclusion, namely the imaginary part ε'' of the data of Chamberlain *et al.* It is easily seen from Fig. 18 that the contribution from an additional absorption process or processes is quite large at 890 GHz, but is negligible at the lower frequencies. We consider that the corresponding contribution at 890 GHz to ε' is sufficiently small to warrant including the data in a constraint fitting of a single relaxation. However, it will be seen from discussion in Section 12 that regression analysis over a wider range of frequency to two or more relaxation processes would yield more accurate values of ε_∞. Furthermore, the recent submillimeter refractive index measurements of Chamberlain *et al.*[176] will, when they are completed over the necessary temperature range, considerably influence the situation.

8. KINKS?

Dielectric measurements can throw light upon the possible exsistence of "kinks" in the temperature-dependent functions of the physical properties of water. Discontinuities in functions might indicate the existence of "higher-order phase transitions"; but it is the current opinion of workers in this field* that such effects are attributable to interfacial phenomena (see also Chapter 1). The static dielectric constant measurements of Rusch and Good[930] were undertaken in order to confirm the conclusion drawn from the work of Malmberg and Maryott[725] as to the absence of a kink at 15°C. This conclusion was confirmed. The static dielectric constant appears to be a smooth function of temperature. Work has been in progress in the author's laboratory upon the temperature dependence of ε' and ε'' at 3 GHz frequency. At this frequency it is clear from the Debye equations that ε' is very close to ε_s and ε'' is very close to the product $\varepsilon_s\tau$; therefore, the separation of the two parts of the dielectric constant enables a separation of possible kinks in ε_s and τ temperature-dependence functions to be made.

Following some observations and Fourier analysis[1080] at 0.5°C temperature intervals between 20 and 70°C, it has become apparent that smaller intervals of temperature would be necessary to uncover the possible existence

* W. Drost-Hansen, private communication (1969).

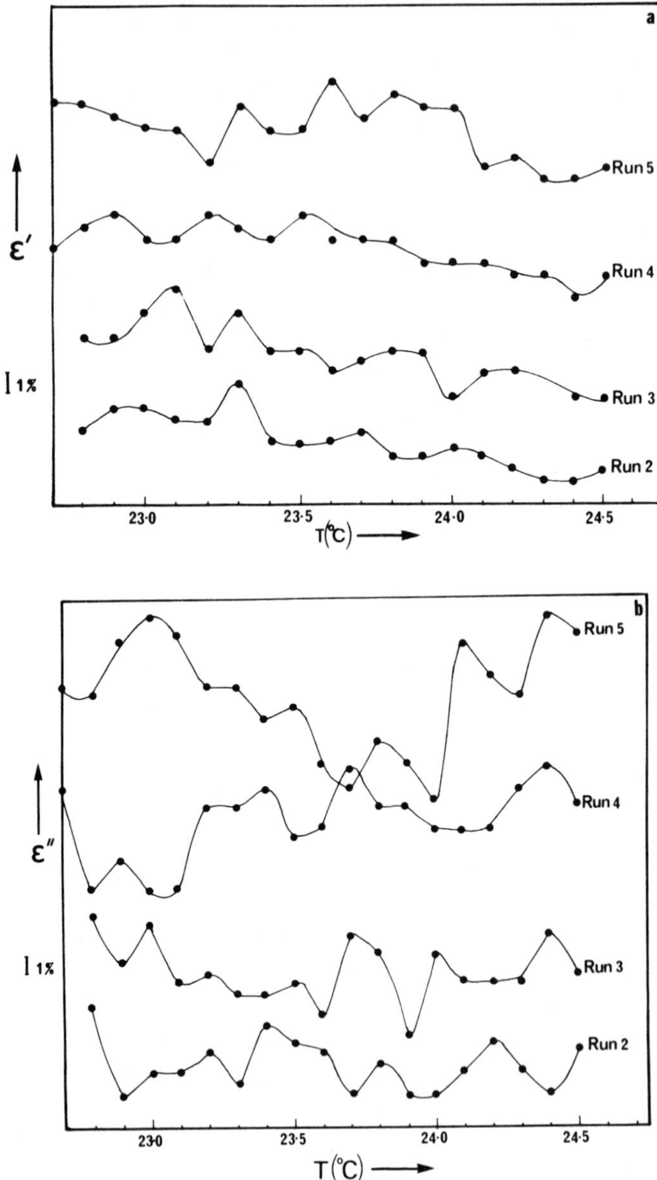

Fig. 11. Comparison of four sets of measurements at 3 GHz of (a) ε' and (b) ε'' for water over a short range of temperature. Possible consistent anomaly in ε'' at 23.25°C, otherwise random errors in data.

of kinks. Measurements are in progress[160] at 0.1°C temperature intervals, using ultrapure water* in a fused silica capillary in a TM_{010} cylindrical cavity. Comparison of a series of runs taken over a small range of temperature is shown in Fig. 11. This range was chosen for initial measurements because the Fourier analysis of previous data looked interesting. It will be seen that the accuracy of the present experiments is not yet adequate to substantiate the existence of small kinks. Furthermore, the errors in ε' appear to be completely random, so that there is no evidence for any kinks in ε_s in this small range of temperature. However, the consistency among four ε'' runs in the region of 23.3°C makes it just possible that structure might exist here in the dependence of the temperature function of the relaxation time. The evidence at this stage is, in effect, negative over this small range of temperature. The diameter of the capillary tube (1.5 mm) is sufficiently large for it to be unlikely that interfacial phenomena are involved, although the only way to be certain of this would be to compare data obtained with different tube diameters.

9. SUBMILLIMETER MEASUREMENTS

In Fourier spectroscopy radiation from a continuum source is directed onto a two-beam interferometer. The optical constants of a material may then be found by placing a specimen within one arm of the interferometer. For the study of heavily absorbing materials such as water this is done by replacing the mirror of one arm by a plane surface of the specimen. An extended interferogram is recorded as a function of path difference, and from the Fourier transform of the interferogram the refractive index n and absorption coefficient $\varkappa$ are derived over a wide range of frequency.[175,176,273,1204] The modular Michelson interferometer developed at the British National Physical Laboratory for this type of measurement is shown in Fig. 12.

The movement of the interferometer mirror M increases the path difference x from $-D$ to $+D$ and provides a slowly varying signal, the variable part of which is called the interferogram. If the mirror M is jittered with an amplitude $\bar{\lambda}/8$, where $\bar{\lambda}$ is the mean wavelength detected, and synchronous amplification is used, only the variable part of the interferogram is recorded. This record is called the phase-modulated interferogram. In the absence of a specimen, the background interferogram $G_{0\phi}(x)$ is recorded; when the specimen is introduced a different interferogram

* D. J. G. Ives, private communication.

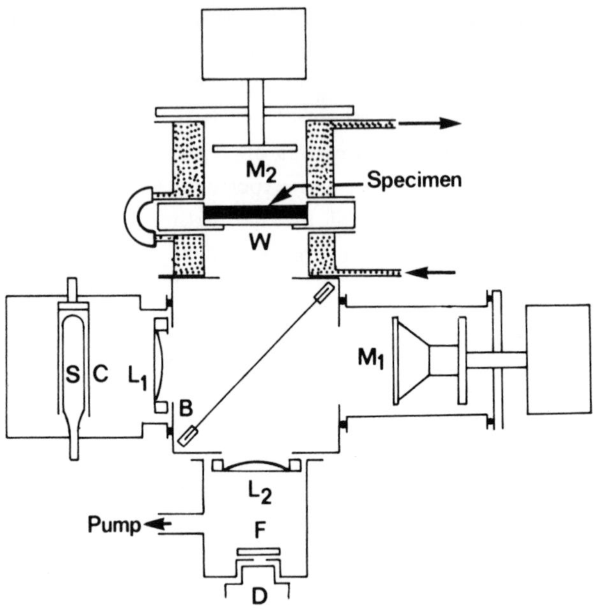

Fig. 12. Schematic diagram of modular cube Michelson interferometer used for submillimeter measurements of complex refractive index for water. S, source; C, chopper; L_1, L_2, lenses; B, beam-divider; M_1, M_2, mirrors; F, filter; D, detector; W, TPX window.

$G_\phi(x)$ is observed. This is the convolution of the impulse response function $V(x)$ of the specimen with the background interferogram:

$$G_\phi(x) = V(x) * G_{0\phi}(x) \tag{44}$$

From the properties of Fourier transforms it follows that

$$\hat{S}_\phi(\nu) = \hat{R}(\nu)\hat{s}_{0\phi}(\nu) \tag{45}$$

where

$$\hat{R}(\nu) = [n_w + \hat{n}_L(\nu)]/[n_w - \hat{n}_L(\nu)] \tag{46}$$

is the (complex) reflection factor of the specimen, $\hat{n}_L(\nu)$ is the complex refractive index of the liquid, n_w is the refractive index of the transparent window, and $\hat{s}_{0\phi}(\nu)$ and $\hat{S}_\phi(\nu)$ are the computed (complex) power spectra evaluated, respectively, for the absence and presence of the specimen.

Submillimeter and far-infrared data are summarized in Figs. 13–18. An absorption band has been found maximizing at 193 cm^{-1}, which is a

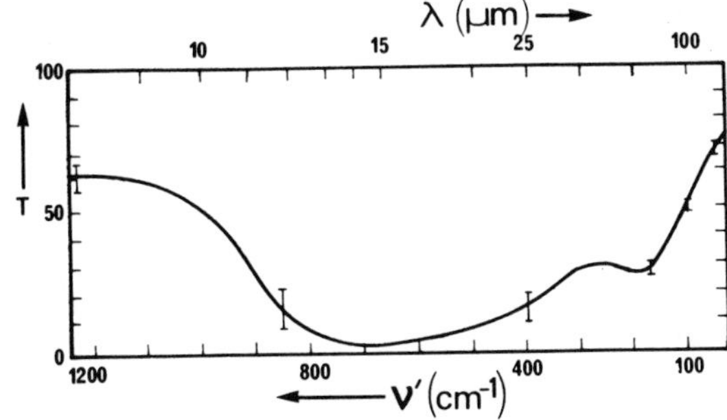

Fig. 13. Transmission function $T(\nu')$ for water (~20°C) in the submillimeter band reported by Draegert *et al.*[273]

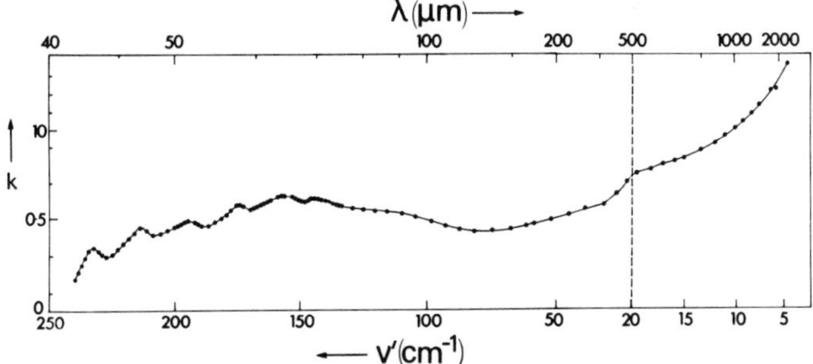

Fig. 14. Transmission function $k(\nu')$ for water (~20°C) in the submillimeter band reported by Stanevich and Yaroslavskii.[1010] Broken line indicates change of ν' scale. The authors' definition of k corresponds to $k = 2\varkappa$.

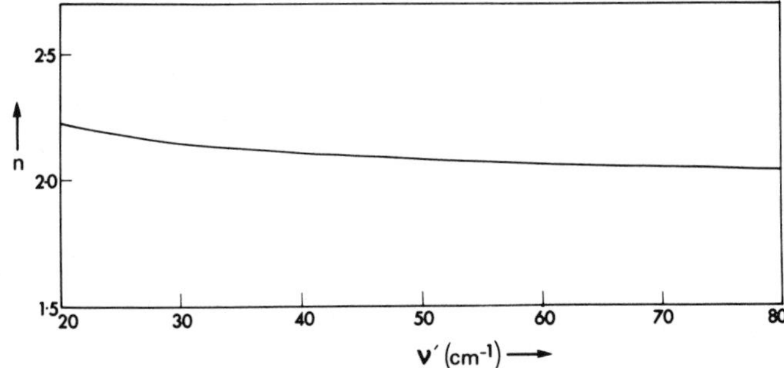

Fig. 15. Refractive index for water (~20°C) in the submillimeter band reported by Chamberlain *et al.*[176]

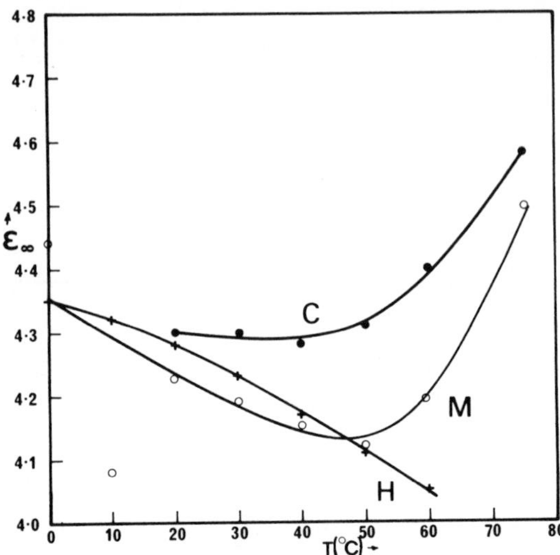

Fig. 16. Temperature variation of high-frequency dielectric constant: *C* (closed circles), ε' measured at 890 GHz by Chamberlain *et al.*;[175] *M* (open circles), ε_∞ deduced from constraint fitting of microwave data; *H* (crosses) ε_∞ required by calculations of Hill.

shoulder (Fig. 13) on a well-known strong band[1116] maximizing at 683 cm^{-1}. The 193 cm^{-1} band is supposedly identical with the band reported by Cartwright[168–170] at 65 μm (154 cm^{-1}).

The fall of dielectric constant with increasing frequency is rather slow[176] in the region 20–100 cm^{-1}, but is probably faster at the maximum of the 193-cm^{-1} band, as appears from Fig. 18.

In the 1000-cm^{-1} region there are some recent refractive index measurements by Querry *et al.*[876] which yield values of ε' a good deal lower than

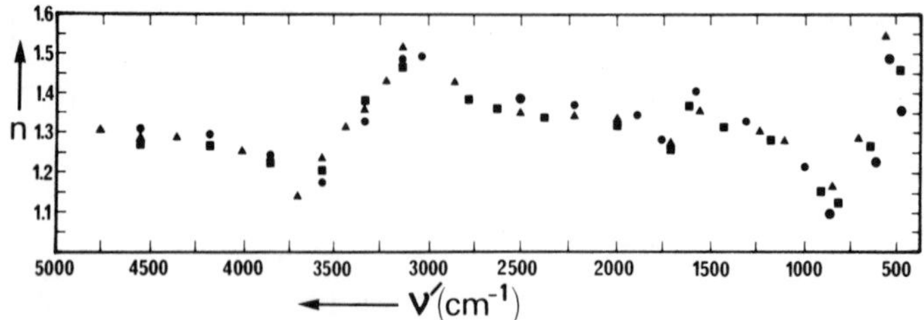

Fig. 17. Infrared refractive index of water (~20°C) reported by Querry *et al.*[876]

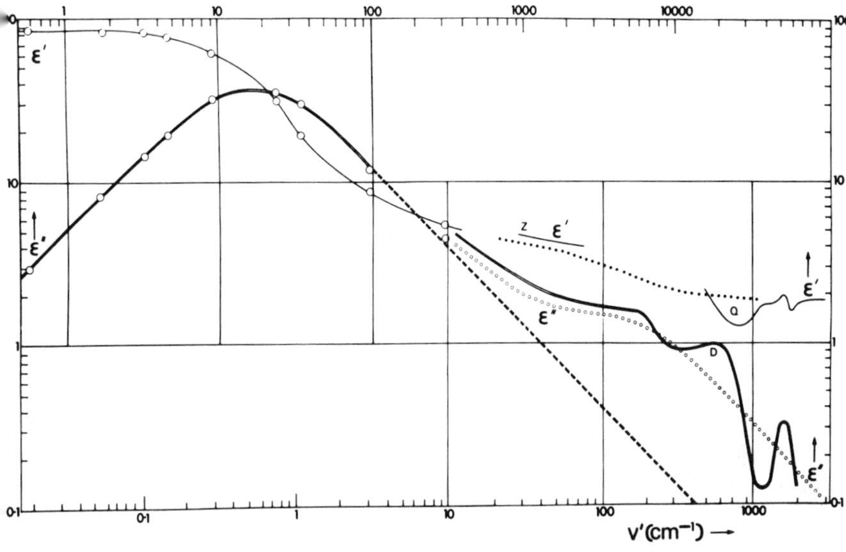

Fig. 18. Logarithmic plot of ε' (single-thickness line) and ε'' (double-thickness line) against frequency ν for water at 20°C; included are data of Table V (large open circles), Fig. 13 (*D*), Fig. 15 (*Z*), Fig. 17 (*Q*). Broken line shows ε'' expected on the basis of single relaxation process. Small circles show calculations of ε' and ε'' from proposed second relaxation process with $\lambda_s = 0.008$ cm.

those at 100 cm^{-1}; the function contains anomalous dispersion features (Fig. 17). It appears that in the unexplored region between these two sets of results a fall in ε' occurs; the frequency variations of both ε' and ε'' are shown in Fig. 18; further data are awaited with interest.

10. THEORY OF THE DIELECTRIC CONSTANT IN A TIME-VARYING FIELD

The theory starts from an assumption about the time variation of the polarization in response to a step function removal of field. The simplest assumption is of an exponential decay:

$$P(t) = P_0 \, e^{-t/\tau} \tag{47}$$

where τ is the relaxation time. The physical meaning of such a decay is that the rate of change of polarization is proportional to the difference between the instantaneous polarization and its final value. This relaxation assumption is often used in the theory of the return to equilibrium of a perturbed system.

For the application of a harmonically time-dependent field of angular frequency $\omega = 2\pi\nu$ described by

$$E(t) = E_0 e^{j\omega t} \tag{48}$$

the frequency-dependent complex dielectric constant $\hat{\varepsilon}$, which is related to the refractive index n and absorption coefficient $\varkappa$ by the equation

$$\hat{\varepsilon} = \varepsilon' - j\varepsilon'' = (n^2 - \varkappa^2) - j(2n\varkappa) \tag{49}$$

was shown to be

$$\hat{\varepsilon} = \varepsilon_\infty + [(\varepsilon_s - \varepsilon_\infty)/(1 + j\omega\tau)] \tag{50}$$

where ε_s is the "static" dielectric constant for time-invariant field and ε_∞ the residual dielectric constant appropriate to frequencies much higher than that of the relaxation process; at sufficiently high frequencies permanent dipoles make no contribution to the dielectric constant and only atomic and electronic polarization contribute.

This expression was first obtained by Debye[245] using a very specialized model, that of spherical molecules in a viscous fluid; a quantity containing ε_s, ε_∞, the viscosity of the liquid, and other factors replaces τ. Nevertheless, the derivation does not require the specialized model; all that is required is the assumption of exponential relaxation.

Rationalizing eqn. (50) yields

$$\begin{aligned} \varepsilon' &= \varepsilon_\infty + [(\varepsilon_s - \varepsilon_\infty)/(1 + \omega^2\tau^2)] \\ &= \varepsilon_\infty + \{(\varepsilon_s - \varepsilon_\infty)/[1 + (\lambda_s/\lambda)^2]\} \end{aligned} \tag{51}$$

$$\begin{aligned} \varepsilon'' &= (\varepsilon_s - \varepsilon_\infty)\omega\tau/(1 + \omega^2\tau^2) \\ &= (\varepsilon_s - \varepsilon_\infty)(\lambda_s/\lambda)/[1 + (\lambda_s/\lambda)^2] \end{aligned} \tag{52}$$

where λ_s is the relaxation wavelength ("Sprungwellenlänge") corresponding to τ. These are known as the Debye equations[245]; they may be expressed graphically in the form of Fig. 19, in which $\log_{10}(\varepsilon')$ and $\log_{10}(\varepsilon'')$ are plotted against $\log_{10}(\omega\tau)$. For $\omega = \omega_s = \tau^{-1}$, when ε' has fallen by half of its total decrease $\varepsilon_s - \varepsilon_\infty$, ε'' has its maximum value, equal to $\frac{1}{2}(\varepsilon_s - \varepsilon_\infty)$. This equality is a good test that the polarization has a simple exponential decay function of time constant $\tau = \omega_s^{-1}$.

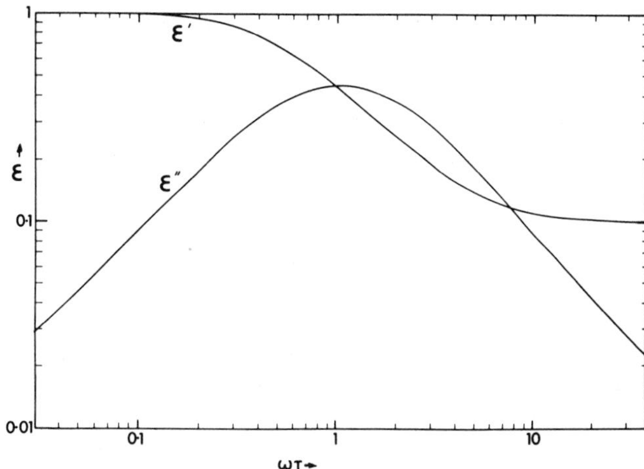

Fig. 19. Variation of $\log_{10}(\varepsilon')$ and $\log_{10}(\varepsilon'')$ versus $\log_{10}(\omega\tau)$ for single Debye relaxation between $\varepsilon_s = 1$ and $\varepsilon_\infty = 0.1$

Another graphic expression of the equations is a plot of ε'' versus ε' for corresponding values of ω. It was shown by Cole and Cole[196] that this plot is a semicircle with center on the ε' axis at $\varepsilon' = \frac{1}{2}(\varepsilon_s - \varepsilon_\infty)$. For the Debye equations to be shown to be valid, it is necessary that the points be distributed around the semicircular "Cole–Cole plot" in the manner prescribed by eqns. (51) and (52). A family of such plots for water is shown in Fig. 20.

The "microscopic" relaxation time τ_μ of the individual dipole is not necessarily the same as that of the dielectric (macroscopic relaxation time

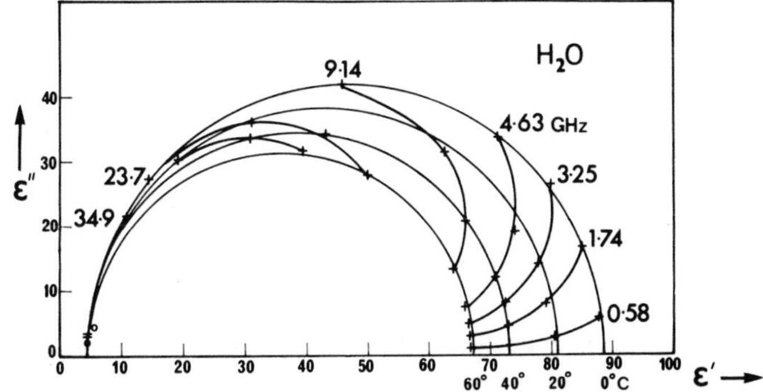

Fig. 20. Cole–Cole representation $\varepsilon'(\varepsilon'')$ of data for water at different temperatures (°C) and frequencies (GHz).

τ_M). Debye treated this problem in terms of molecules experiencing a Lorentz field

$$E_L = \tfrac{1}{3}[\varepsilon'(\omega) + 2]E \tag{53}$$

and rotating against a frictional force proportional to their angular velocity. This leads to Debye eqns. (51) and (52), with

$$\tau_M = [(\varepsilon_s + 2)/(\varepsilon_\infty + 2)]\tau_\mu \tag{54}$$

However, the Lorentz field is a very poor approximation to the instantaneous value of the "internal" field which is actually experienced by the molecule. A better approximation is given in the theory of the static dielectric constant due to Onsager.(808) This theory has been generalized by several authors, the calculations of Collie *et al.*(201) and of Hill(508) being the most free from arbitrary assumptions. The results of this calculation are

$$\hat{\varepsilon} - n^2 = \frac{4\pi N_0}{V} \frac{3(2\varepsilon_s + n^2)}{2\varepsilon_s + 1} \frac{\mu^2}{3kT(1 + j\omega\tau)} \frac{\hat{\varepsilon}}{2\hat{\varepsilon} + 1} \tag{55}$$

Modern theories take account of the phase lag in the reaction field produced at the molecule by the libration of the neighboring molecules. In the model of Fatuzzo and Mason,(324,325) the dipole moments of the molecules are supposed to librate at the frequency of the field through a small angle about their equilibrium positions in zero applied field. The frequency dependence of the dielectric constant is shown to be

$$\frac{\varepsilon_s(\hat{\varepsilon} - \varepsilon_\infty)(2\hat{\varepsilon} + \varepsilon_\infty)}{\hat{\varepsilon}(\varepsilon_s - \varepsilon_\infty)(2\varepsilon_s + \varepsilon_\infty)} = \left[1 + j\omega\tau_\mu + \frac{(\varepsilon_s - \varepsilon_\infty)(\varepsilon_s - \hat{\varepsilon})}{\varepsilon_s(2\hat{\varepsilon} + \varepsilon_\infty)}\right]^{-1} \tag{56}$$

However, a sounder basis for the calculation of frequency-dependent dielectric constant without ignoring the lag in the reaction field is to make use of the Kubo theory of nonequilibrium statistical mechanics. Glarum(388) applied this theory, and subsequently Fatuzzo and Mason criticized the treatment and arrived at the equation:

$$\frac{\varepsilon_s(\hat{\varepsilon} - 1)(2\hat{\varepsilon} + 1)}{\hat{\varepsilon}(\varepsilon_s - 1)(2\varepsilon_s + 1)} = \frac{1}{1 + j\omega\tau_\mu} \tag{57}$$

for rigid, unpolarizable dipoles, for which $\varepsilon_\infty = 1$. This is in fact the identical expression to that obtained from the librating (unpolarizable) dipole theory. It does not give rise to a precisely semicircular Cole–Cole plot, but to the slightly skewed plot shown in Fig. 21. It has not yet been possible to

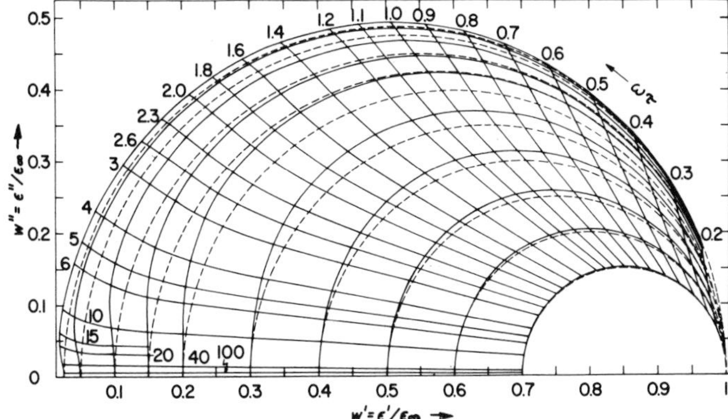

Fig. 21. Plots of $(\varepsilon''/\varepsilon_s)(\varepsilon'/\varepsilon_s)$ computed from eqn. (57).

demonstrate whether the experimental data for any liquid reproduce this slight departure from semicircularity. According to Fatuzzo and Mason, the apparent macroscopic relaxation time of the dielectric constant is equal to the microscopic relaxation time, but according to earlier treatments, it is longer by a factor $\frac{3}{2}$.[388,867]

Although the theories of the frequency variation of the dielectric constant have been worked out for single time of relaxation governing the exponential decay, it can be shown that a general description of the time variation of polarization is possible by means of a superposition of a continuous distribution of relaxation processes with characteristic times varying over a wide range. Such a distribution of relaxation times is often encountered. It gives rise to a Cole–Cole plot which is a shallow arc rather than a semicircle. An empirical expression [eqn. (43)] is based on this construction: If the center of the arc lies on a line below the ε'' axis, making an angle α with the point $\varepsilon' = \varepsilon_\infty$, $\varepsilon'' = 0$, then it can be shown[196] that

$$\hat{\varepsilon} = \varepsilon_\infty + \{(\varepsilon_s - \varepsilon_\infty)/[1 + (j\omega\tau)^{(1-\alpha)}]\} \tag{58}$$

The parameter α is taken as a relaxation-time spread parameter. The deduction of its value for water has been considered above.

The theory of rate processes[395] has been applied to the temperature variation of a single relaxation time, in the form

$$\omega_s = (kT/h)\exp(\Delta S^\ddagger/R)\exp(-\Delta E^\ddagger/RT) \tag{59}$$

where $\omega_s = 1/\tau$ is the angular relaxation frequency, $\Delta E^\ddagger$ is the activation energy of the relaxation process, $\Delta S^\ddagger$ is its entropy change, R is the gas

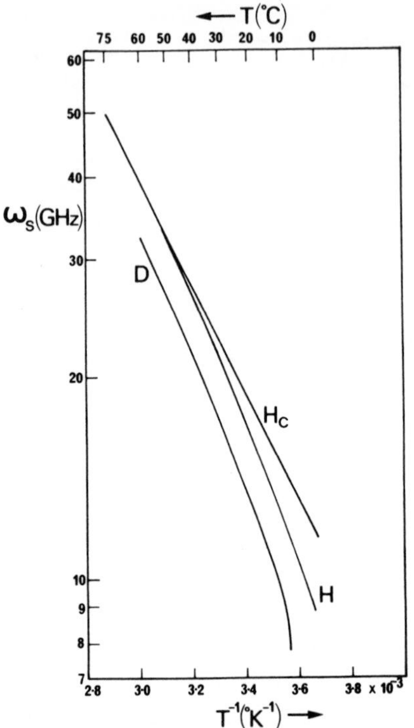

Fig. 22. Plot of $\log_{10} \omega_s$ versus (T^{-1}) for liquid H_2O and D_2O, together with H_2O values corrected for temperature variation of n_3 (respectively, curves H, D, and H_c).

constant, k is the Boltzmann constant, and h is Planck's constant. Thus, a plot of $\log \omega_s$ against T^{-1} should be linear, with slope governed by the activation energy of the process. Such a plot for the principal relaxation of H_2O and D_2O is given in Fig. 22.

11. INTERPRETATION OF THE RELAXATION TIMES AND ABSORPTION BANDS

The experimental data for liquid water in the microwave and submillimeter bands can be interpreted as follows: There is a small spread of relaxation processes in the microwave band, with relaxation times τ tabulated as a function of temperature in Table Va. The approximately exponential temperature dependence of the relaxation time (Fig. 22) demonstrates an activation energy of relaxation of $\sim$5 kcal mol^{-1} for both H_2O and D_2O; this is believed to be of the order of the energy of the OH$\cdots$O hydrogen bond. The dipole orientation process follows eqn. (43), and it has always been thought that it is likely to be related to the breakage of a single hydrogen bond; perhaps the bending of bonds also contributes.

At higher frequencies there is further absorption, accompanied by further depression of real dielectric constant, as shown in Fig. 18. The maximum absorption (20°C) is at 190 cm^{-1}, and there is a further absorption band maximizing at 685 cm^{-1}. These features have been attributed to "vibrations of the hydrogen-bonded liquid lattice," but there are differences of interpretation,[(443,721)] and discussion will be given below in terms of a second relaxation process.

We first discuss the principal relaxation process, which accounts for more than nine-tenths of the static dielectric constant. Its interpretation depends on the model which is proposed for liquid water, and must be different according to whether the breakage or the bending of hydrogen bonds is regarded as important. The essentially single relaxation time implies that the dipole orientation process is unique, i.e., that the reorienting dipoles have essentially identical surroundings. This is in itself difficult to reconcile with a bond-bending model, and must be regarded as evidence against it. However, the small spread of relaxation times might arise from a distribution of environments in which the bond-bending orientation occurs. The relaxation time spread can also be interpreted in terms of bond breaking, as is done below.

A consideration of the processes of reorientation in a bond-breaking model[(443)] is capable of accounting for the principal relaxation. Four-bonded molecules (i.e., molecules bonded to four neighbors) would for reorientation require the breakage of three hydrogen bonds; a defect must be created or destroyed in the process; this is presumably the principal mechanism of relaxation in ice I (ΔE_{act} is found to be $\sim$13 kcal mol^{-1}). Three-bonded molecules require for reorientation the breakage of two hydrogen bonds, two-bonded molecules require the breakage of one, and one- and zero-bonded molecules are free to rotate without breakage. Suppose that all types of molecule are present in the liquid. The faster bond-breaking processes make the slower (high-energy) processes very rare, so that three- and four-bonded species do not contribute relaxations, even though they are probably present in large proportions. It is presumably the two-bonded molecules which contribute the principal reorientation process, the activation energy of which is found to be of the magnitude necessary for the breakage of one hydrogen bond. The zero- and one-bonded molecules may well contribute a higher-frequency process, but there are so few of these species present that the process contributes only a small fraction of the static dielectric constant. Furthermore, this high-frequency process takes place without bond breakage, so that it does not interfere with the two-bonded process by destroying the structure in the way in which the two-bonded process interferes with possible three- or four-bonded processes.

Haggis *et al.*[443] proposed a refinement of this situation, namely that it is the "asymmetrically two-bonded" molecules (Fig. 23) which dominate. It was claimed that the orientation of an H_2O whose oxygen atom is exposed to only one neighbor HO can take place more easily than that of one whose oxygen is exposed to two HO (but with one of its own hydrogens unbonded). There are two centers of minimum potential energy on the oxygen atom, and it may be possible for a hydrogen to pass from one to the other with less of a barrier to overcome than would be the case if the orientation took place by the removal of a hydrogen from an H_2O molecule and its replacement by another.

Whether such a distinction is realistic or not, the two-bonded process may be treated quantitatively as follows. The factor governing the reorientation of four- and three-bonded molecules which make up the bulk of the liquid is the rate of formation of molecules in the two-bonded state, which

Fig. 23. Rotation of the "asymmetrically two-bonded water molecule."

are free to rotate at a higher frequency than the rate at which they are formed. The rate of formation depends on the probability of breaking of a bond, $f_1(T)\exp(-\Delta E/RT)$, and on the number of bond types from which the state can be formed:

$$\begin{aligned} 1/\tau &= (2m_{33} + m_{34} + m_{31} + m_{32}) f_1(T)\exp(-\Delta E/RT) \\ &= 2n_3 f_1(T)\exp(-\Delta E/RT) \end{aligned} \tag{60}$$

The logarithmic plot of relaxation time ($\tau^{-1} = \omega_s$) against the reciprocal of temperature is slightly curved (Fig. 22). This might well be due to the temperature variation of n_3, if variation of $f_1(T)$ with temperature can be neglected. We do indeed obtain a linear plot for $\log(\omega_s/n_3)$ against T^{-1}, which is shown as the line H_c in the figure; the values of n_3 are calculated from eqns. (34)–(38). The value of ΔE corresponding to this corrected plot is $\sim$3 kcal mol^{-1}, which is lower than the bond energy of the OH$\cdots$O bond. This lends some support to the idea that the "asymmetrically two-bonded" molecules can rotate with less energy than the OH$\cdots$O bond energy.

But the asymmetrically two-bonded and symmetrically two-bonded species might well be expected to have different relaxation times; since, according to the bond-breaking model, both are presumed to be present in the liquid, two relaxation times should be observed. This is consistent with the observed relaxation time spread $\alpha \simeq 0.015$. Assuming that the two species are equally populated, and that α is due to a sum of two relaxation times τ_1 and τ_2 rather than to a spread, then it is readily shown that $\tau_1 \simeq 0.8\tau$ and $\tau_2 \simeq 1.2\tau$. The relative insensitivity of α to temperature is consistent with this approach. It may seem surprising that a 1.5% spread parameter should be accounted for by two so widely different relaxation times; but the process of dielectric relaxation is so broad that observed data are relatively insensitive to spreading of characteristic times. The interpretation of α in this way is only hypothetical, but is particularly attractive because it was several years before the discovery of finite α that Haggis *et al.* proposed the difference between symmetrically and asymmetrically two-bonded species, without facing up to the fact that two fairly similar relaxation times might be expected. Grant independently arrived at this conclusion, but did not publish it, although it is mentioned in his thesis.

It is probably an oversimplification to relate the probabilities and energies of "bond breaking" only to the molecular species. The concept that the process is cooperative, and takes place by a mechanism of "flickering clusters" (Chapter 14), is of importance.

The comparison of relaxation time temperature functions for H_2O and D_2O is of interest (see Fig. 22). It is unfortunate that the curvature of the graphs prevents an accurate comparison of the two activation energies to be made without a computer program (not yet written). However, it is easy to see that they are approximately equal. The hydrogen bond energies for the two species are probably almost identical, since the enthalpies of sublimation of H_2O ice and D_2O ice are respectively 11.316 and 11.925 kcal mol^{-1}. It is the activation energies and entropies of relaxation which together determine the ratio of relaxation times of liquid H_2O and D_2O at any temperature [eqn. (59)]. This is a contrasting situation to that obtaining for vibration and rotation bands, which should have a frequency ratio of $\sqrt{2}$, being governed by nuclear mass; this ratio should be insensitive to temperature.

The Debye relaxation theory relates the relaxation to the viscosity as follows:

$$\tau = 8\pi\eta a^3/2kT \tag{61}$$

where a is the molecular radius and η is the viscosity. Values of the ratios of τ for D_2O and H_2O are very close[(201)] to the ratios of η for D_2O and H_2O, as can be seen from Table VII. Recent D_2O measurements confirm this picture[(422)]—an impressive and unique vindication of the relatively crude Debye theory of relaxation as a reorientation of a spherical polar molecule. Moreover, the molecular radius, calculated as one half the inter-oxygen distance, $a = 1.40$ Å, yields an accurate value of $\tau = 0.9 \times 10^{-11}$ sec. This is relevant to the question of identity of "observed" and "molecular" relaxation times; we have seen that according to the treatment which includes the lag of the reaction field,[(324,325)] the two times should be identical; but it is claimed that the relaxation of a dipole in a polarized medium will always be slower than that in an unpolarized medium, and a factor of $\frac{3}{2}$ has been proposed.[(388,867)]

TABLE VII. Comparison of Relaxation Times and Viscosities of H_2O and D_2O

T, °C	$\tau(D_2O)/\tau(H_2O)$	$\eta(D_2O)/\eta(H_2O)$
10	1.30	1.29
20	1.27	1.25
30	1.24	1.22
40	1.21	1.19

The reconciliation of the spherical molecule relaxation theory with the bond-breaking theory has been attempted by Grant.[418] Both relaxation time and viscosity are exponentially proportional to temperature [see eqn. (59)]:

$$\tau = (A/T)\exp(\Delta E_\tau/kT) \tag{62}$$

$$\eta = B\exp(\Delta E_\eta/kT) \tag{63}$$

with A and B constants. It is readily seen that the two energies are approximately equal. Using equations similar to (34), Grant showed that τ was proportional to $\eta/n_2(1-p)$. Since the calculations in Table II show that $T/n_2(1-p)$ is constant over the range 298–373°C, it follows that $\tau \propto \eta/T$.

12. INTERPRETATION OF DATA IN THE SUBMILLIMETER BAND. CONCLUSIONS

Absorption measurements in the submillimeter region have been made by Draegert *et al.*[273] (Fig. 13), Stanevich and Yaroslavskii[1010] (Fig. 14), and Chamberlain *et al.*[175] All these were at $T \simeq 20$°C. Chamberlain *et al.* also report laser measurements at $\nu' = 29.7\ \text{cm}^{-1}$ of both refractive index n and $\alpha = 4\pi\varkappa/\lambda$ over the temperature range 20–70°C; these are shown as ε'' in Fig. 16. The refractive index at 23°C has recently been measured by Chamberlain *et al.*[176] in the range $30 \leq \nu' \leq 85\ \text{cm}^{-1}$, and is shown in Fig. 15. At much shorter wavelengths, refractive index measurements are available, in particular, the recent data of Querry *et al.*[876] (Fig. 17).

In order to display the dispersions and absorptions in a form suitable for understanding in dielectric terms, Fig. 18 illustrates log ε' and log ε'' versus log ν' at 20°C; it is more usual to display ε' and ε'' versus log ν, but the fully logarithmic graph is employed in order to emphasise the weaker dispersions at higher frequencies. The form of the principal Debye relaxation is abundantly clear from the figure.

The temperature variation of the static dielectric constant is considerable, but although the temperature variation of ε' and ε'' at the high-frequency end of the spectrum has not yet been fully investigated it is clear from the 29.7-cm^{-1} measurements,[175] illustrated in Fig. 16, that the temperature variation in this region is much smaller than that of the static dielectric constant.

The relatively slight frequency dependence of ε' in the region of 1000 GHz (30 cm^{-1}) makes it feasible to consider the consequences of assuming

ε_∞ for the first Debye relaxation to be governed by this value of ε'. The ε' data of Chamberlain *et al.* were included in the constraint fitting program by which ε_∞ was calculated in Section 7. The Debye equations (51) and (52) applied with data as in Table V at 20°C are consistent with a small spread of relaxation times. However, the value of ε'' calculated from this relaxation process would be ~1.4 at 1000 GHz, whereas the experimental value is close to 2.3. Figure 18 shows the approximate 45° linear extrapolation to high frequencies of the log ε'' versus log ν' function (double broken line), which demonstrates this effect. There is thus a very broad region of absorption additional to that arising from the microwave Debye process. It commences around 10 cm^{-1} and extends to around 1000 cm^{-1}. It is arguable that this absorption, accompanied by an (incompletely measured) decline of ε', includes a dielectric relaxation process. However, there are two regions of maximum absorption, at about 200 cm^{-1} and 700 cm^{-1}; the latter is almost certainly a resonance rather than a relaxation process, and the former lies suspiciously close to the 175 cm^{-1} Raman band characterized as an intermolecular absorption by Walrafen[1134,1135]; this band decreases in intensity by a factor of seven when the temperature is raised from 0 to 60°C, a fact which is supposed to indicate the rapid disruption of structure in this range. The submillimeter absorption band (and also the real dielectric constant) shows only a limited weakening with temperature, and therefore it must arise from, or at least contain, a different process. At present the only published measurements supporting this statement are at the single frequency $\nu' = 30$ cm^{-1}[175]; it was shown that the absorption was in excess of the value predicted from single relaxation, equally so at 70 as at 20°C.

The criteria for distinguishing between resonances and relaxations are as follows:

1. The temperature dependence of the frequency of maximum absorption is small for a resonance process; but for a relaxation process it is exponentially proportional to $\Delta E^{\ddagger}/kT$.

2. When the vibration of a proton is involved, the ratio of H_2O to D_2O resonance frequencies is usually $\sqrt{2}$, but the ratio of frequencies of the maxima in the two relaxation processes is not necessarily $\sqrt{2}$; this was discussed in the previous section.

3. The breadth of an isolated absorption peak arising from a single relaxation process is more than one order of magnitude (full-width at half-maximum), and will be larger if a spread of relaxation times is involved. That of a single resonance process is usually much narrower although

there can be several resonance processes close together in frequency, or even a spread.

4. The behavior of ε' with increasing frequency in the region of maximum absorption is different for a relaxation and for a resonance process. For a relaxation there is a broad decline from $\varepsilon' = \varepsilon_s$ to $\varepsilon' = \varepsilon_\infty$, but for a resonance process there is an "anomalous dispersion," typified by the farinfrared features in Fig. 18; in general, this contains both a minimum and a maximum.

There are some discrepancies among the absorption measurements, but there seems to be no doubt about the very broad absorption in the region of 200 cm^{-1}. The structured peak (Fig. 14) reported by Stanevich and Yaroslavskii[1010] has been found to be entirely absent by Draegert *et al.*,[273] who suppose that it arises from water or other vapor in the optical path. There is nevertheless a fairly broad maximum in this region which, when represented as $\alpha(\nu)$, appears only as a shoulder on the 685 cm^{-1} band. The temperature variation of the frequency of maximum absorption is not as yet known, and the frequency of the corresponding absorption in D_2O is approximately the same as that in H_2O. The structured band has been omitted from Fig. 18.

The 685-cm^{-1} band shifts by about 100 cm^{-1} in a 70° rise of temperature; this is a small shift when compared with that of the microwave relaxation process, which is proportionately four times greater. The frequency of the corresponding D_2O absorption peak is 1.4 times lower than that for H_2O. The refractive index (and ε') corresponding to this absorption shows an anomalous dispersion feature. All the evidence points to this band arising from a resonance process, and it has been interpreted as a vibration of the hydrogen atoms against the hydrogen bonds, since the band is not found in the vapor phase.

The data for the ~200-cm^{-1} maximum are not consistent with a resonance interpretation. There could be a resonance process superposed on a relaxation process, but the data seem to be dominated by the relaxation process. It is very broad, it is associated with a broad decline of dielectric constant ε', and the D_2O data are not different in frequency by $\sqrt{2}$. What little is known about the temperature variation of the frequency of maximum absorption is consistent with a relaxation of activation energy much lower than that of the principal relaxation.

It is worthwhile performing a crude calculation using the Debye equations (51) and (52), of the magnitude of ε' and ε'' contributed by a second relaxation process; parameters are selected as follows: ε_s (second relaxation) = ε_∞ (first relaxation) = 4.1; ε_∞ (second relaxation) = n^2 = 1.8; λ_s (second relaxation) = 0.008 cm. To the ε'' values so calculated are added the calculated contributions from the extreme high-frequency end of the first relaxation process. Both ε' and ε'' values resulting from this calculation are shown as dotted lines in Fig. 18. The results are encouraging; the extra absorption arising from the 693-cm^{-1} band is obviously attributable to a further process, as has already been mentioned. The asymptotic behavior of ε'' at high frequencies is a 45° line somewhat above the line corresponding to the first relaxation. Around 1100 cm^{-1} the experimental ε'' falls below this; an inertial effect may be involved.

Given this interpretation, it is for the present necessary in the calculation of the static dielectric constant to employ the Fröhlich equation (30) with n^2 = 1.8 and μ_0 = 1.84 D. At 20°C the correlation parameter g required to fit the experimental ε_s = 80.4 has the value g = 2.82. It will be seen from Table II that this is almost exactly the value that would be expected in a "four-bonded" water environment. It is therefore rather higher than might be expected, but a small elevation of n would be sufficient to lower g to a more realistic value. The exact measurement of n over the entire frequency range is necessary for any accurate application of the Fröhlich–Kirkwood theory. In particular, coverage of the missing band at 20–100 μm in the refractive index is important; the anomalous dispersion features obscure the true ε_∞ for the second relaxation. Analysis of the entire 1–10^5 GHz range in terms of two relaxations is possible; the optimum $\varepsilon_\infty(1) = \varepsilon_s(2)$ cannot be chosen without this. An educated guess might be: $\varepsilon_\infty(1) = \varepsilon_s(2)$ = 4.2, $\varepsilon_\infty(2)$ = 2.0.

The temperature variation of $\varepsilon_\infty(1) = \varepsilon_s(2)$ is of great importance from the point of view of knowledge of the proportions of "free" (zero- and one-bonded) H_2O molecules, which are here considered to be responsible for process 2. Haggis *et al.* calculated this proportion, and predicted that $\varepsilon_\infty(1)$ would rise with increasing temperature; as can be seen from Fig. 16, this is not found except at temperatures above 50°C. It is likely that the proportions are sensitive to other factors than those assumed by these authors. The proportion corresponding to the data of Fig. 16 is (4.1 − 1.8) ×100/(80.4 − 1.8) = 2.9% zero- and one-bonded H_2O molecules at 20°C. However, this interpretation is speculative, and its usefulness could be merely that of stimulating further study.

The static dielectric constant ε_s = 92 for ice[(37)] will also be very sensitive to the value of $\varepsilon_\infty = n^2$. A value of 3.05 was measured by Lamb[(648)] at 10 GHz, but the value in the submillimeter region is probably lower. An infrared refractive index of approximately n^2 = 2 would satisfactorily fit the data with $g = 2.91$. For the principal low-frequency relaxation process,[178,648,1116] $\tau = 1.4 \times 10^{-4}$ sec at -5°C. This relaxation process would give rise to very little absorption in the microwave region, $\tan \delta = 0.17 \times 10^{-4}$ at 10 GHz; but Lamb measured $\tan \delta = 0.45 \times 10^{-3}$, and concluded that there was possibly a second absorption maximum at a frequency well above 10 GHz. Such a second relaxation should shift exponentially to lower frequencies as the temperature is lowered, thus raising $\tan \delta$ exponentially with T^{-1}. But $\tan \delta$ actually falls as temperature is lowered, so that another factor must be involved in the temperature dependence. The fall is by no means exponential (rounded $\log(\tan \delta)$ versus $1/T$ plot), and the most likely explanation is that free or interstitial molecule populations rise very rapidly with rising temperature in the region -50 to 0°C.

It is possible that there is a relaxation process for ice in the submillimeter and far-infrared, very similar to that proposed for liquid water. The static dielectric constant for ice[37] is very close to that for water at 0°C (Table II), and the conclusion drawn from this by Haggis *et al.*[443] was that the amount of dipole correlation is almost the same, that is, 9% bond breaking in water at 0°C.

ACKNOWLEDGMENTS

The author wishes to acknowledge valuable criticism and discussions with Drs. Peter Mason, Nora Hill, Ted Grant, and John Chamberlain, the last-named contributing much of Section 9.

CHAPTER 8

Liquid Water: Scattering of X-Rays

A. H. Narten and H. A. Levy

Chemistry Division
Oak Ridge National Laboratory
Oak Ridge, Tennessee

1. INTRODUCTION

The long-range periodicity of a crystal lattice is not a necessary requirement for the production of diffraction effects. Even in a dilute polyatomic gas the interference between waves scattered from different atoms within a molecule is sufficient to produce a characteristic diffraction pattern from which the molecular structure can be deduced. The short-range positional and orientational correlation between molecules, negligible in the gas phase, is sufficient in a liquid to cause additional interference effects between waves scattered from atoms in different molecules. The diffraction pattern of a liquid contains information about these correlations.

2. EXPERIMENTAL METHODS

The X-ray scattering by an isotropic liquid is symmetric about the incident beam, and the accessible range of scattering angles is $0° \leq 2\theta \leq 180°$. At large scattering angles the interference pattern is determined mainly by interactions with a range of the order of the X-ray wavelength (~1 Å). At small angles the interference is due primarily to interatomic distances which are considerably larger than the X-ray wavelength. Because of the different requirements for measuring the X-ray scattering at large and at small angles, the two methods will be discussed separately. The diffractometers described in Sections 2.1 and 2.2 are those used at Oak Ridge National Laboratory for the collection of the X-ray data on water that are presented in Section 4.2.

2.1. Large-Angle Scattering

The arrangement of the X-ray tube X, the sample HH, and the detector arm with monochromator M and detector D is shown schematically in Fig. 1. The arm containing the X-ray tube and the arm containing the monochromator and detector are caused to rotate at the same angular rate and in opposite sense about a horizontal axis A lying on the surface HH of the liquid sample and perpendicular to the plane of the paper. The slits S_1, S_2, and S_3 serve to define the incident and scattered beams and to determine their divergence δ. This arrangement, with the mean angle α between the incident beam and the sample surface always equal to the mean angle β between the scattered beam and the sample surface, makes possible the use of a divergent beam and thus considerably increases the intensity of the scattered radiation over that achieved with a finely collimated beam.

The crystal monochromator M (Fig. 1) is of the focusing type[432] and is set to reflect the characteristic $K\alpha$ radiation of the source. The detector D, set to receive the final beam as reflected by the monochromator, consists of a scintillation counter in conjunction with a linear amplifier, a single-channel analyzer, and a count scaler-timer. The analyzer is set to reject any pulses which might arise from the second-order half-wavelength radiation reflected by the monochromator.

The described geometry and slit arrangement completely eliminate sample holder absorption and scattering; it also reduces the background of stray radiation, particularly at small scattering angles. Instrument resolution is determined primarily by the extent of penetration of the beam into the

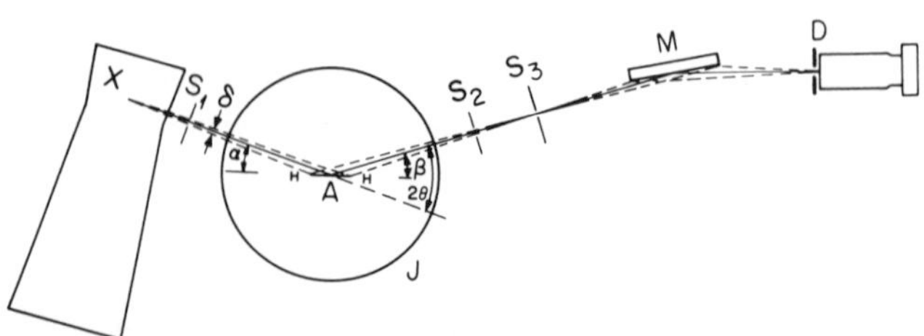

Fig. 1. Reflection geometry used in large-angle scattering experiments with water.

sample, the size of the focal spot in the X-ray tube, the size of the receiving slit S_3, and the mosaic spread of the monochromator crystal. Angular resolution is limited in practice by the requirement for adequate intensity of scattered radiation. For the described diffractometer the slits S_3 and D were chosen approximately the size of the focal spot[678] and the resolution is almost entirely determined by the perfection of the monochromator crystal. The wavelength acceptance can be estimated from the measured rocking curve of the crystal through the Bragg equation[432] as

$$\Delta\lambda/\lambda \simeq \text{ctn}(\phi\Delta\phi) \tag{1}$$

with ϕ the reflecting angle of the monochromator crystal for radiation of wavelength λ. The water data described in Section 4.2 were measured using a sodium chloride crystal set for plane (200) to reflect Mo $K\alpha$ radiation; for this case the angle $\phi = 14.5°$, and the mosaic spread (half-width at half-height) of the crystal was measured as $\Delta\phi = 0.13°$. Equation (1) yields for the wavelength resolution a value $\Delta\lambda/\lambda \simeq 0.01$. This resolution is sufficient to greatly reduce the intensity of Compton-modified radiation passed by the monochromator.

The measured intensity of scattered X-radiation has to be corrected for background radiation, absorption in the sample, polarization of the X-ray beam, multiple scattering, and Compton-modified radiation. For the described diffraction geometry[678] all of these corrections are either negligibly small or can be calculated with good accuracy, except for the Compton scattering, which, in the case of water, is the most important single source of uncertainty in the desired final result, namely the total coherently scattered intensity. Because of its importance in the present case we shall discuss the Compton correction in some detail.

The intensity of Compton-modified radiation[432] as a function of the scattering variable $s = (4\pi/\lambda)\sin\theta$ rises rapidly to its asymptotic value, while the intensity of coherently scattered radiation has its maximum at a value of $s \simeq 2$ (for water) and is then rapidly decreasing (see Fig. 2). All of the previously reported diffraction data on water (see Section 4.1) were collected with instruments in which all of the Compton-modified radiation reached the detector. In these cases the intensity of Compton-modified radiation is equal to the intensity of coherently scattered radiation at $s \simeq 5.5$, and exceeds the coherent intensity by a factor of seven at $s \simeq 16$, thus making the measurement of reliable data at large scattering angles extremely difficult. For the described diffraction geometry the intensity of Compton-modified radiation rises, as predicted by theory,[432] at small scattering angles. However, with increasing angle the exciting source for the Compton scattering shifts from the characteristic line passed by the monochromator into the continuum. As a result, the Compton scattering decreases from a maximum at intermediate angles to a small fraction of the coherent intensity at large angles. The wavelength shift of the Compton-modified radiation is,[432] for Mo $K\alpha$ with $\lambda = 0.7107$ Å.

$$\Delta\lambda/\lambda = 0.0342(1 - \cos 2\theta) \tag{2}$$

and the crystal monochromator discriminates against the Compton scattering at angles for which $\Delta\lambda/\lambda \gtrsim 0.01$, the wavelength resolution of the system described. From eqn. (2) we estimate this to be the case for $2\theta \simeq 45°$, corresponding to a value of $s \simeq 7$. Figure 2 shows that the NaCl monochromator passes about 50% of the Compton scattering at $s \simeq 5.5$, and less than 5% at large scattering angles, qualitatively in good agreement with the above estimate.

Curve C of Fig. 2 was determined by an iterative procedure,[678] making use of the fact that the intensity of Compton-modified radiation must equal the theoretical curve B at small scattering angles, and can be determined experimentally at high angle through the use of Zr filters.[678] Once the discrimination function for a given monochromator has been determined in this way, the Compton scattering curve for any liquid can be

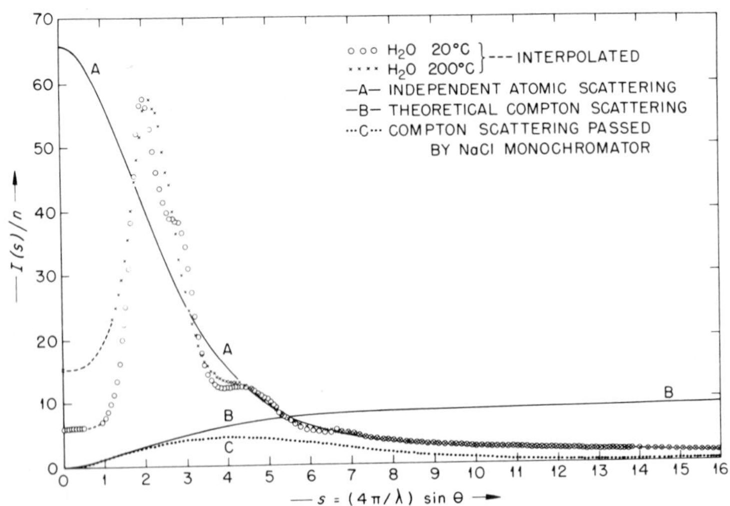

Fig. 2. Total scattered intensities of X-radiation for water at two temperatures; units are electron units per water molecule.

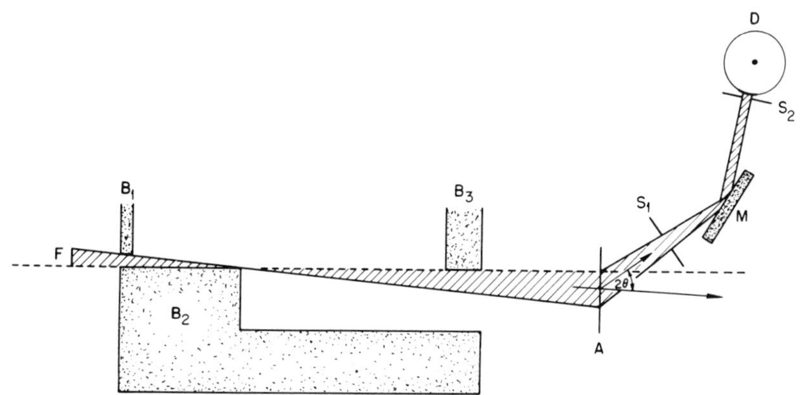

Fig. 3. Transmission geometry used in small-angle experiments with water.

calculated from the tabulated[205,221]* atomic Compton scattering amplitudes.

The absolute scale is established by normalizing the asymptote of the corrected intensity function to the theoretical independent atomic scattering (see Section 3.1) at large scattering angles. An alternate method[883] for the scaling of diffraction data from liquids is less suitable in the case of water.[782]

2.2. Small-Angle Scattering

Scattering measurements at very small angles require the use of a beam with extremely fine collimation. The scattering of a beam transmitted through a thin sample is measured, since surface reflection analogous to the large-angle case would require an extremely accurate adjustment of the reflecting surface. The main difficulty in small-angle scattering experiments is to achieve a sufficiently fine collimation and still have adequate intensity.[433]

The geometry shown schematically in Fig. 3 is that of a commercially available instrument designed by Kratky.[635] The breadth and divergence of the direct beam are limited by the upper face of an optically flat metal block B_2, the underside of an entrance slit B_1, and the lower face of an optically flat bridge B_3. This collimating arrangement is geometrically equivalent to a pair of long slits; when precisely machined and aligned, it prevents practically all parasitic scatter originating from the collimating apertures from reaching the detector and at the same time permits intensity

* Also D. T. Cromer, private communication (Compton scattering factors for hydrogen).

measurements to scattering angles as small as $\sim 20''$ of arc.[635] Its disadvantage is the asymmetry of the system with respect to the direct beam, and therefore the inability to measure the scattering curve on both sides of zero angle. It is essential for optimum performance that the faces of B_2 and B_3 be perfectly plane and that they coincide precisely with one and the same horizontal plane indicated in Fig. 3 by the broken line. The breadth of the beam can be adjusted by varying the width of the entrance slit B_1. The sample A is contained between thin windows held together by a metal frame. The scattered beam passes through a receiving slit S_1, is reflected from the surface of a plane monochromator crystal M (hot-pressed pyrolytic graphite[1007]) through a scatter slit S_2 into the proportional counter D. In the instrument used at Oak Ridge the entire X-ray path is in a vacuum,[491] thus reducing substantially the background arising from vacuum windows and air scatter which are present in the commercial units.

The intensity is measured as a function of the distance from the receiving slit S_1 to the centroid of the direct beam. Since the need for sufficient scattered intensity requires use of collimating systems which average the scattering over an appreciable range of scattering angles, the measured intensity is often very different from the curve for perfect collimation. Deconvolution methods[492] are available to correct for this instrumental effect. However, for most liquids at temperatures and pressures far from the critical point, the observed X-ray scattering in the small-angle region is nearly constant with scattering angle; in such a case, very little collimation correction is required. Corrections for polarization and Compton-modified scattering are very small,[433] and the only appreciable corrections are for background, sample absorption, and multiple scattering, the latter contributing about 10% to the observed intensity in the case of water.[185]

In small-angle experiments the absolute scale is usually[433] established through the use of standards. In the case of water and other fluids[490] the small-angle data can also be scaled by extrapolation to zero scattering angle and comparison with a value calculated for the zero-angle intensity from bulk thermodynamic properties (see Section 3.1).

2.3. Data Reduction

The Oak Ridge X-ray data on water were collected at eight temperatures between 4 and 200°C. The sample pressure was atmospheric at temperatures below 100°C and equal to the vapor pressure above this temperature. Except for the data at 20°C, intensity and radial distribution functions were derived from the raw data previously published.[783] The large-angle data cover an angular range $1 \leq s \leq 16$. Unlike the previously published[783–785] intensities and radial distribution functions, the present curves were derived using more accurate values[205,221] for the atomic scattering amplitudes, and results from small-angle X-ray scattering[490] in the range $0.04 < s \leq 0.5$ were included[782] in the curves up to 75°C. At temperatures above 75°C the total intensity functions (Fig. 2) were extrapolated to the calculated value of the intensity at zero angle [eqns. (5a) and (5b) below].

All details of the data correction and reduction have been described elsewhere and the intensity and radial distribution functions discussed in Section 4.2 are available in tabulated form.[782]

3. ANALYSIS OF DIFFRACTION DATA

In this section we shall outline the type of information which can be extracted from diffraction data on polyatomic liquids. The formal approach to this problem was introduced and developed by Zernike and Prins[1197,1198] and Debye and Mencke[246] for monatomic fluids, and extended to heteroatomic systems by Waser and Schomaker.[1150] A basic assumption in this treatment is that the scattering is due to spherically symmetric atoms, deviations from spherical symmetry due to bonding effects between atoms within a molecule being ignored.

An alternate and more rigorous approach to the problem has been developed by Steele and Pecora,[1015] who derived general expressions for scattering from a fluid composed of nonspherical molecules. This approach, which requires precise knowledge of the distribution of electron density within the molecules, has never been applied to scattering data from real liquids. Preliminary calculations by Blum[96] using Pecora and Steele's expansion for the X-ray intensity function indicate that, in the case of water, this approach may yield useful results in the future, complementing but not significantly modifying the conclusions drawn from the simpler theory.

3.1. Scattering by Heteroatomic Liquids

Consider a macroscopically isotropic diffracting object in which the relative positions of the atoms are fixed. The intensity of a coherently scattered wave is given by the Debye equation[244]

$$I(s) = \sum_{i=1}^{N} \sum_{j=1}^{N} f_i(s) f_j(s)(\sin sr_{ij})/sr_{ij} \tag{3}$$

with summation for both i and j over all N atoms in the system. The scattering variable s has been defined as $(4\pi/\lambda)\sin\theta$, where λ is the X-ray wavelength and 2θ is the scattering angle. The distance between atoms i and j is r_{ij}. The atomic scattering amplitudes $f(s)$ are the Fourier transforms of the radial distribution functions for the electron density in the individual atoms, accessible from *ab initio* calculations. It is convenient to separate the terms in eqn. (3) for $i = j$, which are not dependent on a separation distance r_{ij}. The total intensity corrected for these terms (a reduced intensity function) is

$$I(s) - \sum_{i=1}^{N} f_i^2(s) = \sum_{i=1}^{N} \sum_{i \neq j=1}^{N} f_i(s) f_j(s) (\sin sr_{ij})/sr_{ij} \tag{4}$$

In a real material, and particularly in a fluid, relative atomic positions are not rigidly maintained. We represent by $4\pi r^2 \varrho_{\alpha\beta}(r)$ a function giving the radial density of distinct pairs of atoms of kind α, β separated by a distance r. The functions $\varrho_{\alpha\beta}$ represent the average distribution of pairs both over time and the volume of the sample.

In these terms, the scattered intensity becomes

$$\begin{aligned} i(s) &\equiv I(s)/n - \sum_{i=1}^{m} f_i^2(s) \\ &= \sum_{\alpha=1}^{m} \sum_{\beta=1}^{m} f_\alpha(s) f_\beta(s) \int_0^\infty 4\pi r^2 \varrho_{\alpha\beta}(r)[(\sin sr)/sr]\, dr \end{aligned}$$

in which a structural unit containing m atoms is visualized as representative of the whole sample, which contains n such units. The term $\sum_{i=1}^{m} f_i^2(s)$ is the independent atomic scattering, and the reduced intensity $i(s)$ is the structurally sensitive part of $\mathbf{I}(s)/n$, the measured intensity scaled to one structural unit. In order to apply the Fourier integral theorem, it is advantageous to add and to subtract from the above expression the double sum

$$\sum_{\alpha=1}^{m} \sum_{\beta=1}^{m} f_\alpha(s) f_\beta(s) \int_0^\infty 4\pi r^2 \varrho_0[(\sin sr)/sr]\, dr$$

where ϱ_0 is the bulk number density of structural units. The added double sum corresponds to a delta function at $s = 0$ which does not contribute to the observable scattered intensity and may therefore be neglected. The expression for the reduced intensity becomes

$$i(s) = \sum_{\alpha=1}^{m} \sum_{\beta=1}^{m} f_\alpha(s) f_\beta(s) \int_0^\infty 4\pi r^2 [\varrho_{\alpha\beta}(r) - \varrho_0][(\sin sr)/sr]\, dr \tag{5}$$

with summation over all distinct pairs in the structural unit.

The pair density functions $\varrho_{\alpha\beta}(r)$ for a fluid should approach closely the bulk density ϱ_0 beyond a critical distance $r = r_c$, which may be taken as a measure of the extent of short-range order in the liquid. If it is assumed that all pair density functions $\varrho_{\alpha\beta}(r)$ become precisely equal to ϱ_0 at large r, it may be shown that $I(s)$ approaches zero at $s \ll 1/r_c$. In practice it is found that $I(0)$ is finite. This implies that there are persistent long-range variations of $\varrho_{\alpha\beta}(r)$ from ϱ_0. For a liquid composed of one well-defined molecular species which need not be monatomic, the asymptotic small-angle scattering conforms to the equation[432]

$$I(0)/nF^2 = \langle(\nu - \langle\nu\rangle^2\rangle/\langle\nu\rangle \tag{5a}$$

where ν is the instantaneous and $\langle\nu\rangle$ the average number of molecules in a volume whose dimensions are large compared to r_c, F is the number of electrons per molecule, and the averages are over time and the volume of the sample. The critical length r_c is always several times larger than molecular dimensions; for water at room temperature it is estimated (see Section 4.2) to be about 8 Å. For some exceptional systems (for example, fluids near the critical point) it approaches macroscopic dimensions, and the limiting intensity $I(0)$ is reached only at unobservably small scattering angles.

A particular application of eqn. (5a) concerns random equilibrium fluctuations in the local number density in a system containing one kind of molecule, in which the average density is also the most probable one, and only small fluctuations are present. Statistical mechanical considerations[516] yield for the relative variance in particle number which appears on the right side of eqn. (5a)

$$\langle(\nu - \langle\nu\rangle)^2\rangle/\langle\nu\rangle = kT\varrho_0\varkappa \tag{5b}$$

where $\varkappa$ is the isothermal compressibility, k is the Boltzmann constant, and T is the absolute temperature. Random equilibrium density fluctuations are, of course, present in any system but may be only one source of such fluctuations.

In the special case in which the scattering system consists of one kind of atom, the double sum of eqn. (5) reduces to one term. It is possible to obtain the pair density functions $\varrho(r)$ or the pair correlation functions $g(r) = \varrho(r)/\varrho_0$ by Fourier transformation. In a similar manner, correlation functions for molecular centers can be obtained from eqn. (5) if the system consists of one kind of molecule, and if the molecules are spherically symmetric and randomly disposed.[874] In general, however, the corresponding functions $\varrho_{\alpha\beta}(r)$ and $g_{\alpha\beta}(r)$ are not obtainable individually. It is nevertheless useful to construct a modified correlation function by Fourier transformation, namely

$$G(r) = \sum_{\alpha=1}^{m} \sum_{\beta=1}^{m} G_{\alpha\beta}(r) = 1 + (1/2\pi^2\varrho_0) \int_0^\infty si(s)M(s)\sin(sr)\,ds \tag{6}$$

with

$$M(s) = \left[\sum_{\alpha=1}^{m} f_\alpha(s) \right]^{-2} \tag{6a}$$

for $s < s_{\max}$, the maximum value of s accessible in scattering experiments, and $M(s) = 0$ otherwise. Introduction of this modification function into eqn. (6) makes the product $f_\alpha(s)f_\beta(s)M(s)$ nearly independent of s and thus removes from the resulting radial distribution function (RDF) the average breadth of the distribution of electron density in the atoms. The relation between component X-ray pair correlation functions $G_{\alpha\beta}(r)$ and the true atom pair correlation functions $g_{\alpha\beta}(r)$ is one of convolution[1150]

$$G_{\alpha\beta}(r) = (1/r) \int_{-\infty}^{\infty} ug_{\alpha\beta}(u)T_{\alpha\beta}(u - r)\,du \tag{6b}$$

with u a variable of integration and

$$T_{\alpha\beta}(r) = (1/\pi) \int_0^\infty f_\alpha(s)f_\beta(s)M(s)\cos(sr)\,ds \tag{6c}$$

The function $G(r)$, accessible from X-ray scattering data, is thus a kind of linear combination of true atom pair correlation functions, which are convoluted with the known functions $T_{\alpha\beta}(r)$ given by eqn. (6c). The factor $T_{\alpha\beta}(r)$ may be visualized as a shape function into which a pair distribution function is transformed by the combined effects of the inherent electron distribution and treatment of the X-ray data.

It is not possible to decompose the function $G(r)$ uniquely into components $G_{\alpha\beta}(r)$, which can then be deconvoluted into components $g_{\alpha\beta}(r)$, unless, for a liquid containing m different types of atoms, $m(m + 1)/2$ independent diffraction experiments are performed. For water, three different experiments, involving different values for the scattering amplitudes f, would be necessary. Such a study, although possible in principle by the use of neutron diffraction data on isotopically substituted waters, has not been undertaken. The only recourse open for water and other polyatomic liquids is to calculate the atom pair correlation functions from proposed models and then to construct intensity and radial distribution functions which can be compared with those derived from diffraction data.

3.2. Scattering by Model Liquids

Although a model which scatters X-rays in exactly the same way as a real liquid cannot be proved to be unique, this agreement is necessary for the model to be tenable. A model that is to be tested against diffraction data must have a number of properties and must meet certain conditions in order to be useful, realistic, and tractable in this context. We have discussed this subject in detail elsewhere[785] and shall only list these properties and conditions:

1. The model must specify all interactions, properly weighted, to a distance $r \simeq r_c$, where r_c measures the extent of order in the liquid.
2. It must do so in terms of a small number of independent parameters.
3. It must ensure that structurally equivalent atoms have the same distance spectra.
4. The average density of interactions must be consistent with the density of the liquid.
5. The distance spectra of different atoms must be geometrically compatible.

Conditions 1–5 are sufficient for the construction of a one-phase model in which only short-range correlation exists. In the case of water, there has been much discussion[301] of models of a different sort which possess two or more distinct "phases." By definition, such a model contains qualitatively different local arrangements; the connecting zones between them, whether gradual or abrupt, are necessarily different in local arrangement from the regions that are connected.[570,786] The distance spectra for such zones must therefore be included in a valid comparison with radial distribution functions, and the following additional details specified: (6) the average arrangerangement of molecules within each phase; (7) The relative proportions of two or more phases; (8) The average arrangement of molecules around the boundaries of a phase region; (9) The average separation and relative orientation of pairs of such regions. Conditions 8 and 9 need not be given in detail if the regions are large compared with r_c and randomly disposed. The assumption of such large domains, however, has important implications with regard to small-angle scattering.

The only way known to us of fulfilling the listed conditions is to assume a certain near-neighbor configuration around one or more origin atoms and to repeat this structural unit according to some specified rule. An approach of this kind involves a space lattice. Unlike the crystalline case, however, repetition of this structural unit is imperfect and therefore is accompanied by rapid, progressive loss of positional correlation between atoms in adjacent "primitive cells." This loss of correlation is visualized as the result of continuous, small variations in local environment; it may be embodied in the model in the form of probability distributions with mean-square variations from average separations which increase rapidly with radial distance, and the distance spectrum characteristic of a particular model can be closely approximated by a uniform distance distribution (continuum) a few molecular radii away from any starting point.

For such a description, Eq. (5) takes the form

$$\begin{aligned} i(s) = & \sum_{i=1}^{m} \sum_{j} \exp(-b_{ij}s^2) f_i(s) f_j(s) (\sin sr_{ij})/sr_{ij} \\ & + \sum_{\alpha=1}^{m} \sum_{\beta=1}^{m} \exp(-b_{c\alpha\beta}s^2) f_\alpha(s) f_\beta(s) \\ & \times 4\pi\varrho_0 [sr_{c\alpha\beta} \cos(sr_{c\alpha\beta}) - \sin(sr_{c\alpha\beta})]/s^3 \end{aligned} \tag{7}$$

where the summations are over the structural unit for α, β, and i, and over all atoms in the discrete structure for j. The first double sum in eqn. (7) arises from carrying out the integration in eqn. (5) over the discrete structure, that is for the range $0 < r \leq r_c$, and averaging the Debye equation (3) over discrete Gaussian distributions with average separation r_{ij} and mean-square variation $2b_{ij}$. The second double sum in eqn. (7) results from integration over the continuum, that is, for the range of $r \geq r_c$ in which $\varrho_{\alpha\beta} = \varrho_0$. The boundary between the two regions need not be sharp; in eqn. (7) $r_{c\alpha\beta}$ represents the mean and $2b_{c\alpha\beta}$ the variance of this boundary for the pair density function $\varrho_{\alpha\beta}$.

Fourier transformation of the reduced intensity function (7) according to eqn. (6) yields the X-ray pair correlation function. This procedure has the corollary advantage that possible effects on the features of the function $G(r)$ arising from the truncation of the Fourier integral [eqn. (6)] at finite values of the variable s will be exhibited by both the experimental correlation function and by the curve calculated for the model.

4. DIFFRACTION PATTERN OF LIQUID WATER

The diffraction pattern of liquid water is usually presented as a plot of the total scattered intensity of X-radiation against the variable s (Fig. 2). The total intensity function rises monotonically from its predicted level

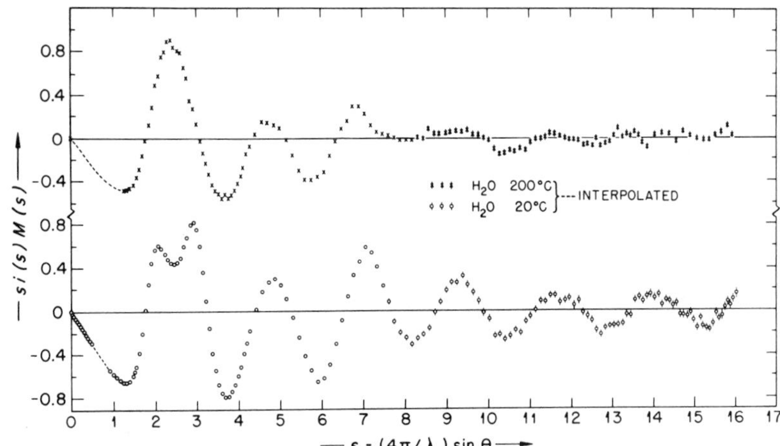

Fig. 4. Kernels of the Fourier integral in eqn. (6) represent the structurally sensitive part of the scattered intensity of X-radiation for water at two temperatures.

[eqns. (5a) and (5b)] at zero scattering angle through a pronounced maximum, and falls off rapidly with the independent atomic scattering curve around which it oscillates with decreasing amplitude. The intensity curves of Fig. 2 for water at widely different temperatures are not very different from each other below a value of $s \simeq 8$, and apparently indistinguishable from each other and from the independent atomic scattering curve above this value; this fact may serve to illustrate two points: First, the customary way of presenting diffraction data on liquids in the form of total intensity curves (as in Fig. 2) is quite insensitive to structural detail and therefore almost pointless. Second, X-ray data of considerable precision, particularly at large scattering angles, are a necessary prerequisite for the resolution of structural differences in a liquid at different temperatures.

The first point makes a comparison among sets of X-ray data on water reported in the literature difficult. The second point is further illustrated in Fig. 4, which shows the kernels of the Fourier integral in eqn. (6) for the same two temperatures. The functions $si(s)M(s)$ represent the structurally sensitive part of the scattered intensity, and in this form of presentation the differences between the curves for water at 20 and 200°C are evident. It can also be seen from Fig. 4 that the functions $si(s)M(s)$ are oscillating around zero to values of the scattering variable $s \simeq 16$ and beyond; this means that the diffraction pattern of liquid water shows interference throughout this large range of scattering angles.

4.1. Survey of Experimental Data

The first X-ray diffraction patterns from liquid water were obtained by Meyer,[745] Stewart,[1024] and Amaldi.[12] Katzoff[577] was the first to apply to water the method of Fourier analysis. Morgan and Warren[761] measured and analyzed the X-ray scattering from liquid water at five temperatures between the melting and boiling points. Between 1938 and 1962 various X-ray studies of water were reported.[107,331,980,1088,1089] The most extensive study of water by X-ray diffraction has been carried out at Oak Ridge National Laboratory[490,782–785]; this work, started in 1961 and still continuing, has considerably improved and extended upon Morgan and Warren's[761] classical study both in resolution and temperature range.

Small-angle scattering data on water have been reported by a number of authors.[185,385,490,676,963,1065] The results of the very careful work of Levelut and Guinier,[676] Chonacky and Beeman,[185] and Hendricks[490] do not differ significantly.

All published large-angle X-ray work on water, although of varying quality, is in essential agreement. An apparent discrepancy between one set of results[484,1088,1089] and those of other workers now seems resolved[98] as due to differences in the presentation and interpretation of the data. With the exception of the Oak Ridge work,[490,782–785] approximations and assumptions were made in the reduction and presentation of all published large-angle X-ray data on water, and this makes the detailed comparison of different studies difficult. We shall base the following discussion exclusively on the Oak Ridge data, and we shall compare these results only with Morgan and Warren's work. Although Morgan and Warren[761] in their data reduction made the assumption that "since the scattering by hydrogen is almost completely Compton-modified radiation, it is justified to treat water as a one-atom substance," their presentation is sufficiently detailed to make such a comparison meaningful.

4.2. Interpretation of the Diffraction Pattern

The diffraction pattern of liquid water (Figs. 2 and 4) contains information about the positional correlation of oxygen and hydrogen atoms, averaged both over time and the volume of the sample. These atom pair interactions contribute to the diffraction pattern approximately as $f_\alpha f_\beta$, the product of the scattering amplitudes for atoms of kind α, β. In the case of X-rays, the scattering is predominantly due to oxygen atom pairs, but the contribution from O···H interactions is $\sim$12% and hence not negligible;

the scattering due to H···H atom pairs is only ~2% compared to O···O interactions. In contrast, in D_2O the neutron scattering amplitudes for oxygen and deuterium atoms are of nearly equal magnitude, and one would expect the neutron diffraction pattern of water to be quite different from that obtained with X-rays; this matter is discussed in more detail in Chapter 9.

Throughout the following discussion, it should be borne in mind that the average spatial arrangement of atoms and molecules cannot be derived uniquely from the one-dimensional diffraction data for a liquid. It is necessary to make assumptions about the short-range order in liquid water and test the validity of these models by comparison with the X-ray data.

4.2.1. *Radial Distribution and Short-Range Order*

The X-ray atom pair correlation functions $G(r)$ for liquid water between 4 and 200°C (Fig. 5) are significantly different from those previously published[783–785] only in the region of radial distances $r < 2.5$ Å. The earlier curves were derived from intensity functions subjected to an empirical correction procedure which resulted in the loss of all structural information at these small distances; no such empirical correction was used in the derivation of the curves shown in Fig. 5.

The large maximum around 1 Å must be ascribed to the intramolecular O—H interaction. Since the structure of an isolated water molecule is known from spectroscopic[63] and gas diffraction[971] studies, this part of the correlation functions can be calculated, and comparison with the curves derived from scattering data on the liquid yields a valuable criterion for their accuracy. This comparison is shown in Fig. 6 for two temperatures, and the agreement is very gratifying. The disagreement between calculated and observed curves below 0.5 Å is the result of accumulated systematic errors in the intensity function. Fourier transformation of this difference below 0.5 Å indicates that the low-frequency error in the functions $si(s)M(s)$ amounts to less than 1%, an uncertainty smaller than that expected from the estimated uncertainties in the calculated scattering amplitudes used in the data reduction.[221,205]* The region below 2.5 Å is not shown in the radial distribution functions of Morgan and Warren.[761]

The maximum around 2.9 Å in the functions $G(r)$ of Fig. 5 must be ascribed to interactions between oxygen atoms from neighboring water molecules. This near-neighbor distance increases from 2.84 Å at 4°C to 2.94 Å at 200°C. The values reported by Morgan and Warren are larger,

* Also D. T. Cromer, private communication (Compton scattering factors for hydrogen).

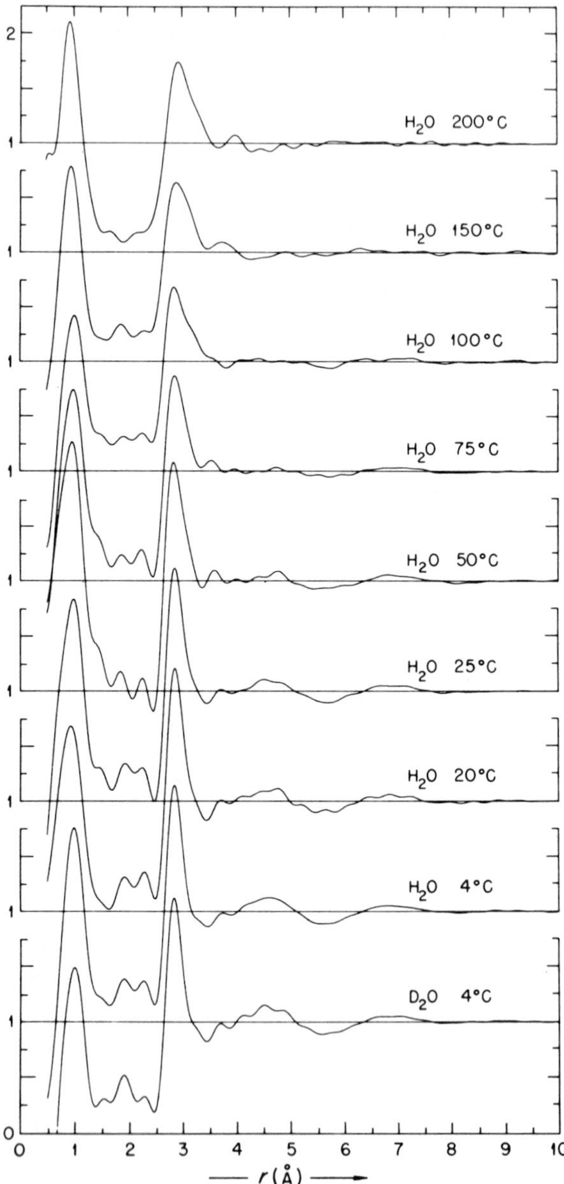

Fig. 5. X-ray correlation functions $G(r)$ for water. Curves represent a superposition of modified atom pair correlation functions $g(r)$ descriptive of O—O, O—H, and H—H interactions.

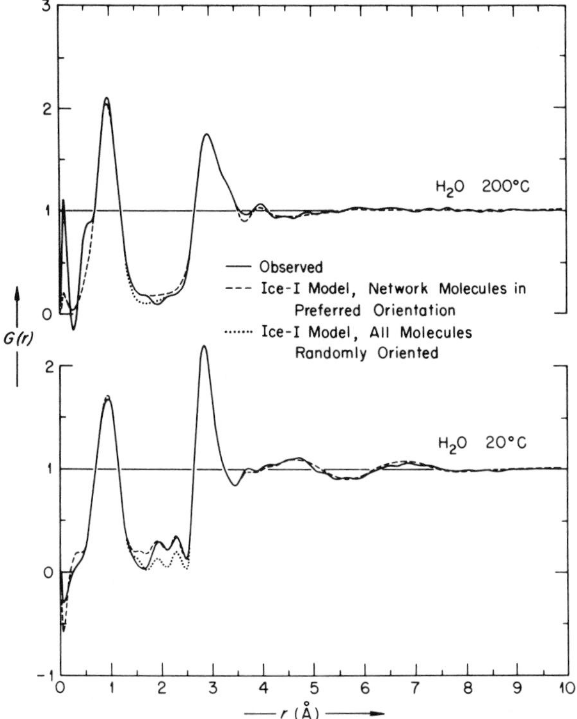

Fig. 6. Interpretation of correlation functions for water at two temperatures. Calculated curves for distances $r < 1.5$ Å are model-independent and based on gas diffraction results.

and this is primarily because of the lower value of s_{max} of their experiments (10–12): It has been shown[782] that termination of the Fourier integral in eqn. (6) at decreasing values of the upper limit results in radial distribution functions with maxima shifted toward longer distances. In Fig. 5 the maximum around 2.9 Å is flanked on both sides by a series of ripples, which are another consequence of the mentioned termination effect and therefore of no structural significance. The first of these ripples on the long-distance side of the peak around 2.9 Å occurs between 3.5 Å and 4 Å; it is barely visible in the curves of Morgan and Warren, primarily because these authors used a modification function different from the one defined in eqn. (6a).

Two broad maxima around 4.5 Å and 7 Å in the curves of Fig. 5 correspond to second- and higher-neighbor interactions. The gradual disappearance of these features with increasing temperature indicates a decrease of the critical distance r_c which measures the extent of short-range order from about 8 Å near the melting point to about 6 Å at 200°C.

While there is no reasonable doubt about the preceding interpretation of the radial distribution curves in Fig. 5, quantitative interpretation of the curves at distances larger than 1.5 Å is quite uncertain without additional assumptions. The evaluation of areas in terms of average coordination numbers, in particular, is never unambiguous.(845) In this case, the lower bound of the maximum around 2.9 Å is fairly well resolved, but the upper bound is not. Interactions with second and higher neighbors contribute significantly to the area under this peak, and proper resolution into individual components is quite uncertain even at low temperatures, where this maximum is relatively sharp. Morgan and Warren(761) derived a coordination number of 4.4 from their curves for water at and below room temperature by placing the mid-position at the observed maximum of the peak and drawing the long-distance side symmetric with the short-distance side. They also showed that it is possible to interpret the peak around 2.9 Å as due to three discrete interactions and a continuous, uniform density starting at about 2.7 Å; this interpretation is, however, quite arbitrary. A less arbitrary way of deriving coordination numbers from diffraction data is the following(785): Let us assume that the maximum around 2.9 Å is characteristic of the interaction of one oxygen atom with N oxygen atom neighbors separated by a Gaussian distribution of distances centered at r_0. The contribution of longer distances can be estimated as arising from a uniform distance distribution starting at r_c, the radius of a sphere of volume $(N+1)/\varrho_0$, ϱ_0 being the bulk density of liquid water. Using eqn. (7) and adjusting by least squares the distance r_0 and the mean-square variations in both r_0 and r_c, we find that a value of $N = 4.4$ gives the best agreement with the experimental curves for water at all temperatures. But the agreement of this "model" (4.4 discrete O—O interactions surrounded by a uniform distance distribution) with the radial distribution functions of water (see Fig. 1 of Ref. 785) is poor. The short-distance part of the peak around 2.9 Å is reproduced reasonably well, but the long-distance part is not. Since, for the predominantly tetrahedral coordination indicated by this number, interactions with second neighbors would be centered at the considerably longer distance of around 4.5 Å (where a broad maximum is observed in Fig. 5), the conclusion cannot be avoided that the first coordination sphere around a water molecule is complex; the same conclusion was reached by Morgan and Warren.(761)

The only significant difference in the $G(r)$ curves for light and heavy water at 4°C is in the region of radial distances $1.5\text{ Å} < r < 2.5\text{ Å}$. This region of distances is of great interest because it contains information about the relative orientation of pairs of water molecules through the O···H interaction between atoms from neighboring molecules. Unfortunately, this region is most difficult to interpret because the satellite ripples of the O···O peak around 2.9 Å [arising from the termination of the Fourier integral in eqn. (6) at finite values of the variable s] are most pronounced below 2.5 Å. It is therefore necessary to reproduce at least the major features of the $G(r)$ functions in Fig. 5 by curves calculated for a model structure before an interpretation of the distance region between 1.5 Å and 2.5 Å can meaningfully be attempted; we postpone this discussion until part 4.2.3 of this section.

4.2.2. *Long-Range Density Fluctuations*

The fact that the correlation functions $G(r)$ in Fig. 5 differ significantly from unity to radial distances which correspond to about five molecular radii at room temperature must be ascribed to positional correlation between water molecules, that is, "structure" in liquid water. If the positional correlation between water molecules were of a long-range nature (as might be the case if the molecules formed large clusters held together by hydrogen bonds, or if regions containing qualitatively different local arrangements or "phases" were present) the intensity functions shown in Fig. 2 should exhibit interference maxima and minima in the region of small scattering angles.[385] The fact that the scattered intensity in the region $0.04 < s < 0.5$ for water below 100°C is found[185,490,676] to be constant with respect to s precludes any significant variation in the pair density not included in the curves of Fig. 5. Furthermore, the value of the relative variance of the particle number derived from extrapolated small-angle scattering [eqn. (5a)] does not differ significantly from that predicted by eqn. (5b). The random fluctuations in density expected for a liquid consisting of a single phase and a single chemical species are thus sufficient to explain the observed small-angle scattering and the small, long-range variations from unity of the functions $G(r)$ implied by it.

4.2.3. *Models*

Ordinary ice (ice Ih) melts with a 10% increase in bulk density. Since the average distance between nearest neighbors in liquid water near the melting point exceeds that found for the solid[55,653,834] by 3%, the packing scheme for water molecules must be different in the liquid than in the solid; the 10% increase in the coordination number upon melting is in agreement with this view. On the other hand, the location of maxima and minima near 4.5, 5.5 Å, and 7 Å in the correlation functions for water at temperatures below 100°C (Fig. 5) roughly coincides with regions of high and low atom-pair concentrations calculated for the ice I structure. In particular, the location of the maximum near 4.5 Å, when compared with a nearest-neighbor distance of 2.84 Å, is a strong indication that the *average* deviation from ideal tetrahedral coordination in the liquid near the melting point must be quite small. For this reason the greater density of the liquid cannot be the result of shorter distances between second-nearest neighbors, as is known to be the case in most of the high-pressure ice polymorphs.(569) Morgan and Warren(761) suggested that the complexity of the first coordination shell of a given water molecule might be caused by "filling in" the space between first and second neighbors in tetrahedral coordination by additional molecules "at a variety of distances."

Since a detailed comparison of proposed water models is presented in Chapter 14, we shall confine ourselves to a brief description of the ice I model proposed by Samoilov(938) and specified in detail by Danford and Levy.(226) This model has been shown(226,784) to reproduce the correlation functions of liquid water (Fig. 5) at all temperatures, the temperature dependence of the model parameters being related to the thermodynamic properties of water in a physically reasonable way.(784) Other proposed models are either incompatible with observed X-ray scattering or insufficiently defined for adequate testing.(785)

The ice I model describes liquid water as an extensive three-dimensional hydrogen-bonded network, the details of which are short-lived. Over short distances from any origin molecule this network is closely related to a slightly expanded ice I lattice. The average structure of this network is very open, with spaces between groups of molecules in tetrahedal coordination sufficiently large to accommodate additional "cavity" molecules. The complexity of the first coordination shell is described by the model in terms of distinctly different *average* environments of network and cavity molecules. Important characteristics of the model, descriptive of a fluid rather than a crystalline solid, are absence of long-range correlation, random occupancy of cavities, and random network vacancies.(570,784–786)

In Fig. 6 correlation functions calculated for the ice I model are compared to the experimental curves of liquid water at two temperatures; the agreement is equally good at all other temperatures.(784) Details of the calculations have been given elsewhere(782,784) and only some conclusions which complement the discussion of Section 4.2.1 will be stated. The model describes the peak around 2.9 Å as arising from 4.4 near-neighbor interactions at all temperatures, the same result as obtained in Section 4.2.1 by a different method. Two calculated curves are shown in Fig. 6; they differ only in the distance region 1.51 Å $< r <$ 2.5 Å. Both calculated curves include the model-independent distances (0.976 and 1.559 Å) and variances (0.067 and 0.115 Å) for the intramolecular O—H and H—H interactions, taken from gas diffraction results.[971] The dotted curves incorporate all intermolecular O···H and H···H interactions as uniform distance distributions; this corresponds to random orientation of all water molecules with respect to each other. In the dashed curves of Fig. 6 it has been assumed that the network molecules of the ice I model are connected to their four nearest neighbors through relatively straight hydrogen bonds, resulting in a discrete O···H distance at ~2 Å; the cavity molecules are assumed to be randomly oriented. The dashed curves, based on the latter description, agree much better with the experimental ones at 20°C, but at 200°C the dotted lines for random orientation agree equally well. The rms displacements associated with the shortest intermolecular O···H distance are larger for H_2O than for D_2O (as might be inferred from a comparison of the curves for 4°C in Fig. 5) and they increase rapidly with temperature.

5. CONCLUSIONS

The positional correlation between molecules, that is, "structure" in liquid water, extends only over relatively short distances. A few molecular radii away from any starting point only the random fluctuations in density expected for a fluid consisting of a single phase and a single chemical species occur. In this respect water is no different from simple monoatomic liquids.

Near the melting point the average separation between oxygen atoms from neighboring water molecules is only slighlty larger than that found in ordinary ice (2.77 Å).[653] This near-neighbor distance increases gradually from 2.84 Å at 4°C to 2.94 Å at 200°C, and the corresponding maximum in the correlation functions broadens progressively with increasing temperature. The changes in position and shape of the O···O peak reflect the weakening of the O···H—O hydrogen bond at high temperatures.

The average coordination number in liquid water is independent of temperature from 4 to 200°C and is slightly larger than four, indicating predominantly tetrahedral coordination. The distance ratio for the maxima in the correlation functions which correspond to first- and second-neighbor O···O interactions indicates that the *average* deviation from ideal tetrahedral coordination in the liquid must be quite small. On the other hand, the peak corresponding to near-neighbor O···O interactions cannot be described by a single Gaussian distance distribution, and this complexity of the first coordination shell of a given water molecule must be due to the presence in the liquid of molecules with at least two types of near-neighbor configurations. However, these "species" exist in environments which are distorted from the average, so that on a scale of many molecules (sampled in small-angle scattering experiments) liquid water is "single phase," homogeneous, and isotropic.

The distribution of instantaneous and local near-neighbor O···O and O···H distances about their respective mean values is much larger in the liquid than in the solid, and we conclude that there must also be a wide distribution of O···H—O angles about the mean value. This distribution is sharper in D_2O than in H_2O at 4°C. The breadth of the distribution of hydrogen bond angles increases rapidly with temperature and, at 200°C, becomes so large that the molecules in liquid water are "seen" by X-rays as randomly oriented.

Any realistic model of water must accommodate (or at least avoid conflict with) these features, which can be deduced more or less directly from the X-ray data. Of the structures which have a sufficiently detailed basis to permit calculation of radial distribution functions, only the ice I model has been shown to give agreement with data from both large- and small-angle X-ray scattering. For the few other models for which radial distribution functions have been calculated, the apparent agreement with the experimental curves for liquid water is meaningless because the computed curves do not correctly describe radial distribution in these cases.

CHAPTER 9

The Scattering of Neutrons by Liquid Water

D. I. Page
Atomic Energy Research Establishment
Harwell, England

1. INTRODUCTION

Neutrons are a unique probe for the study of molecular dynamics. Their mass is comparable to the mass of the nucleus and their velocities undergo a Doppler shift when they are scattered from moving nuclei. The associated energy change is easily measurable if the initial velocity of the neutrons is small enough. Nuclear reactors provide a source of suitable neutrons, a typical energy E being 25 meV with associated de Broglie wavelength of 1.8 Å. This wavelength is comparable to the interatomic distances in condensed matter and so the change in momentum on scattering is sensitive to the structure of the scattering system. Thus, the distribution as a function of angle of neutrons scattered from a system of nuclei gives information on the spatial distribution of the nuclei, while the distribution of neutron energies at any one angle gives information on the motion of the nuclei.

Typical energy changes ΔE are in the range 0.1–100 meV (0.8–800 cm^{-1}) and typical changes in wave vector $\varkappa$ are 0.3–10 $Å^{-1}$. The corresponding frequency components of the motion ($\omega = \Delta E/\hbar$, where $\hbar$ is Planck's constant divided by 2π) are 1.5×10^{11} to 0.3×10^{14} rad sec^{-1} and these are related to relaxations of the structure over periods of the order of $2\pi/\omega$, i.e., 4×10^{-11} to 2×10^{-13} sec. Structure is observed over regions of dimensions $2\pi/\varkappa$, which is of the order of 0.6–20 Å. Neutron scattering is at present the only method capable of revealing details of dynamic processes on such short space and time scales. However, because neutron spectra contain both sets of information, they are usually complex and require considerable analysis in their interpretation.

This chapter deals with the information available in the spectra of neutrons scattered by liquid water and the experimental results obtained so far. No attempt is made to describe the experimental techniques involved, as these are well-covered in the literature.[(39,137,192)] Section 2 outlines the principles of neutron scattering and relates the experimentally observed quantities to correlation functions in both time and space. The presentation is simplified to allow those not engaged in neutron physics to achieve a comprehensive picture of the neutron scattering technique. Section 3 deals with the interpretation of neutron data in terms of structural properties of water and how they can be combined with data from other sources. In section 4 the diffusive motions (i.e., behavior over long time) of the water molecule are considered, while in Section 5 the vibrational motions (i.e., behavior at short times) of the atoms in the water molecule are discussed.

2. THE NEUTRON SCATTERING METHOD

2.1. Nuclear Scattering of Neutrons

The cross section for the scattering of neutrons by nuclei has been discussed many times in the literature and readers are referred for detailed treatment to, for example, Turchin[(1078)] and Sjolander.[(997)] The probability of a neutron being scattered when passing through a system of nuclei is $N\sigma_s$, where N is the number of nuclei and σ_s is defined as the scattering cross section of a nucleus. We consider the neutron beam as a primary plane wave incident on the nuclei, which then act as the origins of secondary wavelets. The principal steps in the calculation of scattering cross sections of thermal neutrons (i.e., energy of $\sim$0.025 eV) are as follows:

(a) For low-energy neutrons the neutron wavelength is much greater than the nuclear diameter, so only *S*-wave scattering is important in the spherical harmonic expansion for the scattered wavelets.

(b) *S*-wave scattering cross sections for an isolated, stationary nucleus are spherically symmetric and independent of neutron energy.

(c) The nuclear size is much smaller than the interatomic distances and so we may write for the nuclear potential $V_i(r)$ of the ith nucleus

$$V_i(r) = \text{const} \times \delta(r - r_i) \tag{1}$$

where r_i is the position of the ith nucleus.

(d) The overall perturbation is small (i.e., although the nuclear potential is very strong within its range, the intensity scattered by a single nucleus is very small), so that the Born approximation is valid.

(e) The constant in (c) is determined by fitting the Born cross section for epithermal neutrons to the measured "free-atom" cross section. The constant is written

$$2\pi\hbar^2 b/m \tag{2}$$

where m is the neutron mass and b is a length called the bound-atom scattering length. The bound-atom cross section (which is usually defined as the scattering cross section for condensed matter) is then given by

$$\sigma_b = 4\pi b^2 \quad (\equiv \sigma_s) \tag{3}$$

and the free-atom cross section σ_f is

$$\sigma_f = 4\pi b^2[A/(A+1)]^2 \tag{4}$$

where A is the ratio of nuclear to neutron masses. σ_f is the cross section measured by the transmission of fast neutrons through matter.

The value of b cannot be calculated and varies randomly for different elements and for different isotopes of an element. A list of recommended values for scattering lengths has been published by the Neutron Diffraction Commission.[792]

The scattering process in condensed matter may or may not involve a change in energy of the neutron, cases termed inelastic and elastic scattering, respectively. Neutron scattering experiments have one or more neutron detectors, each of which detects the neutrons scattered through an angle θ into a small solid angle $d\Omega$. If the energy of the scattered neutron is not resolved, then the scattered intensity as a function of the angle θ gives the differential scattering cross section $d\sigma_s/d\Omega$; if the change in energy ΔE of the neutron is measured, then the double differential scattering cross section $d^2\sigma_s/d\Omega\, dE$ is obtained.

2.2. Coherent and Incoherent Scattering

If the nuclei in a scattering specimen are all identical and have zero spin, then all the scattering lengths are identical and the scattered waves are in phase; they can then interfere with the incident wave. This interference gives information on the nuclei acting together in a cooperative manner and the scattering is said to be *coherent*. However, if there is more than one scattering length per nucleus or per chemical element, there will be a component of *incoherent* scattering, that is, nuclear scattering which is not coherent with that from another nucleus and cannot produce interference effects. Incoherent scattering gives information on nuclei acting individually. The total scattering cross section is the sum of the two components.

Incoherent scattering can arise in two ways:

(a) Isotopes of the same element have different scattering lengths. If the element has two isotopes of abundances c_1 and c_2 and corresponding scattering lengths b_1 and b_2, then the average or coherent scattering length b_{coh} is

$$b_{\mathrm{coh}} = c_1 b_1 + c_2 b_2 \tag{5}$$

which determines the coherent scattering cross section σ_{coh},

$$\sigma_{\mathrm{coh}} = 4\pi(c_1 b_1 + c_2 b_2)^2 \tag{6}$$

The total scattering cross section is given by the sum of the scattered intensities,

$$\sigma_s = 4\pi c_1 b_1^2 + 4\pi c_2 b_2^2 \tag{7}$$

and the incoherent cross section σ_{inc} is

$$\begin{aligned} \sigma_{\mathrm{inc}} &= \sigma_s - \sigma_{\mathrm{coh}} \\ &= 4\pi c_1 c_2 (b_1 - b_2)^2 \end{aligned} \tag{8}$$

Thus, isotopically pure samples must be used to avoid isotopic incoherence.

(b) If the nucleus has a spin I, then it may combine with a neutron (spin $\pm\frac{1}{2}$) to form one of two alternative compound nuclei (spins $I \pm \frac{1}{2}$). Different scattering lengths (b_+, b_-) are associated with these, giving rise to spin incoherence. If we replace the isotopic abundances in the previous paragraph by the statistical weighting factors, we obtain

$$\sigma_{\mathrm{coh}} = 4\pi\{[(I + 1/)(2I + 1)]b_+ + [I/(2I + 1)]b_-\}^2 \tag{9}$$

$$\sigma_s = 4\pi\{[(I + 1)/(2I + 1)]b_+^2 + [I/(2I + 1)]b_-^2\} \tag{10}$$

$$\sigma_{\mathrm{inc}} = 4\pi[I(I + 1)/(2I + 1)^2](b_+ - b_-)^2 \tag{11}$$

An important case of this kind is that of hydrogen. Here, the products of the scattering amplitudes and weighting factors for parallel and antiparallel spins are almost equal in magnitude but of opposite phase. Thus, constructive interference cannot occur and the scattered wave from each nucleus escapes from the specimen without being modified by interference effects.

The presence of incoherent and coherent scattering means that the experimentally observed cross sections consist of two parts:

$$d^2\sigma_s/d\Omega\, dE = (d^2\sigma_{\text{inc}}/d\Omega\, dE) + (d^2\sigma_{\text{coh}}/d\Omega\, dE) \tag{12}$$

The coherent term contains information from all the nuclei acting together; the incoherent term contains information from the atoms acting individually. It is convenient to write (see, for example, Lomer and Low[(695)])

$$d^2\sigma_{\text{inc}}/d\Omega\, dE = (k/k_0)b_{\text{inc}}^2 S_s(\boldsymbol{\varkappa}, \omega) \tag{13}$$

$$d^2\sigma_{\text{coh}}/d\Omega\, dE = (k/k_0)b_{\text{coh}}^2 S(\boldsymbol{\varkappa}, \omega) \tag{14}$$

where b_{inc} is the incoherent scattering length for the system and $\mathbf{k}_0$ and $\mathbf{k}$ are the initial and final wave vectors of the scattered neutron, corresponding to energies E_0 and E. These are related to the energy transfer ΔE and momentum transfer $\hbar\varkappa$ by

$$\begin{aligned} \hbar\omega &= E_0 - E = \Delta E \\ &= (\hbar^2/2m_n)(k_0{}^2 - k^2) \\ \boldsymbol{\varkappa} &= \mathbf{k}_0 - \mathbf{k} \\ \hbar^2\varkappa^2/2m_n &= E_0 + E - 2(E_0E)^{1/2}\cos\theta \end{aligned} \tag{15}$$

where m_n is the mass of the neutron. Figure 1 shows the locus in $\varkappa$, ω space followed by measurements at two constant angles of scatter ($\theta = 20$ and $90°$) for three different values of E_0. $S_s(\boldsymbol{\varkappa}, \omega)$ and $S(\boldsymbol{\varkappa}, \omega)$ are called the incoherent and coherent *scattering laws*, respectively, for the system. The scattering law is independent of the properties of the neutron and the details of the scattering process and depends only on the properties of the scattering system. $S_s(\boldsymbol{\varkappa}, \omega)$ contains information about the behavior of a single nucleus, while $S(\boldsymbol{\varkappa}, \omega)$ contains the information from the nuclei acting collectively. For example, in a simple liquid, diffusive motion of the atoms (self-diffusion) will show up in $S_s(\boldsymbol{\varkappa}, \omega)$, while information on the viscosity will show up in $S(\boldsymbol{\varkappa}, \omega)$.

2.3. Isotopic Substitution

The fact that different isotopes of the same element have different scattering lengths often enables more than one set of scattering data to be obtained for the same substance. This is mainly of interest when working with molecular samples (see Section 3.1) for it allows, in theory, the relative

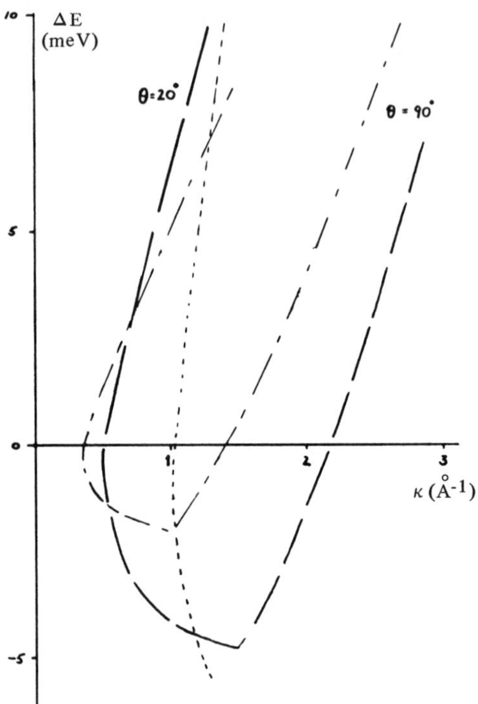

Fig. 1. Locus in $\varkappa$, ω space followed by a measurement at constant angle of scatter θ for $\theta = 20°$ and 90°. Three incident neutron energies are shown: 2 meV (— - - —); 5 meV (— —); and 20 meV (- - -). The curve for 20 meV shows how the use of high incident energies tends toward constant-$\varkappa$ loci.

contributions of the various atoms in the molecule to be varied, bringing each into prominence in turn. For example, Table I gives the neutron scattering cross sections for hydrogen, deuterium, and oxygen. We see the scattering from light water will be mainly by the hydrogen nuclei and will be almost totally incoherent. Thus, only the incoherent scattering law [eqn. (13)] can be measured and this will give information on the motion of single hydrogen atoms. The scattering from heavy water will be from both the deuterium and oxygen and will show strong interference effects. Thus, any information on the structure of water must come from the data on heavy water. In contrast, X-rays will be scattered almost wholly by the oxygen atoms in the water and give almost no information about the hydrogen atoms. The substitution of deuterium for hydrogen is of great im-

TABLE I. Coherent and Incoherent Scattering Cross Sections for the Atomic Components of Light and Heavy Water

Isotope	σ_{inc}, b	σ_{coh}, b
^{1}H	79.7	1.79
^{2}H	2.2	5.4
^{16}O	0.0	4.24

portance in the investigation of organic substances, where one wishes to remove the predominant incoherent scattering from hydrogen.

2.4. The Scattering Law and Correlation Functions

We now consider the Fourier transformation of the scattering law, and to illustrate this, we consider the simple case of radiation being Bragg-reflected by a crystal. Bragg's law relates the wavelength of the radiation λ to the spacing d of the lattice planes in the crystal:

$$\lambda = 2d \sin(\theta/2) \tag{16}$$

This should be compared with eqn. (15) for elastic scattering ($\omega = 0$, $k = k_0$), which gives

$$\varkappa = 2k \sin(\theta/2) = (4\pi/\lambda) \sin(\theta/2) \tag{17}$$

It can be seen that the variable $\varkappa$ behaves like the reciprocal of the spacing, and intuitively one sees that Fourier-transforming with respect to $\varkappa$ will provide information about the spacings between atoms in the scattering system.

Similarly, the variable ω behaves like the reciprocal of a time. We can imagine a simple harmonic oscillator of frequency ω and period of oscillation τ; if we Fourier-transform the motion of the oscillator (i.e., the function $\cos \omega_0 t$), we obtain a sharp peak at a frequency $\omega_0 = 1/\tau$. Thus, the Fourier transform represents the atomic modes of motion. Combining these conclusions, we see the double Fourier transform of the scattering law gives information about the position of the atoms as a function of time. This theorem was first proved by van Hove,[1087] who gave

$$S(\boldsymbol{\varkappa}, \omega) = (1/2\pi) \int_v d\mathbf{r} \int_{-\infty}^{\infty} d\tau [G(\mathbf{r}, t) - \varrho] \exp[i(\boldsymbol{\varkappa} \cdot \mathbf{r} - \omega t)] \tag{18}$$

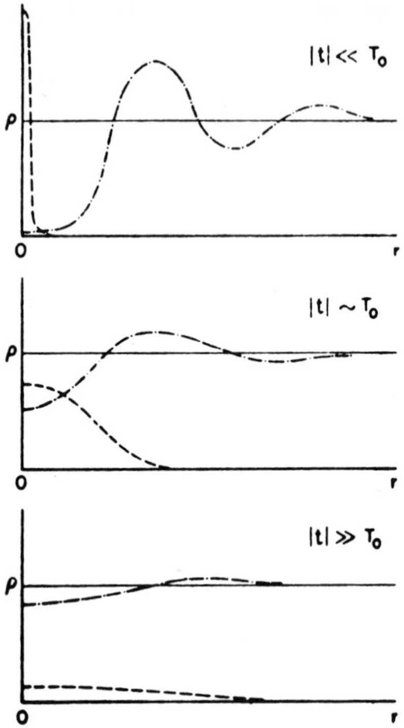

Fig. 2. Dependence of the van Hove correlation functions $G_s(\mathbf{r}, t)$ (— —) and $G_d(\mathbf{r}, t)$ (—·—) on $\mathbf{r}$ for three values of t. The solid line corresponds to the average density of the system whose relaxation time is T_0.[1087]

where ϱ is the average density of the scattering nuclei. The function $G(\mathbf{r}, t)$ is called the space–time correlation function and for a classical system has a simple physical meaning. If there is an atom at position $\mathbf{r}_1$ at a time t_1, then $G(\mathbf{r}, t)$ is the probability of finding an atom at position $\mathbf{r}_2$ at a time t_2. The quantities $\mathbf{r}$ and t are the space and time differences. $G(\mathbf{r}, t)$ can be divided into two parts, $G_s(\mathbf{r}, t)$, which measures the probability of finding the *same* atom each time, and $G_d(\mathbf{r}, t)$, which is the probability of finding a different atom each time.

$$G(\mathbf{r}, t) = G_s(\mathbf{r}, t) + G_d(\mathbf{r}, t) \tag{19}$$

Figure 2 shows the behavior of $G_s(\mathbf{r}, t)$ and $G_d(\mathbf{r}, t)$ as a function of time for a simple liquid with a structural relaxation time T_0.

We note that $G_s(\mathbf{r}, t)$ describes the motion of a single atom, while the total function describes the dynamic behavior of all the atoms. This can be compared with the incoherent and coherent scattering laws in Section 2.2.

A useful function often used in the analysis of data is the intermediate scattering function $I(\varkappa, t)$, which is defined by

$$I(\varkappa, t) = \int_v [G(\mathbf{r}, t) - \varrho] \exp(i\varkappa \cdot \mathbf{r})\, d\mathbf{r}$$
$$= \int_{-\infty}^{\infty} S(\varkappa, \omega) \exp(i\omega t)\, d\omega \tag{20}$$

i.e., the temporal transformation of $S(\varkappa, \omega)$ has been performed so that $I(\varkappa, t)$ contains the effects of the scatterer dynamics.

Another function which often occurs in the literature is the symmetrized scattering law $\tilde{S}(\varkappa, \omega)$, which is defined by

$$\tilde{S}(\varkappa, \omega) = [\exp(-\hbar\omega/2k_{\mathrm{B}}T)]S(\varkappa, \omega) \tag{21}$$

where k_{B} is Boltzmann's constant and T is the absolute temperature of the specimen. It arises because $G(\mathbf{r}, t)$ is a complex function of time and so its transform $S(\varkappa, \omega)$ is a nonsymmetric function of ω. This lack of symmetry is due to the detailed balancing of the cross sections for energy loss and gain, and inclusion of the Boltzmann factor gives a symmetric form of the scattering law which is often more convenient for calculations.

It is usual to describe liquid "structure" as meaning the instantaneous structure and this is reflected in the pair distribution function $g(\mathbf{r})$, which denotes the relative position of a pair of atoms at the same instant of time, i.e., $G(\mathbf{r}, 0)$. From eqn. (18), the liquid structure factor $S(\varkappa)$ is given by

$$S(\varkappa) = \int_{-\infty}^{\infty} S(\varkappa, \omega)\, d\omega$$
$$= 1 + (1/2\pi) \int_v d\mathbf{r}\, [G(\mathbf{r}, 0) - \varrho] \exp(i\varkappa \cdot \mathbf{r}) \tag{22}$$
$$= 1 + (1/2\pi)\varrho \int_v d\mathbf{r}\, [g(r) - 1] \exp(i\varkappa \cdot \mathbf{r})$$

Figure 3 shows the qualitative behavior of $S(\varkappa)$ and $S(\varkappa, \omega)$ for a simple liquid. The uppermost curve shows the structure factor $S(\varkappa)$; the remaining curves show the behavior of $S(\varkappa, \omega)$ at several values of $\varkappa$. At low values of $\varkappa$, $S(\varkappa, \omega)$ splits into three components, where the displaced lines correspond to sound wave modes and their size and shape may be described

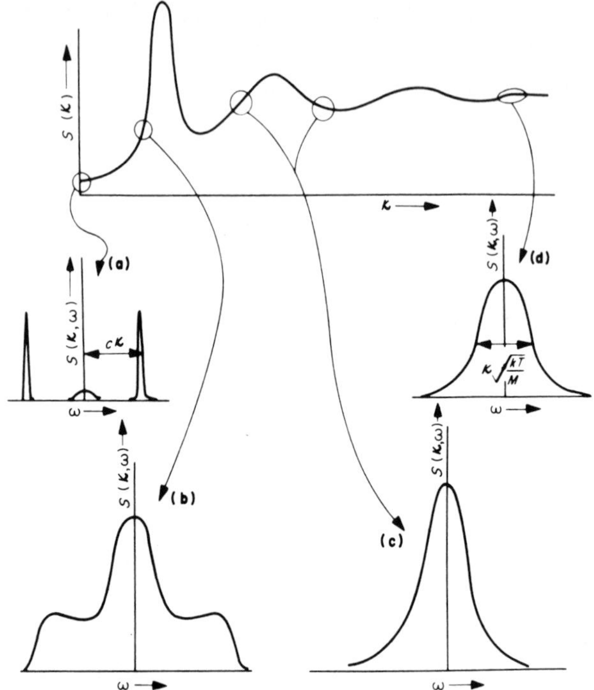

Fig. 3. Qualitative behavior of $S(\varkappa, \omega)$ for several values of $\varkappa$. The upper curve is the structure factor $S(\varkappa)$ and the remaining curves show the spectral shape at several fixed values of $\varkappa$. At low and high values of $\varkappa$ the widths can be calculated from simple considerations.[288]

by simple hydrodynamic theory; the central peak corresponds to the thermal damping of the sound waves or entropy fluctuations in the system. At high values of $\varkappa$, $S(\varkappa, \omega)$ has a shape appropriate to a perfect gas because one is considering atomic movement on a very small time scale. This will be gaslike, as the atom will travel in a straight line for a very small distance until it feels the field of force generated by its neighbors. At intermediate values of $\varkappa$ the cooperative modes merge into single-particle modes.

2.5. Neutron Spectra from Water

Figure 4 shows three energy spectra of neutrons of low incident energy ($E_0 \simeq 4$ meV) scattered from light water through angles of 30, 60, and 90°. These are typical of the slow neutron spectra used to obtain dynamic information about water. Neutrons of low incident energy are used to make the fractional energy change on scattering $\Delta E/E$ as large as possible, because the detecting system has a finite energy resolution.

Note that for low incident energies, neutrons can only gain in energy; for higher incident energies, where both energy loss and gain are possible, the inelastic region is reflected on the energy-loss side of the quasielastic peak but is modified by the Boltzmann factor discussed in the last section.

Integration of the area under each curve would not give a point on the structure factor curve, as the values of $\varkappa$ vary with ω at a fixed angle (Fig. 1) and the structure factor is the integral of the scattering law at constant $\varkappa$ [eqn. (22)]. Structure factor measurements are made with as high an incident neutron energy as possible to approximate the constant-$\varkappa$ condition. These structure measurements are discussed in Section 3.

The spectra in Fig. 4 can be split into two parts: (a) the large quasi-elastic peak around zero energy transfer, which gives information on the

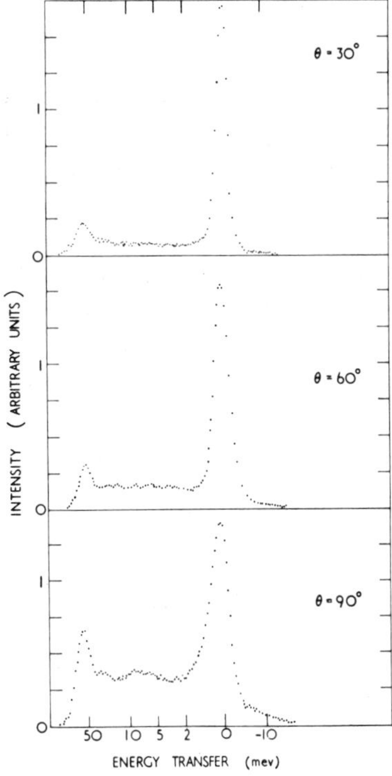

Fig. 4. Typical energy spectra observed from the scattering of slow neutrons (E_0 = 4 meV) by light water at 22°C.

long-time behavior of the atoms, and (b) the inelastic region extending to 100 meV, which represents energy changes due to the interaction of the neutron with the vibrational modes of the system.

We can see how these two regions arise if we consider the situation for a solid. At time zero the position of an atom is given by definition, and on the average it is moving with substantially the velocity it would have as a gas atom. Because the atom is bound, a "thermal cloud" is rapidly built up about its mean position and thereafter changes slightly. The formation and fluctuations of the cloud give rise to the inelastic component; the scattering from the cloud as a form factor gives rise to an elastic component, elastic because the cloud endures "forever." The two energy components reflect the quick formation of the cloud and its long life when formed. The decrease in intensity with angle of the elastic component measures the spatial extent of the cloud. This decrease is measured by the *Debye–Waller* factor, which is usually written as $\exp\langle -\frac{1}{2}(\mathbf{u} \cdot \boldsymbol{\varkappa})^2 \rangle$, where $\mathbf{u}$ is the displacement of the atom. The Debye–Waller factor is the Fourier transform of the thermal cloud. For spherical symmetry the factor becomes $\exp(-u^2\varkappa^2/6)$ due to directional averaging.

If we now consider a liquid, by analogy, the two components in the liquid energy distribution refer to the formation of a "thermal cloud" in the liquid and the subsequent duration of the cloud associated with it, which now has a finite lifetime; this leads to an energy broadening of the elastic peak. Diffusive motion will show up in its effects on the thermal cloud and thus on the quasielastic components. The Debye–Waller factor for a liquid is no longer strictly meaningful because the size and existence of the "thermal cloud" is a function of time. It is often quoted for a liquid, however, as giving an estimate of the mean-square displacement of the atom from its equilibrium position.

Data from the elastic and inelastic regions will be discussed separately in Sections 4 and 5.

3. STRUCTURAL MEASUREMENTS

3.1. Structure Factor for Water

Radiation scattering cannot yet give a satisfactory insight into the "structure" of water. In this section the current data, method of analysis, and preliminary results are presented, but they by no means give a definitive picture of the spatial interrelation of the water molecules. In Section 2.4

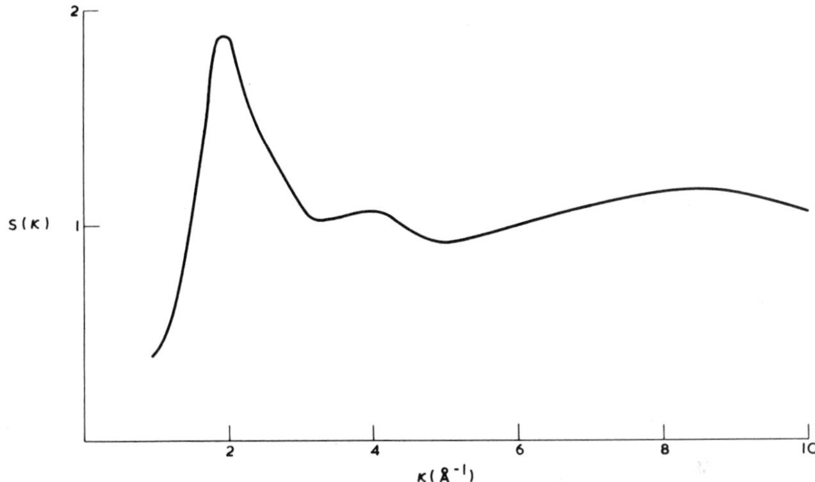

Fig. 5. Structure factor $S(\varkappa)$ for heavy water at 22°C determined by neutron scattering. Scattering from both deuterium and oxygen atoms contribute to the curve.

we saw that the structure factor $S(\varkappa)$ is the Fourier transform of the pair distribution function $g(r)$. For a diatomic molecule AB there are three pair distribution functions corresponding to correlations between atoms of type A, atoms of type B, and correlations between atoms of A and B. To determine these three functions unambiguously requires three independent structure-factor determinations. This can be done by isotopic substitution (Section 2.3) so that the scattering by one type of atom becomes smaller than, equal to, and larger than the scattering by the other type of atom in turn. Work of this type has been done by Enderby *et al.*,[(309)] who studied the alloy Cu_6Sn_5. When we consider water, however, two problems arise. First, of the three isotopes of hydrogen, only deuterium is suitable, as hydrogen scatters incoherently and tritium is scarce and radioactive; second, oxygen-17 and oxygen-18 are scarce and their scattering lengths are almost identical to that of oxygen-16 (5.79, 6.01, and 5.77 fm, respectively). Thus, only the structure factor for $^2H_2{}^{16}O$ is measurable by neutron scattering. This has been done in several laboratories over various ranges of $\varkappa$.[(128, 820,1009)]* Figure 5 shows $S(\varkappa)$ for D_2O at room temperature after all corrections have been applied to the observed differential cross section. These corrections have been discussed by North *et al.*† and include absorption,

* Also R. N. Sinclair and D. Day, private communication (1968).

† North, D. M., Enderby, J. E., and Egelstaff, P. A., *J. Phys. Chem.* **1**, 784 (1968).

multiple scattering, and incoherent scattering by the specimen and the so-called Placzek corrections, all except the last being reasonably straightforward to apply. Placzek corrections arise from the factor k/k_0 in eqn. (14) when the equation is integrated at constant angle (and not constant $\varkappa$) to give $d\sigma/d\Omega$. The problem was considered originally by Placzek[852] and later by Vineyard[1112] and Ascarelli and Caglioti.[34] The correction is small for heavy atoms and if the energy transfers allowed by the system are negligible compared with the energy of the impinging radiation. This is not the case for water, and consequently it is difficult to assign an absolute scale to the measurements. The Placzek corrections are smoothly varying, however, and the scale of Fig. 5 has been fixed by the condition $S(\varkappa) = 1$ at large values of $\varkappa$, using the data of Sinclair and Day between $\varkappa = 10$ to 25 Å^{-1} measured on the Harwell LINAC.

It is interesting to note that Forte and Menardi[345] looked for the existence of large molecular aggregates in D_2O by measuring $d\sigma/d\Omega$ at $0.1 < \varkappa < 0.5$ Å^{-1}, but no coherent effect corresponding to an aggregate structure was found. Only limited information is available from the single structure factor curve, but the information can be enhanced by combining it with data from other sources.

3.2. Molecular Structure Factors

An alternative approach was suggested by Zachariasen,[1194] who proposed using the structure of the molecule as a basic piece of information. He showed that the diffraction data could be represented by

$$S_m(\varkappa) = f_1(\varkappa) + f_2(\varkappa)[S_c(\varkappa) - 1] \tag{23}$$

where $S_c(\varkappa)$ is the theoretical structure factor for molecular centers, $S_m(\varkappa)$ is the experimentally measured structure factor, $f_1(\varkappa)$ is the structure factor for a single molecule, and $f_2(\varkappa)$ is an orientational correlation function. We see that the measured diffraction pattern is a combination of the diffraction pattern for single molecules plus that given by the distribution of molecular centers modified by the correlation between the molecules. Rigorously, this correlation must extend to infinity. If the molecular structure is known, $f_1(\varkappa)$ can be calculated and eqn. (23) then involves only two unknown functions. Neutron data are still insufficient for this purpose; however, X-rays are scattered almost solely by the oxygen atoms in water and the X-ray diffraction data give information on the oxygen–oxygen distribution function. If we assume the distribution is the same as for

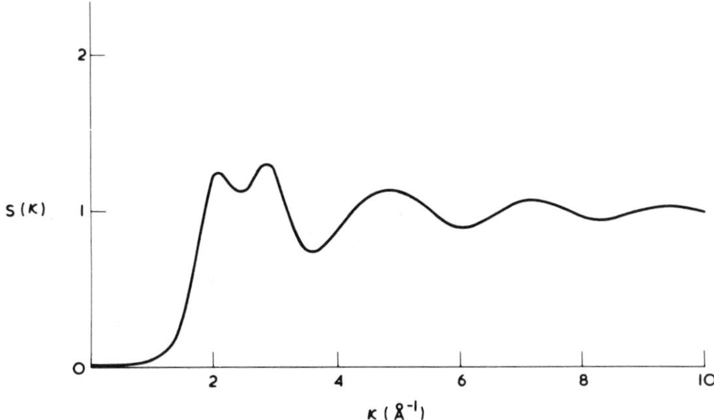

Fig. 6. Structure factor $S(\varkappa)$ for light water at 25°C determined by X-ray scattering. The scattering is mainly by the oxygen atoms.

molecular centers, then X-ray measurements give $S_c(\varkappa)$. Figure 6 shows this structure factor for light water measured by Narten *et al.*[783] Equation (23) is derived in the Appendix together with formulas for $f_1(\varkappa)$ and $f_2(\varkappa)$.

For orientational independence of the molecule D_2O with a hydrogen bond angle of 109° and O—D bond length (r_{OD}) of 0.96 Å, eqns. (A.3) and (A.5) give

$$f_1(\varkappa) = (b_O + 2b_D)^{-2}\{b_O^2 + 2b_D^2 + 4b_O b_D[(\sin \varkappa r_{OD})/\varkappa r_{OD}] + 2b_D^2[(\sin \varkappa r_{DD})/\varkappa r_{DD}]\} \tag{24}$$

$$f_{2u}(\varkappa) = (b_O + 2b_D)^{-2}\{b_O + 2b_D[(\sin \varkappa r_{OD})/\varkappa r_{OD}]\}^2 \tag{25}$$

where b_D and b_O are the scattering lengths for deuterium and oxygen, respectively, r_{DD} is the distance between deuterium atoms in a molecule, and $f_{2u}(\varkappa)$ denotes the function $f_2(\varkappa)$ for molecules completely uncorrelated.

Figure 7b shows the calculated functions $f_1(\varkappa)$ and $f_{2u}(\varkappa)$ and Fig. 7b shows a calculated $S_m(\varkappa)$ using eqn. (23) and the values of $S_c(\varkappa)$ in Fig. 6. This should be compared with the measured neutron diffraction pattern in Fig. 5. There is broad agreement between the measured and calculated structure factors, but significant differences in detail, which are probably due to some limited ordering. However, preliminary calculations* indicate that large-scale ordering, as in ice, of the water molecules leads to results in disagreement with the experimental data.

* D. I. Page and J. G. Powles, *Mol. Phys.* **21**, 901 (1971).

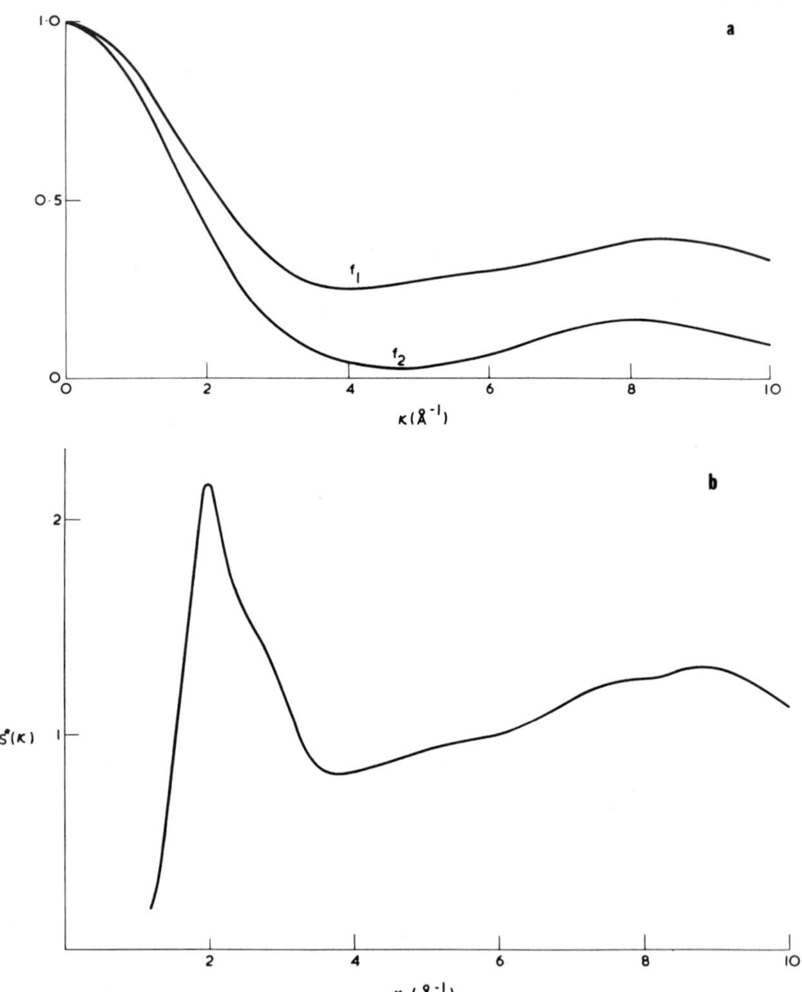

Fig. 7. (a) Calculated structure factors for light water. The function f_1 is the structure factor for a single molecule, f_2 is the orientational correlation function for random molecular orientation. (b) Plot of $S^*(\varkappa)$, the total neutron structure factor calculated from f_1, f_2, and the data in Figure 6.

There are as yet not sufficient calculations using this method of analysis to draw specific conclusions about water structure. However, it is a powerful method which will yield much information; it is particularly pleasing to see different experimental techniques being combined in such calculations to give apparently meaningful results.

4. QUASIELASTIC SCATTERING OF NEUTRONS BY WATER

4.1. Debye–Waller Factor

In Section 2.5 we saw that the energy spectrum of neutrons scattered from water can be treated in two parts: the quasielastic region, which is related to the long-time behavior of the scattering atom (i.e., hydrogen) and the inelastic region, corresponding to a much shorter observation time scale. The falloff with angle of the intensity of the quasielastic component measures the spatial extent of the thermal cloud associated with the vibrating atom and is usually written as $\exp(-\varkappa^2 u^2/6)$, by analogy with the Debye–Waller factor for a solid.

The angular variation of the quasielastic intensity has been measured for water at room temperature by several workers.[128,365,632,656]* Table II gives their values of the root mean square displacement $\bar{u}$ of the vibrating atom and the incident neutron energies for the experiments.

These measurements indicate that the probable value of $\bar{u}$ is between 1.0 and 1.5 Å, but appears to depend on the value of E_0, the value of u decreasing as E_0 increases. This is shown clearly in the results of Stirling and White,* whose three measurements were made under similar conditions on the same spectrometer, only the incident energy being varied. These discrepancies will be discussed in Section 6.

TABLE II. Root Mean Square Displacement $\bar{u}$ of the Hydrogen Atoms in Light Water Deduced from the Debye–Waller Factor[a]

Experiment	E_0, meV	$\bar{u}$, Å
Brockhouse[128]	42	0.9
Larsson *et al.*[656]	5.1	0.9
Kottwitz and Leonard[632]	100	1.0
	147	1.0
	247	0.5
Franks *et al.*[365]	3.1	1.4
Stirling and White, Private communication	2.1	1.5
	3.0	1.3
	4.6	1.0

[a] Here, E_0 is the initial energy of the neutrons.

* Also G. C. Stirling and J. White, private communication (1970).

4.2. Diffusive Motions

Information about the diffusive behavior is contained in the shape of the quasielastic peak, i.e., in the function $S_s(\boldsymbol{\varkappa}, \omega)$ at small values of ω (Section 2.2). For molecular liquids, rotational processes will also contribute to the quasielastic peak, but for small values of $\varkappa$ [O(1Å^{-1}] only the few lowest-order spherical harmonics of the rotational correlation function are important.(960) The corresponding relaxation time obtained for water from NMR experiments(497) is 10^{-11} sec, which corresponds to an unresolved peak in the neutron spectrum. Thus, for data from slow neutron scattering, the rotational effects may be considered as a minor additional broadening of the quasielastic peak to that given by simple diffusion. A treatment of rotational diffusion in molecular liquids has been given by Egelstaff and Harris.(289)

Analysis of $S_s(\boldsymbol{\varkappa}, \omega)$ usually proceeds through the use of a model for $G_s(\mathbf{r}, t)$ in which the motion of the scattering atom is assumed to be harmonic so that $G_s(\mathbf{r}, t)$ is of Gaussian form in $\mathbf{r}$(997):

$$G_s(\mathbf{r}, t) = [2\pi w^2(t)]^{-3/2} \exp[-r^2/2w^2(t)] \tag{26}$$

where $w(t)$ is the width function and is related to the mean-square displacement of the scattering atom at time t. Several diverse systems (e.g., ideal gas, harmonic oscillator) yield a Gaussian function of $\mathbf{r}$ for $G_s(\mathbf{r}, t)$, and Vineyard(1112) suggested that this *Gaussian approximation* should apply for real liquids, the essential details of the motion being contained in $w(t)$. Analysis of neutron scattering data involves selecting the "best" width function.

For example, Fick's law, which defines the macroscopic translational diffusion constant D, can be written

$$D \nabla^2 G_s(\mathbf{r}, t) = (\partial/\partial t) G_s(\mathbf{r}, t) \tag{27}$$

and the solution of this equation corresponding to $r = 0$ at $t = 0$ is

$$G_s(\mathbf{r}, t) = (4\pi D t)^{-3/2} \exp(-r^2/4Dt) \tag{28}$$

Comparison of eqns. (26) and (28) shows

$$w^2(t) = 2Dt \qquad \text{(Fick's law)} \tag{29}$$

Fourier transformation of eqn. (28) gives the scattering function for small $\boldsymbol{\varkappa}$ and ω:

$$S_s(\boldsymbol{\varkappa}, \omega) = (1/\pi)\{D\varkappa^2/[\omega^2 + (D\varkappa^2)^2]\} \tag{30}$$

Thus, $S_s(\varkappa, \omega)$ has a Lorentzian shape in ω with a full energy width at half maximum height $(\Delta E)_{1/2}$ of

$$(\Delta E)_{1/2} = 2D\varkappa^2 \tag{31}$$

In principle, the shape of the quasielastic peak gives most information on the form $S_s(\varkappa, \omega)$. In practice, the peaks at high values of $\varkappa$ contain an inelastic component, while for small values of $\varkappa$ the resolution function of the detectors is the major contribution to the peak shape; both of these effects make shape analysis unreliable at present. Thus, most of the work to date has been concerned with the energy widths of the quasielastic peaks.

Brockhouse[128,129] studied the quasielastic intensity for light water in a temperature range 4–70°C, as well as for ice and heavy water. He found his experimental lines to be considerably narrower than those calculated from eqn. (31) making use of tracer measured values for D, and three examples are shown in Fig. 8. He then considered a jump diffusion model in which it is assumed that diffusion occurs in large jumps corresponding to an activation process in a solid, i.e., the molecule vibrates about an equilibrium site for a time C_0 and then jumps to a new site in a time small compared with C_0. The vibrational behavior at any given site will generate a "thermal cloud" (Section 2.5). This model gives an energy width equal to h/C_0 and independent of the scattering angle. From dielectric relaxation time measurements on water at 25°C, Brockhouse obtained a value of

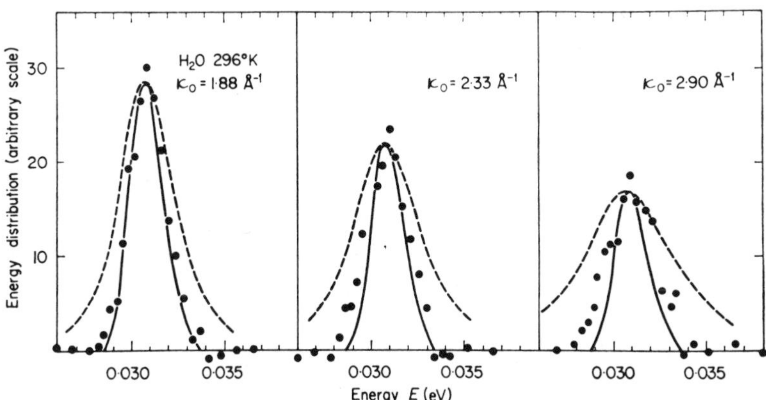

Fig. 8. Quasielastic components of neutron spectra observed from light water at 23°C. The resolution functions measured with vanadium are shown as heavy lines; the dashed lines are calculated from the diffusion coefficient assuming "simple" diffusion. $\varkappa_0$ is the momentum transfer for $\Delta E = 0$.[129]

8.3×10^{-12} sec for C_0. This gave calculated linewidths much smaller than the measured ones and it was concluded that the true diffusion process in water is a mixture of both simple and jump diffusion. These results were substantially confirmed by a third series of measurements.[541,936] Hughes *et al.* studied the 90° scattering from water at 22 and 45°C and surprisingly failed to observe any diffusion broadening of the elastic peak; they drew the conclusion that the effective mass of the scatterer must be much larger than that corresponding to a single molecule. However, Cribier and Jacrot[219] observed diffusive broadening from water at 30 and 43°C, fitted eqn. (30) to their data, and found the best values for D to be 2.7 and 3.4×10^{-5} cm² sec⁻¹, respectively.

It should be noted that the process of jump diffusion lies outside the scope eqn. (26), as it does not generate a Gaussian self-correlation function. If the amplitude u of the thermal cloud is negligible compared with the jump length l, a self-consistent treatment of the process can be given.[288] If, owing to the random nature of a liquid, we assume that l is distributed to le^{-l/l_0}, where l_0 is the average jump length, then

$$(\Delta E)_{1/2} = 2D\varkappa^2/(1 + D\varkappa^2 C_0) \tag{32}$$

Singwi and Sjolander[983] suggested a modification of the form

$$(\Delta E)_{1/2} = 2\{D\varkappa^2 + [1 - \exp(-u^2\varkappa^2/6)]/C_0\}/(1 + D\varkappa^2 C_0) \tag{33}$$

where $\exp(-u^2\varkappa^2/6)$ is the Debye–Waller factor. As $\varkappa \to 0$, this expression will give the correct asymptote if $D \gg u^2/6C_0$. Since random walk theory requires $D = l^2/6C_0$, we find the required condition for eqn. (33) to be $l^2 \gg u^2$, which is also the condition under which eqn. (32) was derived. If $u \simeq l$, then eqns. (32) and (33) should be modified.

Larsson *et al.*[655,656] carried out a series of experiments on light and heavy water and found line broadening of the same order as that of Brockhouse. They found their data at 22°C could be fitted by eqn. (33) with $D = 1.85 \times 10^{-5}$ cm² sec⁻¹ and $C_0 = 1.5 \times 10^{-12}$ sec. Safford *et al.*[933] also fitted eqn. (33) to their quasielastic data; they found optimum values for D and C_0 to be 2.7×10^{-5} cm² sec⁻¹ and 2×10^{-12} sec, respectively.

Franks *et al.*[365] used a model due to Egelstaff and Schofield[292] which uses in eqn. (26) a width function given by

$$w(t) = 2D[(t^2 + C_0^2)^{1/2} - C_0] \tag{34}$$

The form of this width function was introduced for mathematical convenience; it gives the correct limits as $\varkappa \to 0$ and as $\varkappa \to \infty$. The form of $S_s(\varkappa, \omega)$ is

$$S_s(\varkappa, \omega) = \frac{\exp(D\varkappa^2 C_0)}{\pi} \frac{D\varkappa^2 C_0}{[\omega^2 + (D\varkappa^2)^2]^{1/2}} K_1\{[\omega^2 + (D\varkappa)^2]^{1/2} C_0\} \quad (35)$$

where K_1 is a Bessel function of the second kind. There is no analytic form for the linewidths and $(\Delta E)_{1/2}$ must be evaluated numerically for different values of D and C_0. Franks *et al.* noted that previous work had sometimes used unrealistic values of D and of the Debye–Waller factor; they preferred to use the macroscopic diffusion coefficient(1072) and the measured Debye–Waller factors in fitting their data, leaving C_0 as the only disposable parameter. For H_2O at 21°C they obtained a value of $C_0 = 4.7 \times 10^{-12}$ sec for eqn. (35). They also compared thair data with those of Safford and found fair agreement; thus, the calculated values of C_0 depend critically on the model which is used.

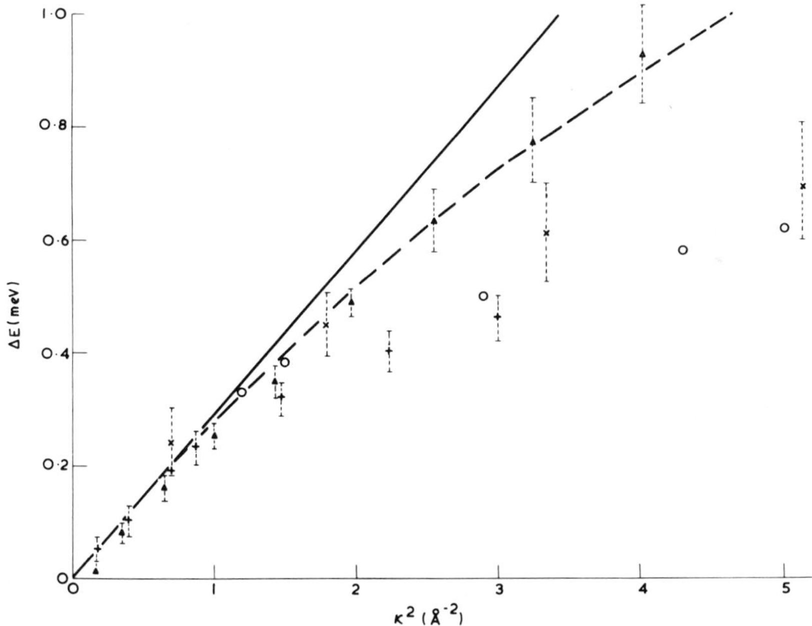

Fig. 9. Variation of the widths of the quasielastic peaks in light water at 22°C as a function of $\varkappa^2$. The data are those of Larsson *et al.*(656) (×), Sakamoto *et al.*(936) (△), Safford *et al.*(933) (○), and Franks *et al.*(365) (+). The dashed line represents the mean of 27 measurements at various incident energies by Stirling and White (private communication).

Stirling and White* found that, unlike the Debye–Waller factor in Section 4.1, their linewidths were independent of incident neutron energy and noticeably greater at the higher values of $\varkappa$ than the other published data, except those of Sakamoto *et al.*[936] Their data departed only slightly from the line predicted by simple diffusion theory.

Figure 9 shows a plot of all these diffusion widths as a function of $\varkappa^2$. The simple diffusion line corresponding to a temperature of 22°C ($D = 2.2 \times 10^{-5}\ \mathrm{cm^2\ sec^{-1}}$) is also shown.

The significance of these results is discussed in Section 6.

5. INELASTIC SCATTERING OF NEUTRONS BY WATER

5.1. Spectral Density Function

Examination of a spectrum of inelastically scattered neutrons will usually reveal a series of peaks corresponding to specific energy gains or losses, and frequently these peaks are identified with a mode for the system already identified by other techniques. For complete analysis of a spectrum a more fundamental approach must be adopted. The space–time correlation function $G_s(\mathbf{r}, t)$ gives a physical picture of the behavior of a single atom, but it is difficult to see the effects of other atoms on that atom; we therefore consider the velocity correlation for an atom when the effect of all the modes of the system will show up through the behavior of a single particle.

The most convenient way of introducing the atomic motions into the incoherent scattering cross section seems to be through a generalized frequency distribution. We again consider the analogy of a classical crystalline solid. The atomic motions can be analyzed into a spectrum of thermal vibrations which travel as waves through the crystal. The energy associated with these waves is quantized and each quantum of energy is regarded as a phonon. Each phonon has a specific wavelength and energy and an infinite lifetime. Inelastic scattering arises from the generation and annihilation of phonons. The frequency distribution function $\varrho(\omega)\, d\omega$ is defined as the number of phonons with energies in the interval between ω and $\omega + d\omega$. For a crystalline solid it can be shown[695] that

$$S_s(\boldsymbol{\varkappa}, \omega) = [\exp(-\mathbf{u}^2 \cdot \boldsymbol{\varkappa}^2)]\{\delta(\omega^2) + (kT/M)\varkappa^2[\varrho(\omega)/\omega^2] + [(kT/M)\varkappa^2]^2 \times \int [\varrho(\omega - \omega')\varrho(\omega)/2(\omega - \omega')^2(\omega')^2]\, d\omega' + O(\varkappa^6)\} \quad (36)$$

* G. C. Stirling and J. White, private communication (1970).

where M is the mass of the scattering atom. The first term in the curly brackets corresponds to elastic scattering, while the second term corresponds to the one-phonon interaction, and the detailed structure of the phonon spectrum can be seen in this term. The third term corresponds to the two-phonon interaction and gives a broad spectrum due to the convolution integral. Finally, the higher-order terms corresponding to multiphonon interactions give even broader spectral distributions.

In general, molecular systems undergo nonharmonic motions to which eqn. (36) does not apply, but the above result may be generalized by considering the velocity correlation function for a single atom. This function is the thermally averaged correlation between the velocity $v(0)$ of an atom at time zero and its velocity $v(t)$ at a later time t. It is usual to discuss the spectral density of this function, since then particular modes of motion will appear as peaks whose areas are proportional to the amplitude of the modes. In the case of a harmonic crystal the spectral density is equivalent to the frequency spectrum.

The velocity correlation for a classical system is simply related to the scattering law by[287]

$$z(\omega) = (1/\pi) \int_0^\infty (\cos \omega t) \langle v(0) v(t) \rangle \, dt = \omega^2 [S_s(\varkappa, \omega)/\varkappa^2]_{\varkappa \to 0} \tag{37}$$

where $z(\omega)$ is the spectral density of the velocity correlation function $\langle v(0)v(t) \rangle$. Thus, most inelastic scattering experiments on water are directed toward the determination of part of $z(\omega)$, usually that part lying between 5 and 100 meV.

It should be noted that every atomic motion makes a contribution to $z(\omega)$, and to isolate one type of motion within the whole spectrum requires some additional information. A knowledge of the approximate frequency and shape of a given rotational or vibrational contribution to $z(\omega)$ may be sufficient for its identification.

5.2. Experimental Results

The Raman and infrared spectra of water (see, for example, Walrafen[1134] and Chapter 5) show several levels attributable to the rotational and vibrational modes of the H_2O molecule. Movements of the molecule as a whole can be seen as broad bands of frequencies up to 120 meV with peaks at 22 meV, attributed to hindered translation of the molecule, and 55 and 67 meV, attributed to hindered rotation or libration of the bonds. Movement of the hydrogen atoms relative to the oxygen atom gives a bending mode at 205 meV and two vibrational modes, seen as a broad band of frequencies between 370 and 450 meV. These measurements cannot be used directly to predict the scattering of neutrons because the intensity of scattering depends upon the amplitude of motion of the hydrogen atom. For example, the relative weight given to each mode cannot be deduced from the amplitudes of the observed infrared peaks because the intensity is related to the Raman and infrared selection rules for the excitation of the levels. The linewidths also cannot be readily evaluated because in infrared and Raman scattering these levels appear in combination with the rotational levels of the water molecule. However, the position of the levels can be used as a guide when interpreting the neutron data.

Early data using low-incident-energy neutrons were obtained;[408,541,656,1025] additionally, spectral density functions have been deduced.[290,655,853] These measurements have been reviewed by Larsson.[654] All these workers observed, in varying degrees, broad spectra with some distinct energy transfers. Figure 4 shows typical spectra. Larsson and Dahlborg[655] split the spectrum into parts corresponding to modes with the molecule as a rigid unit ("lattice modes") and intramolecular modes (Fig. 10). All the results quoted above agree with this and identify the prominent peak at ~65 meV as due to the librational mode. None of the experiments referred to included the vibrational (intramolecular) modes, as neutrons of high incident energy would be required for this. In detail, however, the results of different workers do not agree. Hughes *et al.*[541] observed other distinct energy transfers at 5, 8, and 21 meV and smaller peaks between 0.5 and

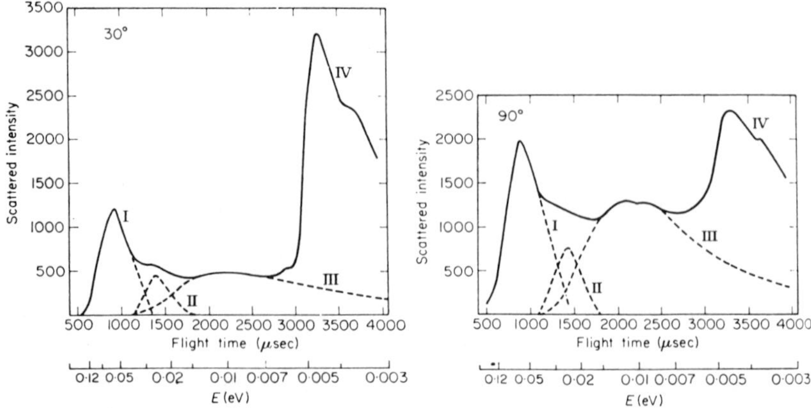

Fig. 10. Time-of-flight spectra of cold neutrons scattered from light water at 20°C at scattering angles of 30° and 90°. The dashed lines represent separation of the spectra into various inelastically (I, II, and III) and quasielastically (IV) scattered parts.[654]

0.7 meV. Larsson and Dahlborg alone noted an asymmetry in the librational peak, which they attributed to the presence of a second component at 1.5–2 times the frequency of the main component. They also claimed that the spectra of water at +2°C and ice at −3°C were identical; Golikov *et al.* have disagreed with this and find[408] the spectrum of water at +1°C to be identical with that of water at 22°C. As noted in Section 4.2, the variations of the widths of the quasielastic peaks with $\varkappa^2$ also differed.

The situation since 1965 has scarcely improved; the uncertainty in the spectra below 50 meV has not been resolved. Burgman *et al.*[151] studied the scattering from ice and water near the melting point, in order to resolve the differences between Larsson *et al.* and Golikov *et al.*, and they concluded that the results of Golikov *et al.* were substantially correct, but they found even more pronounced differences between ice and water. For water at room temperature they observed low-energy peaks at 6.5, 9.6, 11.8, 14.8, 19, 26, and 34 meV. Brugger[138] observed the inelastic scattering at small values of $\varkappa$ and noted that "peaks in the cross-section data, other than the 60 meV hindered rotational band, were not observed," while Prask and Boutin[869] and Safford *et al.*[933] have claimed detailed structure in their spectra and have assigned specific energy transfers to many small peaks. Franks *et al.*[365] prefer to consider the region between 5 and 40 meV as a single region, although "fine structure" was observed but could not be assigned consistently to specific energy transfers. Stirling and White* reached a similar conclusion. Prask and Boutin resolved the librational peak *in ice* into three components at 69, 81, and 100 meV, but could not resolve the peak in water in the same way. Safford *et al.* partially resolved the librational peak in water into three maxima at 57, 75, and 108 meV, which agree with the broad torsional components observed by Raman[1135] and hyper-Raman[1060] studies; the frequencies of the lower maxima in their spectra are in good agreement with those of Burgman *et al.* Table III compares the positions of the maxima in the spectra observed in the various experiments. It is interesting to note that Stafford *et al.* also observed maxima at very low energies ($\lesssim$1 meV), which were due to instrumental effects (cf. Hughes *et al.*, who used a similar apparatus).

The best determination of the spectral density function is that of Haywood,[482] whose result is shown in Fig. 11. The bound part of the function is due to translational and rotational motions of the molecule and was observed experimentally with neutrons; two of the peaks are identified as the librational peak at 70 meV ($\beta = \Delta E/kT = 2.7$) and the

* G. C. Stirling and J. White, private communication (1970).

TABLE III. Summary of the Observed Peak Positions in the Spectrum of Neutrons Scattered Inelastically from Light Water at ~22°C

Author	Lattice modes, meV										Librational modes, meV
Larsson and Dahlborg[655]	—	—	—	—	—	—	20	—	—	—	59
Burgman *et al.*[151]	6.5	—	9.6	11.8	14.8	19	—	—	—	—	60
Prask and Boutin[869]	6.2	7.8	9.5	12.9	—	18	23	27	30	35	60
Haywood[482]	—	—	—	12.5	—	—	22	—	—	—	70
Brugger[138]	No peaks										60
Safford *et al.*[933]	Agreement with Burgman *et al.*										57, 74, 107
Franks *et al.*[365]	"Fine structure" not identifiable with peaks										67
Stirling and White, Private communication	"Fine structure" not identifiable with peaks										63

hindered translational peak at 22 meV, while the peak at 12.5 meV is not identified. The δ-functions at ~200 meV ($\beta = 8.1$) and ~400 meV ($\beta = 16.8$ and 17.7) represent the vibrational modes; they were observed in a later neutron scattering experiment,[463] but their positions were deduced by Haywood from infrared data. They are included to allow the calculation of the scattering cross sections, and hence the assignment of relative weighting factors to the different parts of $z(\omega)$. Haywood estimated the weights to be in the ratios of 0.5 : 0.17 : 0.33 for the bound region and lower- and higher-energy δ-functions, respectively. The vibrational modes were observed by Harling[463] to be at 206 meV (with a full-width at half-height of 23 meV), ~450 meV, and ~530 meV. The value of 206 meV for the bending mode is in good agreement with the infrared data,[1134] but the higher-frequency band has higher average energies and widths than the optical results. Harling suggested this may be due to multiphonon interactions. Several other measurements have been made of the scattering law for both H_2O and D_2O at several temperatures,[483,462,611] with the aim of determining $S(\boldsymbol{\varkappa}, \omega)$ over a wide range of $\varkappa$ and ω for neutron thermalization calculations and not for the examination of low-energy detail. Apart from noting the similarity between the spectral density functions for both H_2O and D_2O, these results will not be discussed.

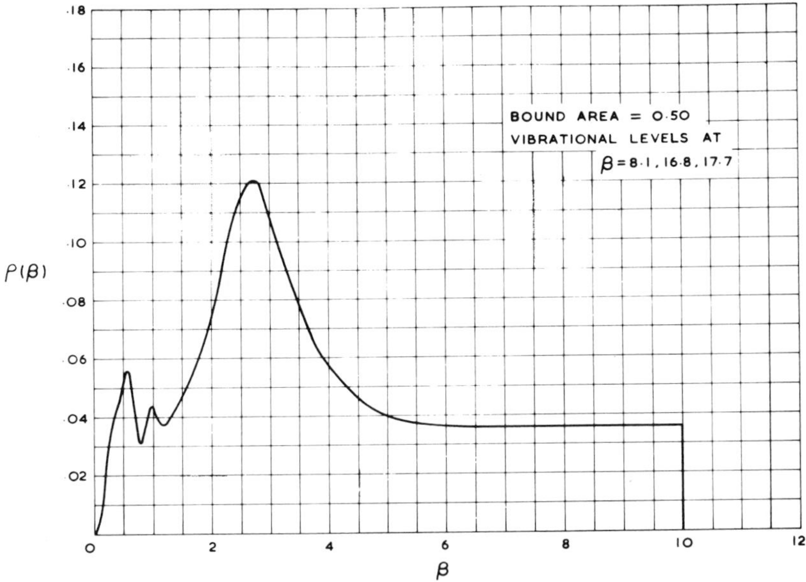

Fig. 11. The spectral density function for light water at 22°C.[482] $\beta = \Delta E/kT$.

6. DISCUSSION

The results presented in the previous sections appear to be at variance on several points, namely the data on the quasielastic peak widths, the presence of small peaks in the inelastic spectrum at low values of the energy transfer, and the value for the Debye–Waller factor. These differences should be resolved before too much credence is given to the various models which have been proposed. Two possible sources of error are as follows:

(a) Inelastic background. Figure 9 shows that the width data from the quasielastic peaks agree up to values of $\varkappa^2 = 2$ Å^{-2}. The inelastic component of the spectra extends into the quasielastic region (Fig. 10) and must be subtracted before the widths are determined. This inelastic component increases as $\varkappa^2$ [eqn. (36)], so that the correction will be much smaller at low values of $\varkappa^2$ than at high values. It seems reasonable that some of the discrepancies at higher values of $\varkappa^2$ could be due to the correction procedure.

The presence of an inelastic component could also explain discrepancies in the Debye–Waller factors because, as the incident neutron energy rises, the absolute resolution of the detector worsens and the amount of inelastic scattering included in the quasielastic measurements increases. This leads to smaller values of the Debye–Waller factor for higher incident energies, and inspection of Table II shows this to be roughly correct.

(b) Incident spectrum of neutrons. The results were mainly obtained from experiments using two types of incident neutron fluxes—a full, cold neutron spectrum using the beryllium filter cutoff, or a monokinetic beam, velocity selected from a Maxwellian spectrum.[138] Figures 10 (beryllium filter) and 4 (velocity selector) show the difference in the spectra obtained; this is particularly noticeable in the quasielastic region. The experiments using the velocity-selected technique appear to see much less pronounced fine structure in the inelastic region (Table III).

Allowing for these differences in the data, the neutron scattering technique confirms that water molecules certainly have (a) vibrational modes at 206, ~450, and ~530 meV, (b) a pronounced hindered rotational mode at 65 meV, and (c) a hindered translational mode at 22 meV. In addition, there are a large number of low-energy modes, but the assignment of precise energy transfer values to specific modes is speculative. These facts are well known from other sources and the neutron scattering method makes its main contributions in fixing the relative displacements of the H atoms vibrating in their different modes and in observing the displacement of the hydrogen atoms at times of the order of 10^{-12} sec. It shows that the hydrogen atoms are vibrating with a root mean square displacement of ~1 Å and, in the long-time limit, the self-diffusion coefficient determined by tracer and NMR methods accounts for the diffusive motion observed by neutron scattering, i.e., the neutron data can be interpreted without recourse to a rotational diffusion contribution. This implies that there is very little rotation of the molecule as it translates from site to site. At these long times, the motion can be described by random walk theory and Fick's law applies. On a shorter time scale, however, simple diffusion no longer applies and the process appears to be a combination of simple plus jump diffusion. The time between jumps has often been calculated, but this must be interpreted with care because it usually implies a jump length which is not negligible compared with the thermal cloud diameter (Section 4.1), thus invalidating a necessary assumption in the theory.

Structural measurements are still at a preliminary stage. The lack of suitable isotopes means that the three partial structure factors cannot be determined uniquely. The neutron scattering data must be combined with those from other sources, and it is from such cooperative analysis of data, as well as the refinement of the scattering experiments, that definitive information will come. High-resolution, fully corrected data need to be obtained to explore the diffusion process more fully, not only by determination of the half-widths of the quasielastic peaks, but by analysis of their shape. Clearly, the ranges of $\varkappa$ and ω which can be covered by experiments must be extended. The present data are dominated by energy-gain scattering of cold neutrons and experiments must be performed to explore the high-energy transfer region of $S(\varkappa, \omega)$ by means of high incident neutron energies and energy-loss scattering. Ideally, the whole of the scattering function $S(\varkappa, \omega)$ should be determined; a single experiment can only measure part of this, and a clear theoretical estimate of the information contained in the part to be measured should be made before the experiment is carried out. In this way neutron scattering can make many contributions to the understanding of the spatial and dynamic behavior of water molecules.

APPENDIX. CALCULATION OF ORIENTATIONAL CORRELATION FACTORS FOR MOLECULES

We define the general structure factor $S_m(Q)$ for neutron scattering by a system of nuclei by

$$S_m(Q) = [N/(\sum_i \bar{b}_i)^2]\langle | \sum_i \bar{b}_i \exp(i\mathbf{Q} \cdot \mathbf{r}_i) |^2\rangle \tag{A1}$$

where $\bar{b}_i$ is the isotopic average scattering length for the ith nucleus. If we assume that the nuclei comprise molecules and the molecular structure is known, then

$$S_m(Q) = [1/(\sum_i \bar{b}_i)^2]\langle | \sum_n b_n \exp(i\mathbf{Q} \cdot \mathbf{r}_{cn}) |^2\rangle + [1/N(\sum_i \bar{b}_i)^2] \times \langle \sum_{i \neq j} [\exp(i\mathbf{Q} \cdot \mathbf{r}_{cij})] \sum_{n_i n_j} b_{n_i} b_{n_j} \exp[i\mathbf{Q} \cdot (\mathbf{r}_{ci} - \mathbf{r}_{cj})]\rangle \tag{A2}$$

where n labels an atom in the molecule and in the second term n_i labels the nth atom in the ith molecule; $\mathbf{r}_{cn}$ is the intermolecular distance of the atom n from the molecular center c; $\mathbf{r}_{cij}$ is the intermolecular distance between the centers of the molecules i and j. We put

$$f_1(Q) = [1/(\sum_i \bar{b}_i)^2]\langle | \sum_n b_n \exp(i\mathbf{Q} \cdot \mathbf{r}_{cn}) |^2\rangle \tag{A3}$$

$f_1(Q)$ is the molecular form factor which is observed for a single molecule, i.e., in the gas at low density, where the second term in (A2) is negligible. $f_1(Q)$ can be calculated for a given molecule but the second term in eq. (A2) cannot be simplified without making assumptions because the orientation of the molecule $\mathbf{r}_{cn_i}$ depends on that of the molecule $\mathbf{r}_{cn_j}$ and on their separation $\mathbf{r}_{cij}$. If $\mathbf{r}_{ij}$ is large, $\mathbf{r}_{cn_i}$ and $\mathbf{r}_{cn_j}$ are surely statistically independent. However, the main contribution to $S_m(Q)$ of interest comes from short $\mathbf{r}_{ij}$. The second term in eq. (A2) can be written for infinite-ranged correlations:

$$[1/(\sum_i \bar{b}_i)^2]\langle \sum_{n_i n_j} b_{n_i} b_{n_j} \exp[i\mathbf{Q} \cdot (\mathbf{r}_{cn_i} - \mathbf{r}_{cn_j})]\rangle \times (1/N)\langle \sum_{i \neq j} \exp(i\mathbf{Q} \cdot \mathbf{r}_{cij})\rangle \tag{A4}$$

and we put

$$f_2(\varkappa) = [1/(\sum_i \bar{b}_i)^2]\langle \sum_{n_i n_j} b_{n_i} b_{n_j} \exp[i\boldsymbol{\varkappa} \cdot (\mathbf{r}_{cn_i} - \mathbf{r}_{cn_j})]\rangle \tag{A5}$$

By analogy with atomic liquids we define the structure factor for molecular centers $S_c(\varkappa)$:

$$S_c(\varkappa) = (1/N)\langle \sum_{i \neq j} \exp(i\boldsymbol{\varkappa} \cdot \mathbf{r}_{ij})\rangle + 1 \tag{A6}$$

Hence,

$$S_m(\varkappa) = f_1(\varkappa) + f_2(\varkappa)[S_c(\varkappa) - 1] \tag{A7}$$

CHAPTER 10

Thermodynamic and Transport Properties of Fluid Water

G. S. Kell

Division of Chemistry
National Research Council of Canada
Ottawa, Canada

1. INTRODUCTION

The engineers of remotest antiquity provided water for their cities. In the Middle Ages the use of water mills flourished in Europe. In these activities practical knowledge of the properties of water—density, volatility, and viscosity—preceded the development of scientific understanding. Similarly, Watt patented the condenser, making the steam engine practical, in 1769, before the discovery of the compound nature of water, shown by Cavendish in 1781, and well before the development of thermodynamics in the nineteenth century.

Today, the properties of water play a key role in, for example, oceanography and limnology, hydraulics, biochemistry, and physical chemistry. Usually, it is the thermodynamic and transport properties that have this practical importance. Indeed, for many applications, values of properties are needed with high precision, even though their relation to structure on a molecular scale is poorly understood. Accordingly, many measurements and tabulations have been made for particular purposes, the most noteworthy being steam tables for the design of power plants.

In the present chapter the thermodynamic and transport properties of the fluid phases, liquid and vapor, are examined from a molecular point of view when this is available—macroscopic thermodynamics alone seems inadequate for the study of ions in solution or of the hydration of proteins—

experimental work is surveyed, and reliable values, including those suitable for calibration, are indicated. The liquid and vapor are considered over their range of stability at temperatures up to the critical point at 374°C. It is true that observations have been made on the liquid at 24°C below the freezing point,[448] and at negative pressures of 270 bars,[124] but the liquid and vapor will be considered here primarily in their regions of stability.

One school of thought sees liquid water as composed of distinguishable species, with the properties reflecting equilibria among those species. This theory has existed since about 1900, when the components were called *hydrol*, *dihydrol*, etc. The mathematics has evolved remarkably little. Another school[590] sees the properties of the liquid as reflecting distributions of oxygen–oxygen distances and of oxygen–oxygen–oxygen angles. Both explanations are consistent with the thermodynamic properties of the liquid; accordingly thermodynamics does not choose between them, and structural discussions based on these properties are idle. On the other hand, for the low-pressure vapor the forces between molecules can be determined with sufficient precision that it is natural to phrase the treatment in molecular terms.

The needs of international trade in the field of thermal power generation led representatives of a number of nations to agree, in 1934, to a skeleton table of values of the thermodynamic properties of water and steam; the detailed tables to be used in design work were required to be within the tolerances specified by the skeleton table at the points it gave. In 1963 the Sixth International Conference on the Properties of Steam, meeting in New York, adopted a revised and enlarged skeleton table giving thermodynamic values from 0 to 800°C and at pressures to 1 kbar,[995] as well as viscosity and thermal conductivity[996] over narrower ranges. It also established an International Formulation Committee (IFC) to obtain, for computer use, formulations that were within the tolerances of the skeleton table.[547,548] It is a need of the thermal power industry that the design tables should remain unchanged for long periods—as from 1934 to 1963. Political considerations are also said to influence the method of arriving at the values and tolerances of the skeleton table. Thus, *steam tables should not be considered critical compilations for other purposes* and, in particular, they should be differentiated only with caution. Their utility for others is that they tabulate properties over a wide range of conditions and are available in various engineering units.*

* Tables agreeing with the 1963 skeleton table: The N.E.L. "Steam Tables 1964"[45] give temperatures in degrees Celsius, pressures in bars, specific volumes in cubic cen-

The major compilation of data on pure water remains that of Dorsey,[269] which gives thorough coverage of work done before 1940; it remains an essential reference book. There are also more limited surveys.[123,132,301,617,660] Kennedy and Holser[595] have given a table of volumes of fluid water up to 1000°C, and Burnham *et al.*[155] give the thermodynamic properties of water at intervals of 20°C and 100 bars from 20°C and 100 bars to 1000°C and 10 kbars with a standard error of about 0.1%.

For heavy water and waters of other isotopic compositions, extensive compilations include those by Kirshenbaum,[612] Shatenshtein *et al.*[967] Kazavchinskii *et al.*,[583] Rabinovich,[881] and Dirian,[265] and there are shorter review articles.[706,915,1162] The thermodynamic properties of heavy water have been tabulated by Elliott[305] and Kazavchinskii *et al.*[583]

2. THERMODYNAMIC PROPERTIES

2.1. The Macroscopic Viewpoint

Classical thermodynamics is based on laws of bulk matter, such as the conservation of energy, laws which are independent of the structure of matter. Hence, measurements of thermodynamic properties have a permanent value independent of dispute over the detailed structure of the liquid. Of course, thermodynamic properties often play a central part in developing models, even if the models do not derive solely from thermodynamic information.[528,580,940] The Gibbs energy G (or equally well, the Helmholtz energy) may be chosen as the primary thermodynamic function from which other equilibrium properties are obtained by standard thermodynamic relations[431]:

Volume	$V = (\partial G/\partial P)_T$
Entropy	$S = -(\partial G/\partial T)_P$
Enthalpy	$H = G + TS$
Internal energy	$U = H - PV$

timeters per gram, and energies in joules per gram. Schmidt has given "Properties of Water and Steam in SI-Units"[954] with pressures in bars, specific volumes in cubic meters per kilogram, and energies in kilojoules per kilogram; this table was obtained from the 1967 IFC "Formulation for Industrial Use."[547] The E.R.A. "1967 Steam Tables"[303] derive from the same formulation, but use English units, that is, temperatures in degrees Fahrenheit, pressures in lbf in.$^{-2}$, specific volume in cubic feet per pound, and energies in Btu lb^{-1}. The "Steam Tables" of Keenan *et al.*,[586,587] which were published in two editions, one in English and one in metric units, were derived from a different equation.

Other important properties may be obtained by further differentiation:

Isopiestic (isobaric) thermal expansivity	$\alpha_P = [\partial(\ln V)/\partial T]_P$
Isothermal compressibility	$\varkappa_T = -[\partial(\ln V)/\partial P]_T$
Isopiestic specific heat	$C_P = (\partial H/\partial T)_P$
Isochoric specific heat	$C_V = (\partial U/\partial T)_V$

The microscopic approach to thermodynamic properties starting from molecular properties and applying either statistical mechanics[677,1158] or computational methods[51] does not yet reproduce experimental values for liquid water very closely.

In liquid or gaseous mixtures of H_2O and D_2O the equilibrium

$$H_2O + D_2O \rightleftharpoons 2HDO \tag{1}$$

gives a certain concentration of HDO but prevents the preparation of pure HDO. If the hydrogens and deuterium were distributed randomly on the oxygens, the equilibrium constant for eqn. (1) would be 4, the concentration of HDO being given by the rule of the geometric mean. The value in the vapor, which is well understood, in nearer 3.5 because of differences in the energy levels of the species. It has been reported that the equilibrium constant for the liquid is significantly less than 4, (3.76 at 25°C)[368] though Salomon,[937] in a recent report, finds that the equilibrium constant for eqn. (1) in the liquid is close to 4.0, and calculates the heat of mixing as 33.2 cal mol^{-1} compared with an experimental value of 32 cal mol^{-1}. In any event, the usual assumption that the abundance of HDO is given by the geometric mean and that H_2O and D_2O give an ideal solution is a good approximation for ordinary work.[5]

2.2. Volume Properties

2.2.1. *Vapor*

Steam has been used for power generation for over two centuries, and the volume properties are now known over a considerable range. Experimental work may be found through the references given by Bruges.[132] Data at low pressures is most satisfactorily compared with molecular theory by the virial equation[734]

$$P/\varrho RT = 1 + B^{(2)}\varrho + B^{(3)}\varrho^2 + \cdots \tag{2}$$

where P is the pressure, ϱ is the molar density, R is the gas constant, T is

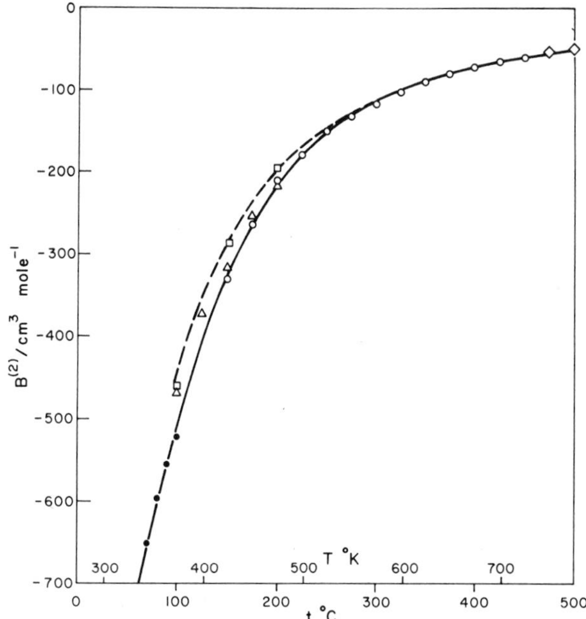

Fig. 1. Second virial coefficient of steam. Above 400°C, agreement is excellent. The solid line is drawn to pass through the values of Kell *et al.* at high temperatures, and of McCullough *et al.* at low temperatures. The dotted line is from an equation given by Keyes *et al.*[203,603] (○) Kell *et al.*;[593] (●) McCullough *et al.*;[712] (◇) D_2O, Kell *et al.*;[592] (□) latent heat, Curtiss and Hirschfelder;[223] (△) latent heat, Kell *et al.*[593]

the absolute temperature, and $B^{(n)}$ is the nth virial coefficient. The second virial coefficient $B^{(2)}$ is a measure of two-body interactions in the vapor, the third $B^{(3)}$ includes three-body interactions as well, etc.

The determinations[203,223,592,593,603,712] of $B^{(2)}$ for H_2O are shown in Fig. 1, except for the values arrived at by Vukalovich *et al.*[1120] which fall on top of the ones shown. Above 400°C agreement is excellent. Even at lower temperatures the difference between the results of Keyes and his co-workers[203,603] and those of Kell *et al.*[593] may be acceptable in view of the experimental difficulties in working with a highly polar substance such as water,[602] but the differences are great enough that the determination of a molecular property such as the quadrupole moment or the polarizability from the data is quite uncertain. Kell *et al.*[592] have shown that $B^{(2)}$ is very nearly the same for H_2O and D_2O, although it appears slightly more negative for D_2O at low temperatures.

The second virial coefficient $B^{(2)}$ of a dipolar molecule may be expressed in terms of the intermolecular potential energy Φ integrated over all positions and configurations,[516]

$$B^{(2)} = \tfrac{1}{4}N_A \int_0^\infty \int_0^{2\pi} \int_0^\pi \int_0^\pi (1 - e^{-\Phi/kT}) r^2 \sin\phi_1(1) \sin\phi_1(2) \times d\phi_1(1)\, d\phi_1(2)\, d\phi_3\, dr$$

where N_A is the Avogadro number, k is the Boltzmann constant, r is the distance between the centers of molecules 1 and 2, $\phi_1(2)$ is the angle between r and the dipole axis of molecule 2, and ϕ_3 is the aximuthal angle about r between the dipoles. The potential energy can be expressed as the sum of a nondirectional Lennard-Jones $(m, 6)$ potential energy,

$$\Phi_{LJ} = 4\varepsilon[(\sigma/r)^m - (\sigma/r)^6]$$

plus an electrostatic energy involving the permanent and induced multipoles of the molecules.* The simplest such potential is that of Stockmayer,[1030] which considers only the permanent dipoles. The Stockmayer potential Φ_S is given by

$$\Phi_S = \Phi_{LJ} + \mu^2 r^{-3}[2\cos\phi_1(1)\cos\phi_1(2) + \sin\phi_1(1)\sin\phi_1(2)\cos\phi_3] \quad (3)$$

where μ is the dipole moment. If the observed value of μ of 1.84 D is accepted and the parameters ε and σ adjusted, the Stockmayer potential represents observed values of $B^{(2)}$ for steam nearly to experimental precision, giving $\varepsilon/k \simeq 350°K$ and $\sigma \simeq 2.70$ Å. If meaningful parameters are to be obtained from $B^{(2)}$, precise data are needed over a wide temperature range.[614,685] Higher permanent moments, such as the quadrupole, which is important for water, can be included, as well as the inductive effects produced by the polarization of a molecule in the electrical field of its neighbors. For these higher polar forces a closed expression is not normally written for $B^{(2)}$ but, following Buckingham and Pople,[144] the second virial coefficient is expressed in terms of a function $H_k(y)$ introduced for this purpose; it was tabulated by them for the Lennard-Jones (12–6) potential as the central part, and for (18–6) and (27–7) by Saxena and Joshi.[944] However, it is doubtful how far this classical approach may be pushed, as McCarty and Babu[710] have found that the translational and rotational quantum cor-

* Carrà and Zanderrighi[166] have expressed the potential as a nondirectional Morse potential plus polar forces.

rections to $B^{(2)}$, using the Stockmayer potential, are greater than the experimental error.[593]

$B^{(2)}$ for a mixture of mole fraction x_1 of species 1 and x_2 of species 2 is given by

$$B^{(2)} = x_1{}^2 B_{11}^{(2)} + 2x_1x_2B_{12}^{(2)} + x_2{}^2 B_{22}^{(2)}$$

where $B_{11}^{(2)}$ is the second virial coefficient for pure 1. Insofar as the potential Φ_{12} can be described by a central part and polar forces, the molecular moments and polarizabilities remain the same as for the pure gases, but new parameters ε_{12} and σ_{12} are needed to describe the central part of the potential. In this respect water and other polar gases seem no more difficult than nonpolar ones.

Dimers and Trimers. An important question for liquid water is the geometry and energy of neighboring water molecules. Thorough calculations are only available for clusters of a few molecules, so they strictly apply only to the vapor, but, as the favored configurations are close to those for the ices, the results likely apply to the liquid as well. The formation of an n-mer is described by

$$(H_2O)_{n-1} + H_2O \rightleftharpoons (H_2O)_n \tag{4}$$

the equilibrium constant K_n for this polymerization reaction can be related to the virial coefficient. The present treatment follows Kell and McLaurin.[591] The equilibrium constant is

$$K_n = P_{(H_2O)_n}/P_{(H_2O)}P_{(H_2O)_{n-1}}$$

where $P_{(H_2O)_n}$ is the partial pressure of species $(H_2O)_n$. Writing $B_*^{(n)}$ for the difference

$$B_*^{(n)} = B^{(n)} - B_f^{(n)}$$

the equilibrium constants are related by

$$B_*^{(2)} = -K_2RT$$

$$B_*^{(3)} = (4K_2{}^2 - 2K_2K_3)(RT)^2$$

$$B_*^{(4)} = (-20K_2{}^3 + 18K_2{}^2K_3 - 3K_2K_3K_4)(RT)^3$$

The contribution $B_f^{(n)}$ made by free molecules, that is, by molecules whose collision affects $B^{(n)}$ but does not lead to a polymer, must be estimated. For water, Hirschfelder *et al.*[517] took $B_f^{(2)}$ to be about 40 cm^3 mol^{-1}. Some authors, for example Woolley,[1184] have taken $B_f^{(n)}$ to be zero. These equi-

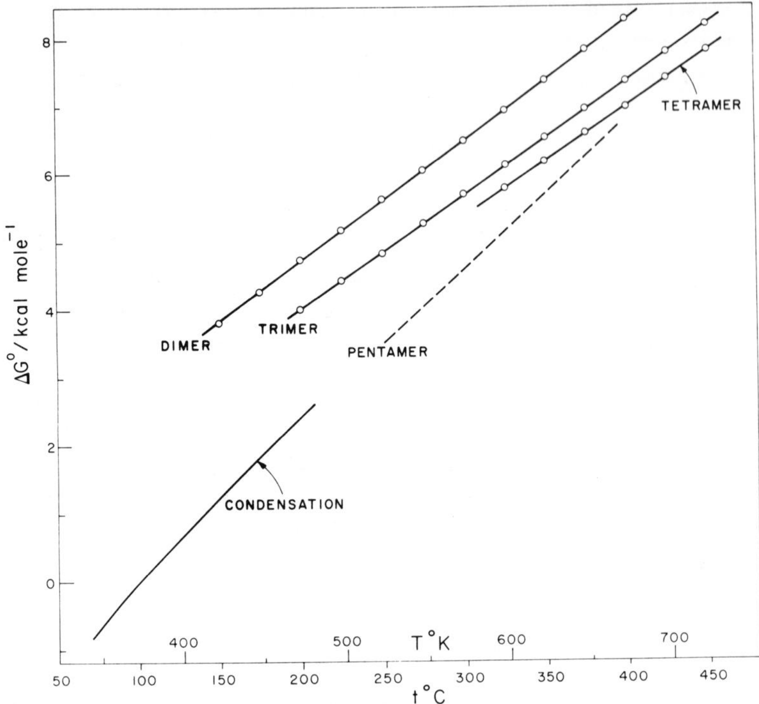

Fig. 2. The Gibbs energy of polymerization in water vapor. The $\Delta G_n{}^\circ$ shown is for the reaction given by eqn. (4), and data are from Kell *et al.*[591,593] Values for dimerization, trimerization, and tetramerization are experimental. The value shown for pentamerization is estimated for the formation of a cyclic species, and the agreement of the slopes for dimer, trimer, and tetramer at a different value shows that a negligible fraction of the populations of those species can be cyclic.

librium constants can be related to the thermodynamic parameters for polymerization by the relations

$$\Delta G_n{}^\circ = -RT \ln K_n = \Delta H_n{}^\circ - T\,\Delta S_n{}^\circ$$

Figure 2 shows $\Delta G_n{}^\circ$ for $n = 2$, 3, and 4 as determined from the virial coefficients of Kell, McLaurin, and Whalley. ΔG° is slightly lower for $n = 3$ and 4, indicating that the higher polymers are slightly more stable than the dimer, but all three lines are close together. This shows that cyclic structures are below the limits of detectability in the trimer and tetramer populations. As Table I shows, in each case ΔH° for the addition of a monomer is near -3.5 kcal mol^{-1}, and ΔS° about -17 cal mol^{-1} deg^{-1}. The pentamer can be cyclic without strain, and estimated values of $\Delta G_5{}^\circ$ are shown in Fig. 2,

TABLE I. Stabilization Energy of Water Polymers in the Vapor[a]

	$-\Delta E_i$, kcal mol^{-1}	$-\Delta H_i^\circ$, kcal mol^{-1}	$-\Delta S_i^\circ$, cal mol^{-1} deg^{-1}
Dimer: Theoretical			
Del Bene and Pople[(250,251)]	6.09	—	—
Hankins *et al.*[(454)]	4.72	—	—
Morokuma and Winnick[(764)]	6.55	—	—
Kollman and Allen[(619)]	5.94	—	—
Bollander *et al.*[(99)]	5.2	—	—
Dimer: Experimental			
O'Connell and Prausnitz[(803,871)]	—	4.6–5.2	—
Bollander *et al.*[(99)]	—	3.0	—
Kell and McLaurin[(591)]	—	3.8	17.9
Green *et al.*[(423)]	—	6.5	26
Trimer: Theoretical			
Del Bene and Pople[(250,251)]			
Trimer I	3.24	—	—
Trimer II	4.05	—	—
Trimer III	8.57	—	—
Trimer IV	10.75	—	—
Trimer V	12.45	—	—
Trimer: Experimental			
Kell and McLaurin[(591)]	—	3.20	16.8
Tetramer: Experimental			
Kell and McLaurin[(591)]	—	3.96	16.3

[a] Here, $i = 2, 3, 4$ for dimer, trimer, and tetramer, respectively.

taking $-\Delta H_5^\circ = 8$ kcal mol^{-1} and $-\Delta S_5^\circ = 22$ cal mol^{-1} deg^{-1}; the line is clearly quite different from the others, supporting the view that the lower forms are linear. Independent information about polymers in the vapor can be obtained by mass spectrometry, either to obtain the population of polymers in the bulk gas[(423,664)] or to study ionized species.[(584,585)] Greene preferred values of $-\Delta H_2^\circ = 6.5$ kcal and $-\Delta S_2^\circ = 30$ cal mol^{-1} deg^{-1}, but as the values for condensation, which correspond to the formation of two hydrogen bonds, are $-\Delta H^\circ_{\text{cond}} = 9$ kcal mol^{-1} and $-\Delta S^\circ_{\text{cond}} = 26$ cal mol^{-1} deg^{-1}, these values seem distinctly too high. O'Connell and Prausnitz(803,871) have reviewed the older evidence that the dimer occurs in the open-chain form, rather than as a bifurcated or cyclic dimer, possibilities that are shown in Fig. 3. They concluded that $-\Delta H_2^\circ$ could have values up to 6 kcal mol^{-1}. *A priori* theoretical calculations have been made with wave functions.(99,250,251,454,619,764) The quantity reported is ΔE, which is the potential energy change on formation of the polymer from monomers. The values of ΔE calculated for dimerization are about -6 kcal mol^{-1}, which may be compared with a value of ΔU_2°—in the perfect gas approximation, eqn. (4) gives $\Delta U = \Delta H - RT$—nearer -4 kcal mol^{-1}. The results of Del Bene and Pople(250,251) appear most far-reaching. They indicate that the preferred structure for the dimer is clearly the open-chain dimer in the *trans* form as shown in Fig. 3, rather than in a *cis* form, and that the po-

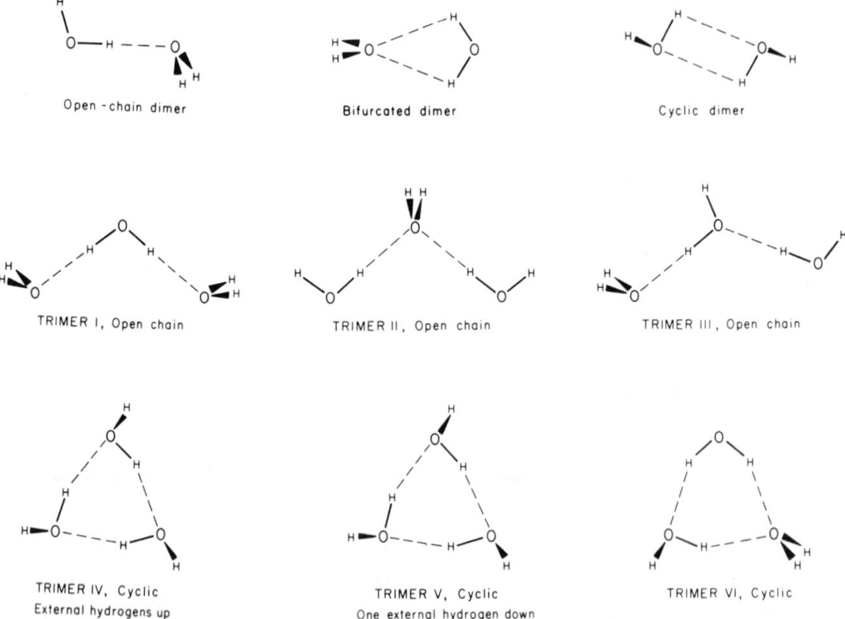

Fig. 3. Structures for polymers in steam. Dimers appear to occur only in the open-chain form. Of the open-chain trimers, form III appears more stable because in it the hydrogen bonds alternate, H—O...H—O...H—O. The evidence on the stability of cyclic trimers in comparison with the open-chain form is not clear, but trimer VI must be less stable because the bonds do not alternate in it.

tential energy has only one minimum corresponding to this form. (For more details see Chapter 3.) In the case of the higher polymers, they show that the repeated unit O—H...O—H...O—H, with a head-to-tail arrangement, is preferred over O—H...O...H—O or O...H—O—H...O, so that, in the nomenclature of Fig. 3, open-chain trimer III is more stable than I or II. On this ground alone, cyclic trimer VI is to be predicted as less stable than IV or V. Bolander *et al.*[99] point out that as ΔE is the potential energy, inclusion of the zero-point energy nearly accounts for the difference from the experimental value of ΔU. For comparison, the change in internal energy U on condensation is 5 kcal mol^{-1} [341,590] and the change in potential energy is about 6 kcal mol^{-1}. The significant feature to be taken from the theoretical calculations is that the distances and angles in vapor-phase polymers are not too far different from those in the ices.

2.2.2. *Volume of Liquid Water*

Many of the determinations of density at atmospheric pressure have been made in terms of the 1901 liter, one (1901) liter = 1.000028 dm^3. The liter, which will not be used here, was originally defined so that water at its maximum density at atmospheric pressure had a density of 1 kg liter^{-1}. Unfortunately, densities in g cm^{-3} and g ml^{-1} and specific volumes in cm^3 g^{-1} and ml g^{-1} are all near unity, which often leads to confusion.[740] In the liquid range, confusion between densities and specific volumes can be avoided by the use of SI units. Thus, the density of water at its maximum at 4°C is 999.972 kg m^{-3}, and its specific volume is $\bar{V} = 1.000028 \times 10^{-3}$ m^3 kg^{-1}.

a. Volume at Atmospheric Pressure. The most precise measurements of the density of water in the range 0–40°C, culminating extensive work at the end of the nineteenth century, appear to be those of Chappuis[180]; these measurements, precise to 1 ppm from 0 to 20°C, and with a scatter of about 5 ppm in the higher part of the range, were the basis of the table given by Tilton and Taylor.[1067] As far as can be reconstructed[742]—for isotopes had not yet been discovered—the water used by Chappuis had the maximum density 999.972 kg m^{-3} given in the preceding paragraph. The measurements of Thiesen *et al.*[1062] over the same range are nearly as precise; these authors and Chappuis agree well in the lower part of the range, as they must, since the density of water at its maximum was taken as unity, but Chappuis's densities are 7 ppm higher at 40°C. The International Critical Tables[1033] gave an arithmetic mean of the tables from this paper and from that of Chappuis. Bigg[83] reanalyzed both sets of data together and gave a table for this range. Good agreement and high precision were possible because the density was defined to be unity at the temperature of maximum density, near 4°C, which decreases the importance of knowledge of the isotopic composition. The work is of modern value because the temperature scale used was sufficiently close to the International Practical Scale of 1948 that uncertainties are said by Bigg to cause an error of less than 1 ppm at 40°C.

The thermal expansion between 50 and 100°C was measured by Thiesen(1061) to a few parts in 10^5, and from 45 to 85°C by Owen *et al.*(819) —their reproducibility was 0.2 ppm, but Kell and Whalley(594) suggested that their higher points were systematically in error by a few parts per million. Steckel and Szapiro(1013) measured the expansion of H_2O, as well as of waters of other compositions, up to 78°C. Kell and Whalley measured the thermal expansion from 80 to 150°C at 5 bars, and calculated the values at 1 atm using their measured compressibilites; their estimated error at 100°C is 10 ppm; above 100°C, their values for atmospheric pressure refer to the metastable superheated liquid. (It should be noted that some tables combine values for 1 atm below 100°C along with values of saturation at higher temperatures, without indicating the change.) A table covering the range −20 to 110°C at atmospheric pressure was given by Kell(588); in the range 0–40°C, agreement with the table of Bigg is to 1 ppm.

Measurements in the metastable region between 0 and −13°C, typically with a precision of a few parts in 10^5, have been reviewed by Kell.(588) In this range, the specific volume is nearly symmetric about the minimum volume at 4°C, the volume at $4 - x$ degrees being close to that at $4 + x$ degrees; this contradicts the conclusion of Mohler(759) that the specific volume rises much faster with temperatures decreasing below 4°C than with temperatures rising above 4°C. It is not clear whether measurements at lower temperatures(235,772,973–975,1199) relate to bulk water or show surface effects; it appears that at these lower temperatures, the volume does rise more rapidly with decreasing temperature. Extrapolation to lower temperatures, where vitreous ice could be expected, is uncertain.

The volumes of D_2O and T_2O, reviewed by Kell,(588) are shown in Fig. 4. As the temperature increases from 10 to 100°C, the molar volumes of H_2O, D_2O, and T_2O approach each other, but at higher temperatures, the values diverge again.

Mendeleef(743) represented the density of water from −10 to 200°C by a rational function with four constants

$$\bar{\varrho}/(\text{g ml}^{-1}) = 1 - [(t - t_m)^2/(A + t)(B - t)C]$$

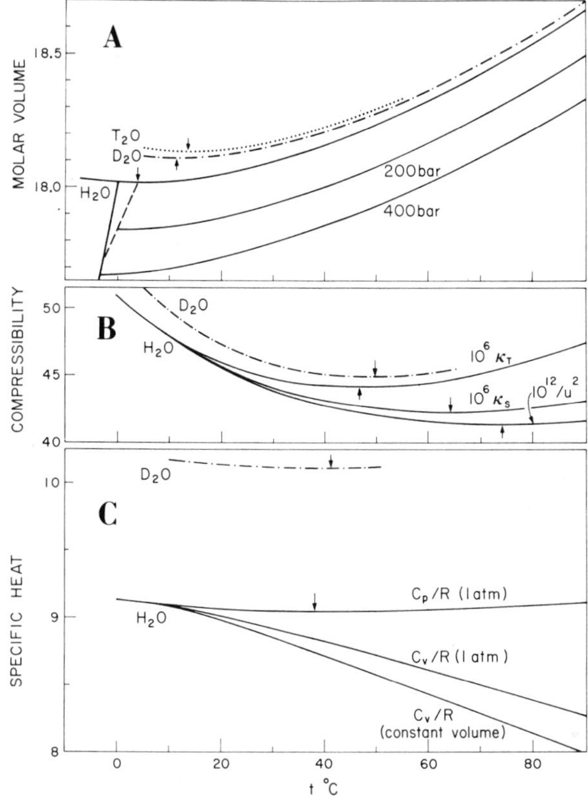

Fig. 4. Thermodynamic properties of liquid water. The vertical arrows indicate the temperatures of extrema. (A) The molar volume of water at atmospheric pressure and at low pressures. The dashed line at lower left shows the temperature of maximum density as a function of pressure. The heavy line to its left marks the freezing line to ice I*h*. (B) Functions related to the compressibility. The unit for β_S and β_T is bar^{-1}, and for $10^{12}/u^2$ it is $\text{cm}^{-2}\ \text{sec}^2$. (C) The specific heat at constant pressure C_P of H_2O at atmospheric pressure, the specific heat at constant volume C_V at atmospheric pressure, and C_V at a density of 1000.0 kg m^{-3}; the three coincide at the temperature of maximum density.

where t_m is the temperature of maximum density, and A, B, and C are adjusted to fit the data; this equation is still sometimes used.[636,772] Chappuis[180] used three cubic polynomials over the range 0–40°C. For the same range, Thiesen *et al.*[1062] used a rational function with four parameters

$$\bar{\varrho}/(\text{g ml}^{-1}) = 1 - [(t - t_m)^2(t + C)/B(t + D)]$$

TABLE II. Coefficients Representing the Density of Water, for the Rational Function Given by Eqn. (5)[a]

Coefficient, kg m^{-3}	H_2O	D_2O	$H_2{}^{18}O$	$D_2{}^{18}O$	T_2O
a_0	999.8396	1104.690	1112.333	1215.371	1212.93
a_1	18.224944	20.09315	13.92547	18.61961	11.7499
$10^3 a_2$	−7.922210	−9.24227	−8.81358	−10.70052	−11.612
$10^6 a_3$	−55.44846	−55.9509	−22.8730	−35.1257	—
$10^9 a_4$	149.7562	79.9512	—	—	—
$10^{12} a_5$	−393.2952	—	—	—	—
$10^3 b$	18.159725	17.96190	12.44953	15.08867	9.4144
Range of function, °C	0–150	3.5–100	1–79	3.5–72	5–54
Standard error, ppm	0.2–10	3	2	8	20
Estimated accuracy, ppm	0.4–20	10	50	100	200
Temperature of maximum density, °C	3.984	11.185	4.211	11.438	13.403
Maximum density, kg m^{-3}	999.972	1106.00	1112.49	1216.88	1215.01

[a] Ref. 588. Copyright American Chemical Society.

and this was the form employed by Tilton and Taylor.[(1067)] Kell[(588)] found the best fit to the data for H_2O from 0 to 150°C was given by a rational function with seven parameters,

$$\bar{\varrho} = (a_0 + a_1 t + a_2 t^2 + a_3 t^3 + a_4 t^4 + a_5 t^5)/(1 + bt) \tag{5}$$

with smaller numbers of parameters needed for the less extensive data of other isotopic forms; the coefficients obtained are given in Table II. The data for H_2O have been expressed in terms of Tchebychev polynomials by Gibson.[(377)]

b. Isotopic Variations. The volumes of H_2O and D_2O are approximately equal on a molar basis, as shown by Fig. 4, but laboratory and engineering work is usually done on a mass basis, so the isotopic composition of the water must be considered. There are two stable isotopes of hydrogen, ^1H and ^2H (D, or deuterium), as well as radioactive ^3H, (T, tritium), with half-life of 12.26 years, whose presence in waters of natural origin is negligible for thermodynamic purposes—the abundance of tritium in rainfall is related to thermonuclear testing in the atmosphere, and is at present about 2×10^{-16} mole fraction of hydrogen.[(434,950)] The atomic weights and abundances are shown in Table III.

TABLE III. Isotopic Masses and Natural Abundances[a]

	Weight	Natural abundance	
		Mole %	Weight %
H (1969)	1.008_0	—	—
^{1}H	1.007825	99.985	99.970
^{2}H (D)	2.01410	0.015	0.030
^{3}H (T)	3.01605	—	—
O (1969)	15.999_4	—	—
^{16}O	15.99491	99.759	99.732
^{17}O	16.99914	0.037	0.039
^{18}O	17.99916	0.204	0.229

$H_2O = 18.015_4$; $D_2O = 20.027_6$; $T_2O = 22.031_5$

[a] Masses are on the scale $^{12}C = 12.00000$. The 1969 atomic weights for hydrogen and oxygen indicate less precision than the 1961 values, as variations in the isotopic abundance of terrestrial samples limit the precision. The masses of the individual isotopes are known with greater precision.

The volumes of isotopic mixtures are calculated assuming additive molar volumes.[612] If N_i and V_i refer to the mole fraction and molar volume of the ith species, the molar volume of the solution is

$$V = \sum N_i V_i$$

and its density is

$$\varrho = (\sum N_i M_i)/V = (\sum N_i M_i)/\sum N_i V_i$$

where M_i is the molecular weight of the ith species, and $\sum N_i M_i$ is the molecular weight of the solution. As the volume is linear in the mole fraction, the density is not. The treatment when the isotopic compositions of both hydrogen and oxygen vary is given by Girard and Menaché[383,384] and Shatenshtein.[967]

A measurement of the density alone does not give the deuterium concentration, for enrichment processes change not only the hydrogen ratio $^2H/^1H$, but also, to a lesser extent, the oxygen ratios $^{17}O/^{16}O$ and $^{18}O/^{16}O$. Thus, after the deuterium concentration has been changed, it is necessary to *normalize* the composition of the oxygen isotopes, for example, by equilibration with natural CO_2.[1012] For a normalized binary solution of ^1H_2O and D_2O at 25°C and 1 atm, the mole fraction of D_2O is related to the density by

$$N_{D_2O} = (9.2735 \times 10^{-3}\, \Delta\bar{\varrho})/(1 - 3.29 \times 10^{-5}\, \Delta\bar{\varrho})$$

where $\Delta\bar{\varrho}$ (kg m^{-3}) $= \bar{\varrho} - \bar{\varrho}_{H_2O}$; that is, an increase of 1 ppm in density corresponds to an increase of 9.27 ppm in the mole fraction of deuterium. Thus, ordinary water, with 150 ppm mole fraction deuterium, is 16 ppm denser than ^1H_2O. According to Shatenshtein,[967] $^1H_2{}^{16}O$ is 262 ppm lighter than ordinary water with, in addition to the 16 ppm due to deuterium, 224 ppm due to ^{18}O and 22 ppm due to ^{17}O.

In ordinary water, the four major molecular species occur in the proportions[741]

$$^1H_2{}^{16}O/^1H_2{}^{18}O/^1H_2{}^{17}O/^1HD^{16}O = 997280/2000/400/320$$

The fractionation, both of the hydrogen and of the oxygen isotopes, during purification is usually much greater than the variation in natural sources; the density difference between the first and last fractions of a single distillation can be 20 ppm.[187]

Craig[217] introduced "standard mean ocean water" (SMOW) with a definite isotopic composition, and this will, no doubt, come to be the composition intended for "ordinary" water. Cox *et al.*[216] have proposed that water for the most precise uses should be that from 2000 m in the western Mediterranean Sea, triply distilled by a rigidly prescribed procedure which produces no significant fractionation in the hydrogen or oxygen isotopes.

c. Compressibility at Atmospheric Pressure. Like the volume at atmospheric pressure, the isothermal compressibility $\varkappa_T$,

$$\varkappa_T = -[\partial(\ln V)/\partial P]_T = [\partial(\ln \varrho)/\partial P]_T$$

goes through a minimum, in this case near 46.5°C. Direct determinations of volume changes, whether caused by temperature or pressure, are made relative to a vessel, and the volume change of the vessel must then be determined in a separate experiment. Among the better direct determinations,[261,594,752] the differences, of about 0.1×10^{-6} bar^{-1}, in $\varkappa_T$ probably arise primarily from errors in the determination of the compressibility of the vessel, which will be about 3×10^{-6} bar^{-1} for glass and 0.7×10^{-6} bar^{-1} for steel.

The isothermal compressibility of fluids can also be obtained from the velocity of sound u by the relation

$$\varkappa_T = (V/u^2) + (T/VC_P)(\partial V/\partial T)_P{}^2$$

The velocity of sound itself can be determined with high precision, but for most fluids, the errors in $(\partial V/\partial T)_P{}^2/VC_P$ are greater than the errors in the determination of the compressibility of a vessel, and the direct method is to be preferred. However, as $(\partial V/\partial T)_P$ is zero at the temperature of maximum density, the velocity of sound is the preferred method for water at temperatures not too far removed from the density maximum. There are four determinations of the velocity of sound in ordinary water that agree well and cover the range from 0 to 75°C.(52,164,252,718) The earlier three of these, along with some higher-temperature direct values,(594) have been analyzed by Kell,(589) who gives the equation

$$\begin{aligned} 10^6\varkappa_T\ (\text{bar}^{-1}) = {} & (50.88630 + 0.7171582t \\ & + 0.7819867 \times 10^{-3}t^2 + 31.62214 \times 10^{-6}t^3 \\ & - 0.1323594 \times 10^{-6}t^4 + 0.6345750 \times 10^{-9}t^5) \\ & \div (1 + 21.65928 \times 10^{-3}t) \end{aligned} \tag{6}$$

with a standard error of 0.0021×10^{-6} bar^{-1}. This standard error is low enough that in this temperature range, water is suitable for the calibration of the compressibility of pressure vessels.

The isothermal compressibility $\varkappa_T$ has its minimum at 46.5°C, but it should not be inferred that this temperature has some deep significance in properties of water, for simple transformations give other functions of equivalent physical content with extrema at other temperatures. Thus, for the velocity of sound u itself, i.e., for $(\partial\varrho/\partial P)_S$, the extremum is at 74°C, for the isentropic compressibility $\varkappa_S = -[\partial(\ln V)/\partial P]_S$, it is 64°C, and for $(\partial V/\partial P)_T$, it is 42.3°C, all of these differing from the 46.5°C for $\varkappa_T$. The existence of these temperatures is related to the maximum density, but it is clear that there is no single temperature above which water is a *normal* liquid and below which it is peculiar.

The isothermal compressibility of D_2O was determined from 5 to 65°C by Millero and Lepple,(754) who gave

$$\begin{aligned} 10^6\varkappa_T\ (\text{bar}^{-1}) = {} & 53.61 - 0.4717t + 0.9703 \times 10^{-2}t^2 \\ & - 1.015 \times 10^{-4}t^3 + 0.5299 \times 10^{-6}t^4 \end{aligned}$$

with a standard error of 0.02×10^{-6} bar^{-1}. As Fig. 4 shows, the compressibility of D_2O is larger than that of H_2O in about the same pattern as its volume is larger, that is, at lower temperatures, where $V(D_2O)$ is markedly larger than $V(H_2O)$, $\beta(D_2O)$ is markedly larger than $\beta(H_2O)$.

d. Temperature of Maximum Density. At atmospheric pressure, the temperature of maximum density, that is, the temperature at which

$$(\partial \varrho/\partial T)_P = 0$$

is 3.98°C. At higher pressures, the temperature of maximum density moves to lower temperatures, with the dependence given by

$$\left(\frac{\partial T}{\partial P}\right)_{\varrho \max} = -\frac{[(\partial/\partial T)(\partial \varrho/\partial P)_T]_P}{(\partial^2 \varrho/\partial T^2)_P}$$

Kell[589] evaluated this from the temperature dependence of the compressibility and of the thermal expansion, and found $(\partial T/\partial P)_{\varrho \max} = -0.0200 \pm 0.0003$ deg bar^{-1}. This line reaches the ice–liquid equilibrium line at -4°C and 400 bars. This slope disagrees with that reported by Bridgman,[116] but agrees with the data reviewed by Tammann and Schwarzkopf.[1053] For D_2O—volumes from Kell[588] and compressibilities from Millero and Lepple[754]—the corresponding value is -0.0178 deg bar^{-1}. The position of this line is a measure of the peculiarity of liquid water, and with D_2O, this is not only higher at atmospheric pressure, 11.44°C versus 3.98°C, but does not move to lower temperatures as quickly with increasing pressure.

e. PVT Relations. As knowledge of densities is a prerequisite for most other studies of water, many measurements of liquid densities are available The precise ones published by Amagat[11] in 1893 remain among the best available, as his scales of pressure and temperature were close to the present ones. Other studies include those by Bridgman,[116,117] Smith and Keyes,[1001] Kennedy *et al.*,[521,596] Kell and Whalley,[594] Grindley and Lind,[428] Wilson and Bradley,[1176] Franck *et al.*,[630,723] Vukalovich *et al.*,[1121,1122] Vedam and Holton,[1102] and Burnham *et al.*[154] Of these, the work of Kell and Whalley, 0–1 kbar, 0–150°C, appears the most precise, and that of Burnham *et al.*, 1–8.9 kbar, 20–900°C, appears the most extensive. Measurements are also available on D_2O.[728]

Figure 5 shows the stable region at temperatures up to the critical point. Three general areas may be distinguished. At lower pressures, near the critical point, the behavior is essentially like that of other materials. At high pressures, e.g., at 10 kbar, the behavior is also very regular. It is at low temperatures and pressures that the peculiar features of water, such as

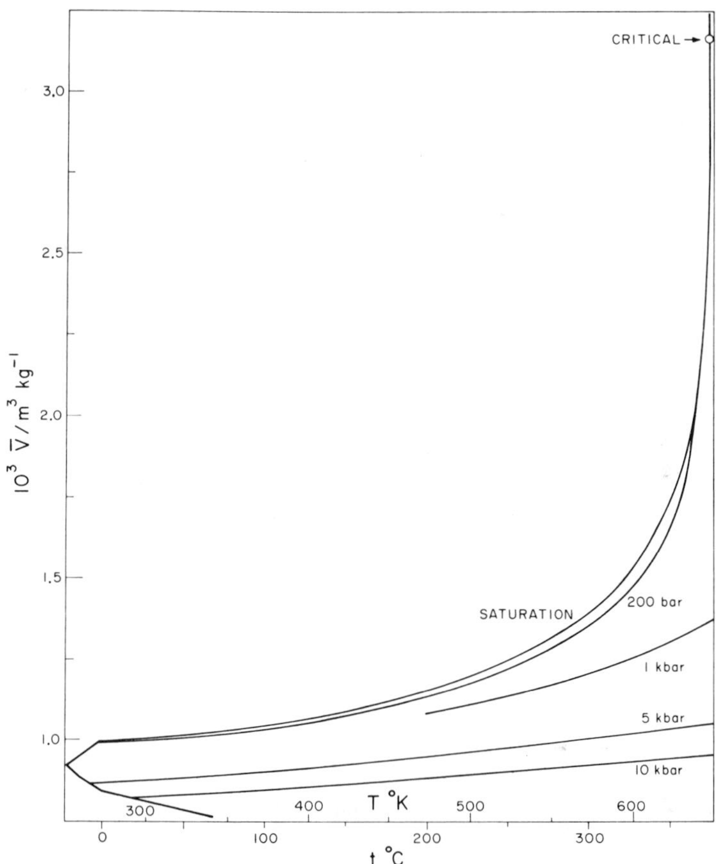

Fig. 5. Volume of liquid H_2O. Here, the range of densities and temperatures is greater. The field ends at the lower left at the freezing curves of the various ices. It is only at low temperatures and low pressures that water is strikingly different from other liquids.

the minimum compressibility and maximum density, are found. The *PVT* relations of this region are indicated in Fig. 4(a). At these lower pressures, the liquid can afford the greater volume entailed by less deformation of the intermolecular angles.

The most striking feature of the *PVT* properties in the low-pressure region is not apparent from data on the liquid alone, but can be seen in comparison with the ices. Liquid water is 10% denser than ordinary ice I*h*, which means that the mean oxygen–oxygen distance is smaller. However, X-ray diffraction(784) and the infrared spectrum(1132) agree that the oxygen–

oxygen distance to the nearest neighbor is larger in the liquid. Hence, the distances to higher neighbors must be smaller, and this is likely the case for second neighbors, implying deformation of oxygen–oxygen–oxygen angles. It is the balance between first-neighbor distances increasing with temperature, which would increase the volume, and angular deformation increasing with temperature, which would decrease the volume, that produces the maximum density and the minimum compressibility.

f. Equations of State. Many equations have been proposed to represent the isothermal behavior of liquid water, but only a few can be mentioned here. The observations can be represented rather closely by any of a number of equations with three parameters, one of which is usually the volume at low pressure. One such equation,

$$(V_0 - V)/V_0P = A/(B + P) \tag{7}$$

where V_0 is the volume at zero pressure (sometimes approximated by the volume at atmospheric pressure), and A and B are adjusted to fit the data, was published by Tait[1044,1046] in 1889. This equation is identical with[1045,1047]

$$(P + B)(V - V_\infty) = C \tag{8}$$

where $C = ABV_0$, $V_\infty = V_0(1 - A)$, and V_∞ would be the volume if the equation held at infinite pressure. Equation (8) is often called the Tumlirz equation,[1077] though the form used by Tumlirz had DT in place of C. Equation (8) was used for water by Eckart.[284] The equation that bears Tait's name,

$$(-1/V_0)(\partial V/\partial P) = A/(B + P)$$

or

$$V = V_0\{1 - A \ln[(B + P)/B]\} \tag{9}$$

where A and B differ from the values in eqn. (7) for the ordinary range of data, was proposed by Tammann[1050] in 1895 and was named the Tait equation by him. Equation (9) has been widely used for a number of liquids, including water,[681] but Hayward[479,480] finds that eqn. (7) is to be preferred.

The representation of the isothermal equation of state by a power series in the pressure,

$$V = V_0 + a_1P + a_2P^2 + \cdots$$

is less satisfactory, as the coefficients have alternating signs and many terms are needed to represent data over a wide range. I have noted that if P is replaced by $z = P/(\Pi + P)$, where Π is given a fixed value of 5–10 kbar, then the series $V = V_0 + a_1'z + a_2'z^2 + \cdots$, where the first two terms are equivalent to eqn. (7), converges well and does not have alternating signs. Alternatively, the pressure P could be represented as a series in V or in $V - V_\infty$. In practice, a compromise must be made between, on the one hand, precision and convenience of the representation and, on the other, intelligibility of the coefficients and of their temperature dependence.(713) There does not appear to be a simple isothermal equation that represents both vapor and liquid adequately.

Many equations of state have been used to represent volumes or the other thermodynamic properties as a function of both temperature and pressure. Bain(44) has described the use of orthogonal functions. The formulations of the Sixth International Conference on the Properties of Steam(547,548) were mentioned in the introduction. Gibson and Bruges(380) and Walker(1130) deal with the region from 0 to 150°C and to 1 kbar.

In our laboratory some years ago, we represented the volume data of Kell and Whalley(594) over this range by the equation

$$V = V_A(1 + a_1x + a_2x^2)/(1 + bx)$$

where V_A is the volume of water at atmospheric pressure (1.013 bars) as obtained from the function for H_2O given by eqn. (5) using the coefficients of Table II, $x = P - 1.013$, where P is the pressure in bars, and a_1, a_2, and b are functions of the IPTS-48 temperature t given by

$$\begin{aligned}10^6a_1 &= 55.71454 + 2.262165t - 1.391722 \times 10^{-2}t^2 \\ &\quad - 2.120850 \times 10^{-4}t^3 + 3.453244 \times 10^{-6}t^4 \\ &\quad - 1.103666 \times 10^{-8}t^5\end{aligned}$$

$$\begin{aligned}10^9a_2 &= 2.656276 - 1.603340 \times 10^{-1}t + 2.690604 \times 10^{-3}t^2 \\ &\quad - 1.910151 \times 10^{-5}t^3 + 4.074224 \times 10^{-8}t^4\end{aligned}$$

$$\begin{aligned}10^6b &= 106.6390 + 1.895276t - 6.887601 \times 10^{-3}t^2 \\ &\quad - 2.745977 \times 10^{-4}t^3 + 3.790356 \times 10^{-6}t^4 \\ &\quad - 1.171620 \times 10^{-8}t^5\end{aligned}$$

The form of this function was chosen because it provides a good fit, is close enough in form to eqn. (7) that it can probably be extrapolated some distance with confidence, and can be integrated in closed form to give the change in Gibbs energy with pressure. It has not been adjusted to bring the 1-atm compressibilities in line with eqn. (6).

An equation of state is used to calculate each steam table, and that used by Keenan *et al.*[586,587] is a single equation to represent the Helmholtz energy over the whole range, both liquid and vapor, with a total of 55 adjustable constants. It contains unsatisfactory features in the critical region, as will be noted below, but is otherwise to be considered the best comprehensive equation of state ever established for any substance.

Selected values of the equilibrium properties at atmospheric pressure are given in Table IV, and volumes under pressure are given in Table V.

TABLE IV. Thermodynamic Properties of Ordinary Water at Atmospheric Pressure[a]

t_{68}, °C	t_{48}, °C	Vapor pressure, bar	Density, kg m^{-3}	Isothermal compressibility, bar^{-1}	Isobaric specific heat, J g^{-1}
0	0.0000	0.00611	999.840	50.886	4.2174
5	5.0023	0.00872	999.964	49.170	4.2019
10	10.0043	0.01228	999.700	47.810	4.1919
15	15.0060	0.01705	999.100	46.735	4.1855
20	20.0074	0.02338	998.204	45.894	4.1816
25	25.0085	0.03169	997.045	45.249	4.1793
30	30.0094	0.04245	995.648	44.773	4.1782
35	35.0100	0.05627	994.032	44.443	4.1779
40	40.0104	0.07382	992.215	44.243	4.1783
45	45.0105	0.09591	990.212	44.158	4.1792
50	50.0104	0.12346	988.033	44.179	4.1804
55	55.0101	0.15754	985.691	44.297	4.1821
60	60.0096	0.19936	983.193	44.504	4.1841
65	65.0089	0.25027	980.548	44.797	4.1865
70	70.0081	0.31181	977.761	45.171	4.1893
75	75.0071	0.38569	974.839	45.622	4.1925
80	80.0059	0.47379	971.788	46.149	4.1961
85	85.0046	0.57821	968.610	46.751	4.2002
90	90.0032	0.70123	965.311	47.428	4.2048
95	95.0016	0.84534	961.892	48.180	4.2100
100	100.0000	1.01325	958.357	49.007	4.2157

[a] Temperatures are on the International Practical Temperature Scale of 1968, with corresponding values on IPTS-48 indicated. To the precision given, only vapor pressure and density change with the change in temperature scale. 1 atm ≡ 1.01325 bar. Sources: Temperature scales, Bedford and Kirby.[59] Vapor pressure, Bridgeman and Aldrich.[113] Density, Kell.[588] Isothermal compressibility, Kell.[589] Isobaric specific heat, de Haas.[247]

TABLE V. Specific Volume $\bar{V}$ of Ordinary Water under Pressure[a]

P, kbar	t, °C								
	0	25	50	75	100	125	150	175	200
0.5	0.97666	0.98185	0.99135	1.00421	1.02008	1.03887	1.0607	1.0860	1.1149
1.0	0.95656	0.96344	0.97321	0.98552	1.00016	1.0171	1.0363	1.0578	1.0826
1.5	0.9385	0.9471	0.9572	0.9696	0.9839	0.9995	1.0174	1.0366	1.0588
2.0	0.9241	0.9341	0.9447	0.9561	0.9681	0.9822	0.9989	1.0166	1.0368
2.5	0.9117	0.9218	0.9321	0.9430	0.9544	0.9675	0.9824	0.9987	1.0172
3.0	0.9005	0.9106	0.9209	0.9317	0.9426	0.9550	0.9689	0.9839	1.0004
3.5	0.8901	0.9002	0.9105	0.9212	0.9320	0.9440	0.9572	0.9707	0.9860
4.0	0.8805	0.8906	0.9009	0.9114	0.9220	0.9338	0.9465	0.9595	0.9733
4.5	0.8715	0.8815	0.8918	0.9022	0.9128	0.9241	0.9361	0.9485	0.9617
5.0	0.8627	0.8727	0.8829	0.8933	0.9038	0.9148	0.9264	0.9385	0.9511
5.5	0.8551	0.8649	0.8749	0.8851	0.8952	0.9059	0.9171	0.9289	0.9410
6.0	0.8481	0.8577	0.8675	0.8773	0.8872	0.8976	0.9084	0.9199	0.9316
6.5	(0.8417)	0.8510	0.8605	0.8701	0.8798	0.8901	0.9004	0.9116	0.9229
7.0	—	0.8451	0.8543	0.8636	0.8730	0.8829	0.8929	0.9037	0.9147
7.5	—	0.8393	0.8483	0.8574	0.8666	0.8772	0.8871	0.8971	0.9071
8.0	—	0.8339	0.8427	0.8516	0.8606	0.8709	0.8805	0.8902	0.8999

[a] The values are based primarily on Kell and Whalley[594] and Burnham *et al.*[154] and have been smoothed to avoid discontinuities at the edges of data sets. These data agree with those for 1 atm in Table IV. The table gives values of $10^3\bar{V}$ in $m^3\ kg^{-1}$.

2.3. Thermal Properties

A thermodynamic description of a substance involves a thermal quantity, such as the enthalpy, as well as the volume. As the enthalpy is needed in the design of steam plants, many measurements have been made, and steam tables give values of this quantity. In the present chapter, specific heat will be considered, either at constant pressure, $C_P = (\partial H/\partial T)_P$, or at constant volume, $C_V = (\partial U/\partial T)_V$. C_V is always less than C_P, and the relationship between them involves only the isothermal compressibility and the isobaric thermal expansion, that is, only volume properties:

$$C_P = C_V + T\frac{(\partial V/\partial T)_P{}^2}{(\partial V/\partial P)_T}$$

Isothermal changes of enthalpy with pressure, or of internal energy with volume, like isothermal changes of any thermodynamic quantity, also depend only on volume properties:

$$\begin{aligned} dH &= C_P\,dT + [V - T(\partial V/\partial T)_P]\,dP \\ dU &= C_V\,dT + [T(\partial P/\partial T)_V - P]\,dV \\ (\partial C_P/\partial P)_T &= -T(\partial^2 V/\partial T^2)_P \\ (\partial C_V/\partial V)_T &= T(\partial^2 P/\partial T^2)_V \end{aligned} \tag{10}$$

It is thus possible to combine (a) a calculation of the enthalpy or specific heat in the dilute vapor by statistical methods applied to spectroscopic data, with (b) a calculation of the change of enthalpy from the equations of state of vapor and liquid, plus (c) measurements of the enthalpy of phase change, to arrive at the values of the enthalpy at high pressure. This may be preferable to direct measurement in some cases. However, it is possible to seek an explanation of the thermal properties of the liquid without reference to the vapor.

The most highly regarded enthalpy measurements of fluid water at saturation are those undertaken at the U.S. Bureau of Standards in the 1930's.(810,813–815) A recent description of the calorimeters is available.(382) The most extensive measurements of C_P are those of Sirota, which will be treated below. A review of determinations of C_P and C_V has been given by Bruges.(132)

A great number of units, often with rather similar names, have been used as units of heat, and the values have, on occasion, varied with time and country.(269,1028) The recommended unit is now the absolute joule, and this is meant here by the word "joule." Before 1948, measurements based on electrical units were in international joules, which, despite the name, varied from country to country. In 1948, the U.S. international joule was 1.000165 absolute joules, but in 1930, it was considered 1.0004 absolute joules, and at that time, there were differences of 0.01% in the international joule as realized in different national laboratories.(269) Many measurements have also been reported in calories, and at least four calories may be distinguished. The IT (*I*nternational Steam *T*able) calorie was defined for early steam tables as one (international)kilowatt hour = 860 IT calories. The values of the heat units are shown in Table VI. The situation was sufficiently confused that Dorsey did not try to reduce heat measurements to a common unit.

2.3.1. *Vapor*

The energy or specific heat of dilute steam may easily be calculated at not too low temperatures in the approximation that the molecule is a har-

TABLE VI. Units of Heat

1 international (U.S. 1948) joule	= 1.000165 absolute joules
1 (thermochemical) calorie	= 4.184 (exactly) absolute joules
1 (15°C) calorie	= 4.18580 absolute joules
1 (20°C) calorie	= 4.18190 absolute joules
1 (mean) calorie	= 4.19002 absolute joules
1 (IT) calorie	= 4.1867 absolute joules
	or 4.1868 absolute joules

monic oscillator. There are contributions from the translational, rotational, and vibrational degrees of freedom.[498] The three translational degrees of freedom give

$$U_{\text{tr}} = \tfrac{3}{2}RT$$

Water is nonlinear, so there are three rotational degrees of freedom, which give, classically,

$$U_{\text{rot}} = \tfrac{3}{2}RT$$

The vibrational energy may be summed to give

$$U_{\text{vib}} = \tfrac{1}{2}N_A hc \sum \nu + RT \sum [\xi/(e^{\xi} - 1)] \tag{11}$$

where N_A is Avogadro's number, h is Planck's constant, c is the velocity of light, ν is the frequency in cm^{-1}, $\xi = hc\nu/kT$, where k is Boltzmann's constant, and the summations are over the three internal vibrations of the molecule, which are at $3657\ \text{cm}^{-1}$, $1595\ \text{cm}^{-1}$, and $3756\ \text{cm}^{-1}$ for H_2O. The total U_{vapor}, relative to the energy of the molecule without motion, is given by the sum

$$\begin{aligned} U_{\text{vapor}} &= U_{\text{tr}} + U_{\text{rot}} + U_{\text{vib}} \\ &= 3RT + U_{\text{vib}} \end{aligned}$$

Also,

$$\begin{aligned} H_{\text{vapor}} &= U_{\text{vapor}} + RT \\ &= 4RT + U_{\text{vib}} \end{aligned}$$

and

$$C_V = 3R + R \sum [\xi^2 e^{\xi}/(e^{\xi} - 1)^2]$$

$$C_P = 4R + R \sum [\xi^2 e^{\xi}/(e^{\xi} - 1)^2]$$

In D_2O, all three frequencies shift downward because of the increased mass, and $\frac{1}{2}N_A hc \sum \nu$ therefore decreases, but the terms $RT \sum [\xi/(e^\xi - 1)]$ and $R \sum [\xi^2 e^\xi/(e^\xi - 1)^2]$ are greater. (This simple picture of the specific heat will also be used below in the discussion of the liquid.) By 1934, Gordon[411] had removed many of the approximations of the model just given. A full calculation of the various isotopic species of water, considering low- temperature rotational corrections, and considering the molecules as nonrigid, rotating, anharmonic oscillators has been given by Friedman and Haar[367] and McBride and Gordon.[708]

Measurements include those of Keyes *et al.*,[603] Havlicek and Miskovský,[476] Vukalovich *et al.*,[1123,1124] and Angus and Newitt.[18] Those of Sirota *et al.*,[984–989,994] and Amirkhanov and Kerimov[13,14] have mostly been at higher temperatures and pressures, including much work in the supercritical region. It should be noted that in the supercritical region, Sirota finds a maximum in C_P—corresponding to a maximum in $(\partial V/\partial T)_P$—indicating that, although the saturation line ends at the critical point, the locus of the maximum of C_P represents an extension into the supercritical region that continues to divide the T–P plane into a gaslike and a liquidlike region.[990,992] Amirkhanov and Kerimov found maxima in C_V giving a similar locus, though slightly different from the one of C_P.

2.3.2. *Specific Heat of the Liquid at Atmospheric Pressure*

Before electrical measurements became usual, heat capacity measurements were made in terms of the specific heat of water at atmospheric pressure. At present, the temperature dependence of C_P at atmospheric pressure is taken from the work of Osborne *et al.*[814] of the U.S. Bureau of Standards. The best values of C_P at 1 atm are given by

$$C_P(1\text{ atm}) = 4.1855\{0.996185 + 0.0002874[(t + 100)/100]^{5.26} + 0.011160 \times 10^{-0.036t}\} \qquad (12)$$

In this equation the expression in curly brackets gives the temperature dependence found by NBS. The leading multiplicative constant, the specific heat at 15°C, was derived by de Haas[247] (*cf.* Stimson[1028]) on behalf of the International Bureau of Weights and Measures from a consideration of the best determinations at 15°C.[557,646,814]

In a solid or in a low-density vapor, thermodynamic functions such as specific heat are most easily understood at constant volume, for the functions at constant pressure contain a contribution due to the work of expanding the system against its own internal forces as well as the work of expansion against the external pressure. In fluids at higher densities, thermodynamic functions at constant volume need not be simple. In water, the distributions of oxygen–oxygen distances on one hand, and the angular correlations given by the distribution of oxygen–oxygen–oxygen angles on the other, can vary independently, so that the configuration can vary on a molecular level even at constant volume. It is sometimes convenient to visualize a vitreous form of water—whose properties can, though not readily in practice, be observed at low temperatures[17,590] or high frequencies—in which angular correlations do not change with temperature and for which properties at constant volume are intelligible in terms of simple statistical mechanical models.

This vitreous form has been called "vibrational,"[590] or "glass,"[238] and the additional part needed to make up the observed properties has been called "potential,"[590] "relaxational,"[238] or "configurational."[301] The vibrational part can be understood by eqn. (11), applied now to all vibrations, and the configurational part is the difference between eqn. (11) and experiment. In the liquid at room temperature, the translational and rotational and rotational degrees of freedom of the vapor are replaced by three hindered rotations ν_R with a maximum at about 685 cm^{-1} in the infrared—this band is particularly broad—and three hindered translations ν_T with a maximum at about 190 cm^{-1}.[273] Ford and Falk[341] considered that these values overweight the high frequencies and that it is preferable to use $\nu_R = 450$ cm^{-1} and $\nu_T = 115$ cm^{-1} in Einstein calculations; these lower values were used in the calculations that follow. The description of changes in the vibrational spectrum of liquid water with change of temperature is a subject of sharp controversy. Some, illustrated by work found elsewhere in this volume, see the spectrum as composed of a number of components whose relative intensity changes with temperature; the change of configurational energy with temperature is then considered as produced by change in the proportions of the species present in the liquid. I prefer to regard the shift of the absorption maxima as typical of vibrations in the bands. In this case, eqn. (11) suggests two contributions to the specific heat:

$$C_{V,\mathrm{vib}} = \tfrac{1}{2}N_A hc \sum (\partial \nu/\partial T) + R \sum [\xi^2 e^\xi/(e^\xi - 1)^2] \tag{13}$$

the first arising from the shift in zero-point energy because of the shift in frequency, and the second being the Einstein specific heat of harmonic oscillators. Summation is over the three internal vibrations, the three hindered rotations, and the three hindered translations. If the vibrations did not shift in frequency and were all classically excited, C_V/R would be 9. However, the second term of eqn. (13) gives $C_V/R \approx 0.00$ for each of the two internal vibrations near 3300 cm^{-1}, ~0.01 for the vibration near 1650 cm^{-1}; ν_R gives ~0.65 for each vibration, and ν_T gives ~0.97. That is, the internal vibrations are scarcely excited, ν_T is practically classically excited, and only the contribution from ν_R changes significantly with temperature or pressure in the liquid. The sum of all nine vibrations for this second term gives $C_V/R \approx 4.87$, leaving a contribution of 4.13 to be explained by the change of zero-point energy (the ν_R band appears to be the one that shifts most rapidly with temperature) and the change of configurational energy. It seems highly probable that, no matter what model is chosen, the change of configurational energy with temperature is difficult to understand at present.

The specific heat C_P of D_2O is larger than that of H_2O on a mass basis,[(191)] and hence more than 10% greater on a mole basis, as can be seen from Fig. 4. In D_2O, the ν_R vibrations appear to lie nearly 100 cm^{-1} lower than in H_2O, and the other frequencies are lower, too. This effect, due to greater isotopic mass, only accounts for an increase of 3% in the specific heat of D_2O, so the higher specific heat of D_2O largely represents increased configurational energy.

2.3.3. *Specific Heat under Pressure*

The specific heat falls with increasing pressure by an amount that can be calculated by eqn. (10). Hence, relatively few measurements have been made of the liquid under pressure at lower temperatures, and most values have been obtained from equations of state based on volume measurements. (It has been recognized that in the critical region, equations of state present difficulties, and that independent measurements are needed there.) Among the direct measurements are those by Koch[(616)] for H_2O and Rivkin and Egorov[(907)] for D_2O. The ratio of heat capacities $C_P(D_2O)/C_P(H_2O)$ is a function of temperature but not of pressure to 800 bars in the range investigated. It also appears[(729)] that C_V becomes almost independent of temperature at 800 bars, though these calculations are not too precise.

Sirota and Shrago[(991,993)] have shown that at 0°C, the values of ΔC_P calculated from *PVT* measurements are discordant, and suggest that calorimetric measurements would be desirable. Their finding possibly reflects no more than the difficulty of twice differentiating the volume with respect to the thermodynamic temperature, a difficulty that is partly mathematical because of the large temperature intervals between observations, and partly due to experimental scatter.

The ν_R band, which will give information about the contributions to the specific heat, has not yet been measured under pressure.

2.4. Liquid–Vapor Equilibria

At room temperature, water vapor and the liquid are clearly different phases, but they are related by the fact that a system that is clearly vapor can be taken to one that is clearly liquid by paths around the critical point that do not involve any phase transition. A fluid, on the other hand, cannot transform to a crystal without a phase change. Further, which crystalline phase nucleates to establish equilibrium with a fluid is partly subject to chance, but there appears to be only one vapor and one liquid for water (if reports of polywater[257,691] can be excluded) so that the liquid–vapor equilibrium properties are well-defined.

As the temperature increases toward the critical point, the latent heat, surface tension, and the difference in liquid and vapor densities—the maximum density of the liquid being unimportant in this regard—all decrease, and the vapor pressure rises. As will be clear from Section 2.5 on the critical point, most efforts to fit equations to include the critical region that were made before 1965 are no longer theoretically acceptable because it was not realized that the Helmholtz function is nonanalytic at the critical point.

2.4.1. *Latent Heat*

The measurements made at the Bureau of Standards[813–815] appear most reliable over the whole range from 0°C to the critical temperature, but the published functional representations are not satisfactory near 100°C. Using Tchebychev polynomials, Gibson and Bruges[379] have presented a systematic treatment of the saturation properties using the NBS and more recent data.

2.4.2. *Vapor Pressure*

Gibson and Bruges[379] included the vapor, or saturation, pressure in their treatment. The vapor pressure has been represented by a rational function of the temperature by Tanishita and Nagashima.[1055] Using modified Tchebychev polynomials, Bridgeman and Aldrich[113,114] also gave an expression for the vapor pressure of ordinary water:

$$\log_{10} P\ (\text{bar}) = A + Y_1 - B(1 + Ct)Y_2 \qquad (14)$$

where $A = 1.06994980$, $Y_1 = 4.16385282(t - 187)/(t + 237.098157)$, $B = 1.0137921$, $C = 5.83531 \times 10^{-4}$, and

$$Y_2 = [3\sqrt{3}/(2 \times 1.87^3)][0.01(t-187) - 0.01\alpha]\{187^2 - [0.01(t - 187) - 0.01\alpha]^2\}$$

$$\alpha = Z^2(187^2 - Z^2)/[0.30231574(1 + 3.377565 \times 10^{-3}t)]$$

$$Z = -1.87 + 3.74\{1.152894 - 0.745794 \cosh^{-1}[654.2906/(t + 266.778]\}$$

and t is the IPTS-48 Celsius temperature. This equation represents the vapor pressure and its derivative $(\partial P/\partial T)_{\text{sat}}$ with high precision from 0°C to nearly critical, covering a wider range than previous equations by Osborne and Meyers,(811) Gerry,* and Goff and Gratch,(405) and is based on the experimental values of Osborne *et al.*(812) It is now recognized that the vapor pressure curve shows a singularity at the critical point which none of these equations can accommodate. There are other experimental papers.(746,766,1002,1029)

The vapor pressure is, perhaps, the best-investigated property of waters of other isotopic compositions.(881) Data are available on D_2O,(679,749,805,903) T_2O,(565,859,860,872) $H_2{}^{18}O$,(558,904,1041,1081) and $D_2{}^{18}O$, $D_2{}^{16}O$, and $H_2{}^{17}O$.(1041) The vapor pressure of H_2O–D_2O mixtures, i.e., the vapor pressure of HDO, was found by Zieborak(1201) to be slightly higher than would be the case for an ideal solution.

It is convenient to describe the vapor pressure of the isotopic waters in terms of the partition coefficient α, illustrated in Fig. 6, defined by

$$\alpha = P_X/P_{H_2O} \tag{15}$$

where P_X is the vapor pressure of the heavier isotopic molecule. At low temperatures, α is less than unity, which accounts for the fact that the oceans are richer in heavier isotopes than is rain. This is the normal effect. For some substances, including water, α becomes greater than unity at higher temperatures, which is the inverse effect. A theory, given by Bigeleisen(82) and extended by other authors,(527,1179) calculates the change in free energy produced by shifts in the vibration frequencies. However, the greater part of the effect is due to the shifts in the libration bands ν_R, which, as already mentioned in connection with the specific heat, are broad and difficult to locate precisely. Hence, while the theory seems qualitatively correct, it is not yet satisfactorily exact. O'Neill and Adami(807) have studied the oxygen

* H. T. Gerry, quoted in Ref. 815.

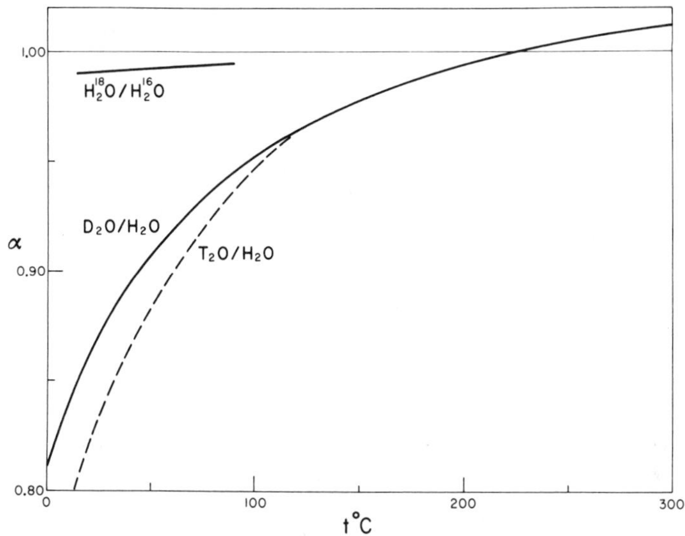

Fig. 6. Partition coefficient α for isotopic waters. At low temperatures, where α is much less than unity, the composition of D_2O–H_2O and T_2O–H_2O mixtures changes strikingly on distillation. Sources of data: D_2O/H_2O, Whalley(1162); T_2O–H_2O, Jones(565); $H_2{}^{18}O/H_2{}^{16}O$, Szapiro and Steckel.(1041)

isotope partition function for water and conclude that their results are incompatible with many mixture theories of the structure of water.

2.4.3. *Surface Tension*

Below 100°C, surface tension usually refers to the interface between water, saturated with air, and air, saturated with water, though the air can be replaced by other gases, or pure vapor alone can be present with little change. Above 100°C, tabulations are for an interface between water and steam at their saturation pressure. All the experimental methods are difficult,(461) as impurities in the parts per billion range can affect the values. A number of corrections must be made, and these are often inadequately reported. The consistent results obtained by Gittens(386) by two methods, between 5 and 45°C, appear more reliable than those of Moser(765) or Teitel'baum *et al.*(1058) and are slightly higher than those results. Caskey and Barlage(171) have shown that at 25°C, the surface tension does not differ significantly from its equilibrium value when the surface is as young as 0.05 sec. Hacker(441) has measured the surface tension down to −22°C; it follows smoothly from the values at higher temperatures. The surface tension falls approximately linearly with increasing temperature from a value of 75.6 dyn cm^{-1} at 0°C to zero at the critical point. Grigull and Bach[426] have given an equation for the values over the whole range. Claussen[188] has shown that, rather than the surface tension σ itself, the function $\sigma/\varrho^{2/3}$ gives a straight line against the temperature from 0 to 50°C.

The surface tension of D_2O is about 0.1% lower than that of H_2O at 25°C,[562] and varies linearly with density in mixtures. Costello and Bowden[208] and Heiks *et al.*[486] have measured the temperature dependence of the surface tension of D_2O. With increasing temperature, the value for D_2O falls faster than that for H_2O, and it is several per cent lower at 200°C.

2.5. Critical Point

The critical parameters of water are considered in a subsequent section. This section sketches the current approach to the critical point based on the recognition that there is a singularity there. Until evidence to the contrary is obtained, we naturally assume that the properties of a physical system are to be represented by analytic functions, that is, by ones having derivatives of all orders. In particular, this means that the properties in the neighborhood of a point may be expressed by a Taylor series. Thus, if we write T_c, P_c, and ϱ_c respectively for the temperature, pressure, and density at the critical point—the point at which the liquid–vapor saturation line ends—we naturally assume that we can write a Taylor expansion of the pressure $\Delta P = P - P_c$ (or alternatively, of the Helmholtz energy) in terms of $\Delta\varrho = \varrho - \varrho_c$ and $\Delta T = T - T_c$:

$$\Delta P = a_{10}\,\Delta\varrho + a_{01}\,\Delta T + a_{20}(\Delta\varrho)^2 + a_{11}\,\Delta\varrho\,\Delta T + a_{02}(\Delta T)^2 + a_{30}(\Delta\varrho)^3 + \cdots \tag{16}$$

At the critical point, the following conditions hold:

$$(\partial P/\partial\varrho)_{T_c} = 0 \tag{17}$$

$$(\partial^2 P/\partial\varrho^2)_{T_c} = 0 \tag{18}$$

From condition (17), $a_{10} = 0$, and from (18), $a_{20} = 0$. Rewriting eqn. (16) to keep only the leading nonzero terms gives

$$\Delta P = a_{01}\,\Delta T + a_{11}\,\Delta\varrho\,\Delta T + a_{30}(\Delta\varrho)^3 + \cdots \tag{19}$$

For the critical isotherm, $\Delta T = 0$, so that in the limit as the critical point is approached along that isotherm,

$$\Delta P \propto | \Delta \varrho |^3$$

provided a_{30} is not zero. It is then a matter of experiment to determine δ in the expression

$$\Delta P \propto | \Delta \varrho |^\delta$$

for the critical isotherm to see if $\delta = 3$. In fact, the exponent δ appears to have a value greater than 4 for nonconducting fluids, and to have a nonintegral value. This means that the *PVT* properties of fluids are not analytic at the critical point, but that there is, rather, a singularity. As the Ising model in two dimensions, which can be solved rigorously, has a singularity at the critical point, the apparent singularity that is observed in physical systems[(291)] is considered real rather than due to our failure to reach the asymptotic region. Like ΔP, the Helmholtz energy is also nonanalytic at the critical point.

In 1964, Fisher[(334)] showed that analytic equations of state for which the leading coefficients of eqn. (19) are nonzero—the van der Waals equation, for example, being one—give the exponent δ the value 3 for the asymptotic behavior as the critical point is approached, and this is accordingly called the *classical* value.

A number of exponents are introduced to cover the singularities in various properties.[(335)] For the specific heat at constant volume C_V,

$$1/C_V \propto | \Delta T |^\alpha$$

α has a classical value of 0, but experiment gives a value between 0 and 0.1. For the difference in density between liquid and gas, there is

$$\varrho_L - \varrho_G \propto | \Delta T |^\beta$$

Classical theory gives $\beta = \frac{1}{2}$, but the experimental value is approximately $\frac{1}{3}$. For the isothermal compressibility $\varkappa_T$,

$$1/\varkappa_T \propto | \Delta T |^\gamma$$

γ has a classical value of 1, but the experimental value lies nearer 1.1. δ has been defined above. For the surface tension σ,

$$\sigma \propto | \Delta T |^\mu$$

μ has a classical value of 1.5, but the experimental value lies nearer 1.2.

Near the critical point, the correlations between molecules extend to much greater distances than the range of intermolecular forces. Hence, not only is the critical region essentially the same for all molecular fluids, but also for metals and for the coexistence curves of binary systems; in magnetic systems, critical points occur at the Curie temperature, at which a ferromagnetic material becomes disordered, and at the Néel temperature, at which an antiferromagnet becomes disordered. The details of intermolecular forces will have some effect on the critical phenomena but, to the precision of the data available to her, Levelt Sengers[673] found the indices for water $\alpha = 0.1$, $\beta = 0.35$, $\gamma = 1.2$, and $\delta = 4.4$ to be within the range for nonpolar fluids.[675,1110,1111] The aim of research at present should be to obtain values of the indices α, β, etc. with sufficient accuracy that they may be compared with those for other substances, such as CO_2, He, Ar, and Xe, for which good values are available.

A theory is available, the scaling-law theory, originally due to Widom[1171] and Griffiths,[425] which gives relations among the exponents, and indicates how a satisfactory equation of state may be constructed. In the regions that are physically accessible, a satisfactory equation of state must be singular only at the critical point, and must incorporate the power laws properly. Goodwin[409] has offered an equation of state for hydrogen that meets at least some of these requirements.

The concern with the values of the exponents in the asymptotic limit means that $T - T_c$ must be pushed to small values. As this requires that the values of T and T_c of experimentalist and correlators must be consistent and convertible to the absolute scale without ambiguity, we digress to temperature scales.

2.5.1. *Temperature Scales*

In the present chapter, t (or in this section, t_{48}) denotes the Celsius temperature on the International Practical Temperature Scale of 1948 (IPTS-48),* while T refers to the thermodynamic temperature, which is defined to have a value 273.16°K at the triple point of water ($t = 0.01$°C) and for which values are known more or less closely at other temperatures. It is, of course, the thermodynamic scale which enters into thermodynamic

* The International Practical Temperature Scale of 1948 succeeded the International Temperature Scale of 1927, which gave the same values as IPTS-48 for temperatures between the freezing point of water and the sulfur point at 444.60°C. Earlier measurements cited in the present chapter are in a temperature range where the scale used was close to IPTS-48.

calculations, but for some purposes, the International Practical Kelvin Temperature T_{48} could be considered sufficient, where

$$T_{48} = t_{48} + 273.15°\mathrm{K}$$

The relation between T_{48} and T, and its effect on equations for the properties of steam and water, has been considered by Bridgeman and Aldrich[114] and Gibson,[378] who give tables showing the relation between T (as then realized) and T_{48}, as well as the derivative dT/dT_{48}. At the critical point of water, according to their figures, $T = T_{48} + 0.113$ or $T = T_{48} + 0.116$, and $dT/dT_{48} = 1.0000$ or 0.99994.

In January 1969, the International Practical Temperature Scale of 1968 (IPTS-68) came into use,[49,59,204] defining t_{68}, and with Kelvin temperature $T_{68} = t_{68} + 273.15$. The zinc point is now at $t_{68} = 419.58°\mathrm{C}$, while the value of t_{48} for this point was 0.075°C lower, and the interpolation formulas are different. The International Practical Kelvin Temperature of 1968 approximates the thermodynamic temperature to within present knowledge. Table IV gives the relation between t_{48} and t_{68} from 0 to 100°C. At the critical point of water,

$$t_{68} = t_{48} + 0.077$$

and with the change of temperature scales, the former estimate of the thermodynamic temperature at the critical point of water is now to be considered 0.039°C too high.

2.5.2. *Critical Parameters for Water*

Equations of state proposed before the recognition of the nonanalytic character of the critical point were not satisfactory in practice, and are no longer theoretically satisfactory. Experimental work needs to be reevaluated, but only a start has been made at this. Most early experimental work has been summarized by Nowak *et al.*[796,798] and later data are available.[14,90,906]

Table VII gives values from various sources of the critical parameters of H_2O and D_2O. The experimental work of Blank[90] is highly regarded, and his results suggest that previous estimates of the critical temperature for H_2O have been too high. As I have no values for D_2O obtained by scaling-law analysis, values by classical analysis are included. It is seen that the critical temperature of D_2O is about 3°C lower than that of H_2O, that the critical pressure is perhaps slightly lower, and that the molar volume is about the same. Figure 7 shows a plot of $\log_{10} |(\varrho - \varrho_c)/\varrho_c|$ versus

TABLE VII. Critical Parameters of Water[a]

Experimentalist (E) or correlator (C)	T_c, °C, IPTS-48	P_c, bars	$\bar{\varrho}_c$, g cm^{-3}	V_c, cm^3 mol^{-1}
Classical analysis: H_2O				
Bridgeman and Aldrich[115] (C)	374.02	220.91	0.317	56.8
Gibson and Bruges[379] (C)	374.15	221.23	0.317	56.7
Nowak *et al.*[796,798] (C)	374.15	221.1	0.305	59.1
Keyes *et al.*[586,587] (C)	374.02	220.89	0.317	56.7
Eck[283] (E)	374.23	221.15	0.326	55.2
Blank[90] (E)	373.91	220.45	—	—
Classical analysis: D_2O				
Elliott[305] (C)	371.1	221.36	0.338	59.2
Whalley[1162] (C)	371.1	218.8	0.328	55.0
Riesenfeld and Chang[902] (E)	371.5	221.2	0.326	55.3
Oliver and Grisard[805] (E)	370.9	218.6	—	—
Eck[283] (E)	371.5	217.2	0.361	54.8
Blank[90] (E)	370.66	216.59	—	—
Scaling law analysis: H_2O				
Levelt Sengers[673] (C) (data of Eck[283])	374.35	—	0.325	55.4
Levelt Sengers[674] (C) (data of Osborne *et al.*[813])	373.77 (t_{68} = 373.85°C)	—	—	—

[a] Only the scaling law analysis is now acceptable theoretically. The results of the classical analysis are given for comparison and because scaling-law values are not available for D_2O.

$\log_{10}|(T - T_c)/T_c|$ for the data of Eck, using $T_c = 374.35$°C (IPTS-48) and $\varrho_c = 0.325$ g cm^{-3}, as given by Levelt Sengers.[673] With the scaling-law analysis, the critical temperature is 0.1°C higher than Eck found by classical analysis of the same data. The critical density is significantly greater than that of other workers, but about what Eck found. The value of exponent β is the asymptotically limiting slope; it has been drawn with $\beta = 0.353$ as given by Levelt Sengers. Clearly, these data are not fitted by a classical value of $\frac{1}{2}$. Levelt Sengers[674] has also obtained a value of the critical temperature of water from the saturation data of Osborne *et al.*[813]

Gregorio and Merlini[424] have discussed defects of the 1967 IFC Formulation for Industrial Use, caused by its failure to conform to scaling-law requirements, but they have not realized that, as we recognize the

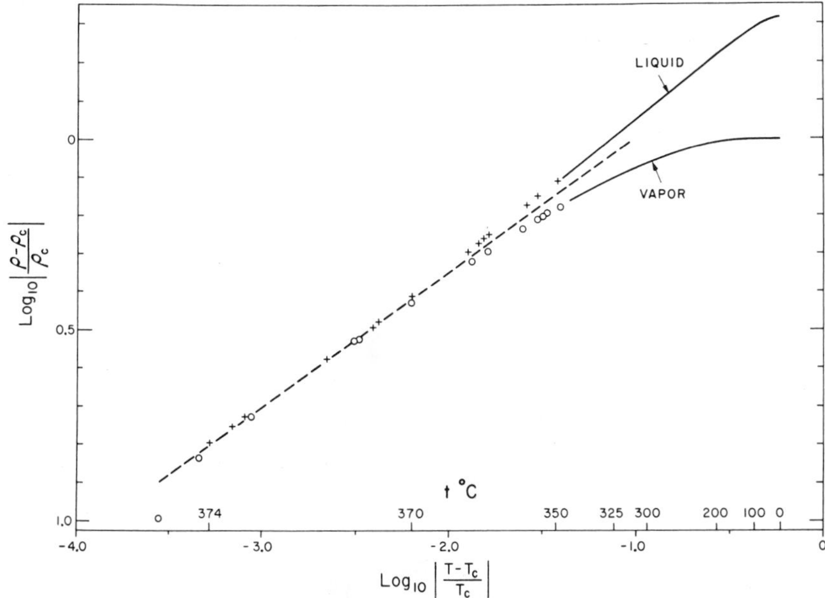

Fig. 7. The scaling-law β for H_2O. The points shown are from Eck.[283] (○) vapor, (×) liquid, plotted using values $\varrho_c = 0.325$ g cm^{-3} and $T_c = 374.35$°C (IPTS-48) determined by Sengers[673] from these data. The error of the function plotted increases toward the left. The slope of the dashed curve is $\beta = 0.353$, which shows that the equation of state of water must be nonanalytic at the critical point.

critical point by the behavior of our system in a region, there must be a reanalysis of experimental data in terms of a satisfactory equation before the critical parameters are chosen.

3. TRANSPORT PROPERTIES

3.1. The Hydrodynamic Viewpoint

Macroscopic hydrodynamics is based on laws of bulk matter, such as the conservation of momentum, on the same level as those employed by classical thermodynamics. Here, however, we consider the coefficients in the linear approximation. The (dynamic or shear) viscosity* η measures the

* There is also a bulk or volume viscosity which is important when the volume changes rapidly, as in a high-frequency acoustic field, but which does not come within the scope of this chapter.[478,823]

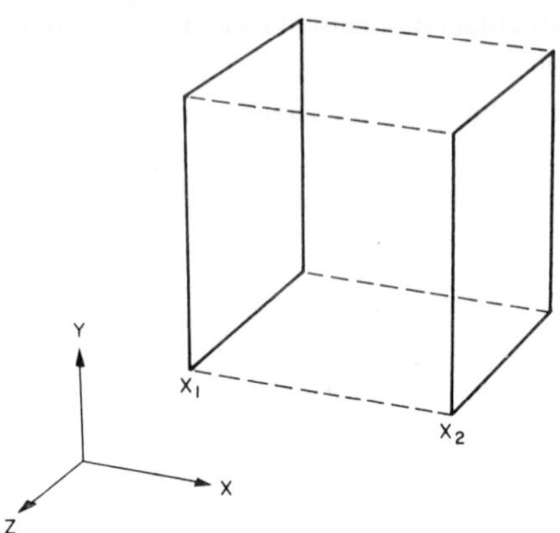

Fig. 8. Coordinate system for transport properties. The unit of thermal conductivity or diffusion is the flux when the planes shown are of unit area at unit separation with unit driving gradient. For viscosity, the plane at X_1 is stationary and that at X_2 is moving in the Y direction with unit velocity.

transport of momentum, the thermal conductivity λ the transport of heat, and the diffusivity $\mathscr{D}$ the transport of mass. Using the coordinate system of Fig. 8, the flux of heat J_q between unit planes at X_1 and X_2 is, according to the Fourier law of heat conduction, proportional to the temperature gradient between them:

$$J_q = -\lambda(\partial T/\partial x)$$

Heat may be conducted by several mechanisms, for example, by the diffusion of species, by phonons, i.e., elastic waves, and by photons. In nonmetallic solids at low temperatures, the principal method is by phonons. Photons become important at high temperatures. In the gas phase, heat is conducted primarily by the movement of molecules between collisions. In the liquid, scattering of phonons is large, rendering the phonon a less useful concept, and both collisional and diffusional processes are important.[716]

If, instead of a temperature gradient, there is a gradient of concentration (chemical potential μ), the flow of mass J_m is given, in the same coordinate system, by Fick's law:

$$J_m = -\mathscr{D}(\partial \mu/\partial x)$$

At liquid densities, diffusion is measured relative to one of the chemical species present, and in the present chapter this will be another isotopic or spin species of water. As intermolecular forces are very nearly the same for isotopic species, this measures the *self-diffusion*, the motion of one water molecule relative to its neighbors.

If the plane at X_1 is fixed and that at X_2 is moving in the Y direction, a velocity gradient is established; the movement of momentum down this gradient is given by Newton's law of viscosity:

$$J_{mv} = -\eta(\partial v_y/\partial x)$$

Alternatively, the viscosity η is the coefficient of proportionality between the velocity gradient and the stress τ_{yz} producing the gradient:

$$\tau_{yz} = \eta(\partial v_y/\partial x)$$

It is through equations of this form that viscosity enters the equations of hydrodynamics. The unit of viscosity used here is the poise, 1 P $\equiv$ 1 dyn sec cm^{-2} $\equiv$ 1 g cm^{-1} sec^{-1}. Viscosity may also be expressed in terms of kinematic viscosity $\nu = \eta/\bar{\varrho}$, for which the unit is the stoke (cm^2 sec^{-1}).

The Sixth (1963) Conference on the Properties of Steam appointed a committee to survey the experimental data on transport properties, of both liquid water and steam, and to arrive at correlating equations. Kestin and Whitelaw[(601)] have reviewed the data considered, which in some regions were quite discordant. More recent data, in a few cases much superior, are noted in their place below.

As the transport properties reflect relaxation processes, it is possible to discuss the corresponding time of relaxation on a molecular scale.[(501)]

3.2. Vapor

3.2.1. *Monatomic Gas Approximation*

In a dilute gas, energy, momentum, and mass are carried from place to place by the molecules themselves. In the case of a hard-sphere, monatomic, ideal gas, the transport coefficients are given simply[(516)]:

$$\lambda \propto (T/M)^{1/2}/\sigma^2 \quad (20a)$$

$$\mathscr{D} \propto (T^3/M)^{1/2}/P\sigma^2 \quad (20b)$$

$$\eta = 4M\lambda/15R = 2M\lambda/5C_V \quad (20c)$$

$$\propto (MT)^{1/2}/\sigma^2 \quad (20d)$$

where σ is the hard-sphere diameter, and M the molecular weight. The features to notice in these formulas are that in each case the coefficient is proportional (or inversely proportional) to the square root of the molecular weight, that the thermal conductivity and viscosity differ only by a multiplicative factor, and that the temperature coefficient is greater in the case of diffusion. For a monatomic gas, the dimensionless Prandtl number, $\mathrm{Pr} = C_P\eta/M\lambda$, has a value $\frac{2}{3}$.

3.2.2. *Collision Integral*

Calculation of the transport properties of the system may be considered as having three stages[733]: the determination of the transport properties from the behavior during molecular collisions; the determination of the behavior during collision from the intermolecular forces; and the determination of the intermolecular forces.

Low-pressure steam differs from a monatomic gas in three important respects. The molecule has rotational energy, it has internal vibrational energy, and it has directional forces. These give the possibility of inelastic collisions in which translational, rotational, and vibrational energies all change. The polarity makes it necessary to consider properly weighted averages over all configurations. The expressions obtained are inherently complicated, and are shown here in the form given for ideal dipolar gases by Monchick and Mason,[760] who considered the Stockmayer potential given by eqn. (3). It is convenient to introduce the collision integral $\Omega^{(l,s)*}$, tabulated by the authors, which is a function of the reduced temperature $T^* \equiv kT/\varepsilon$ and the reduced dipole moment $\mu^* \equiv \mu/(\varepsilon\sigma^3)^{1/2}$. In the polar case, the collision integral appears as $\langle\Omega^{(l,s)*}\rangle$ which has been averaged over all configurations.

For the translational part of the thermal conductivity, Monchick and Mason obtained

$$\lambda_{\mathrm{tr}} = (75k/64)(kT/\pi m)^{1/2}f_\lambda/(\sigma^2\langle\Omega^{(2,2)*}\rangle)$$

where m is the mass of the molecule and f_λ is a correction factor which is near unity. For the viscosity, they obtained

$$\eta = (5/16)(mkT/\pi)^{1/2}f_\eta/(\sigma^2\langle\Omega^{(2,2)*}\rangle) = (4m/15k)(f_\eta/f_\lambda)\lambda_{\mathrm{tr}}$$

and for the self-diffusion,

$$\mathscr{D} = (3/8n)(kT/m)^{1/2}f_{\mathscr{D}}/(\sigma^2\langle\Omega^{(1,1)*}\rangle)$$

where n is the concentration of molecules.

The first feature to note is that λ_{tr} and η are still proportional (except for the small temperature variation of f_η/f_λ). The second is that a different collision integral occurs in $\mathscr{D}$.

3.2.3. *Viscosity*

The reliability of older experimental determinations of the viscosity of gases has been brought in question by the discovery that slip at the walls, which was not taken into account, has made most previous determinations low[43,243,456,457] by perhaps as much as 10% above 300°C. This error in the values relates to the well-known observation that the intermolecular potential parameters that represent the viscosity have been rather different from those which represent other properties, such as the second virial coefficient. However, Monchick and Mason[760] used viscosity data for steam that should not be seriously in error to arrive at potential parameters that seemed unreasonable. They concluded that the Stockmayer potential is too artificial to represent the force field of water in sufficient detail.

In most gases, the viscosity increases with increasing pressure, but for steam below about 300°C, the viscosity decreases. Barua and his colleagues[56,982] have shown that dimerization, as in eqn. (4), can explain the phenomenon, as the viscosity of a dimer is lower than that of two monomers. At lower temperatures, where equilibrium favors their formation, the production of dimers produces a net decrease in viscosity with increasing pressure.

The experimental precision of viscosity measurements of steam, as of liquid water, is about 0.2%. A number of papers report work with steam at higher pressures.[279,555,659,737,908–911] For viscosity, the values and interpolating equations of the Sixth (1963) Conference on the Properties of Steam[996] should be superseded by later correlations[133,134,136,427] that incorporate more recent and more consistent data.

3.2.4. *Thermal Conductivity*

The simplest treatment of the contribution of internal degrees of freedom to the thermal conductivity is that given by Eucken,[316] who assumed that eqn. (20c) still holds for the translational contribution to the thermal conductivity,

$$\lambda_{\text{tr}} = (5C_{V,\text{tr}}/2M)\eta$$

where $C_{V,\text{tr}} = 3R/2$ is the translational part of the specific heat, and that a similar expression holds for the internal contribution $C_{V,\text{int}} = C_V - \frac{3}{2}R$, but with the numerical factor equal to unity. Thus,

$$\lambda = \lambda_{\text{tr}} + \lambda_{\text{int}} = (\eta/M)(\tfrac{5}{2}C_{V,\text{tr}} + C_{V,\text{int}})$$

As for an ideal gas $C_P = C_V + R$, on introducing the ratio of specific heats $\gamma = C_P/C_V$, this becomes

$$\lambda = (\eta C_V/4M)(9\gamma - 5)$$

This equation is not used for practical calculations. The rather difficult problem of choosing between the approximations made in this and in related expressions has been discussed by Sandler.(941)

Brain(108–110) has accompanied his experimental data on the thermal conductivity of steam at atmospheric pressure from 100 to 700°C with a review of the work of Vargaftik *et al.*,(1094,1095,1098–1100) Venart,(1103) and Keyes and Vines.(604) At the higher temperatures, there has been a severe disagreement between the results of different laboratories, reflecting the difficulties of the design of this experiment and amounting to as much as 30%, though Brain finds values in good agreement with those of Vargaftik *et al.*

3.3. Liquid

In dilute vapors, as eqns. (20) show, viscosity and thermal conductivity are nearly proportional and show the same temperature dependence. In liquids, the temperature dependences are quite different. At constant pressure, viscosity decreases nearly exponentially with increasing temperature:

$$\eta = \eta_0 \exp(-\Delta E_\eta/RT) \tag{21}$$

though for water, ΔE_η, the Arrhenius energy of activation for viscous flow (approximately 5 kcal mol^{-1} at 0°C), changes more with temperature than is the case with most liquids. Similarly, the pressure dependence at constant temperature can be represented by

$$\eta = A' \exp(P\,\Delta V_\eta/RT) \tag{22}$$

where the phenomenological volume of activation ΔV_η is positive if the viscosity rises with increasing pressure and negative if it falls. In water, it is negative at low temperatures and positive at high temperatures.

In contrast, the thermal conductivity of liquids is nearly linear with temperature and, over narrow ranges, has traditionally been represented by

$$\lambda = \lambda_0(1 + at) \tag{23}$$

These differing temperature dependences clearly show that there are at least two mechanisms effective in the liquid. As moments of inertia change in different proportion to change of mass on isotopic substitution in water, McLaughlin[715] has shown that rotational motion is more important for viscosity, but translational processes are more important for heat transport.

3.3.1. *Viscosity*

The viscosity of liquid water at atmospheric pressure is needed with good precision because of water's use for calibration and of its use as a reference point for the study of solutions. Experimental technique has been discussed by Swindells *et al.*[1040] Values at atmospheric pressure are all tied to the value* $\eta = 1.0020$ cP at 20°C obtained by Swindells *et al.*,[1039] and older values, such as those of Bingham and Jackson[84] should be multiplied by a suitable factor to bring them to this value at 20°C.

There are a number of papers on experiments at atmospheric pressure,[193,459,460,529,574,626,627,769,1156] including Hallett's virtuoso determination[448] of the viscosity of supercooled water down to −24°C. The viscosity has been measured at close intervals from 0 to 70°C by Korson *et al.*,[628] who found that ΔE_η can be represented over this range by a polynomial in the temperature, and that it is free of fine structure. The experimental error of viscosity measurements is about 0.2% and differences approaching this are found in tables, such as those of Stokes and Mills[1032] or Korosi and Fabuss,[626,627] which claim to represent best values. The viscosity of H_2O and D_2O at atmospheric pressure is illustrated in Fig. 9(a), and some values given by correlators and experimentalists are in Table VIII. This table shows good agreement between the values of Korson *et al.*[628] and those of Weber[1156] over their common range, 0–40°C, but above 70°C, agreement among the sources cited is less good, indicating lower precision in this range.

In deriving the usual equations of hydrodynamics for nonpolar materials, it is postulated that the only forces on a fluid element are surface forces, acting through the surface of the element, and body forces, which are proportional to the mass of the element. However, the water molecule, because of its electric dipole and quadrupole, must be capable of transmitting

* With the change from the temperature scale of 1948 to that of 1968, we must write either

$$t_{48}(20°\text{C}) = t_{68}(19.993°\text{C}), \qquad \eta = 1.0020 \text{ cP}$$

or

$$t_{48}(20.007°\text{C}) = t_{68}(20°\text{C}), \qquad \eta = 1.0018 \text{ cP}$$

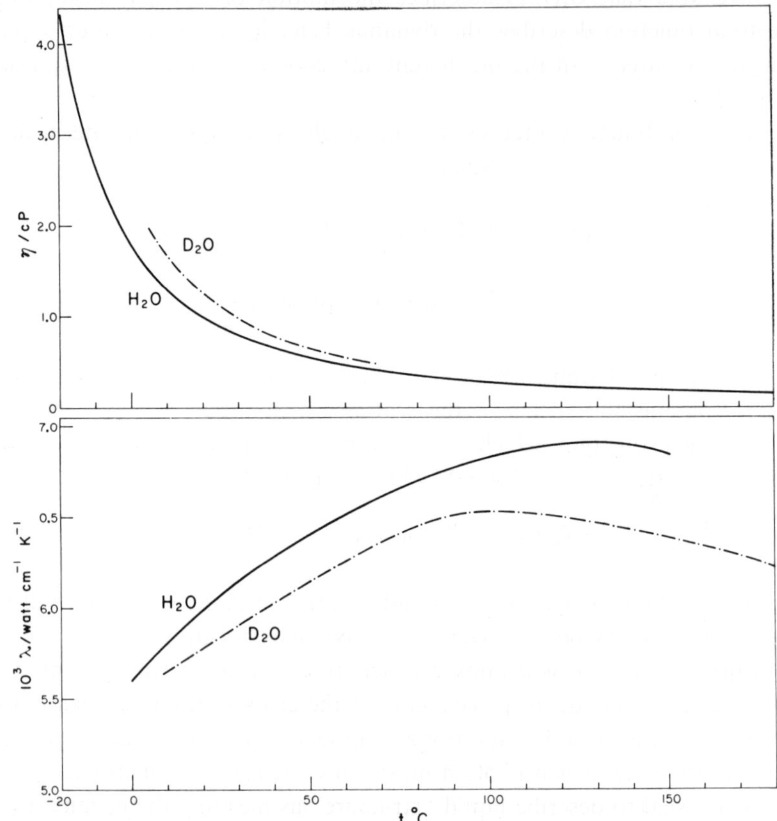

Fig. 9. Transport properties of liquid at atmospheric pressure. (a) Viscosity η. (b) Thermal conductivity λ. For D_2O, the values up to 60°C are the means given by Powell,[866] and those above 75°C are from Ziebland and Burton.[1200]

stress couples and of being subject to body torques.[27,1066] Does bulk water experience such couples over appreciable distances? An increase in the viscosity of liquid water in very thin films, or other non-Newtonian behavior, has sometimes been reported, for example, by Peschel and Adlfinger[833] and Bonarenko,[100] but Hayward and Isdale[481] have reviewed the evidence for rheological abnormalities near solid boundaries and find that there is not yet sufficient reason to believe that earlier results of this sort indicate more than the presence of dust.

For practical purposes, eqn. (21) does not represent the viscosity of water with sufficient precision to be used as an interpolating equation if $-\Delta E_\eta$ is taken as a constant. For such purposes, the viscosity is often

TABLE VIII. Viscosity of H_2O at Atmospheric Pressure[a]

t_{48}, °C	Viscosity, cP				
	Stokes and Mills[1032]	Korosi and Fabuss[626,627]	Korson *et al.*[628]	Weber[1156]	Hardy and Cottington[459,460]
0	1.787	—	1.7916	1.792	—
5	1.516	—	1.5192	1.520	1.5184
10	1.306	—	1.3069	1.3069	—
15	1.138	—	1.1382	1.1383	—
20	1.002	1.0020	1.0020	1.0020	—
25	0.8903	0.8903	0.8903	0.8903	—
30	0.7975	—	0.7975	0.7975	—
35	0.7194	—	0.7195	0.7193	—
40	0.6531	0.6527	0.6532	0.6531	0.6531
45	0.5963	—	0.5963	—	—
50	0.5467	—	0.5471	—	—
55	0.5044	—	0.5042	—	—
60	0.4666	0.4665	0.4666	—	0.4665
65	0.4342	—	0.4334	—	—
70	0.4049	0.4045	0.4039	—	—
75	0.3788	0.3784	0.3775	—	—
80	0.3554	0.3546	0.3538	—	0.3548
85	0.3345	—	0.3323	—	—
90	0.3156	0.3143	0.3128	—	0.3148
95	0.2985	—	0.2949	—	0.2976
100	0.2829	0.2820	0.2783	—	0.2822

[a] All values here are adjusted to a value of 1.002 cP at 20°C.

represented by the equation, due to Vogel,[1113,1114]

$$\eta = \eta_0' \exp[-\Delta E_\eta'/(T + C)] \tag{24a}$$

or

$$\log_{10}\eta = A + [B/(T + C)] \tag{24b}$$

which is sometimes called a *free-volume* equation,[750] and $T = -C$ is the glass transition temperature. Korosi and Fabuss[626,627] give the constants for this equation as $A = -1.64779$, $B = 262.37$, and $C = -133.98$°K, thus estimating the glass-transition temperature of water as 134°K. Equation (24) itself does not represent the viscosity perfectly, and modified or extended forms have been considered by Barna,[53] Fieggen,[330] and Bernini *et al.*[72]

Coe and Godfrey[193] found the rational function

$$\log(\eta/\eta_{20}) = [A'(t-20) + B'(t-20)^2]/(t + C') \qquad (25)$$

to represent their data closely. This equation has also been used by Korosi and Fabuss[626,627] and Korson *et al.*[628] This equation gives about 164°K for the glass-transition temperature, which shows that eqn. (24) and eqn. (25) extrapolate downward rather differently. Kampmeyer[574] represented the viscosity by a power series in the reciprocal absolute temperature, which is a generalization of eqn. (21).

The values of the viscosity of D_2O at atmospheric pressure found by Millero *et al.*[753] are in good agreement with those of Hardy and Cottington[459,460]; the values of Lewis and MacDonald,[680] Timrot and Shuiskaya,[1069] and Heiks *et al.*[486] seem less reliable. At low temperatures, the viscosity of D_2O is 30% higher than that of H_2O, much larger than the ratio (1.05) of the square roots of the masses, but in agreement with other evidence that D_2O has less deformation of tetrahedral angles; at higher temperatures, the viscosities become more similar.

The viscosity of most liquids increases rapidly with increasing pressure, but as Hausen[473] first showed, below about 30°C, the viscosity of water falls with increasing pressure. At higher pressures, it then rises. The data of Bett and Cappi[80] are shown in Fig. 10. They found that $\partial\eta/\partial P$ changes sign at 33.5°C, and are supported by the measurements of Wonham[1181] and Stanley and Batten.[1011] On the other hand, Horne and Johnson[530] and Agaev and Yusibov[2,3] found irregularities near the freezing point; these probably reflect some experimental difficulty. Bruges and Gibson[135] extrapolated their equation for the viscosity of water[133,134] to temperatures below 0°C. If their extrapolation is reliable, the pressure at which $(\partial\eta/\partial P)_T = 0$ moves to higher pressures as the temperature falls from 30 to 0°C, but below −10°C, it moves toward lower pressures, a result that seems unlikely. Rabinovich[881] points out that, although measurements on D_2O are too scattered to show it, the temperature for which $\partial\eta/\partial P = 0$ ought to be higher for D_2O.

The existence of a negative volume of activation at low temperatures and pressures has clear structural implications. If the phenomenological volume of activation defined by eqn. (22) has meaning on a molecular scale, a positive volume suggests that a hole must form into which a neighboring molecule then diffuses. On the other hand, a negative volume can be ex-

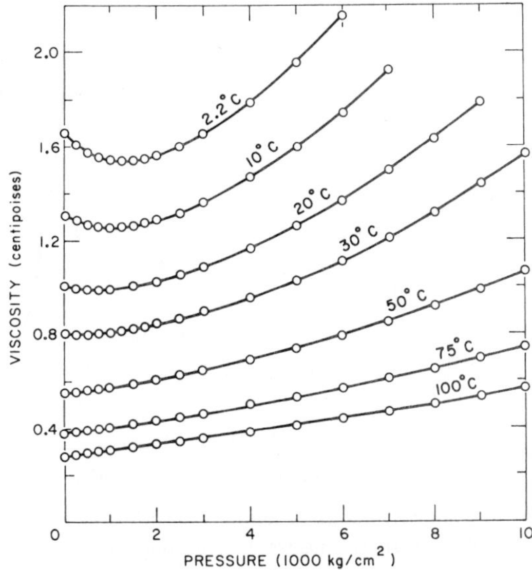

Fig. 10. Viscosity of water under pressure. The data shown indicate that $(\partial\eta/\partial P)_T = 0$ at 33.5°C. Reproduced from Bett and Cappi.[80] (Reprinted with permission from *Nature*, copyright 1965.)

plained for water in terms of a four-coordinated network[590] in which the angles are nearly tetrahedral. In this case, a net deformation of coordination bonds is needed to permit a molecule to start its diffusive motion, and the net deformation produces a decrease in volume.

The viscosity of water under pressure at higher temperatures has also been measured.[501,597,598,737,908,909] Mineev and Zaidel'[755] have suggested that the viscosity of water rises rapidly with increasing pressure in shock waves, but Hamann and Linton[451] have argued that other evidence bearing on the viscosity of shocked water suggests that the value is not greatly different from the normal value, and that the results of Mineev and Zaidel' must result from difficulties in the analysis of their experimental conditions.

The viscosity and self-diffusion are related by the Einstein–Stokes relation which will be discussed below. Further, either the viscosity or the self-diffusion can be related empirically to the dielectric relaxation time. This relation is discussed for H_2O and D_2O by Rizk and Girgis,[912] who express the viscosity by eqn. (24a), and the dielectric relaxation time τ by

$$\tau = (\tau_0'/T)\exp[-\Delta E_\tau'/(T+C)]$$

3.3.2. *Thermal Conductivity*

The thermal conductivity of water rises approximately linearly between 20 and 60°C, in agreement with eqn. (23). It then passes through a maximum at about 130°C. That is, the curve is approximately parabolic. Experimental results have been reviewed by Powell.[866] The scatter in measurements exceeds 1%. Experimental papers on light water include Refs. 118, 174, 661, 892, 955, 1070, 1096, 1103, and on heavy water Refs. 174, 665, 1097, 1200. The thermal conductivity has considerable technical importance, but understanding of the molecular behavior underlying the phenomenon is poorly developed. Figure 9(b) shows the thermal conductivity of H_2O and D_2O. For most liquids, the thermal conductivity falls with increasing temperature, but for water, this is only the case above about 150°C. Accordingly, the thermal conductivity is another property in which liquid water shows an "anomalous" extremum. Adrianova *et al.*[15,16] have given an interpretation of the maximum in terms of Samoilov's picture of liquid water as a distorted ice lattice including interstitial molecules. The equations given in the Supplementary Release on Transport Properties[996] of the Sixth International Conference of the Properties of Steam are the best available for the thermal conductivity.

3.3.3. *Self-Diffusion*

At first glance, the self-diffusion of a liquid, involving as it does the motion of a molecule relative to its neighbors, might seem related to structure and motion on a molecular scale in a more obvious way than is the thermal conductivity or the viscosity, but in fact, the viscosity is rather closely related to the self-diffusion. The mobility u of a spherical particle in a continuum is given by Stokes's law

$$u = 1/6\pi\eta r$$

where η is the viscosity of the medium and r is the radius of the particle. Further, the mobility is related to the diffusion constant $\mathscr{D}$ by the Einstein relation

$$\mathscr{D} = kTu$$

Hence, the Einstein–Stokes relation for the diffusion of a Brownian particle relates the self-diffusion and the viscosity,

$$\mathscr{D} = kT/6\pi\eta r$$

and predicts that for a single diffusing species, $\mathscr{D}\eta/T$ should be a constant. (As the real solvent is not a viscous continuum, the observed coefficient is not 6π.) Then, as eqn. (21) considers the temperature dependence of viscosity to be given by

$$\eta = \eta_0 \exp(-\Delta E_\eta/RT)$$

that for diffusion can be approximated by

$$\mathscr{D} = \mathscr{D}_0 \exp(\Delta E_{\mathscr{D}}/RT)$$

Measurements are available by three methods. Some measurements claim a precision of 1%, but different laboratories sometimes agree only to 10%. The earliest method used isotopic tracers. Values are available for the diffusion of ^2H,[259,697,1144,1147,1149] ^3H,[259,563,1149] and ^{18}O[1148,1149] in H_2O.

The self-diffusion coefficient of heavy water has been determined by Silk and Wade[979] in a pulsed neutron experiment. Like other workers using D_2O, they found $\mathscr{D}(D_2O, 25°C) = 2.0 \times 10^{-5}$ cm² sec⁻¹, while values of $\mathscr{D}(H_2O, 25°C)$ are nearer 2.5×10^{-5} $cm^2\ sec^{-1}$. These are in the ratio of the viscosities and, with the Einstein–Stokes relation, indicate that the diffusing species is the same size in H_2O and D_2O.

The most common method at present for determining the self-diffusion is the spin-echo technique of nuclear magnetic resonance, in which the decay of the spin-echo amplitude of magnetic nuclei is followed. The decay is proportional to $\exp(-\gamma^2 G^2 \mathscr{D} \tau^2/12)$, where γ is the magnetogyric ratio of the nucleus, G is the magnetic field gradient, and $\tau/2$ is the time for field reversal at resonance.[165] Such a method is related to the diffusion constant by a more elaborate theory than is the case for direct determinations of the diffusion constant, but Glasel[392] believes that the results are reliable in the case of water. Perhaps the most interesting experimental measurements are those of McCall *et al.*,[709] who have determined the volume of activation for self-diffusion. Here again, the self-diffusion is approximately inversely proportional to the viscosity. Other measurements include Refs. 190, 270, 629, 981, 1073. Hausser *et al.*[474] have measured the self-diffusion up to the critical point, finding that the Einstein–Stokes relation is well satisfied over the range where viscosity values are available.

Wang[1147] has shown that the product $\mathscr{D}\tau_D$ obtained by multiplying (a) the self-diffusion coefficient of ^{18}O in H_2O by (b) the dielectric relaxation time τ_D from the work of Collie *et al.*[201] is a constant at temperatures from 5 to 25°C. A simple model gives $\mathscr{D}\tau_D = j^2$, where j, the jump distance, is the average distance between equilibrium positions of a diffusing molecule. Wang obtained $j = 3.7$ Å, which between the first and second peaks of the radial distribution function of water.[784] This correlation, like the theory of the spin-echo technique, raises questions about the structure of liquid water on a molecular level that lie outside the scope considered for the liquid phase in the present chapter.

ACKNOWLEDGMENTS

I am indebted to Dr. E. Whalley for helpful discussions on the material of this chapter, and to Dr. J. M. H. Levelt Sengers for comments about the critical region.

CHAPTER 11

Application of Statistical Mechanics in the Study of Liquid Water

A. Ben-Naim

Department of Inorganic and Analytical Chemistry
The Hebrew University of Jerusalem
Jerusalem, Israel

1. INTRODUCTION

"Theories of the statistical mechanics of fluids have either been very formal and based upon first principles or they have been extremely heuristic, depending largely upon models".[898]

Referring to simple fluids, a remarkable progress in both classes of theories has been achieved.[369,509,1059] While the formal approach is based on first principles of statistical mechanics, followed by a well-defined set of approximations, the heuristic approach depends on a successful choice of a suitable model, which reasonably well represents the system on the one hand, and provides, on the other hand, a relatively simple partition function, from which numerical quantities can be obtained.

The situation in the study of water[67,301] is that most, if not all, of the theoretical progress has been through the modelistic approach, and development along formal routes has been very limited. Moreover, until very recently, the prospects for persuing such an approach have been regarded as being very dim. The reason for this can be traced back to three fundamental difficulties. In the first place, virtually nothing is known about the analytical form of the intermolecular potential function between two water molecules, an essential ingredient in any fundamental theory of fluids. In fact, some modeled potential functions have been suggested[301] mainly for the purpose of reproducing the second virial coefficient of water. However, as is well known, this quantity cannot serve as a stringent test of the details of the potential function employed, and different potential functions may produce a reasonable agreement with the experimental values of the second virial coefficient as well as with its temperature dependence.

The second difficulty can be stated as follows: Suppose we have been able to formulate a pair potential which ideally describes the potential energy of interaction for each configuration of a pair of water molecules. Will this function be adequate when employed in the theory of liquid water? For simple fluids, most of the progress has been based, so far, on one fundamental assumption, namely the pairwise additive character of the total potential energy. Symbolically, if $U_{ij}(i, j)$ is the pair potential for particles i and j at configuration (i, j) and $U_N(1, 2, \ldots, N)$ is the potential energy for the system of N particles at configuration $(1, 2, \ldots, N)$, the assumption can be stated as

$$U_N(1, 2, \ldots, N) = \sum_{1 \le i < j \le N} U_{ij}(i, j) \tag{1}$$

Recent studies have shown[454] that nonadditivity of the hydrogen bond energy may be important. Thus, if $U(1, 2, 3)$ is the potential energy for a triplet of water molecules at configuration $(1, 2, 3)$, the magnitude of nonadditivity for this cluster may be expressed by

$$\delta U(1, 2, 3) = U(1, 2, 3) - U(1, 2) - U(1, 3) - U(2, 3) \tag{2}$$

Similar higher-order nonadditivity functions can be defined for larger clusters of molecules. Little is yet known on the analytic form of $\delta U(1, 2, 3)$, but intensive study of this problem is being carried out for some particular configurations that seem to be of special importance in the liquid and in the solid phases.

In a kind of a practical procedure, one may incorporate all nonadditivity effects into the theory by defining an effective pair potential U_{eff} which fulfills the condition

$$U_N(1, 2, \ldots, N) = \sum_{1 \le i < j \le N} U_{\text{eff}}(i, j) \tag{3}$$

This potential may be useful in the elucidation of some aspects of the structure of water when employed in the theory.

This brings us to the third difficulty, which, though technical in essence, imposes severe limitations on the extent to which one can carry out numerical computations, on such a system. The employment of an angle-dependent pair potential function introduces little modification in the formal appearance of the theory, but vastly increases the amount of computational time required to accomplish a project which would normally take a relatively short time in the angle-independent case. This difficulty must be regarded as being temporary, since increases in both the efficiency and the capacity of modern computers may be anticipated in the foreseeable future.

In the following sections, we shall focus our attention on possible ways of computing, and on the information which can be derived from, the pair correlation function. This function provides an important bridge between the molecular structure of the liquid on the one hand, and its bulk thermodynamic properties on the other. Thus, in principle, calculation of the pair correlation function of water could help in the elucidation of some of its anomalous thermodynamic properties.

We shall now list some pertinent definitions of quantities that will be employed in subsequent sections.

In a system of N water molecules contained in a volume V and at temperature T, the probability density of observing a specific configuration of the N water molecules is[67,509]

$$P^{(N)}(\mathbf{X}_1, \ldots, \mathbf{X}_N) = \frac{\exp[-\beta U_N(\mathbf{X}_1, \ldots, \mathbf{X}_N)]}{\int \cdots \int \exp[-\beta U_N(\mathbf{X}_1, \ldots, \mathbf{X}_N)]\, d\mathbf{X}_1 \cdots d\mathbf{X}_N} \tag{4}$$

where $U_N(\mathbf{X}_1, \ldots, \mathbf{X}_N)$ is the total potential energy of interaction among the N molecules at the specific configuration $\mathbf{X}_1, \ldots, \mathbf{X}_N$. Here, $\mathbf{X}_i$ stands for both the positional coordinates $\mathbf{R}_i = (R_{ix}, R_{iy}, R_{iz})$ and the orientational coordinates, which may be selected as the three Euler angles[67,407] $\mathbf{\Omega}_i = (\phi_i, \theta_i, \psi_i)$. The probability density function written in eqn. (4) serves as a source for defining various molecular distribution functions. We shall be interested solely in the pair distribution function, which is the probability density of simultaneously observing two water molecules at configuration $(\mathbf{X}_1, \mathbf{X}_2)$. From eqn. (4) one gets, by integration over the configurations of the remaining $N-2$ molecules,

$$\varrho^{(2)}(\mathbf{X}_1, \mathbf{X}_2) = \frac{N! \int \cdots \int \exp[-\beta U_N(\mathbf{X}_1, \ldots, \mathbf{X}_N)]\, d\mathbf{X}_3 \cdots d\mathbf{X}_N}{(N-2)! \int \cdots \int \exp[-\beta U_N(\mathbf{X}_1, \ldots, \mathbf{X}_N)]\, d\mathbf{X}_1 \cdots d\mathbf{X}_N} \tag{5}$$

The factor $N!/(N-2)!$ is included since we have exactly $N!/(N-2)!$ ways of selecting two out of N molecules to occupy the configuration $(\mathbf{X}_1, \mathbf{X}_2)$. (More precisely, the probability density is referred to the differential elements $d\mathbf{X}_1$ and $d\mathbf{X}_2$.)

The pair correlation function $g^{(2)}(\mathbf{X}_1, \mathbf{X}_2)$ is defined by the relation

$$\varrho^{(2)}(\mathbf{X}_1, \mathbf{X}_2) = (\varrho/8\pi^2)^2 g^{(2)}(\mathbf{X}_1, \mathbf{X}_2) \tag{6}$$

where $\varrho = N/V$ is the number density, and the factor $(8\pi^2)^2$ is a normalization constant. The function $g^{(2)}(\mathbf{X}_1, \mathbf{X}_2)$ approaches unity as the relative separation between the two particles approaches infinity.

It would certainly be desirable to have, either experimentally or theoretically, the full function $g^{(2)}(\mathbf{X}_1, \mathbf{X}_2)$, a knowledge of which is essential to the computation of most of the thermodynamic quantities.[67,509] However, what is actually available from experiments, and what has been so far obtained from theory, is only the angle-averaged pair correlation function, defined by

$$\bar{g}^{(2)}(\mathbf{R}_1, \mathbf{R}_2) = [1/(8\pi^2)^2] \int g^{(2)}(\mathbf{X}_1, \mathbf{X}_2)\, d\mathbf{\Omega}_1\, d\mathbf{\Omega}_2 \tag{7}$$

which is, in effect, a function of the intermolecular distance $R_{12} = |\,\mathbf{R}_2 - \mathbf{R}_1\,|$ only. This function will be referred to as the radial distribution function (RDF), and will be denoted by $g(R)$.

Fortunately, in spite of the averaging over the angles, there is still some important information conveyed by this function. From a mere glance at the RDF of water as a function of the reduced distance $R^* = R/\sigma$, where $\sigma = 2.82$ Å is the effective diameter of a water molecule, one can observe two important features that yield very fundamental information on the local mode of packing of water molecules in the liquid state. Figure 1 displays the experimental RDF of water[783,784] at 4°C and 1 bar (see also Chapter 8) and for argon* at 84.25°K and 0.71 bar as a function of the reduced distance R^*. The first important difference is related to the area under the first peak of $g(R)$, which gives a measure of the average first coordination number around a given molecule. More precisely, since $\varrho g(R)$ is the density of molecules at a distance R from a given molecule, the number of molecules contained within a sphere of radius R_M around a given molecule (excluding the one at the center) is

$$n_{\mathrm{CN}} = \varrho \int_0^{R_M} g(R) 4\pi R^2\, dR \tag{8}$$

There is no universal way of choosing the upper limit of the integral which would be most suitable for defining the first coordination sphere.† A simple and practical choice is the location of the first minimum just after the first peak. Calculations based on this procedure show that the coordination number for liquid argon is about[940] 10 at the melting point and gradually decreases as temperature increases. The corresponding value for water is

* Data provided by Prof. N. S. Gingrich.

† For more detailed discussion of the various methods for calculating n_{CN}, see Pings.[845]

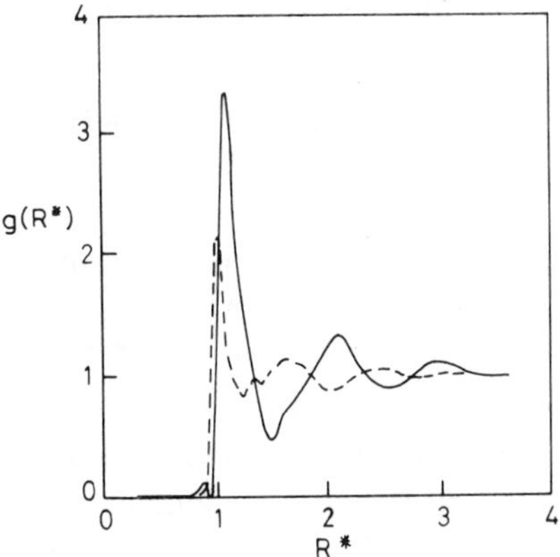

Fig. 1. Radial distribution function for water[783,784] (dashed line) at 4°C and 1 bar and for argon (data provided by Prof. N. S. Gingrich) (solid line) at 84.25°K and 0.71 bar, as a function of the reduced distance $R^* = R/\sigma$, σ being taken as 2.82 Å for water and 3.4 Å for argon.

about 4.4 at the melting point of ice (the ideal number for ice being 4) and *increases* with rising temperature. Figure 2 schematically displays the dependence of n_{CN} upon temperature for these two liquids. This indicates that the local mode of packing of water molecules around a given central molecule is retained to a large extent upon melting and even beyond that.

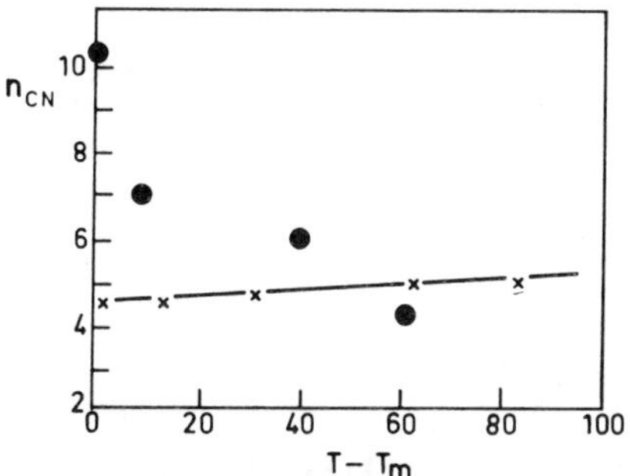

Fig. 2. First coordination number n_{CN} as a function of the temperature above the melting point T_m for argon (dots) and for water (crosses). The drawing is based on data from Samoilov.[940]

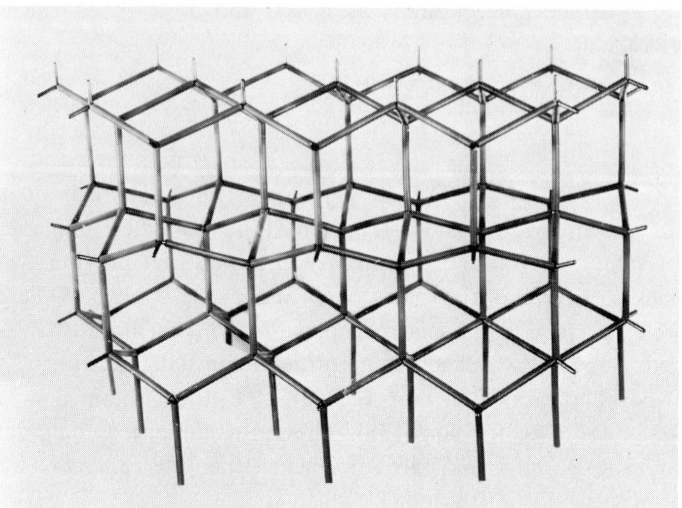

Fig. 3. The arrangement of oxygen atoms in ice I. Each water molecule is surrounded by four nearest neighbors, leaving a large empty space in its immediate vicinity.

(See Chapter 8.) The local geometry around a molecule in ice I is described in Fig. 3. One clearly sees that the open-structure mode of packing leaves vacant spaces which apparently do not get filled to a large extent upon melting. Supplementary information on the second-nearest neighbors is provided by the location of the second peak of the RDF. Referring again to Fig. 3, we see, that in ice I, the distance between second-nearest neighbors is

$$2 \times 2.76 \sin(\theta_T/2) = 4.5 \text{ Å} \tag{9}$$

where θ_T is the angle between the two bonds in a perfect tetrahedral geometry,

$$\theta_T = 109°28' \tag{10}$$

The location of the second peak of the RDF of water is almost exactly at 4.5 Å, which indicates that in liquid water, there is a large concentration of molecular triplets with a geometry similar to the triplets occupying consecutive lattice points in ice. The difference in the mode of packing between a simple fluid and water is demonstrated in Fig. 4. It is clearly seen that the nearly concentric distribution of simple molecules around a central molecule (Fig. 4b) produces a high concentration of molecules at about σ, 2σ, and 3σ. In water, however, the strong directional forces dictate a unique distribution of water molecules around the central one, thereby producing high concentration of molecules at σ and 1.63σ, respectively (Fig. 4a). It must be noted that although the above-mentioned features of the RDF are the most important ones, more information may be extracted: for example, the location of the third peak, the location of the various minima, and the values of the function at the maxima and minima are quantities that carry additional information, the exploitation of which may be possible by indirect methods, i.e., when the RDF is employed in calculations of derived quantities.

In Section 3 and 4 we shall outline possible routes along which one may investigate the structure of water by starting from an assembly of *single* water molecules as the basic entities. It must be borne in mind that most of the theoretical development in the study of water has so far been based on *ad hoc* models. Those models have no doubt provided an insight into the origin of some peculiarities of liquid water. Yet they cannot serve to elucidate the molecular structure of water since all of them in one way or another, assume the two basic ingredients, namely the typical distribution of the first and second nearest neighbors. If this is built into in the theory, then achieving agreement between theory and experiment should be a relatively simple task. On the other hand, the statistical mechanical theory

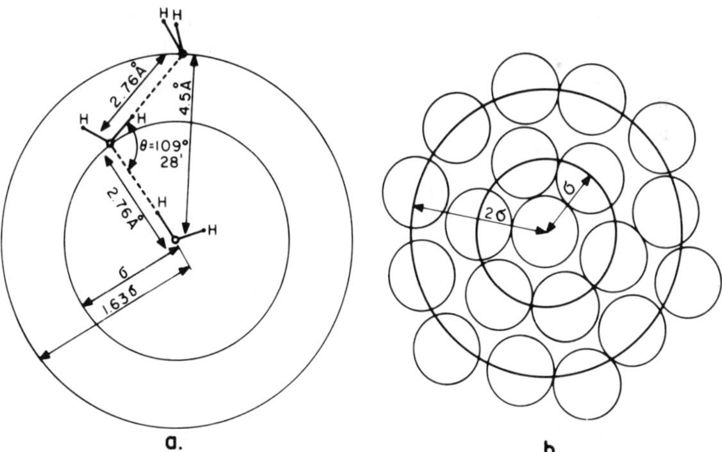

Fig. 4. Schematic description of the distribution of first- and second-nearest neighbors (a) in water and (b) in a simple fluid. The tetrahedral orientation of the hydrogen bond induces a radial distribution of first- and second-nearest neighbors at σ and 1.63σ, respectively, $\sigma = 2.76$ Å being the O—O distance in ice I. The almost equidistant and concentric nature of the packing of particles in a simple fluid produces corresponding neighbors at σ and 2σ.

of fluids has been striving to establish the relation between the pair correlation function and the pair potential. It will be shown in the next sections that employment of a judiciously chosen pair potential may lead to some of the most important features of the RDF of water. A refinement and extension on these lines will, no doubt, be the goal of theoretical research on water in the near future.

In the next section, some general features of a possible pair potential for water molecules in the condensed phase will be outlined. Section 3 reports an attempt to use an approximate Percus–Yevick equation to obtain the radial distribution function for water. Section 4 describes the results of a computer experiment carried out for waterlike particles based on the Monte Carlo technique. Section 5 is devoted to a few derived quantities that provide further information about the molecular structure of this liquid.

2. CHARACTERISTIC FEATURES OF AN EFFECTIVE PAIR POTENTIAL FOR LIQUID WATER

In view of the discussion in the previous section, a pair potential for water in a condensed phase should be chosen in such a way that it will lead to the tetrahedral geometry of packing of the molecules. Thus, the structural information available on both liquid water and ice serves as guidance in designing this function. As we have seen, the tetrahedral geometry is retained upon melting, and therefore we should endeavor to construct a pair potential such that configurations consistent with this geometry will be favored. We shall first outline a possible general pair potential which is expected to fulfill this requirement. A more specific though less elegant, example based on the Bjerrum four-point-charge model, which has actually been tried out in some calculations,[67] will be described later.

The various ingredients from which an analytic expression for a pair potential is constructed must conform with the following requirements:

1. At very short distances, say <2 Å, the two molecules will exert strong repulsive forces on each other, and thereby prevent extensive penetration into each other. As a first approximation, we can choose a Lennard-Jones function of the form

$$U_{LJ}(R_{12}) = 4\varepsilon[(\sigma/R_{12})^{12} - (\sigma/R_{12})^{6}] \tag{11}$$

to take care of this requirement.

Since water and neon are isoelectronic, it will be reasonable to take for ε and σ the corresponding parameters for neon,[516]

$$\varepsilon = 5.01 \times 10^{-15}\,\text{erg} \qquad (7.21 \times 10^{-2}\,\text{kcal mol}^{-1})$$
$$\sigma = 2.82\,\text{Å} \tag{12}$$

2. At large distances, say a few molecular diameters, the interaction between the two molecules can be expressed as arising from a few point electric multipoles, situated in the center of the molecule. The most commonly used is the dipole–dipole term[553]

$$U_{DD}(1, 2) = \boldsymbol{\mu}_1 \cdot \mathbf{T}_{12} \cdot \boldsymbol{\mu}_2 \tag{13}$$

Where $\mathbf{T}$ is the dipole–dipole tensor

$$\mathbf{T}_{12} = (1/R_{12}^3)[\mathbf{I} - 3\mathbf{nn}] \tag{14}$$

with $\mathbf{n}$ a unit vector in the direction of $\mathbf{R}_{12}$. Similarly, one may add dipole–quadrupole, quadrupole–quadrupole interactions, etc. This term, as well as a dipole–quadrupole term, has been employed by Rowlinson[925–927] to compute the virial coefficients of water.

In principle, by considering a sufficient number of terms in the multipole expansion, one may be able to reproduce the characteristics of the desired pair potential. However, our knowledge of the higher moments ends effectively at the dipole moment (see Ref. 301 for more details on this subject).

Moreover, even if we knew a few more multipole moments, it is unlikely that the multipole expansion converges for short distances. Therefore, we must resort to a somewhat artificial *ad hoc* model potential that will produce effectively what the hydrogen bond does in the real situation. This brings us to the third ingredient of the potential.

3. An effective pair potential must account for the short-range tetrahedral orientation of pairs of water molecules. We shall outline here a general procedure for constructing that part of the potential function.

Recognizing the fact that hydrogen bonding occurs only along the four directions pointing to the vertices of a regular tetrahedron, we introduce, in each molecule, four unit vectors along these directions. Figure 5 demonstrates the directions of these four unit vectors, $\mathbf{i}(H_1)$ and $\mathbf{i}(H_2)$ being the directions of the hydrogen bonds formed by molecule i as a donor molecule. These directions are close to, though not necessarily identical with, the direction of the O—H bonds. Similarly, $\mathbf{i}(L_1)$ and $\mathbf{i}(L_2)$ will denote

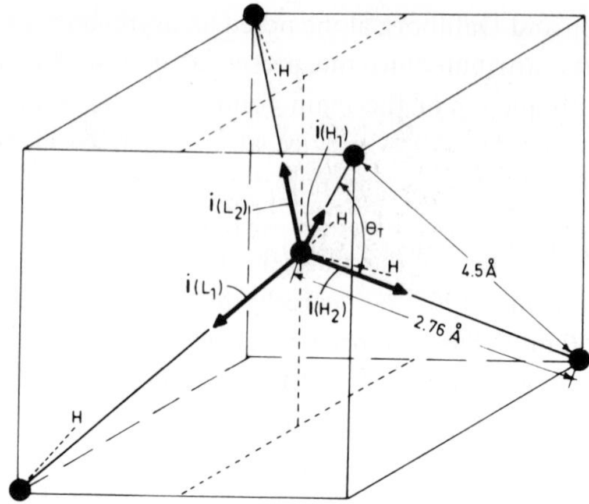

Fig. 5. Four unit vectors attached to a water molecule i are directed toward the vertices of a regular tetrahedron. $\mathbf{i}(H_1)$ and $\mathbf{i}(H_2)$ are along the directions determined by the two O—H bonds. $\mathbf{i}(L_1)$ and $\mathbf{i}(L_2)$ are along the direction of the lone-pair electrons denoted in the text by the O—L direction.

the two unit vectors along the direction of the bonds formed by molecule i, as an acceptor. These directions will be close to the direction commonly assigned to the lone-pair electrons, and will be referred to as the O—L direction.

Let $G_\sigma(x)$ be an unnormalized Gaussian function defined by

$$G_\sigma(x) = \exp(-x^2/2\sigma^2) \tag{15}$$

i.e., a function symmetric about $x = 0$, with maximum height of unity and variance σ^2. A qualitative form of the curve is shown in Fig. 6.

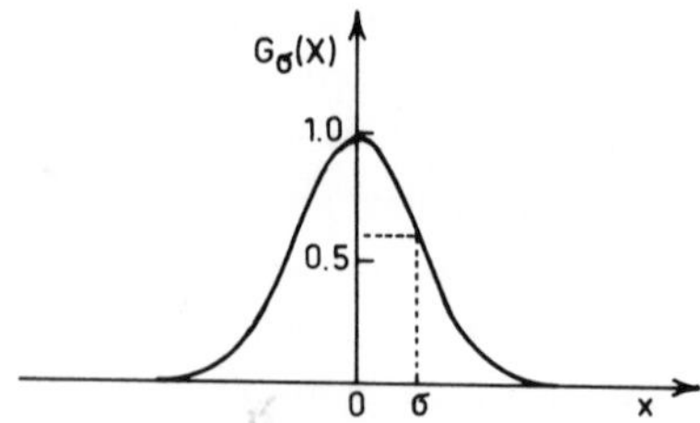

Fig. 6. Gaussian function $G_\sigma(x)$. The value of the function at $x = 0$ is unity. The value drops to $\exp(-0.5) = 0.605\ldots$ at $x = \sigma$.

We now propose the following form of the hydrogen-bond-like part of the effective pair potential, operating between molecule i and j:

$$U_{\mathrm{HB}}(i, j) = \varepsilon_{\mathrm{H}} G_{\sigma'}(R_{ij} - R_{\mathrm{H}}) \left\{ \sum_{\alpha,\beta=1}^{2} G_{\sigma}[\mathbf{i}(H_{\alpha}) \cdot \mathbf{n} - 1] G_{\sigma}[\mathbf{j}(L_{\beta}) \cdot \mathbf{n} + 1] \right.$$
$$\left. + \sum_{\alpha,\beta=1}^{2} G_{\sigma}[\mathbf{i}(L_{\alpha}) \cdot \mathbf{n} - 1] G_{\sigma}[\mathbf{j}(H_{\beta}) \cdot \mathbf{n} + 1] \right\} \tag{16}$$

Here, ε_{H} is an energy parameter which gives the maximum strength of the bond energy, and R_{H} is the intermolecular distance at which the maximum binding energy is attained. The rate at which the energy drops as the distance $| R_{ij} - R_{\mathrm{H}} |$ increases is determined by σ', which in principle can be estimated from quantum mechanical calculations on a pair of water molecules. The conditions relating to the relative angular orientations have been introduced in the brackets of eqn. (16). They require that a bond will form whenever the direction of the O—H bond of one molecule lies along (or nearly so) the direction $\mathbf{n} = \mathbf{R}_{ij}/R_{ij}$, and the direction O—$L$ of the second molecule simultaneously lies along (or nearly so) the direction $-\mathbf{n}$. There are altogether eight terms in these brackets corresponding to all possible configurations for the formation of a bond by the two molecules i and j. One such configuration is depicted in Fig. 7. Again allowance is made for some variation of the bond angle by the choice of σ. The smaller the value of σ, the more rapidly the hydrogen bond energy will fall, as the O—H···O angle deviates from 180°. The overall effect of the particular combination of Gaussians in eqn. (16) is that a bond will form whenever *both* the distance R_{ij} *and* the relative orientations of the two molecules conform with the required conditions discussed in the beginning of this section. Note that the

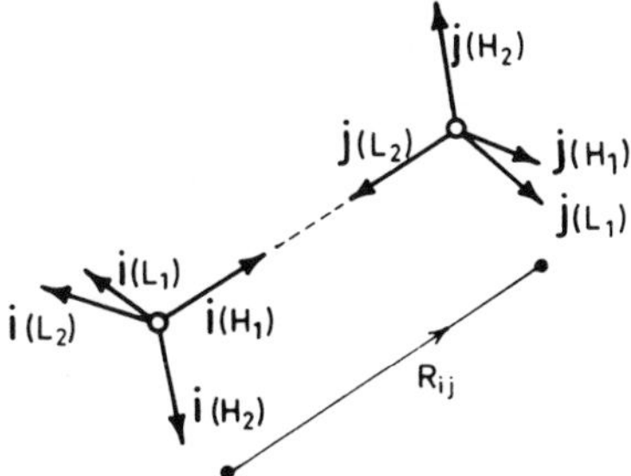

Fig. 7. One possible configuration for hydrogen bonding between molecules i and j. The distance $R_{ij} = | \mathbf{R}_j - \mathbf{R}_i |$ must be about R_{H}. The vector $\mathbf{i}(H_1)$ is pointing along $\mathbf{R}_{ij}$, while the vector $\mathbf{j}(L_2)$ is pointing along $-\mathbf{R}_{ij}$.

angular condition is split into a sum of eight terms, only one of which may be appreciably different from zero at any given configuration of i and j.

The total effective pair potential may now be written as

$$U(i, j) = U_{LJ}(R_{ij}) + U_{EL}(i, j) + U_{\mathrm{HB}}(i, j) \tag{17}$$

For some purpose, the multipole interaction term U_{EL} may be reduced to the dipole–dipole interaction. However, in order to study the peculiar properties of the tetrahedral mode of packing, one may dispense with this term in a preliminary stage. Also, for such a study, one may simplify U_{HB} by identifying the role of O—H and O—L bond directions. This means that two water molecules will form a hydrogen bond at any one of the *sixteen* configurations that brings the unit vectors of two molecules into favorable directions. (In fact, the ultimate effect of U_{HB} is to place four oxygen molecules at the vertices of a *regular* tetrahedron around a given molecule, a configuration that certainly does not depend on our recognition of the difference between an O—H and an O—L bond direction.) The particular choice of the Gaussian functions in eqn. (16), which have some convenient analytic features may prove useful in a theoretical study of the effect of coupling the orientational contribution U_{HB} to the spherical component U_{LJ} of the pair potential. The potential function suggested may also be used as an initial imput in an iterative procedure advocated by Stillinger[(1026)] to compute an optimal effective pair potential for liquid water.

We now turn to a second effective pair potential, based on the Bjerrum[(87)] four-point-charge model for a water molecule. Figure 8 displays the Bjerrum model for a water molecule. Four point charges are situated at the vertices of a regular tetrahedron, the center of which coincides with the center of the molecule. This model induces a pair potential function which has some characteristics of a hydrogen bond potential. It has been employed previously[(66,67)] and will be discussed in the next section. The predominance of the tetrahedral bonding is produced by a combination of Coulombic interactions, due to the point charges at the vertices of the tetrahedron of each molecule, and a Lennard-Jones-type of potential function. Thus,

$$U(1, 2) = U_{LJ}(R_{12}) + S(R_{12})U_{\mathrm{HB}}(1, 2) \tag{18}$$

where U_{HB} comprises the interaction among the 16 pairs of point charges and $S(R_{12})$ is a switching function, the employment of which is necessary to avoid divergence of the potential function when two point charges of opposite signs come close to each other.[(66,67)] The various ingredients of this potential function are shown schematically in Fig. 9.

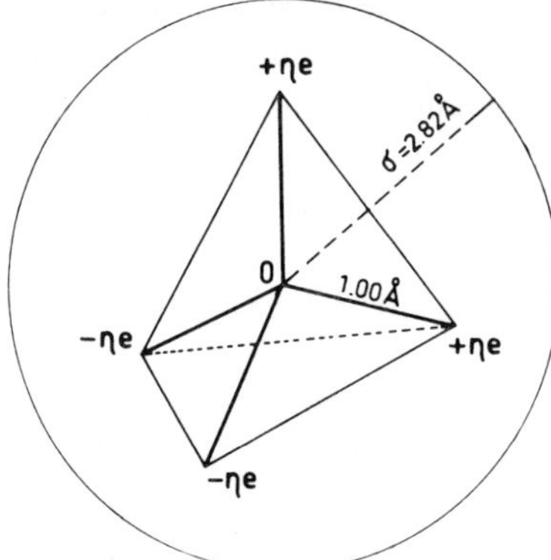

Fig. 8. Bjerrum's four-point-charge model for a water molecule. The oxygen atom coincides with the center of a regular tetrahedron. The fraction of charge $\pm\eta e$ is placed at the vertices of the tetrahedron at a distance of 1.00 Å from the center. The van der Waals diameter of the molecule is 2.82 Å.

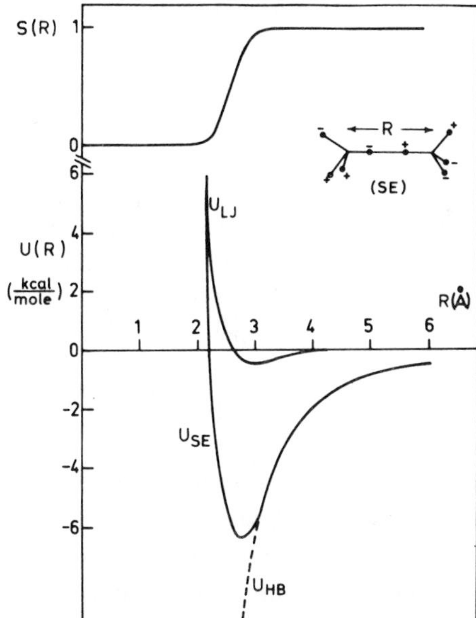

Fig. 9. Schematic description of the various ingredients of the potential function based on the Bjerrum model for water [eqn. (18)]; $U_{\mathrm{SE}}(R)$ is the full potential function for the symmetric eclipse (SE) approach of two water molecules (drawn at the upper right part of the figure). The three components of U_{SE} for the same line of approach are also included.

This function has been employed in an approximate Percus–Yevick equation and will be described in the next section. A very similar potential function has also been employed in the Monte Carlo calculation to be discussed in Section 4.

3. APPLICATION OF THE PERCUS–YEVICK EQUATION

The Percus–Yevick (PY) equation[67,369,832] is an integral equation for the pair correlation function. So far, it has been successfully applied only to simple liquids. Recently,[66,67] an approximate version of this equation has been employed for more complex fluids. The formal generalization of the PY equation to an angle-dependent potential function is a straightforward matter. One defines the two auxiliary functions

$$f(1, 2) = \exp[-\beta U(1, 2)] - 1 \tag{19}$$

and

$$y(1, 2) = g^{(2)}(1, 2) \exp[+\beta U(1, 2)] \tag{20}$$

The PY equation then reads

$$y(1, 2) = 1 + (\varrho/8\pi^2) \int y(1, 3)f(1, 3)[y(3, 2)f(3, 2) + y(3, 2) - 1]\, d(3) \tag{21}$$

which is an integral equation for the function y when f is presumed to be a known function. Solution of this equation should provide an approximate function $y(1, 2)$ and hence $g(1, 2)$. Unfortunately the six-dimensional integration of the r.h.s imposes a severe limitation to the extent to which a numerical integration of this equation can be performed.

Although effective methods are available for computing six-dimensional integrals,[206,470] the number of times these integrals have to be evaluated is exceedingly large. Therefore, it seems unfeasible at present to solve eqn. (21) by direct iterative methods.

An approximate version of this equation has been suggested[66,67] which greatly reduces the computation time, but still preserves some of the features that we expect from the solution.

The approximation can be stated as follows: The potential of average force $W^{(2)}(1, 2)$ is related to the pair correlation function by[509]

$$\begin{aligned} g^{(2)}(1, 2) &= \exp[-\beta W^{(2)}(1, 2)] \\ &= \exp[-\beta U(1, 2) - \beta Q(1, 2)] \end{aligned} \tag{22}$$

where in the last expression on the r.h.s of eqn. (22) the potential of average force has been divided into two parts: the direct pair potential U and an indirect component Q. The approximation involves the assumption that the angular dependence of $W^{(2)}(1, 2)$ is carried only by $U(1, 2)$, while Q is assumed to be a function of distance only. Thus,

$$W^{(2)}(\mathbf{X}_1, \mathbf{X}_2) = U(\mathbf{X}_1, \mathbf{X}_2) + Q(\mathbf{R}_1, \mathbf{R}_2) \tag{23}$$

The integral equation is then transformed into

$$\begin{aligned} y(\mathbf{R}_1, \mathbf{R}_2) = 1 &+ [\varrho/(8\pi^2)^3] \int y(\mathbf{R}_1, \mathbf{R}_3) y(\mathbf{R}_3, \mathbf{R}_1) F_2(\mathbf{R}_1, \mathbf{R}_2, \mathbf{R}_3)\, d\mathbf{R}_3 \\ &+ [\varrho/(8\pi^2)^3] \int y(\mathbf{R}_1, \mathbf{R}_3)[y(\mathbf{R}_3, \mathbf{R}_2) - 1] F_1(\mathbf{R}_1, \mathbf{R}_3)\, d\mathbf{R}_3 \end{aligned} \tag{24}$$

Where now we have integrals over the locational coordinates only, the angular dependence has been averaged out in the definitions of F_1 and F_2,

$$F_1(\mathbf{R}_1, \mathbf{R}_3) = 8\pi^2 \int f(\mathbf{X}_1, \mathbf{X}_3)\, d\mathbf{\Omega}_1\, d\mathbf{\Omega}_3 \tag{25}$$

$$F_2(\mathbf{R}_1, \mathbf{R}_2, \mathbf{R}_3) = \int f(\mathbf{X}_1, \mathbf{X}_3) f(\mathbf{X}_3, \mathbf{X}_2)\, d\mathbf{\Omega}_1\, d\mathbf{\Omega}_2\, d\mathbf{\Omega}_3 \tag{26}$$

Once the functions F_1 and F_2 are computed for a selected value of their arguments, the solution of eqn. (21) by direct iterative methods becomes a straightforward matter. In the actual computation of this equation, it was found[66] that no solution could be obtained for the full potential function $U(1, 2)$, as defined in eqn. (18). Therefore, a procedure in which angle-dependent part U_{HB} is gradually coupled to U_{LJ} has been attempted. Thus, one writes

$$U(1, 2, \lambda) = U_{LJ}(R_{12}) + \lambda S(R_{12}) U_{\mathrm{HB}}(1, 2) \tag{27}$$

where $0 \leq \lambda \leq 1$ is a coupling parameter. Initially, we put $\lambda = 0$. The solution of eqn. (21) for this case corresponds to a system of simple spherical particles. Specifically, in the present case, we have chosen the Lennard-Jones parameters for neon [eqn. (12)], and therefore the solution obtained corresponds to neon at 4°C and at the density of water at this temperature ($\varrho = 0.03346$ Å^{-3}). Figure 10 displays the function $g(\mathrm{R})$ obtained for this case. We note the location of the first two peaks at 2.8 Å and 5.6 Å, respectively. A very flat peak at 8.6 Å exists, but cannot be seen on the scale of Fig. 10. The third peak develops at higher densities, and this is demonstrated in Fig. 11, which corresponds to the same case but for densities of $\varrho = 0.0450$

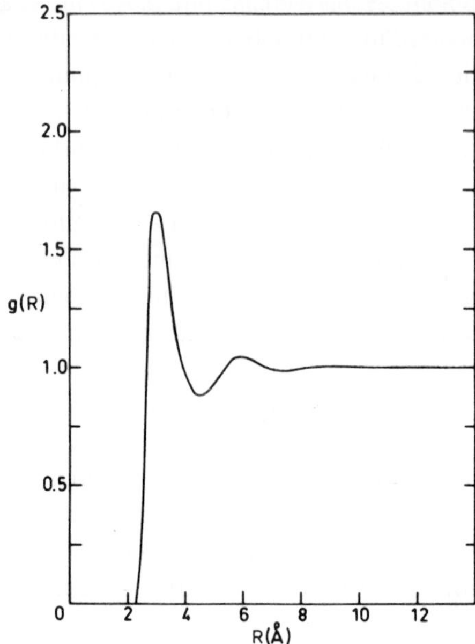

Fig. 10. Radial distribution function $g(R)$ for neon [with U_{LJ} of eqns. (11) and (12) at 4°C and $\varrho = 0.03346$ Å^{-3}] obtained by numerical solution of the PY equation.[66]

and $\varrho = 0.05045$ Å^{-3}. The solution obtained from the case $\lambda = 0$ has been utilized as an initial input in the r.h.s. of eqn. (21) to obtain a solution for $\lambda \neq 0$. This procedure is continued so that in every step the final output for a given value of λ is used as an input for a new value $\lambda + \Delta\lambda$. Such a procedure, though very tedious, reveals some of the most interesting features of the RDF of water. Figure 12 demonstrates the response of the function $g(R)$ to the increase of λ within the approximate scheme employed. The most important result is a gradual shift of the location of the second peak leftward, from an initial value of 5.6 Å for $\lambda = 0$ to about 4.8 Å for $\lambda \approx 0.3$ (for more details, see Ref. 66). Also, a clear gradual shift of the location of the third peak from 8.6 Å to 7.4 Å is observed. (The location of the third peak of the experimental RDF for water at 4°C is at about 6.9 Å.)

Thus, the locations of the first, second, and third peaks clearly tend to the corresponding values of the experimental RDF of water. One cannot expect much more from an approximate PY equation (which itself is an approximate integral equation). An apparent failure of this method is the coordination number, which has been calculated to be about 7.4 instead of the experimental value of about 4.4. It is possible that for the full PY equation [i.e., without introducing the approximation in eqn. (23)], one could have recovered the correct coordination number. It is difficult to speculate about this aspect at present, since no convergent solution could have been obtained in this method for $\lambda \lesssim 0.3$. However, from an examination of the solution of the PY equation for a two-dimensional model[(66a)] of "waterlike" particles it seems that the correct coordination number may be obtainable by the PY equation.

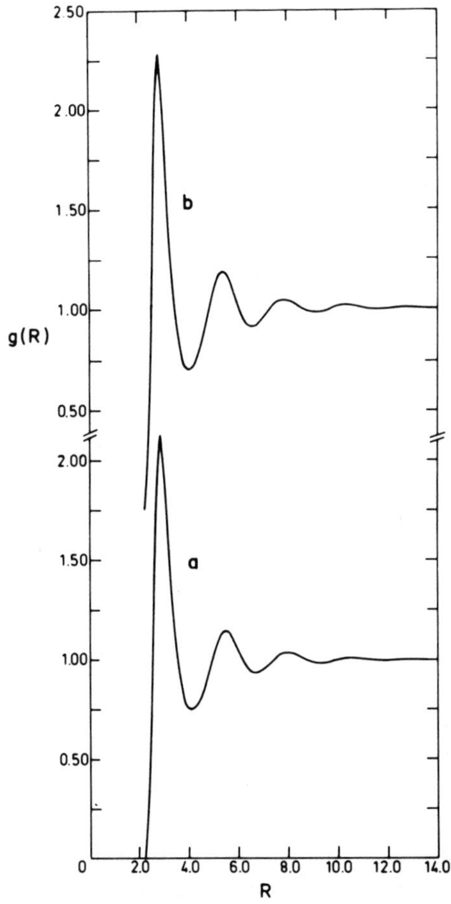

Fig. 11. Radial distribution function $g(R)$ for neon at 4°C and densities (a) $\varrho = 0.0450$ Å^{-3}, (b) $\varrho = 0.05045$ Å^{-3}, obtained by numerical solution of the PY equation (Ben-Naim, unpublished results).

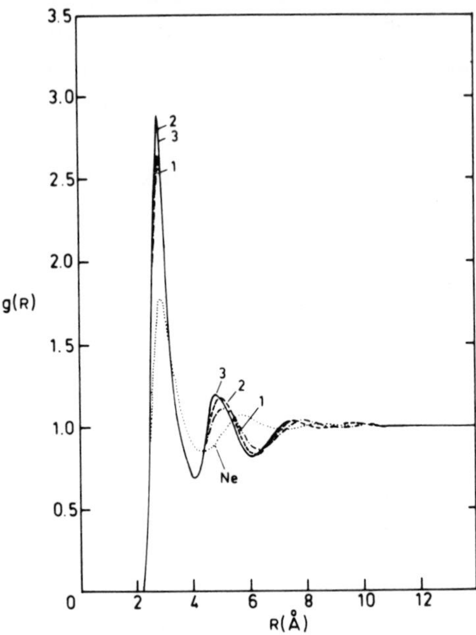

Fig. 12. Radial distribution function $g(R)$ for "waterlike" particles interacting with the potential function described in eqn. (27). The numbers indicate increasing values of the coupling parameter λ between $0 \geq \lambda \geq 0.3$. (For more details, see Ref. 66.)

4. APPLICATION OF THE MONTE CARLO TECHNIQUE

There are essentially two available methods to perform a "computer experiment" on a sample of fluid. The first one, referred to as molecular dynamics, seeks to solve the equation of motion of a set of M particles, where M is of the order of a few hundred. This technique has been extensively used for a system of simple particles.[7] Until now, no published work has appeared on the application of this method to liquid water, but it is expected that such a report will appear in the near future.*

The second method is referred to as the Monte Carlo method.[1182] In this case, the momentum partition function is evaluated analytically and therefore attention is focused on the configurational partition function only.

* F. H. Stillinger, Jr., private communication.

Basically, the idea is to approximate integrals over the configurational phase space of M particles by a sum of a finite number of terms. Thus, if $F(\mathbf{X}_1 \cdots \mathbf{X}_M)$ is a function of the "state" of the system [by state, we mean here the particular spatial and orientational configuration of the classical system of M particles, denoted by $(\mathbf{X}_1, \ldots, \mathbf{X}_M)$], then the statistical mechanical average (in the canonical ensemble, specified by a constant volume V and temperature T)

$$\bar{F} = \frac{\int \cdots \int F(\mathbf{X}_1, \ldots, \mathbf{X}_M) \exp[-\beta U(\mathbf{X}_1, \ldots, \mathbf{X}_M)]\, d\mathbf{X}_1 \cdots d\mathbf{X}_M}{\int \cdots \int \exp[-\beta U(\mathbf{X}_1, \ldots, \mathbf{X}_M)]\, d\mathbf{X}_1 \cdots d\mathbf{X}_M} \tag{28}$$

is approximated by the sum

$$\bar{F} \approx \left[\sum_{i=1}^{N_s} F_i \exp(-\beta U_i)\right] \Big/ \left[\sum_{i=1}^{N_s} \exp(-\beta U_i)\right] = \sum_{i=1}^{N_s} F_i P_i \tag{29}$$

where the summation is carried out over a finite number of states $i = 1, \ldots, N_s$ and P_i is the probability of occurrence of state i. One expects that, by selecting an infinitely large set of states in such a way that each point in the configurational space is arbitrarily closely approached, the sum in eqn. (29) will converge to the average value $\bar{F}$. The art of the Monte Carlo method is to find a judicious recipe for selecting a finite set of states that will be small enough to be amenable for computation, and yet large enough to produce a plausible approximation to $\bar{F}$. An inherent limitation of this method is also imposed by the number of degrees of freedom that can be handled by the computer. Particularly, in the case of water, each particle should be specified by six coordinates (three positions and three angles—assuming that the molecule is rigid). Therefore, the number of particles that can be included in such calculations is severely restricted.

Recently, Barker and Watts[(51)] applied this method to "water-like" particles. The model used was similar to the one described in Section 2, and was based on the four-point-charge model. Also, the intermolecular potential operating between two particles employed in this calculations was similar to the one described in eqn. (18). [A hard-sphere diameter $\sigma = 2.0$ Å was assigned to the particles. The switch from the Coulombic potential to the hard-sphere repulsive potential was therefore more abrupt than the smoothed change accomplished by the function $S(R)$ in eqn. (18).]

The sample consisted of 64 particles at 25°C, which is quite a small number compared with a typical sample of several hundred particles employed for a simple fluid. The resulting RDF obtained by Barker and Watts is reproduced in Fig. 13. As these authors noted, these constitutes preliminary results and a more extended project along the above lines is now being undertaken. The important features of these results are the following: The location of the first peak is clearly to the left of that in the experimental curve. This seems to be easily remedied by replacing the hard cutoff by a smoothly changing function. More important is the location of the second peak, which falls at about 5.8 Å (as may be estimated from Fig. 13), which is at nearly twice the distance of the location of the first peak, and therefore is more typical of a packing of simple spheres than of water. The number of neighbors within a sphere of 3.5 Å was found to be 6.4, which is close enought to the experimental value of about 4.4. Also, some derived thermodynamic quantities, such as the average potential energy and the heat capacity, agree reasonably well with the experimental values. It is to be hoped that more could be learned about the local structure and the thermodynamic properties of water using the Monte Carlo technique with an improved pair potential.

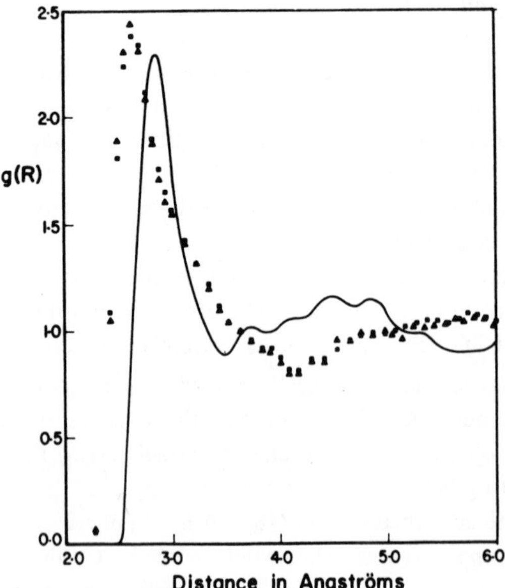

Fig. 13. Comparison between the experimental radial distribution function for water at 25°C (full curve) and the corresponding results from Monte Carlo computations (triangles and squares). Reproduced, by permission, from Ref. 51.

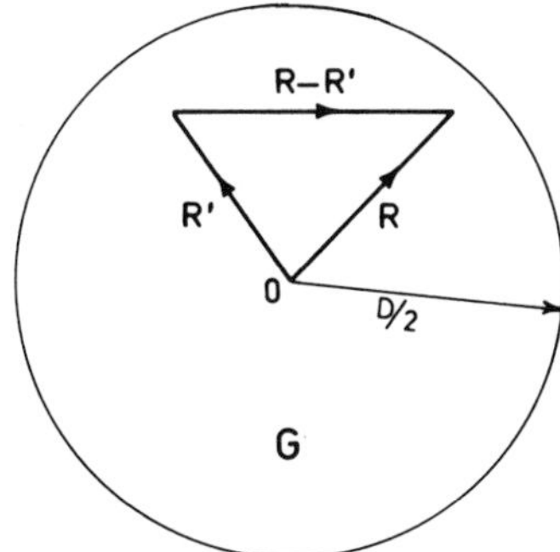

Fig. 14. A spherical region of diameter D for which the density fluctuation is computed by eqn. (30). $\mathbf{R}$ and $\mathbf{R}'$ are two vectors pointing to any pair of points within this region.

5. SOME SPECIFIC APPLICATIONS OF THE PAIR CORRELATION FUNCTION

As noted earlier, the pair correlation function, if known, [i.e., $g^{(2)}(\mathbf{X}_1, \mathbf{X}_2)$, not just the RDF] can be employed in standard statistical mechanical expressions to yield information on the macroscopic properties of the liquid.[67,509] In this section, we shall seek to exploit the available function to derive some more structural information on the *molecular* level.

5.1. Density Fluctuations in Water

Fisher and Adamovich[333] have utilized the experimental RDF of water to obtain some interesting information on local density fluctuations in the liquid.

Let G be a spherical region of diameter D. For any specific configuration of the system, we may find N_G molecules within G; the average value of N_G will be denoted by $\bar{N}_G$. A theorem in statistical mechanics[332] provides a link between the mean fluctuation of the number of particles occupying G and integrals involving the RDF*:

$$\langle(N_G - \bar{N}_G)^1\rangle_{\text{av}} - \bar{N}_G = \varrho^2 \int_{V_G} d\mathbf{R} \int_{V_G} d\mathbf{R}'[g(|\mathbf{R} - \mathbf{R}'|) - 1] \quad (30)$$

where ϱ is the bulk density, V_G is the volume of the spherical region, $\mathbf{R}$ and $\mathbf{R}'$ are two vectors that can be chosen as originating from some fixed origin (Fig. 14), and the integral on the r.h.s of eqn. (30) extends over the

* We shall denote either by bar or by $\langle\ \rangle_{\text{av}}$ average quantities.

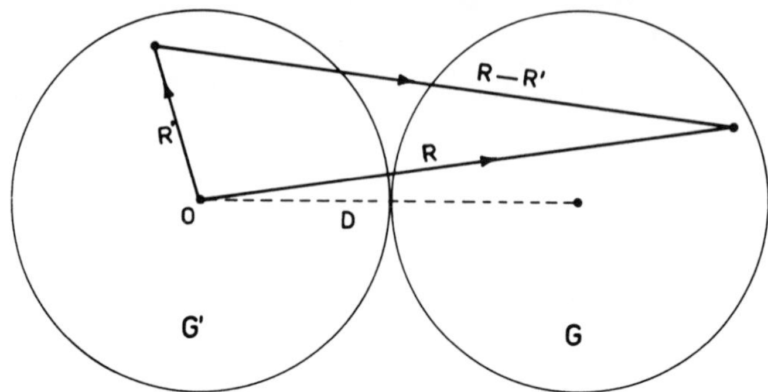

Fig. 15. Two spherical regions of equal diameter at contact, for which the correlation in density fluctuation is computed by eqn. (31). **R** and **R**′ are two vectors over which the integration is carried out over the regions G and G', respectively.

chosen volume V_G. A second quantity of interest is the correlation between the density fluctuations in two adjacent regions G and G', which is given by[332]

$$\langle (N_G - \bar{N}_G)(N_{G'} - \bar{N}_{G'}) \rangle_{\text{av}}$$
$$= \bar{N}_{G\cap G'} + \varrho^2 \int_{V_G} d\mathbf{R} \int_{V_{G'}} d\mathbf{R}' [g(|\mathbf{R} - \mathbf{R}'|) - 1] \tag{31}$$

Here $\bar{N}_{G\cap G'}$ is the average density in the intersection region $G \cap G'$ between G and G'. For two nonoverlapping, adjacent spheres of equal diameters (Fig. 15), $\bar{N}_{G\cap G'}$ vanishes, and both the integrals in eqns. (30) and (31) can be transformed into a one-dimensional integral (this is possible since g is a function of the scalar distance $|\mathbf{R} - \mathbf{R}'|$ only). Using the available experimental RDF for water [supplemented by the asymptotic behavior[332] of $g(R) - 1 \approx (a/R)\cos(\beta R + \delta)\, e^{-\alpha R}$ for $R > 8.2$ Å], Fisher and Adamovich computed both integrals as a function of the diameter of the spheres D.* Note that for large spheres, eqn. (30) leads to the thermodynamic limit

$$C_G(D) = \langle (N_G - \bar{N}_G)^2 \rangle_{\text{av}} / \bar{N}_G \xrightarrow{D\to\infty} \varrho k T \varkappa_T \tag{32}$$

where $\varkappa_T$ is the isothermal compressibility, $\varkappa_T = -V^{-1}(\partial V/\partial P)_T$. For spheres of finite diameter, the relative mean fluctuation in G is always positive and it reaches unity for $D \to 0$. Because of the limiting form of

* The computation of the integrals in eqns. (30) and (31) has been repeated using data from Ref. 783. The curve of Fig. 16 could not be reproduced by this data.

eqn. (32), one may interpret the quantity

$$\varkappa_L = \langle (N_G - \bar{N}_G)^2 \rangle_{\text{av}} / \bar{N}_G \varrho kT \tag{33}$$

as a local compressibility in G. The results depicted in Fig. 16 show that the local compressibility in small regions is considerably larger than the thermodynamic limit $\varkappa_T$. This value falls closely, but not exactly, to the value of $\varkappa_T$ at about $D = 10$ Å.

The second quantity plotted in Fig. 16 is the relative mean correlation between adjacent spherical regions of equal diameter:

$$C_{G,G'}(D) = \langle (N_G - \bar{N}_G)(N_{G'} - \bar{N}_{G'}) \rangle_{\text{av}} / \langle (N_G - \bar{N}_G)^2 \rangle_{\text{av}} \tag{34}$$

For very large diameters $D > 30$ Å, the density fluctuations in adjacent regions become statistically independent. For $D < 8$ Å, the correlation is negative, and small. Fisher and Adamovich have interpreted this behavior in a geometric fashion. A molecule leaving the region G has a high probability of entering G'. Therefore, a negative fluctuation in G is likely to produce a positive fluctuation in G'. In particular, one finds a minimum of $D_{G,G'}(D)$ at $D \approx 3$ Å, which is of the order of magnitude of the molecular diameter of the water molecule. This is consistent with the interpretation given above that the correlation in density fluctuations is dominated by the jump of a single molecule from one region to another.

The most interesting finding, however, is the behavior of $C_{G,G'}(D)$ for spheres of larger diameter, $9 \lesssim D \lesssim 25$ Å. Here, the relative correlation is positive, with a maximum at about $D = 16$ Å. This leads to the following conclusion: A positive value of $C_{G,G'}(D)$ will arise whenever the deviation from the average $\bar{N}_G$ in G has the same sign as the deviation from the average

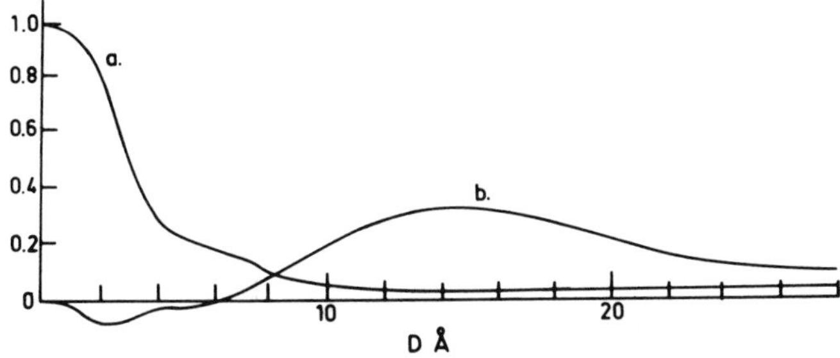

Fig. 16. Density fluctuation (a) in a single region $C_G(D)$ [eqn. (32)] and (b) in a pair of adjacent regions $C_{G,G'}(D)$ [eqn. (34)]. These computed curves refer to water at 25°C and 1 bar (see Ref. 333).

$\bar{N}_{G'}$ in G', which means that G and G' must be in the same "fluctuating region," i.e., contraction or expansion takes place in both G and G' simultaneously. Therefore, these authors conclude that the average diameter of a single fluctuating region in liquid water must be of the order of about 30 Å.

5.2. Local Structure Index for Water

A different application of the compressibility equation

$$\varrho kT\varkappa_T = 1 + \varrho \int_0^\infty [g(R) - 1]4\pi R^2\, dR \tag{35}$$

has been suggested by Stillinger and Ben-Naim.[1027]

The basic idea is to apply eqn. (35) to liquid water, viewed as a mixture of oxygen and hydrogen. In this case, one writes

$$\varrho kT\varkappa_T = 1 + \int [g_{\text{OO}}(R) - 1]4\pi R^2\, dR \tag{36}$$

$$2\varrho kT\varkappa_T = 1 + 2\varrho \int [g_{\text{HH}}(R) - 1]4\pi R^2\, dR \tag{37}$$

According to eqn. (36), the system is regarded as an assembly of oxygen atoms, while the hydrogens merely play the role of providing the potential function for the oxygens. In eqn. (37), the roles of the oxygens and hydrogens are reversed. Note that the total density of the hydrogens is 2ϱ, ϱ being the number density of water molecules. We can now view water as a mixture of doubly charged anions O^{-2}, and singly charged cations H^+. The condition of local electroneutrality imposes the following two relations:

$$1 = \varrho \int [g_{\text{OH}}(R) - g_{\text{OO}}(R)]4\pi R^2\, dR \tag{38}$$

$$1 = 2\varrho \int [g_{\text{OH}}(R) - g_{\text{HH}}(R)]4\pi R^2\, dR \tag{39}$$

The first relation states that the total net charge around a given oxygen must be equal to the net charge (but with opposite sign) on the oxygen, this is seen more clearly if we write eqn. (38) in the form

$$2e = e(2\varrho) \int g_{\text{OH}}(R)4\pi R^2\, dR - 2e\varrho \int g_{\text{OO}}(R)4\pi R^2\, dR \tag{40}$$

where in the r.h.s., the contributions due to hydrogens (with density 2ϱ and charge $+e$) and due to oxygens (with density ϱ and charge $-2e$) are explicitly expressed in the two terms.

Similarly, eqn. (39) states that the total net charge around a given hydrogen must be equal to the net charge (but with opposite sign) on the hydrogen; eqn. (38) can be used to eliminate g_{OO} from eqn. (36) [the same result is obtained by eliminating g_{HH} from eqns. (37) and (39)]. The final equation obtained is

$$kT\varkappa_T = \int_0^\infty [g_{OH}(R) - 1]4\pi R^2\, dR \tag{41}$$

This can obviously be rewritten as

$$kT\varkappa_T = \int_0^{R^*} [g_{OH}(R) - 1]4\pi R^2\, dR + \int_{R^*}^\infty [g_{OH}(R) - 1]4\pi R^2\, dR \tag{42}$$

If we choose R^* such that the sphere of radius R^* about an oxygen will include at most two hydrogens, the first integral on the r.h.s. of eqn. (42) becomes trivial.

We now define the quantity Δ_H, which may be interpreted as the local average excess, or deficiency, of protons beyond the sphere of radius R^* about the oxygen,

$$\Delta_H = 2\varrho \int_{R^*}^\infty [g_{OH}(R) - 1]4\pi R^2\, dR \tag{43}$$

Using eqn. (42) we obtain a practical connection with experimental quantities,

$$\Delta_H = 2kT\varrho\varkappa_T - 2 + [8\pi\varrho(R^*)^3/3] \tag{44}$$

The quantity Δ_H has been calculated (with a choice of $R^* = 1.75$ Å) for water and some nonaqueous liquid, as well as for dilute aqueous solutions of various electrolytes.(1027) Figure 17 demonstrates the difference between water and liquid methane (which can be similarly treated as a mixture of carbon and hydrogen(1027)). The important finding is that Δ_H for water is negative and, at atmospheric pressure, has a negative temperature dependence with an apparent minimum at about 120°C. The corresponding values for liquid methane are positive [in this case, g_{CH} replaces g_{OH} in eqn. (43)]. It is difficult to offer a precise interpretation of these findings. However, in view of a well-known relation between the composition-fluctuation and the pair correlation function for binary systems,(609)

$$\int_0^\infty [g_{\alpha\beta}(R) - 1]4\pi R^2\, dR = V\langle(N_\alpha - \bar{N}_\alpha)(N_\beta - \bar{N}_\beta)\rangle_{av}/\bar{N}_\alpha\bar{N}_\beta \tag{45}$$

we are tempted to interpret Δ_H, as defined in eqn. (43), in terms of the corre-

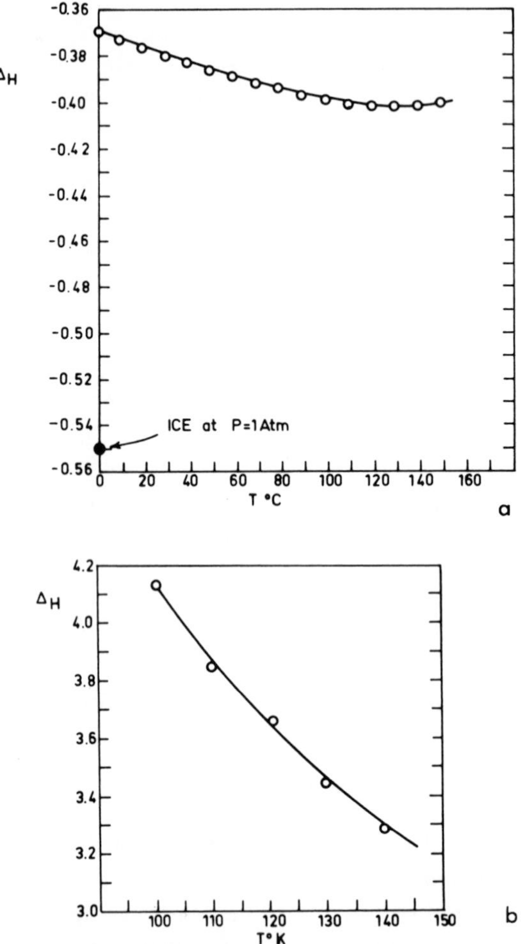

Fig. 17. Dependence of Δ_H on temperature: (a) for liquid water a 1 bar the values (above 100°C were obtained by extrapolation[1027]); (b) for liquid methane.

lation between the density fluctuations of oxygens and hydrogens outside the sphere of radius R^*.

Another interesting finding is the effect of various salts on Δ_H. Ions which are presumed to be "structure breakers" tend to reduce Δ_H, while "structure-former" ions increase Δ_H. This is a strong indication that the quantity Δ_H may serve as an index for local structure around a given water molecule.

5.3. Hole and Particle Distributions in Water

A further quantity which may furnish information on the mode of molecular packing in water can be extracted from the available experimental RDF. We recall that the experimental RDF provides a firm indication that the average local coordination number and the most probable distance for nearest and second-nearest neighbors in water and ice are similar. With this information at our disposal, there are still innumerable ways of packing water molecules consistent with this information; for instance, the hexagonal and the cubic forms of ice represent two possibilities of a regular mode of packing in conformity with the known RDF.

The hole–hole pair correlation function may be used as an additional criterion for accepting or rejecting various models for water.[65] This function is defined in the following manner[510]: Let us observe a small cell in the fluid, the diameter D of which is small enough so that no more than a single molecule can occupy it at any given time.

The probability of finding this cell occupied is evidently

$$P_1 = (4\pi/3)(D/2)^3\varrho \tag{46}$$

where ϱ is the number density. Since the cell under observation is either occupied or vacant, the probability of finding it empty is

$$P_0 = 1 - P_1 \tag{47}$$

The relation between the hole–hole correlation function and the RDF was first derived by Hill.[510] If $P_{HH}(R)$ denotes the probability that two cells at a distance R are simultaneously empty, then the hole–hole correlation function is defined by

$$g_{HH}(R) = P_{HH}(R)/P_0^2 \tag{48}$$

The function $g_{HH}(R)$ has been computed for argon (at 84.25°K and 0.71 bar) and for water (at 4°C and 1 bar). Figure 18 reproduces the plot of $g_{HH}(R^*)$ as a function of the reduced distance R^*. The most interesting finding is the following: there are two peaks for the $g_{HH}(R^*)$ of argon at about $R^* = 1$ and $R^* = 2$ which correspond to the peaks in the RDF. This reflects a kind of symmetry between particle–particle distribution and hole–hole distribution in a simple fluid. On the other hand, the single peak of the function $g_{HH}(R^*)$ for water at $R^* = 1.46$ (or at $R = 4.1$ Å) clearly indicates that no such symmetry exists in liquid water. The location of the peak at 4.1 Å may also indicate which, out of the many possible structures

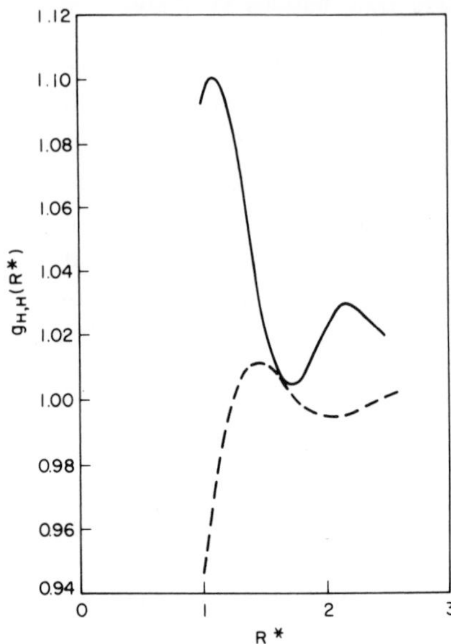

Fig. 18. $g_{HH}(R^*)$ for water (dashed line) at 4°C and 1 bar and for argon (solid line) at 84.25°K and 0.71 bar as a function of the reduced distance R^*.

suggested for water, is consistent with this finding. Unfortunately, the sizes of the holes for which these computations have been performed are too small to serve as conclusive evidence for or against a particular model. It will certainly be worthwhile to extend these calculations to larger cells, an extension that, by the present method, will require information on higher-order correlation functions, which is presently unavailable.

5.4. A Possible Exact Definition of the Structure of Water

It is instructive to end this chapter with a discussion of a possible definition of the structure of water. The present definition is based on the knowledge of the full pair correlation function $g^{(2)}(\mathbf{X}_1, \mathbf{X}_2)$. It was earlier[66a] applied to a two-dimensional system of waterlike particles. This definition is by no means a unique one, and some other possibilities will be indicated below.

We start with the recognition of the fact that hydrogen bonds exist and they are likely to be responsible for the particular structure of water. There is no universal definition of the concept of the hydrogen bond. Following our discussion of the features of the pair potential for water in Section 2, we may recognize water molecules as being hydrogen-bonded whenever their mutual configuration satisfies certain conditions.

Referring to the notation in Section 2, we may define the configuration favoring hydrogen bond formation by*

$$\begin{aligned} -\sigma' < R_{ij} - R_{\mathrm{H}} < \sigma'; \qquad -\sigma < \mathbf{i}(H_\alpha) \cdot \mathbf{n} - 1 < \sigma; \\ -\sigma < \mathbf{j}(L_\beta) \cdot \mathbf{n} + 1 < \sigma \end{aligned} \tag{49a}$$

or

$$\begin{aligned} -\sigma' < R_{ij} - R_{\mathrm{H}} < \sigma'; \qquad -\sigma < \mathbf{i}(L_\alpha) \cdot \mathbf{n} - 1 < \sigma; \\ -\sigma < \mathbf{j}(H_\beta) \cdot \mathbf{n} + 1 < \sigma \end{aligned} \tag{49b}$$

for any pair $\alpha, \beta = 1, 2$. The conditions eqns. (49a, b) require that the distance R_{ij} will be confined between certain limits and that the O—H bond direction of one molecule will be in a favorable orientation relative to the O—L bond direction of the second molecule. The quantities σ and σ' allow for some freedom in stretching and bending of the bond. The set of all configurations $(\mathbf{X}_i, \mathbf{X}_j)$ conforming to the conditions in eqn. (49) will be denoted by $\mathbf{X}_{ij}(\mathrm{HB})$. Thus, for any given configuration of the pair of molecules i and j, we shall say that they are hydrogen-bonded if and only if

$$(\mathbf{X}_i, \mathbf{X}_j) \in \mathbf{X}_{ij}(\mathrm{HB}) \tag{50}$$

Note that $\mathbf{X}_{ij}(\mathrm{HB})$ is a set of configurations, the measure of which depends on the choice of σ and σ'. We now define the characteristic function

$$\begin{aligned} B(\mathbf{X}_i, \mathbf{X}_j) &= 1 \quad \text{if} \quad (\mathbf{X}_i, \mathbf{X}_j) \in \mathbf{X}_{ij}(\mathrm{HB}) \\ &= 0 \quad \text{if} \quad (\mathbf{X}_i, \mathbf{X}_j) \notin X_{ij}(\mathrm{HB}) \end{aligned} \tag{51}$$

According to a general theorem of classical statistical mechanics, the average value of the function B is

$$\bar{B} = (1/2) \int d\mathbf{X}_1 \int d\mathbf{X}_2 B(\mathbf{X}_1, \mathbf{X}_2) \varrho^{(2)}(\mathbf{X}_1, \mathbf{X}_2) \tag{52}$$

Since the total number of pairs of molecules in the system is

$$(1/2) \int d\mathbf{X}_1 \int d\mathbf{X}_2 \varrho^{(2)}(\mathbf{X}_1, \mathbf{X}_2) = N(N-1)/2 \tag{53}$$

* σ and σ' in (49) may be chosen at will. One possibility is to make them equal to σ and σ' of eqn. (16).

it is evident that by introducing the function $B(\mathbf{X}_1, \mathbf{X}_2)$ into the integrand, we shall count only those pairs that are bonded according to eqn. (50), so that $\bar{B}$ may be interpreted as the average number of pairs which are hydrogen-bonded according to our conventions. The average number of bonds per molecule is thus

$$\bar{n}_{\mathrm{HB}} = \bar{B}/N = (1/2N) \int d\mathbf{X}_1 \int d\mathbf{X}_2 B(\mathbf{X}_1, \mathbf{X}_2) \varrho^{(2)}(\mathbf{X}_1, \mathbf{X}_2) \tag{54}$$

This quantity may be used as an index for measuring the degree of structure of the system. It is obvious that this quantity depends on the parameters σ and σ'. It will be larger, the larger are these parameters. The method outlined above may be generalized as follows. A group of M molecules will be said to form a bonded cluster if their configuration $\mathbf{X}_1, \ldots, \mathbf{X}_M$ fulfill certain specified conditions, which by generalization of the notation of (50) and (51), may be written as follows:

Particles $1, \ldots, M$ form a bonded cluster if and only if

$$(\mathbf{X}_1, \ldots, \mathbf{X}_M) \in \mathbf{X}_{12\ldots M}(\mathrm{HB}) \tag{55}$$

and similarly one defines

$$\begin{aligned} B(\mathbf{X}_1, \ldots, \mathbf{X}_M) &= 1 \quad \text{if} \quad (\mathbf{X}_1, \ldots, \mathbf{X}_M) \in \mathbf{X}_{12\ldots M}(\mathrm{HB}) \\ &= 0 \quad \text{otherwise} \end{aligned} \tag{56}$$

and

$$\bar{B}_M = (1/M!) \int d\mathbf{X}_1 \cdots \int d\mathbf{X}_M B(\mathbf{X}_1, \ldots, \mathbf{X}_M) \varrho^{(2)}(\mathbf{X}_1, \ldots, \mathbf{X}_M) \tag{57}$$

where $\bar{B}_M$ is the average number of bonded clusters with M molecules. It is important to note that a bonded cluster with M molecules may be a part of a larger bonded cluster. Furthermore, the set of configurations included in the symbol $\mathbf{X}_{12\ldots M}(\mathrm{HB})$ must be such that each of the molecules in the set $1, \ldots, M$ is bonded, in the sense of eqn. (50), to at least one member of the set.

The definitions outlined in this section have been based on the concept of the hydrogen bond. There are other ways of defining the degree of structure, independent of this concept. It should also be noted that a clear-cut distinction must be made between the formal definition of the structure and the operational definition based on a particular tool of measurement. The latter, though useful from the practical point of view, largely depends on our interpretation of the outcome of the experiments.

CHAPTER 12

Liquid Water—Acoustic Properties: Absorption and Relaxation

Charles M. Davis, Jr., and Jacek Jarzynski
Naval Research Laboratory
Washington, D.C.

1. INTRODUCTION

In general, liquids differ from solids primarily in their ability to flow. While in the process of flowing, however, a liquid is not in thermodynamic equilibrium. On a miscoscopic basis, for flow to occur, the molecules must rearrange relative to each other. In thermodynamic equilibrium, a liquid is completely characterized by a small number of state variables (e.g., the internal energy, volume, number of moles, total magnetic and electric moments, etc.). When a system initially in equilibrium is perturbed into a nonequilibrium state, the method by which equilibrium is restored is known as a relaxation process. Any such nonequilibrium state requires that parameters in addition to the state variables be specified for a complete characterization. In the present chapter, we will be concerned with structural relaxation. This is the process by which the molecules of a system "flow" from a nonequilibrium configuration to a new equilibrium configuration.

Associated with every relaxational process is the accompanying microscopic transport of some extensive property (e.g., mass, energy, etc.). For such a transport process to occur, the corresponding intensive variable must exhibit a gradient; thus, for molecules to flow, a chemical potential gradient must be established. Furthermore, every such transport process results in dissipation, which, in the case of small perturbations from equilibrium, can be described by a relaxation time and some characteristic coefficient, both of which are specified by the equilibrium thermodynamic parameters. The relaxation time and the corresponding characteristic coefficient are not independent of each other. Thus, the dissipation associated with flow may be described in terms of the viscosity, which is in turn related to the structural relaxation time. In most cases, when a system not at equilibrium tends toward equilibrium, more than one transport process is involved. In a real fluid, the pressure depends not only on the density, but also on the temperature. A pressure fluctuation in one part of a system may lead to both density and temperature fluctuations; thus, the flow of both internal energy and mass may occur. The corresponding transport coefficients are the thermal conductivity and the viscosity.

2. ULTRASONIC ABSORPTION

At this point, it will be convenient to restrict our discussion to those relaxation processes associated with density fluctuations. A similar discussion may be given for any other relaxation process one wishes to consider. A low-amplitude, longitudinal sound wave propagating through a liquid results in alternate regions of rarefaction and compression. Thus, a given region in the liquid is alternately subjected to perturbations favoring higher and lower densities. For low-frequency sound waves, these density perturbations occur slowly (relative to the structural relaxation time) and the liquid has time to flow. The corresponding viscosity exhibited is the zero-frequency value characteristic of a liquid. On the other hand, for high-frequency sound (i.e., sound whose period is very short compared to the structural relaxation time), no flow occurs and the viscosity goes to zero. In such a case, the liquid may be said to behave like an amorphous solid.[693]

The nature of the structural relaxation time can be best understood on a microscopic basis. Instantaneously, the particles in a liquid occupy positions relative to their neighbors reminiscent of the positions occupied in an amorphous solid. Unlike the case in an amorphous solid, where the molecules are frozen relative to each other, the particles in a liquid may structurally rearrange. The time for the structure to change is known as the structural relaxation time. The magnitudes of such thermodynamic quantities as the isothermal compressibility $\varkappa_T = -(1/V)(\partial V/\partial p)_T$ and its reciprocal, the bulk modulus K_T, as well as the thermal expansibility $\beta = (1/V)(\partial V/\partial T)_p$, depend on the ability of the volume to change with a change in pressure or temperature; therefore, like the viscosity, these parameters are also frequency-dependent. If a step function of pressure were applied to a liquid, the change in volume could be separated into two parts: first, a rapid, solidlike change (indicated with a subscript ∞); and second, a slower relaxational change (indicated with a subscript r). For a sinusoidally varying pressure, a similar situation results, but instead of discussing instantaneous and relaxational changes, one refers to in-phase and out-of-phase components of stress P and strain S. Thus, for example, for a single relaxation process, the isothermal bulk modulus may be expressed in the complex form[502]

$$K_T = -P/S = K_{T,0} + [K_{T,r}\omega^2\tau_v^2/(1+\omega^2\tau_v^2)] + i[K_{T,r}\omega\tau_v/(1+\omega^2\tau_v^2)] \quad (1)$$

where $K_{T,0}$ is the zero-frequency modulus, $K_{T,r}$ is the relaxation component defined as the difference between the infinite-frequency modulus $K_{T,\infty}$ and $K_{T,0}$, τ_v is the structural relaxation time for constant volume, and $\omega = 2\pi f$, where f is the frequency. The imaginary term in eqn. (1) arises as a result of energy loss and therefore serves to define the volume viscosity as

$$\eta_v = K_{T,r}\tau_v/(1+\omega^2\tau_v^2) \quad (2)$$

An expression similar to eqn. (1) may be written for $\varkappa_T$ as

$$\varkappa_T = \varkappa_{T,0} - [\varkappa_{T,r}\omega^2\tau_p^2/(1+\omega^2\tau_p^2)] - i[\varkappa_{T,r}\omega\tau_p/(1+\omega^2\tau_p^2)] \quad (3)$$

where the parameters in this expression are defined in a manner analogous to the case for K_T (e.g., $\varkappa_{T,r} = \varkappa_{T,0} - \varkappa_{T,\infty}$, and τ_p is the structural relaxation time at constant pressure). The structural relaxation times measured under various conditions can all be shown to be related to each other.[649] Thus, τ_v and τ_p are related by the expression

$$\tau_v = (\varkappa_{T,\infty}/\varkappa_{T,0})\tau_p \quad (4)$$

From the definitions of the various $\varkappa$'s and K's, we may write

$$\begin{aligned} K_{T,r} &= K_{T,\infty} - K_{T,0} = (1/\varkappa_{T,\infty}) - (1/\varkappa_{T,0}) \\ &= (\varkappa_{T,0} - \varkappa_{T,\infty})/(\varkappa_{T,\infty}\varkappa_{T,0}) = \varkappa_{T,r}/\varkappa_{T,\infty}\varkappa_{T,0} \end{aligned} \quad (5)$$

(Both $K_{T,0}$ and $K_{T,\infty}$, as well as $\varkappa_{T,0}$ and $\varkappa_{T,\infty}$, may in principle be experimentally measured, and therefore the relations $K_{T,\infty} = \varkappa_{T,\infty}^{-1}$ and $K_{T,0} = \varkappa_{T,0}^{-1}$ are legitimate thermodynamic expressions. The corresponding relaxational components are defined as differences and therefore cannot be experimentally measured. Furthermore, as shown by eqn. (5), they are not merely reciprocals of each other.)

In the case of water, the structural relaxation times[238] are on the order of 10^{-12} sec; therefore, with present technology, $\omega\tau_v \ll 1$ and eqn. (2) combined with eqns. (4) and (5) simplifies to

$$\eta_v = (\varkappa_{T,r}/\varkappa_{T,0}^2)\tau_p = K_{T,r}\tau_v \tag{6}$$

In a manner similar to eqn. (1), the static shear modulus G is given by the expression

$$G = [G_\infty \omega^2 \tau_s^{\,2}/(1 + \omega^2 \tau_s^{\,2})] + i\omega[G_\infty \tau_s/(1 + \omega^2 \tau_s^{\,2})] = \mathrm{Re}\, G + i\omega\eta_s(\omega) \tag{7}$$

where τ_s is the shear relaxation time and G_∞ is the high-frequency shear modulus. At low frequencies ($\omega\tau_s \ll 1$), $G \to 0$; therefore, there is no low-frequency shear modulus. Equation (7) also serves to define the frequency-dependent shear viscosity η_s:

$$\eta_s(\omega) = G_\infty \tau_s/(1 + \omega^2 \tau_s^{\,2}) \tag{8}$$

and at low frequencies ($\omega\tau_s \ll 1$),

$$\eta_s = G_\infty \tau_s \tag{9}$$

In order to obtain an expression for the sound absorption due to the propagation of longitudinal waves in a liquid, one begins with the wave equation[502]

$$\varrho\, \partial^2 X_1/\partial t^2 = (K + \tfrac{4}{3}G)\, \partial^2 X_1/\partial x_1^{\,2} \tag{10}$$

where X_1 is the displacement at the point x_1 and in the x_1 direction and ϱ is the density. Next, assume that X_1 has the form of a decaying sinusoidal wave,

$$X_1 = X_0 \exp[i\omega t - (\alpha + i\omega/c)x_1] \tag{11}$$

where α is the absorption coefficient and c is the velocity of the longitudinal wave. When $(\alpha c/\omega)^2 \ll 1$ (as in water), and at low frequencies, substituting eqn. (11) into eqn. (10) and solving for α/f^2 yields the expression

$$\alpha/f^2 = (2\pi^2/\varrho c^3)(\eta_v + \tfrac{4}{3}\eta_s) \tag{12}$$

which relates the ultrasonic absorption to η_v and η_s. (When η_v is set equal to zero, one arrives at the shear viscous loss predicted by Stokes.[1031]) Since there are no viscosimeters available which measure η_v, the measurement of ultrasonic absorption plays a unique role in the study of volume viscosity.

3. ULTRASONIC TECHNIQUES

In 1946, Pellam and Galt[830] employed ultrasonic pulses to determine the sound absorption in a number of liquids. The technique consisted in passing a radiofrequency pulse through the liquid over a variable path length and measuring the change in amplitude of the received pulse resulting from a known change in path length. Pinkerton[846] applied this technique to water over the temperature range 0–60°C and at a number of frequencies between 7.37 and 66.1 MHz. Since the periods corresponding to these frequencies are greater by many orders of magnitude than the structural relaxation time, the value of α/f^2 should remain constant, independent of frequency at each temperature [see eqn. (12)]. Since the work of Pinkerton, a number of investigations of sound absorption in water have been carried out as a function of both temperature and pressure.[239,478,692,1048]

The experimental apparatus used by Davis and Jarzynski[239] will be described in some detail.* It consists of: (a) variable-path-length acoustic interferometer; (b) a pressure system composed of a pressure vessel fitted with a turning shaft, a pressure pump, and a pressure gauge all designed for measurements up to 2 kbar; (c) a temperature-control system designed to maintain the contents of the pressure vessel at constant temperature to within 0.1°C with fluctuations in the order of ±0.05°C; and (d) the necessary electronics for transmitting, detecting, and measuring the change in amplitude of an ultrasonic pulse.

3.1. Acoustic System

The acoustic system used is an interferometer similar to the device described by Litovitz and Carnevale.[692] It consists of a stainless steel cylinder into which is fitted a movable piston. Both cylinder and piston are ground to an accuracy of 0.0001-in. The electrical signal excites a quartz crystal (attached to one end of the cylinder) which serves as both the transmitting and receiving crystal. The ultrasonic pulse travels through the liquid, is reflected from a liquid–stainless steel interface (attached to the movable piston), and returns through the liquid to the crystal. The path length between the crystal and the reflecting surface is varied by means of a micrometer screw attached to a turning shaft passing through the plug of the pressure vessel. Thus, the change in path length can be determined to better

* The experimental arrangement used by Davis and Jarzynski is being described because it is the one with which the present author is most familiar.

than 0.001 in. Pressure is transmitted to the water by means of a floating piston inserted into the cylinder.

3.2. Pressure- and Temperature-Control Systems

The temperature of the interferometer is controlled by circulating a water–ethylene glycol mixture from the temperature-controlled reservoir through lead-encased copper coils wound around the pressure vessel. The temperature of the solution is controlled by a Thermistemp (thermistor) temperature controller model 63 to $\pm 0.1°C$. When operating above room temperature, the thermistor controller actuates a heater in the reservoir; when operating below room temperature, the controller actuates a pump that circulates a water–ethylene glycol mixture from the refrigeration unit through a heat exchanger in the reservoir. The temperature in the pressure vessel at a point on the outside wall of the interferometer is measured by a thermocouple calibrated both before and after the measurements against a platinum-resistance thermometer to $\pm 0.02°C$. The thermocouple emf is detected on a Leeds and Northrup K-3 potentiometer and the corresponding temperature variations are less than 0.02°C during a typical experimental run. The limit of error in the absolute temperature determination of the liquid sample being investigated is estimated to be less than 0.1°C.

The pressure system consists of a Blackhawk hand pump model P-228 and intensifier (for fine pressure adjustments). Pressures are read on a Heise gauge (16-in.-diameter dial, 2-kbar capacity). In order to make pressure determinations with errors of less than ± 3 bars, it is necessary to make zero-point corrections after each data run as well as corrections for atmospheric pressure and changes in gauge linkage dimensions with room temperature. The pressure vessel, designed for operation to 3 kbar, has an inside diameter of $2\frac{1}{2}$ in. and an inside length of 20 in. The plug of the pressure vessel has provision for four electrical leads (only three of which were used) and a turning shaft. The pressure seal for the turning shaft is made by means of a Teflon ring sandwiched between two pieces of steel, one of which has a V-shaped face designed to fit into a V-shaped circumferential slot in the Teflon ring. The sandwich arrangement rests on a shoulder in the pressure plug. A nut is tightened down against the other face of the sandwich causing the V-shaped face of the steel ring to be forced into the V-shaped slot of the Teflon ring, spreading it and simultaneously forming a seal with the turning shaft and with the pressure plug.

3.3. Electronics

A block diagram of the electronic apparatus used is given in Fig. 1. A short rf pulse (about 5 μsec duration) containing an appreciable number of oscillations at frequency f (where f is between 45 and 95 MHz, depending on the coil used in the transmitter) is generated by a suitably modulated transmitter and is fed to an X-cut, 5-MHz piezoelectric quartz transducer operated between its 9th and 19th harmonic. The resulting ultrasonic pulse from the transducer passes through the liquid, is reflected by the interface, and returns to the quartz transducer, where it is reconverted to an electrical pulse. After suitable amplification by the receiver, the envelope of the received pulse is displayed on an oscilloscope. Since both the transmitter and receiver are connected to the quartz transducer, a 60-V blanking pulse is applied to the suppressor grids of the rf and i.f. stages of the receiver for the duration of the transmitter pulse. The blanking pulse switches the receiver off and hence prevents excessive overloading. An alternate arrangement using an additional crystal (in this case, mounted on the movable piston) may be used to isolate the transmitter and receiver. The transmitter is connected to the crystal on the movable piston and the receiver to the other crystal. This approach requires either a sliding electrical contact or a flexible wire in order to impress the signal on the added crystal. When

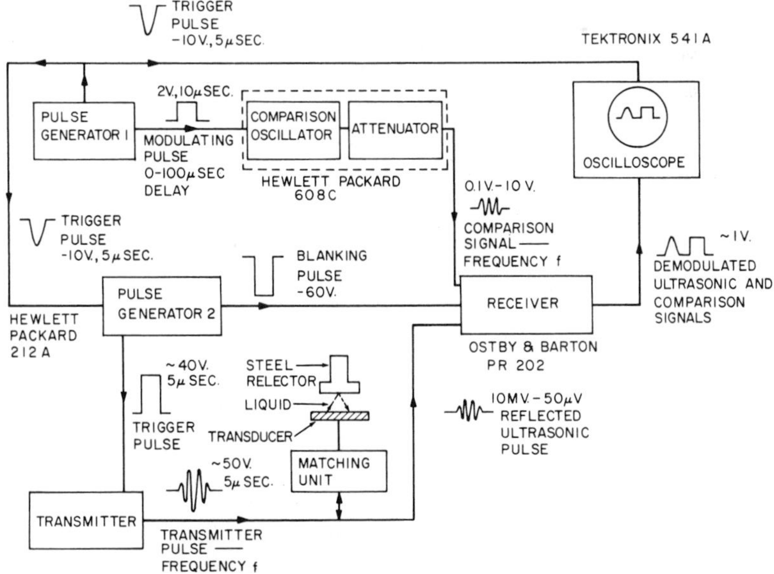

Fig. 1. Electronic arrangement used in measuring the ultrasonic absorption in water.[239]

the piston is moved, either arrangement produces a change in electrical impedance, which in turn affects the amplitude of the received pulse, introducing an error. In highly absorbing liquids, it is only necessary to move the piston very little in order to produce a significant change in amplitude, so that this error is negligible. In the case of the present measurements in water, this error amounted to as much as several per cent and therefore the single-crystal arrangement was used.

The pulse generator, in addition to triggering the oscilloscope and transmitter, provides a second pulse (delayed with respect to the first by up to 100 μsec) to modulate the comparison oscillator. The rf pulse from the comparison oscillator passes through a precision attenuator. It is delayed with respect to the transmitted pulse so that the envelopes of the two pulses are displayed adjacently when they reach the oscilloscope. Both pulses pass through the same amplification path. The advantage of this arrangement is that the gain of the receiver need not remain constant over the period of time required for a measurement (10–15 min).

To ensure efficient transfer of power from the transmitter to the transducer, it is essential to match the output impedance of the former to the input impedance of the latter. The matching network used for this purpose consists of two 50-cm, adjustable stubs (General Radio 874-D50L) separated by a length of coaxial cable. The length of this cable was adjusted at each frequency to obtain optimum impedance matching. After tuning the system at any one frequency, note was made of all electrical settings so that the conditions could be readily repeated at a later date.

Motion of the movable piston produces a change in the path length traversed by the ultrasonic signal through the liquid, and therefore, a change in amplitude of the detected pulse. This change in amplitude is measured by adjusting the attenuator to make the amplitude of the comparison pulse equal to that of the ultrasonic pulse. The absorption coefficient α is then calculated from a plot of amplitude versus path length. When the electrical system was tuned carefully, the values of α/f^2 measured at different frequencies usually deviated by less than 1% from the mean value at a given temperature and pressure.

4. EXPERIMENTAL RESULTS

Measurements of ultrasonic absorption in water have been reported by a number of authors: Pinkerton[(846)] (1947); Litovitz and Carnevale[(692)] (1955); Tait[(1048)] (1957); Davis and Jarzynski[(239)] (1968); and Hawley *et*

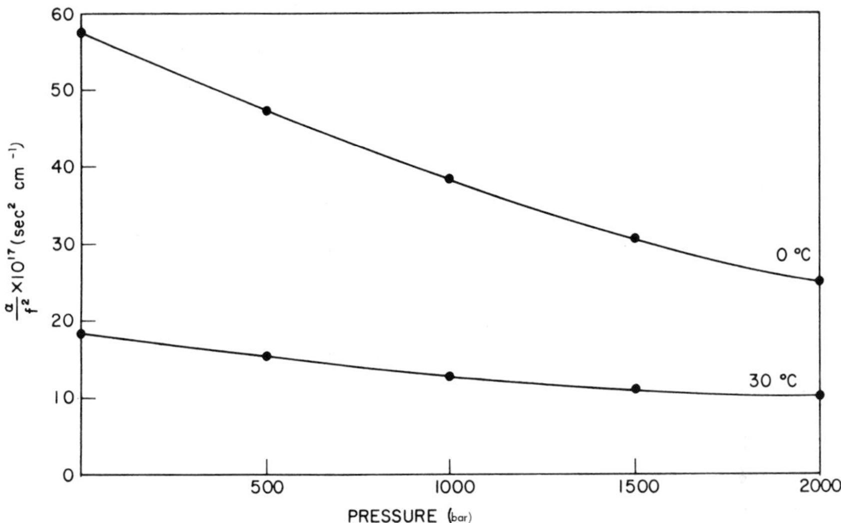

Fig. 2. Values of $\alpha/f^2 \times 10^{-17}$ (sec^2 cm^{-1}) reported for water by Litovitz and Carnevale.[692]

al.[478] (1970). The measurements by Pinkerton, although the most carefully made,* were restricted to atmospheric pressure. The remaining investigations were carried out as a function of pressure, and due to the limitation imposed by such measurements, could not be made with quite the same care as those by Pinkerton. The results of Tait's were confined to a narrow range of temperature (16 and 23°C) and pressure (to 690 bars). In addition, these measurements were made at a single low frequency (12 MHz) and are therefore likely to include diffraction errors. For these reasons, the data obtained by Tait will not be included in most of the following discussions. The measurements by Litovitz and Carnevale (see Fig. 2) were made at 0°C and 30°C in each case at a single frequency (25 MHz and 45 MHz, respectively) and at pressures up to 2 kbar. Hawley *et al.* (see Fig. 3) made their measurements in the same temperature range, but covered the most extended pressure range yet considered (to 4.6 kbar at 0°C and 2.9 kbar at 30°C). Their measurements were also carried out between 22.5 MHz and 31.7 MHz, but at only one frequency for each condition of temperature and pressure. Finally, Davis and Jarzynski (see Fig. 4) have covered a

* Measurements were made in both the Fraunhofer and Fresnel diffraction zones of the quartz crystal; separate measurements were made to ensure that no energy was lost due to side scattering; also, the path length available was several feet, compared to several inches for measurements made as a function of pressure.

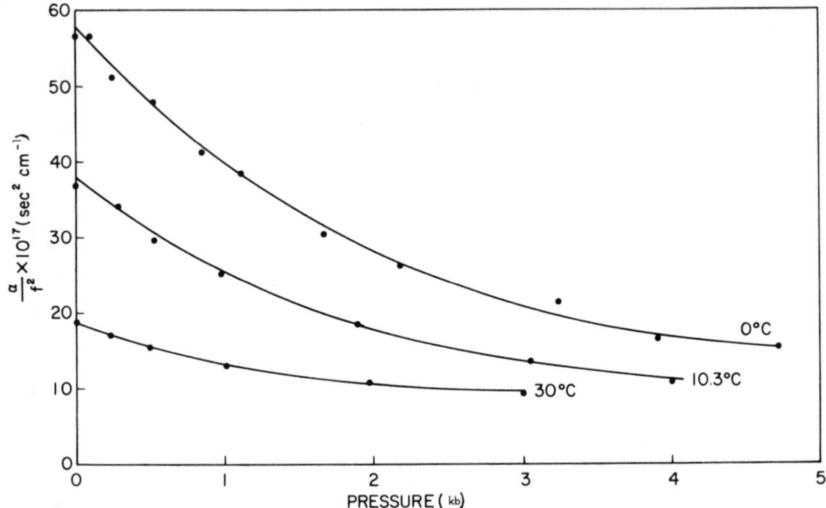

Fig. 3. Values of $\alpha/f^2 \times 10^{-17}$ (sec^2 cm^{-1}) reported for water by Hawley *et al.*[478]

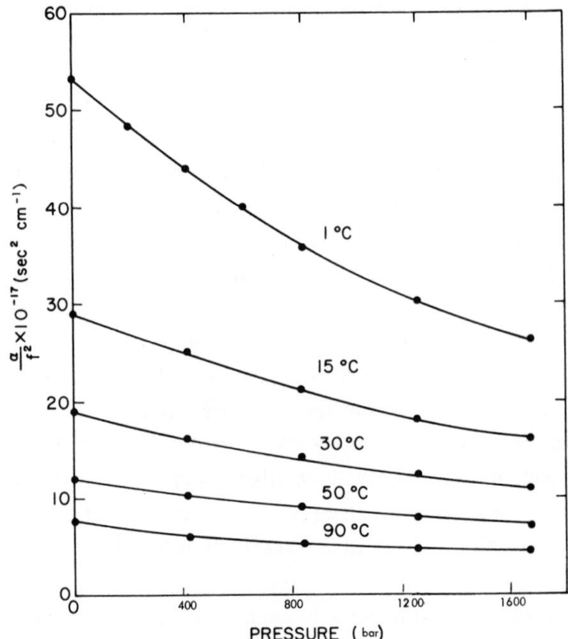

Fig. 4. Values of $\alpha/f^2 \times 10^{-17}$ (sec^2 cm^{-1}) reported for water by Davis and Jarzynski.[239]

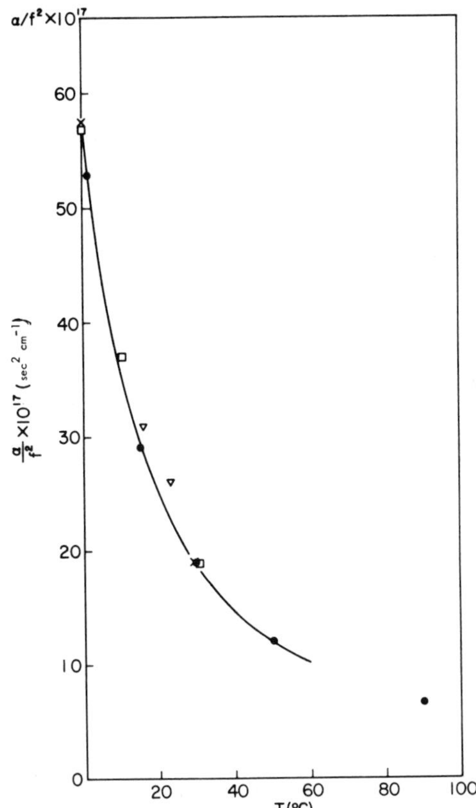

Fig. 5. Comparison of $\alpha/f^2 \times 10^{-17}$ ($sec^2\ cm^{-1}$) reported for water at atmospheric pressure by various investigators: (——) Pinkerton[846]; (×) Litovitz and Carnevale[692]; (▽) Tait[1048]; (●) Davis and Jarzynski[239]; (□) Hawley *et al.*[478]

temperature range from 0 to 90°C at pressures up to 1.6 kbar. At each temperature and pressure, the measurements were made at a minimum of two and most often three frequencies in the range from 45 to 95 MHz, and agreed with each other to within ±1% up to 50°C. Due to the small value of α/f^2, measurements at higher temperatures are more difficult. At 90°C, data corresponding to a given frequency were obtained by averaging the results of many measurements. A spread of ±5% was not unusual. The average values agreed to within ±4%.

Values of α/f^2 obtained at atmospheric pressure are compared in Fig. 5, excluding the data of Tait's, which are is 13% higher at 23°C; the agreement

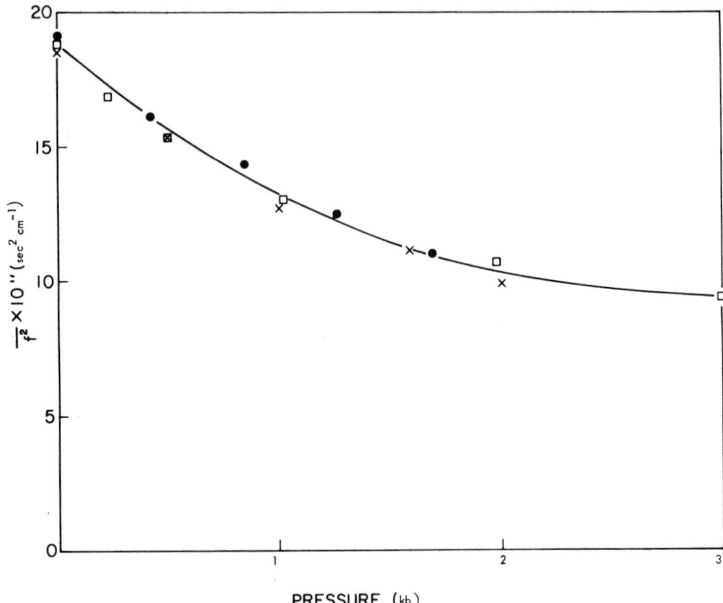

Fig. 6. Comparison of $\alpha/f^2 \times 10^{-17}$ (sec^2 cm^{-1}) reported for water at 30°C as a function pressure by various investigators: (×) Litovitz and Carnevale[692]; (□) Hawley *et al.*[478]; (●) Davis and Jarzynski.[239]

between various investigators is quite reasonable. Measurements made at 30°C and at elevated pressures are compared in Fig. 6. Data obtained by the various investigations agree to within ±4% with a smooth curve drawn through the experimental points.

4.1. Volume Viscosity

In water, the behavior of η_s versus pressure[163] is anomalous (see Fig. 7). For temperatures below approximately 30°C, the value of η_s passes through a minimum as the pressure is increased, while above 30°C, the value of η_s increases in a normal manner. Thus, below 30°C, water becomes more fluid as the pressure, and therefore the density, is increased. This behavior has been observed in no other liquid. Since η_v as well as η_s is associated with structural rearrangement, it is desirable to determine whether η_v behaves in the same manner as η_s. Using measured values of α/f^2, η_v may be calculated by means of eqn. (12). Litovitz and Carnevale[692] interpreted their results as indicating that at 0°C, η_v did not behave in the

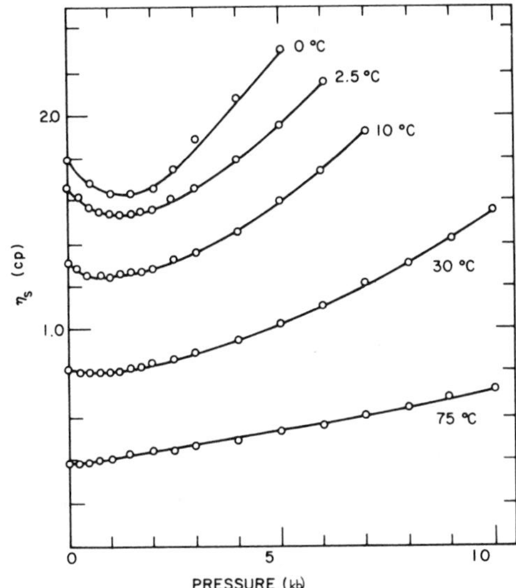

Fig. 7. Shear viscosity (in centipoise) measured as a function of temperature and pressure.[163]

same manner as η_s, but instead exhibited a small maximum.* Davis and Jarzynski,[239] however, did not find this to be the case (see Table I). Instead, in agreement with the variation in η_s, they observed a minimum* in the value of η_v plotted versus pressure for temperatures below 30°C and an increase with pressure at temperatures above 30°C. Moreover, the value of η_v/η_s (see Table II) remained nearly constant as a function of pressure at all temperatures. It is an interesting fact that, while over the entire temperature and pressure range, the measured values of α/f^2 varied by a factor of 13.5 and the calculated values of η_v varied by a factor of 6.5, nevertheless the ratio η_v/η_s remained constant to within 20%. Thus, the variation in the measured sound absorption is almost two orders of magnitude greater than the variation in η_v/η_s. To within the uncertainty of their measurements, the results of Hawley *et al.*,[478] made over a considerably wider pressure range, are in essential agreement with the above observations.

The anomalous decrease in both η_v and η_s with pressure at low temperatures can be explained in terms of a mixture model for water (see Chapter 14). On the basis of such a model, as the pressure is increased, a proposed equilibrium structure of water shifts in the direction of a more close-

* The experimental uncertainty is approximately equal to the observed extrema.

TABLE I. The Variation of η_v with Temperature[239] and Pressure

P, bar	η_v, cP				
	at $T = 1°C$	15°C	30°C	50°C	90°C
1	5.24	3.12	2.24	1.49	0.81
207	5.19	—	—	—	—
414	5.18	3.21	2.21	1.46	0.82
621	5.13	—	—	—	—
828	4.97	3.15	2.26	1.51	0.85
1240	4.84	3.07	2.25	1.50	0.87
1656	4.98	3.12	2.23	1.51	0.92
Spread	8%	4%	1%	3%	14%

packed and less porous structure. The latter structure, however, is assumed to be characterized by a higher degree of hydrogen bonding than the former and therefore to be more resistant to flow. Thus, the minimum in viscosity versus pressure is explained in terms of two competing effects. As the pressure is raised, the density increases (tending to increase the viscosity), but the degree of hydrogen bonding decreases (tending to decrease the viscosity). At 0°C, the latter effect predominates for lower pressures and the former effect predominates for higher pressures, with the result that

TABLE II. The Variation of the Ratio η_v/η_s with Temperature and Pressure[239]

P, bar	η_v/η_s				
	at $T = 1°C$	15°C	30°C	50°C	90°C
1	3.03	2.73	2.80	2.71	2.57
207	3.07	—	—	—	—
414	3.13	2.88	2.76	2.62	2.51
621	3.15	—	—	—	—
828	3.08	2.84	2.81	2.66	2.52
1242	3.03	2.76	2.76	2.58	2.51
1656	3.05	2.77	2.68	2.53	2.56
Spread	4%	5%	5%	7%	2%

Overall decrease in η_v/η_s, 20%

the viscosity passes through a minimum. At temperatures above 30°C, the effect of increasing density predominates at all pressures.

The small temperature dependence of η_v/η_s is observed in many liquids[693] and is an indication that the activation enthalphy for both shear and structural flow is approximately the same. In most liquids, however, the pressure dependence of η_v/η_s is less constant,[478,693] suggesting that the activation volume for the two processes is different. In the case of water, the practically constant value of η_v/η_s versus pressure observed at all temperatures strongly indicates that the activation volumes for shear and structural flow in water are nearly identical. This surprising result has rather important structural implications, as it suggests that both shear and structural flow depend upon the same microscopic mechanism.

Using eqn. (6) and the above values of η_v, the product $\varkappa_{T,r}\tau_p$ may also be calculated. Thus, if either $\varkappa_{T,r}$ or τ_p is known, the other may be obtained. In the sections which follow, we will consider these quantities in more detail.

5. RELAXATIONAL COMPRESSIBILITY

Slie *et al.*[1000] have obtained estimates of $\varkappa_{T,r}$ from ultrasonic measurements in water–glycerol mixtures (containing up to 68 mol % H_2O) where, due to the high viscosity, τ_p could be experimentally determined. The value of $K_{T,\infty}$ was found to be a linear function of water content. By extrapolating to 100% water, an estimate of $K_{T,\infty}$ (and therefore $\varkappa_{T,\infty}$) for water was obtained. The corresponding values obtained for $\varkappa_{T,r}$ and τ_p in pure water at 0°C are 33.5×10^{-12} dyn cm^2 and 4.3×10^{-12} sec, respectively.

The possibility of interpreting the anomalous physical properties of water in terms of a mixture of two or more species in chemical equilibrium has been recognized for many years. (For a detailed discussion of these and other models, see Chapter 14.) Recent experimental evidence, particularly the investigation of Raman spectra (see Chapters 5 and 13), strongly supports the hypothesis that over a wide range of temperature (even extending above the critical temperature at pressures sufficiently high to maintain a density of unity), the properties of water can be accounted for by the simplest of two-state models (i.e., corresponding to the unimolecular reaction $A \rightleftharpoons B$). In this case, the following relation for $\varkappa_{T,r}$ may be obtained[238]:

$$\varkappa_{T,r} = (1 - x)(\Delta V^2)/VRT^2 \tag{13}$$

TABLE III. Relaxational Isothermal Compressibility ($\times 10^{-12}$ cm^2 dyn^{-1})

	T = 0°C	20°C	30°C	40°C	60°C	80°C	100°C
Slie *et al.*[1000]	33.5	—	30.1	—	—	—	—
Davis and Litovitz[240]	30.9	27.8	26.8	26.1	25.3	25.2	25.3
Eucken[317–319]	11.6	7.2	—	4.2	2.0	1.2	0.6
Frank and Quist[360]	22.8	—	16.4	—	—	—	—
Grjotheim and Krogh-Moe[429]	5.2	3.9	—	2.6	1.7	1.1	0.6
Hall[444]	32.8	27.8	26.7	26.2	26.3	27.1	—
Litovitz and Carnevale[692]	36.5	—	30.1	—	—	—	—
Nemethy and Scheraga[788]	7.8	5.8	4.8	4.0	2.5	1.5	0.8
Smith and Lawson[1003]	39.0	35.0	33.0	31.0	26.0	20.0	14.0
Wada[1125]	4.5	3.4	—	—	—	—	—
Walrafen (Chapter 5)	26.4	29.1	28.4	26.7	—	—	—

where ΔV is the difference in mol volume between the two states, x is the mol fraction of one of the states, and R is the gas constant. Values of $\varkappa_{T,r}$ obtained on the basis of various proposed models of water are compared in Table III. The estimates of Davis and Litovitz,[240] Frank and Quist,[360] Hall,[444] Litovitz and Carnevale,[692] Smith and Lawson,[1003] and Walrafen (see Chapter 5) agree reasonably well with that of Slie *et al.* The remaining estimates are quite low, due primarily to the small values of ΔV associated with the corresponding models.

6. STRUCTURAL RELAXATION TIME

As pointed out in Section 4.1, once the values of η_v and $\varkappa_{T,r}$ are known, the value of τ_p may be calculated. The only experimentally determined values of $\varkappa_{T,r}$ are those obtained by Slie *et al.* as discussed in Section 5. In the case of τ_p, however, it is possible to utilize other experimental observations to place limits on the allowable range of values for τ_p.

Litovitz and McDuffie[694] considered the relation between the dielectric relaxation time τ_D in a liquid and τ_p. They observed that in all hydrogen-bonded liquids where both τ_D and τ_p have been measured, τ_D is always the longest. Thus, for these liquids, the structure must break up and reform several times before the dipole is reoriented. Since water is an associated liquid, (i.e., is hydrogen-bonded) it can be expected to behave in this manner, in which case τ_D represents the upper limit for τ_p. Furthermore, in the case of water, dielectric relaxation has been measured[945] over a range of temperatures from 0 to 50°C and found to exhibit a single relaxation time (see, however, Chapter 7). If τ_p were greater than τ_D, water molecules would remain in a given state (or environment) for a time long compared to τ_D. In this case, the ability of a molecule to reorient should depend on the local structure and τ_D would be an average rather than a single time.

The inelastic scattering of neutrons provides another means of estimating the structural relaxation time in water. The time required for a single diffusive step to occur was determined by Safford *et al.*[933,935] from the width of the neutron inelastic scattering peak. In the temperature range 1–25°C, their results were found to agree quite well with a delayed-jump diffusion model*; that is, with a model in which the molecules are assumed to remain in a quasilattice site for a time τ_0 (the residence time) which is much longer than the time τ_1 during which a diffusive jump takes place. In this case ($\tau_0 \gg \tau_1$), the motions observed correspond to the jumping of individual water molecules. At higher temperatures, the approximation $\tau_0 \gg \tau_1$ becomes less valid. In this case, every molecule spends an appreciable amount of time moving between sites. Thus, the probability that neighboring molecules will be simultaneously in motion becomes large; therefore, in addition to the jumping of individual water molecules, correlated motions of groups of water molecules may contribute to structural rearrangements. Safford *et al.*[933,935] point out that such groups of water molecules should not be confused with the "flickering clusters" or "icebergs" which have been proposed for the structure of water.[360,788,828] The value of τ_0 obtained in this manner varied from 3.1×10^{-12} sec at 0°C to 1.7×10^{-12} sec at 25°C. The fact that τ_0 is approximately the same as the value of τ_p obtained by Slie *et al.* is quite significant. The time that an individual molecule remains in a quasicrystalline site before undergoing a diffusive jump to a new site is τ_0. Since τ_p is equal to τ_0, it is reasonable to assume that the rate-controlling step associated with structural rearrangement in water must also be the jump of individual water molecules and cannot be due to the melting of a cluster of molecules, as would occur for models such as have been proposed by Nemethy and Scheraga[788] or Pauling.[828]

A simple flow mechanism such as the uncorrelated jumping of individual molecules provides an explanation for the observed pressure independence of the ratio η_v/η_s. If flow depends on the correlated motions of groups of

* The validity of this and other diffusion models employed in the interpretation of neutron scattering spectra is discussed more fully in Chapter 9.

TABLE IV. Relaxation Times

	Constant-pressure structural relaxation time τ_p, $\times 10^{-12}$ sec						
	at $T = 0°C$	20°C	30°C	40°C	60°C	80°C	100°C
Slie *et al.*[1000]	4.3	2.1	1.8	—	—	—	—
Davis and Litovitz[240]	4.3	2.1	1.7	1.3	1.0	0.8	0.7
Eucken[317–319]	11.6	8.2	—	8.2	12.4	16.1	29.4
Frank and Quist[360]	5.9	—	2.7	—	—	—	—
Grjotheim and Krogh-Moe[429]	25.9	15.1	—	13.2	14.6	17.5	29.4
Hall[444]	4.0	2.1	1.6	1.3	0.9	0.6	—
Litovitz and Carnevale[692]	4.1	—	1.5	—	—	—	—
Nemethy and Scheraga[788]	17.2	10.2	9.3	8.6	9.9	12.8	22.1
Smith and Lawson[1003]	3.4	1.7	1.4	1.1	1.0	1.0	1.3
Wada[1125]	29.9	17.3	—	—	—	—	—
Walrafen (Chapter 5)	5.1	2.0	1.6	1.3	—	—	—
τ_D [945]	18.7	10.1	7.5	5.9	—	—	—
τ_0 [933,935]	2.4	1.7	(25°C)	—	—	—	—

molecules, then it is quite logical to expect some differences in the motions occurring in shear and volume flow. This will result in an activation volume which differs for the two processes. On the other hand, the jumping of individual molecules between quasicrystalline sites is expected to exhibit the same activation volume regardless of whether the final result is shear or volume flow.

Estimates of τ_p obtained from various models are compared with the values of τ_D, τ_0, and Slie's estimates of τ_p in Table IV. The best agreement with the values of Slie *et al.* is obtained for the model of Davis and Litovitz.[240] The values of τ_p obtained from the results of Walrafen (see Chapter 5), Frank and Quist,[360] and Hall[444] (as well as the related theories of Litovitz and Carnevale[692] and of Smith and Lawson[1003]) also agree quite well with the estimates of Slie *et al.*[1000] The remaining estimates yield values of τ_p which exceed the expected values by nearly an order of magnitude, and in addition, are larger than τ_D. Finally, the estimated values of τ_p, plotted versus temperature, obtained from the models of Eucken,[317–319] Grjotheim and Krogh-Moe,[429] and Nemethy and Scheraga[788] pass through an unreasonably large minimum as the temperature increases.

7. CONCLUSIONS

The investigation of ultrasonic absorption in water provides a number of important clues to its structural nature: (a) η_v and η_s exhibit a negative pressure dependence at low temperatures; (b) the ratio η_v/η_s is independent of pressure at all temperatures; and (c) $\tau_p \approx \tau_0$. All of the above observations are consistent with the concept that water is a mixture of two or more distinguishable species whose mechanism of structural rearrangement is the uncorrelated jumping of molecules between quasicrystalline sites. Furthermore, the simplicity of the molecular motions and the existence of a dielectric relaxational process characterized by a single time argue against the existence of large icelike clusters.

CHAPTER 13

Water at High Temperatures and Pressures

Klaus Tödheide

Institut für Physikalische Chemie und Elektrochemie
Universität Karlsruhe
Karlsruhe, Germany

1. INTRODUCTION

This chapter deals with the properties of water in its fluid phases at high temperatures and high pressures. Figure 1 shows the *P–T* projection of the high-pressure portion of the phase diagram of water, with its various modifications of ice indicated by Roman numerals. The temperature and pressure ranges covered in this chapter are limited by the dashed line and the melting pressure curve of ice VII, which has been measured by Pistorius *et al.*(847) up to 200 kbar, where the melting temperature is 442°C. Beyond the upper pressure and temperature limits of 250 kbar and 1000°C, respectively, data on the properties of water are not yet available. The lower limits are chosen somewhat arbitrarily. Therefore, it will sometimes be necessary to report also on the properties of water at lower temperatures and pressures. The immediate neighborhood of the critical point (CP), whose coordinates(798) are $T_c = 374.15°C$, $P_c = 221.2$ bar, and $V_c = 3.28\ cm^3\ g^{-1}$, is treated in Chapter 11.

The properties of water at high temperatures and high pressures have been investigated both for technical and purely scientific reasons. Knowledge of the thermodynamic properties of water at high temperatures and pressures, for example, has become increasingly important in designing power stations, in solving corrosion problems in these and similar units, and in growing synthetic crystals from hydrothermal solutions. The understanding of many geochemical processes is directly dependent on the knowl-

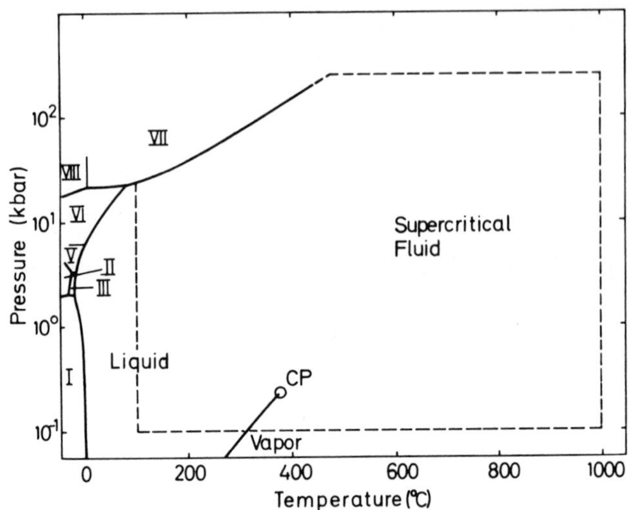

Fig. 1. *P–T* projection of the high-pressure portion of the phase diagram of water according to Bridgman,[116] Brown and Whalley,[130] and Pistorius *et al.*[847]

edge of the properties of water and of aqueous solutions. The application of pressure allows volume and temperature to be changed independently. Thus, the temperature dependence of the properties of water can be separated from their volume dependence. This is especially true in the supercritical region, where temperature and volume can be changed independently and continuously over wide ranges. At 500°C, for example, water may be compressed from zero density to slightly above 2.1 g cm^{-3}, at which density it solidifies to ice VII. During this increase from zero to more than twice its normal density, water undergoes a continuous transition from "vaporlike" to "liquidlike" behavior. The properties of water in all the intermediate states between a dilute vapor and a compressed liquid can thus be studied and the information obtained in this region may even contribute to the understanding of the complex nature of water under ordinary conditions.

2. THERMODYNAMIC PROPERTIES

The thermodynamic properties of water have been measured or calculated over almost the whole area confined within the dashed line in Fig. 1. In the pressure range below 1 kbar, several compilations[45,953,1119] of *PVT* data, based on a large number of accurate experimental determinations, are available. Since they also contain tables for the specific enthalpy, specific entropy, and specific heat,[953] no thermodynamic data will be presented here for that pressure range. The *PVT* measurements of Kell and Whalley[594] and the enthalpy determinations of Newitt *et al.*[794] demonstrate the application of modern experimental techniques at pressures below 1 kbar.

2.1. *PVT* Data and Equation of State

Table I gives a list of *PVT* measurements performed above 100°C and 1 kbar. Most of the measurements fall into the range below 10 kbar. Bridgman[122] and Rice and Walsh[901,1143] reported *PVT* data above 10 kbar.

2.1.1. *Experimental Methods*

The experimental methods for *PVT* data determinations fall into two complementary groups: static and dynamic measurements. The static experiments give more accurate results in the lower and moderate pressure ranges. Dynamic measurements have been carried out at very high pressures, where static methods become complicated and inaccurate, and, finally, impossible.

A number of static techniques may be used at high pressures. One has proved to be particularly suitable for high-pressure application: A fixed amount of water is introduced into a closed system the volume of which can be measured at a given pressure and temperature. Since accurate volume measurements at high temperatures are difficult to perform, part of the closed water system is at, or slightly above, room temperature. Figure 2

TABLE I. Experimental Determinations of *PVT* Data of Water above 100°C and 1 kbar

Author	Year	Temperature range, °C	Pressure range, bars
Tammann and Rühenbeck[1052]	1932	20–650	1–2500
Goranson[410]	1938	800–1200	100–4000
Holser and Kennedy[521,522]	1958–1959	100–1000	100–1400
Jůza *et al.*[566]	1966	80–450	500–4500
Maier and Franck[723]	1966	200–850	1000–6000
Burnham *et al.*[153,154]	1969	20–900	150–8400
Köster and Franck[630]	1969	25–600	1000–10000
Bridgman[122]	1942	25–175	5000–35000
Rice and Walsh[901,1143]	1957	175–1000	25000–250000

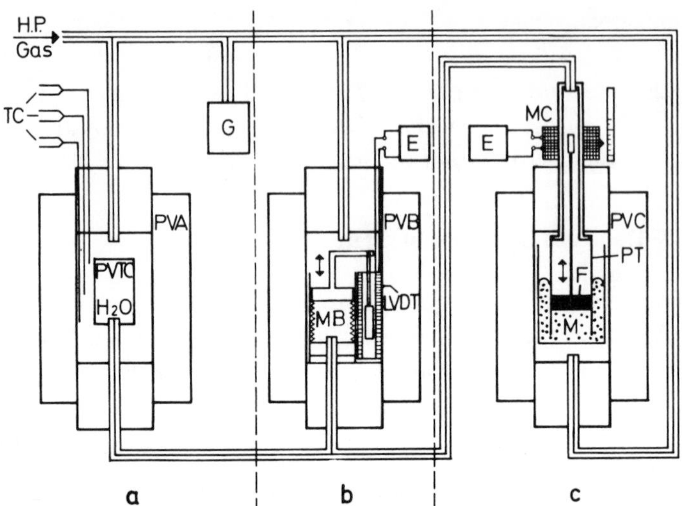

Fig. 2. Schematic diagram of an apparatus for *PVT* measurements of water: (PVA) pressure vessel containing *PVT* cell; (PVB) pressure vessel for volume measurement containing metal bellows and one part of a linear variable differential transformer; (PVC) alternative pressure vessel for volume measurement containing liquid mercury separator and precision tubing; (TC) thermocouples; (G) gauge; (PVTC) *PVT* cell; (MB) metal bellows; (LVDT) linear variable differential transformer; (M) mercury; (F) float; (PT) precision tubing; (MC) movable coil; (E) electrical equipment for displacement measurements of metal bellows and mercury level, respectively.

shows a schematic diagram of an apparatus for pressures of about 10 kbar. In addition to the pressure-generating equipment (not shown in the figure), the apparatus consists of two interconnected pressure vessels PVA and PVB or, alternatively, PVA and PVC. Pressure vessel PVA contains the *PVT* cell PVTC, which is surrounded by the pressure-transmitting medium, usually compressed argon. The *PVT* cell is heated either by a temperature bath outside the pressure vessel or, at higher temperatures, by an electrical furnace inside the pressure vessel. The temperature is measured by different thermocouples TC distributed along the cell. The *PVT* cell is interconnected with the low-temperature part of the system by capillary tubing. Figure 2 shows two alternatives for the low-temperature part. In Fig. 2(b), a metal bellows MB in the pressure vessel PVB provides pressure equilibrium between the pressurizing gas and the water by adjustment of the volume of the closed water system. The resulting volume changes can be read from the calibrated output of a linear variable differential transformer LVDT which is also mounted in the pressure vessel and whose core is linked to the moving top of the bellows. The values obtained have to be corrected for the effects of compression and thermal expansion of the container materials on the volume of the water system. The pressure is measured by a pressure gauge G in the gas system, usually a manganin coil the resistance of which is calibrated as a function of pressure against a dead-weight piston gauge. Burnham *et al.*[154] obtained *PVT* data to 900°C and 8400 bars with a total uncertainty of about 0.3% in specific volume using a combination of pressure vessels PVA and PVB.

Pressure vessel PVA can also be combined with pressure vessel PVC in Fig. 2(c). In this case, pressure equilibrium between the water in the *PVT* cell and the gas outside the cell is established by means of a liquid mercury separator. A precision tube PT dips with its lower open end into the mercury, which is in an open container inside pressure vessel PVC. The volume of the water system is a function of the position of the mercury level in the precision tubing. The position of the level is determined by an assembly of a floating disk F, a vertical wire with a ferromagnetic top, and a movable coil MC. If the ferromagnetic top changes its position with respect to the coil, an ac bridge becomes unbalanced. In order to determine the new position of the top, the coil is readjusted by a rack-and-pinion arrangement, until the bridge is rebalanced. After some calibration, the volume of the water system may be calculated from the final position of the coil. In comparison with the bellows system, additional corrections for the compressibilities of the floating disk and the wire, and for the change in buoyancy with pressure, are necessary. On the other hand, this method is suited for larger volume changes and does not depend on a reproducible linear relationship between the volume of the bellows and the displacement of one of its ends. Köster and Franck[630] used a pressure vessel as shown in Fig. 2(c) for their volume measurements up to 10 kbar.

Above 10 kbar, the technique described is no longer applicable. Since the melting temperature of water exceeds room temperature at these pressures, and consequently water starts to freeze in the cold part of the apparatus, the entire water system must be kept at high temperature. Bridgman[122] used a piston-cylinder apparatus and a differential method in this range. He compressed a water sample which was sealed into a collapsible lead capsule and a very much less compressible standard solid substance, both surrounded by a pressure-transmitting plastic solid, in the same experimental setup in subsequent runs. The compression of water was calculated from the difference in piston displacements in the two runs as a function of pressure and from the known compression of the standard.

In the dynamic experiments, a plane shock wave is generated by a high explosive. The shock front is propagated with a supersonic velocity v_s into the substance to be studied. When a mass element is traversed by the shock front, its thermodynamic state is changed almost discontinuously. The states ahead of and behind the shock front are related through the conservation of mass, momentum, and energy. If the substance in its initial state is at rest ($v_{P_0} = 0$), the conservation relations lead to*

Mass: $$V/V_0 = (v_s - v_{P_1})/v_s \tag{1}$$

Momentum: $$P_1 - P_0 = v_s v_{P_1}/V_0 \tag{2}$$

Energy: $$U_1 - U_0 = \tfrac{1}{2}(P_1 + P_0)(V_0 - V_1) \tag{3}$$

In these equations, P_0, V_0, U_0 and P_1, V_1, U_1 are pressure, specific volume, and specific internal energy ahead and behind the shock front, respectively; v_{P_1} is the particle velocity behind the shock front. Equation (3) defines the loci in the P–V plane of all the states attainable by propagating shock waves with different velocities into the same initial state P_0, V_0. The curve connecting these loci is called the Hugoniot curve centered at P_0, V_0. Thus, measurements of v_s and v_{P_1} suffice to calculate P_1, V_1, U_1 from known values of P_0, V_0, U_0.

Walsh and Rice[(1143)] measured the Hugoniot curve of water centered at 20°C and 1 bar, that is, at $P_0 = 1$ bar and $V_0 = 1.0018$ cm^3 g^{-1}, up to pressures of 400 kbar using an experimental arrangement which is shown schematically in the lower part of Fig. 3. A plane shock wave was propagated into an aluminum plate with several grooves milled into its front surface; some of these were filled with water. When the shock wave arrived at an iron shim in front of the aluminum plate, a small argon gap was rapidly closed. The argon in the gap became luminous because of multiple shock reflection. The light produced was recorded photographically by a moving image camera behind a slit. The resulting trace is shown in the upper part of Fig. 3. From the trace, the difference between the shock velocities in aluminum v_{sa}, and water v_s (behind slit 1), and the difference between the velocity v_{sf} of the free aluminum surface traversing the gaps and the shock velocity in aluminum v_{sa} (behind slit 1 and 3), could be obtained. Since the relationship between v_{sf} and v_{sa} and the equation of state for aluminum were known from previous shock experiments on aluminum, the authors were able to calculate v_s from the observed

* For the derivation of (1)–(3), see, e.g., J. N. Bradley, "Shock Waves in Chemistry and Physics," Methuen, London (1962).

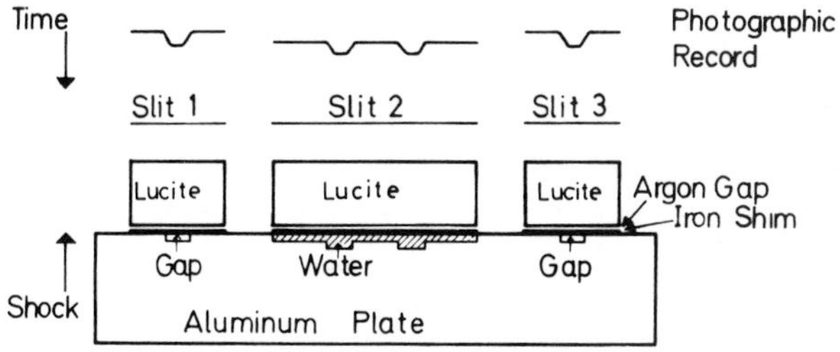

Fig. 3. Schematic diagram of an experimental setup for dynamic measurements of *PVT* data.[1143]

velocity differences, P_1 and v_{P_1} from the boundary condition that P and v_P must be continuous across the aluminum–water interface, and subsequently V_1 and U_1 from eqns. (1)–(3). Using reflected shock waves, Walsh and Rice obtained additional data for high-pressure states in the neighborhood of the Hugoniot curve up to 250 kbar. From these data, the thermodynamic quantity $(\Delta H/\Delta V)_P$ was determined. Assuming that above 25 kbar, $(\partial H/\partial V)_P$ is a function only of pressure, Rice and Walsh[901] derived an equation of state based on their experimental results and calculated complete *PVT* data between 175 and 1000°C and 25 and 250 kbar.

2.1.2. *Results*

The results of the *PVT* measurements for water above 1 kbar are given in Tables II and III in the form of specific volume as a function of pressure and temperature. The values in Table II below 10 kbar were taken from Burnham *et al.*[153] The authors claim an accuracy of 0.3% in specific volume. The specific volumes measured by Holser and Kennedy,[521,522] Jůza *et al.*,[566] Maier and Franck,[723] Köster and Franck,[630] and Bridgman[122] generally agree within about 1% or less with the values given in Table II except close to the upper temperature and pressure limits reached in the respective studies. The values in Table III were taken from Rice and Walsh[901] and are based on their shock wave experiments. The pressure ranges of Tables II and III overlap at 5 and 10 kbar. The agreement at those pressures is better than 1% at 175°C, but only about 5% at 1000°C. The large deviation at 1000°C may be due mainly to inaccuracies arising from the assumption concerning $(\partial H/\partial V)_P$ that had to be made by Rice and Walsh in order to calculate *PVT* data from their experimental Hugoniot curve. These inaccuracies are highest in their high-temperature, low-pressure range, because of its considerable distance from the measured Hugoniot curve centered at 1 bar and 20°C, which, for example, reaches a pressure as high as 169 kbar at

TABLE II. Specific Volume of Water between 100 and 1000°C and 1 and 10 kbar[a]

P, kbar	Specific volume of water, cm³ g⁻¹									
	at T=100°C	200°C	300°C	400°C	500°C	600°C	700°C	800°C	900°C	1000°C
1.0	1.000	1.084	1.213	1.446	1.894	2.667	3.543	4.337	5.095	5.684
1.5	0.984	1.060	1.169	1.330	1.588	1.968	2.442	2.944	3.425	3.868
2.0	0.968	1.037	1.134	1.265	1.450	1.703	2.012	2.352	2.696	3.037
2.5	0.954	1.018	1.104	1.216	1.363	1.552	1.777	2.028	2.293	2.555
3.0	0.943	1.001	1.078	1.176	1.300	1.451	1.627	1.824	2.035	2.236
3.5	0.932	0.986	1.056	1.144	1.250	1.377	1.522	1.684	1.856	2.011
4.0	0.922	0.973	1.038	1.116	1.211	1.321	1.445	1.582	1.726	1.847
4.5	0.913	0.962	1.022	1.094	1.179	1.276	1.385	1.504	1.627	1.727
5.0	0.904	0.951	1.007	1.074	1.152	1.240	1.337	1.442	1.551	1.638
5.5	0.895	0.941	0.994	1.056	1.128	1.209	1.296	1.392	1.489	1.570
6.0	0.887	0.932	0.982	1.041	1.107	1.182	1.262	1.349	1.438	1.517
6.5	0.880	0.923	0.971	1.026	1.089	1.157	1.232	1.311	1.394	1.473
7.0	0.873	0.915	0.961	1.013	1.072	1.136	1.205	1.278	1.355	1.436
7.5	0.867	0.907	0.952	1.002	1.056	1.116	1.180	1.248	1.320	1.402
8.0	0.861	0.900	0.943	0.990	1.042	1.098	1.158	1.221	1.288	1.371
8.5	0.855	0.893	0.935	0.980	1.029	1.081	1.137	1.196	1.259	1.341
9.0	0.849	0.886	0.927	0.970	1.016	1.065	1.117	1.172	1.231	1.312
9.5	0.844	0.880	0.919	0.960	1.003	1.049	1.099	1.150	1.205	1.283
10.0	0.839	0.874	0.911	0.951	0.992	1.034	1.081	1.130	1.180	1.253

[a] From Burnham *et al.*[153]

TABLE III. Specific Volume of Water between 175 and 1000°C and 5 and 250 kbar[a]

P, kbar	Specific volume of water, $cm^3\ g^{-1}$				
	at T = 175°C	250°C	500°C	750°C	1000°C
5	0.934	0.966	1.139	1.350	1.616
10	0.864	0.882	0.975	1.077	1.189
15	0.814	0.830	0.896	0.966	1.040
20	0.777	0.794	0.848	0.903	0.959
25	0.749	0.764	0.814	0.864	0.914
30	0.727	0.741	0.789	0.838	0.886
40	—	0.704	0.749	0.794	0.839
50	—	0.673	0.715	0.757	0.799
60	—	—	0.686	0.725	0.764
70	—	—	0.660	0.697	0.733
80	—	—	0.638	0.672	0.706
90	—	—	0.617	0.649	0.681
100	—	—	0.599	0.628	0.658
110	—	—	0.583	0.610	0.638
120	—	—	0.568	0.594	0.619
130	—	—	0.554	0.578	0.602
140	—	—	0.542	0.565	0.587
150	—	—	0.531	0.552	0.573
170	—	—	0.512	0.530	0.549
190	—	—	0.496	0.512	0.528
210	—	—	0.483	0.497	0.511
230	—	—	0.472	0.484	0.496
250	—	—	0.463	0.473	0.484

[a] From Rice and Walsh.(901)

1000°C. It may be concluded, therefore, that the accuracy of the specific volumes presented in Table III is about 1–3% over most of the range covered.

In the whole temperature and pressure range from 100 to 1000°C and 1 to 250 kbar, respectively, the specific volume of water is a monotonic function of pressure and temperature and does not show any anomaly like the minimum at 4°C and 1 bar. Accordingly, the isobaric coefficient of thermal expansion $\alpha = (1/V)(\partial V/\partial T)_P$ has a positive sign everywhere. Moreover, at constant temperature, it always decreases as the pressure is increased, contrary to the temperature range below 40°C, where the coefficient of thermal expansion increases upon compression at low pressures.(117)

When plotted as a function of temperature for constant pressures, the coefficient of thermal expansion exhibits a maximum, which shifts from about 600°C at 1 kbar toward higher temperatures and flattens out with increasing pressures. At about 25 kbar, it is almost independent of temperature.

As for normal liquids, the isothermal coefficient of compressibility $\beta = (1/V)(\partial V/\partial P)_T$ is a monotonic function of pressure and temperature in the range covered in Tables II und III, whereas it shows a minimum at about 50°C when plotted against temperature for constant pressures below 3 kbar.[117]

2.1.3. *Equation of State*

An equation of state for dense, hot water has not yet been derived from first principles. At low densities, the *PVT* behavior of water can be described by the virial equation of state

$$PV/RT = 1 + [B(T)/V] + [C(T)/V^2] + \cdots \tag{4}$$

where V is the molar volume, R is the gas constant, and $B(T)$ and $C(T)$ are the second and third virial coefficients, respectively. The virial coefficients, which are functions of temperature only, can be calculated from intermolecular potentials by means of statistical mechanics. As the density increases, eqn. (4) becomes more and more inadequate, even if the contributions of the higher virial coefficients are taken into account. At high densities, eqn. (4) does not even converge.

In this range, empirical or semiempirical equations of state are used. They contain a number of parameters which are chosen to fit the experimental data as closely as possible. A large number of parameters is usually necessary to fit the experimental results over wide ranges of the variables with appreciable accuracy.

Equations of state for water in the moderate pressure range (below 1 kbar) are reviewed by Nowak and Grosh.[797] Burnham *et al.*[153] were able to fit their experimental data between 20 and 900°C and between 2 and 10 kbar within 0.1% by a ninth-degree polynomial of the form

$$PV = \sum_{j=0}^{9} \sum_{i=0}^{9-j} a_{ij} T^i P^j \tag{5}$$

where P is the pressure in kbar, T is the temperature in hundreds of degrees centigrade, and V is the specific volume in $cm^3\ g^{-1}$. The 55 coefficients a_{ij} are tabulated in Table IV. In order also to represent the experimental data

TABLE IV. Coefficients a_{ij} in Eqn. (5) Taken from Burnham et al.[153]

i	$a_{ij} \times 10^6$									
	$j = 0$	$j = 1$	$j = 2$	$j = 3$	$j = 4$	$j = 5$	$j = 6$	$j = 7$	$j = 8$	$j = 9$
0	−38106.590	1056466.5	−76454.683	709.56600	10454.935	−4488.7845	905.86480	−98.313225	5.5147402	−0.12547995
1	66431.473	−42654.409	82193.439	−46017.503	15484.386	−3080.9442	344.62611	−19.805288	0.45366798	—
2	−89413.365	8922.2030	1971.9748	−4879.6036	1901.7140	−271.55343	15.518783	−0.29266747	—	—
3	92076.993	4380.5877	317.79963	−1411.1684	190.87901	−3.3016835	−0.20239777	—	—	—
4	−47887.922	−618.91487	1904.3321	−111.53384	−12.650068	0.60544695	—	—	—	—
5	13798.565	−1323.3046	−149.43777	25.569347	−0.34096098	—	—	—	—	—
6	−1988.6789	239.12344	−7.5721292	−0.60823080	—	—	—	—	—	—
7	150.64451	−12.791592	0.57844948	—	—	—	—	—	—	—
8	−6.0211303	0.17501650	—	—	—	—	—	—	—	—
9	0.10682110	—	—	—	—	—	—	—	—	—

between 1 and 2 kbar over the whole temperature range, and between 900 and 1000°C for all pressures, with the same standard deviation of 0.1%, the authors needed two additional polynomials of eighth and ninth degree.

Jůza *et al.*[566] developed an equation of state for water valid from 0 to 1000°C and from 1 bar to 100 kbar. They started with the van der Waals equation of state

$$P = [RT/(V - b)] - (a/V^2) \tag{6}$$

In order to make this equation more flexible, a and b are no longer restricted to be constants. Jůza *et al.* assumed that a is a function of volume and temperature and may be expressed as

$$a = \sum_{n=1}^{8} f_n(V)g_n(T) \tag{7}$$

whereas b is a function of volume only. With these assumptions, eqn. (6) can be rearranged to its final form:

$$PV = RTF(V) - (a/V) \tag{8}$$

with a given by eqn. (7). It is beyond the scope of this chapter to present the 17 functions $F(V)$, $f_n(V)$, and $g_n(T)$ explicitly. In order to obtain a good fit to the experimental data, Jůza *et al.* had to divide the whole density range into four regions and to determine different sets of functions $f_n(V)$ and $g_n(T)$ for the various regions.

2.2. Thermodynamic Functions

As mentioned above, accurate values for the thermodynamic functions of water are compiled for pressures below 1 kbar.[45,953,1119] They are derived from direct measurements of specific heat and enthalpy or from *PVT* data fitted by an empirical equation of state and the thermodynamic functions of water in the ideal gas state which are calculated from spectroscopic data. No measurements of the thermal properties of water have been performed above 1 kbar. Consequently, all values of the thermodynamic functions in this range are based on *PVT* data and some reference value in the lower pressure range.

2.2.1. *Gibbs Free Energy*

The molar Gibbs free energy of water above 100°C and 1 kbar is tabulated in Table V. The values refer to a standard state $G^\circ = 0$ at 25°C

TABLE V. Gibbs Free Energy of Water above 100°C and 1 kbar[a]

P, kbar	Gibbs free energy, kJ mol^{-1}									
	at T=100°C	200°C	300°C	400°C	500°C	600°C	700°C	800°C	900°C	1000°C
1.0	1.18	−1.29	−5.36	−10.8	−17.6	−25.6	−34.7	−44.6	−55.1	−66.2
1.5	2.07	−0.33	−4.28	−9.56	−16.0	−23.6	−32.1	−41.4	−51.4	−62.0
2.0	2.95	0.62	−3.25	−8.39	−14.7	−22.0	−30.1	−39.1	−48.7	−58.9
2.5	3.82	1.54	−2.24	−7.27	−13.4	−20.5	−28.5	−37.1	−46.5	−56.4
3.0	4.67	2.45	−1.26	−6.19	−12.2	−19.2	−26.9	−35.4	−44.5	−54.3
3.5	5.52	3.34	−0.29	−5.15	−11.1	−17.9	−25.5	−33.8	−42.8	−52.4
4.0	6.35	4.22	0.65	−4.13	−9.95	−16.7	−24.2	−32.3	−41.2	−50.6
4.5	7.18	5.10	1.58	−3.14	−8.88	−15.5	−22.9	−31.0	−39.7	−49.0
5.0	8.00	5.96	2.49	−2.16	−7.83	−14.3	−21.7	−29.6	−38.2	−47.5
6	9.61	7.65	4.28	−0.26	−5.80	−12.2	−19.4	−27.1	−35.5	−44.7
7	11.2	9.31	6.03	1.59	−3.83	−10.1	−17.1	−24.8	−33.0	−42.0
8	12.8	11.0	7.74	3.39	−1.93	−8.14	−15.0	−22.5	−30.6	−39.5
9	14.3	12.6	9.43	5.16	−0.08	−6.19	−13.0	−20.4	−28.4	−37.1
10	15.8	14.1	11.1	6.89	1.73	−4.30	−11.0	−18.3	−26.2	−34.8
15	23.1	21.6	18.7	14.8	9.92	4.26	−2.10	−9.13	−16.7	−24.8
20	30.1	28.8	26.1	22.4	17.8	12.3	6.23	−0.54	−7.76	−15.6
30	—	42.4	40.1	36.7	32.4	27.3	21.4	15.2	8.22	0.89
40	—	55.2	53.1	50.0	46.0	41.1	35.8	29.7	22.6	15.8
50	—	—	65.6	62.7	59.0	54.3	49.1	43.5	36.5	30.0
60	—	—	77.7	75.0	71.2	67.0	62.2	56.9	50.0	43.7
70	—	—	—	86.9	83.7	78.7	74.2	69.0	63.4	56.2
80	—	—	—	97.9	94.9	91.3	85.9	81.0	75.6	68.6
90	—	—	—	109	106	103	97.5	92.8	87.6	80.8
100	—	—	—	120	117	114	109	104	98.5	93.0

[a] Standard state: $G^{\circ} = 0$ at 25°C and 1 bar. The tabulated values may be referred to a standard state $G^{*\circ} = 0$ at the triple point of water ($P^{*\circ} = 6.1$ mbar; $T^{*\circ} = 273.16$°K) using the relation $G^{*} = G + (1.89 - 6.62 \times 10^{-3}T)\ J\ mol^{-1}$.

and 1 bar. After conversion of units and choice of a new standard state, the values of Burnham *et al.*[153] are tabulated below 10 kbar. Burnham *et al.* calculated the Gibbs free energy at a pressure P and a temperature T from the relation

$$G(P, T) = G(1 \text{ kbar}, T) + \int_{1 \text{ kbar}}^{P} V(P, T)\, dP \tag{9}$$

where V is the molar volume. They used eqn. (5) and two other equations of similar type as equations of state for the computation of the integral. The $G(1 \text{ kbar}, T)$ values were calculated from $G(1 \text{ kbar}, T) = H(1 \text{ kbar}, T) - TS(1 \text{ kbar}, T)$, with enthalpy and entropy values at 1 kbar taken from the NEL Steam Tables.[45] The accuracy of the values below 10 kbar in Table V is estimated to be about 1%.

The values above 10 kbar were calculated from enthalpies and entropies computed by Jůza *et al.*[566] with the aid of eqn. (8). The accuracy of these values is considerably lower than for those below 10 kbar.

The Gibbs free energy decreases with rising temperature and increases with pressure over the whole range. It has a negative sign at low pressures and high temperatures.

2.2.2. *Entropy*

Values for the molar entropy of water above 100°C and 1 kbar are compiled in Table VI. They refer to a standard state $S^\circ = 0$ at 25°C and 1 bar. Again, the values below 10 kbar are taken from Burnham *et al.*[153] They were calculated from the relation

$$S(P, T) = S(1 \text{ kbar}, T) - \int_{1 \text{ kbar}}^{P} [\partial V(P, T)/\partial T]_P\, dP \tag{10}$$

with eqn. (5), and two similar equations, as equations of state. The $S(1 \text{ kbar}, T)$ values were taken from the NEL Steam Tables.[45] The entropy values above 10 kbar were computed by Juza *et al.*[566] using eqn. (8).

As required by the criterion for thermal stability, $0 < C_P = T(\partial S/\partial T)_P$, the entropy increases with temperature. It decreases monotonically with increasing pressure. This is a consequence of the positive sign of the isobaric thermal expansion coefficient over the whole range.

2.2.3. *Enthalpy and Molar Heat Capacity*

The molar enthalpy of water above 100°C and 1 kbar is listed in Table VII. The standard state chosen is $H^\circ = 0$ at 25°C and 1 bar. The values up to

P, kbar	Entropy, J mol^{-1} °K^{-1}									
	at T=100°C	200°C	300°C	400°C	500°C	600°C	700°C	800°C	900°C	1000°C
1.0	15.7	33.1	47.8	61.4	74.3	86.2	95.3	102	108	112
1.5	15.1	32.3	46.5	59.1	70.4	80.7	89.4	96.7	103	108
2.0	14.6	31.5	45.4	57.5	68.0	77.5	85.7	93.0	99.5	104
2.5	14.2	30.8	44.5	56.2	66.5	75.6	83.0	90.0	96.7	101
3.0	13.7	30.1	43.7	55.2	65.1	73.9	81.1	87.9	94.6	99.5
3.5	13.3	29.5	42.9	54.3	64.0	72.6	79.6	86.2	92.9	98.2
4.0	12.9	29.0	42.2	53.4	63.0	71.5	78.3	84.8	91.5	97.2
4.5	12.5	28.5	41.6	52.7	62.1	70.5	77.2	83.6	90.3	96.5
5.0	12.1	28.0	41.1	52.0	61.3	69.6	76.2	82.6	89.3	95.9
6	11.4	27.1	40.0	50.8	60.0	68.1	74.6	80.8	87.6	94.8
7	10.6	26.3	39.1	49.8	58.7	66.8	73.2	79.3	86.1	93.5
8	9.94	25.5	38.2	48.8	57.7	65.7	72.0	78.1	84.7	91.8
9	9.28	24.8	37.5	48.0	56.8	64.7	71.0	77.0	83.5	89.9
10	8.65	24.1	36.7	47.2	56.0	63.8	70.0	76.0	82.5	87.9
15	6.41	22.1	34.4	44.4	53.0	60.4	67.0	72.9	78.2	83.1
20	4.28	20.1	32.3	42.2	50.6	58.0	64.4	70.2	75.4	80.3
30	—	16.9	29.0	38.8	47.1	54.3	60.7	66.3	71.5	76.2
40	—	14.4	26.3	36.1	44.3	51.5	57.8	63.4	68.5	73.2
50	—	—	24.0	33.7	41.9	49.1	55.4	61.0	66.1	70.7
60	—	—	22.0	31.7	39.8	46.9	53.2	58.8	63.9	68.6
70	—	—	—	29.8	37.9	45.0	51.2	56.8	61.9	66.6
80	—	—	—	28.3	36.3	43.2	49.4	55.0	60.0	64.7
90	—	—	—	27.0	34.8	41.7	47.8	53.3	58.3	63.0
100	—	—	—	25.9	33.6	40.3	46.4	51.8	56.7	61.3

[a] Standard state: $S^{\circ} = 0$ at 25°C and 1 bar. The tabulated values may be referred to a standard state $S^{*\circ} = 0$ at the triple point of water ($P^{*\circ} = 6.1$ mbar; $T^{*\circ} = 273.16$°K) using the relation $S^{*} = S + 6.62$ J mol^{-1} °K^{-1} and to the absolute entropy scale ($S^{**\circ} = 69.99$ J mol^{-1} °K^{-1} at 25°C and 1 bar) by the relation $S^{**} = S + 69.99$ J mol^{-1} °K^{-1}.

TABLE VII. Enthalpy of Water above 100°C and 1 kbar[a]

P, kbar	Enthalpy, kJ mol^{-1}									
	at T=100°C	200°C	300°C	400°C	500°C	600°C	700°C	800°C	900°C	1000°C
1.0	7.03	14.4	22.0	30.5	39.9	49.6	58.0	65.1	71.4	77.0
1.5	7.72	14.9	22.4	30.2	38.4	46.8	54.9	62.3	69.4	75.2
2.0	8.42	15.5	22.8	30.3	37.9	45.7	53.3	60.7	68.0	73.7
2.5	9.11	16.1	23.3	30.6	38.0	45.5	52.3	59.4	67.0	72.8
3.0	9.80	16.7	23.8	30.9	38.1	45.4	52.0	58.9	66.4	72.5
3.5	10.5	17.3	24.3	31.4	38.4	45.5	51.9	58.7	66.2	72.6
4.0	11.2	17.9	24.9	31.8	38.7	45.7	52.0	58.7	66.2	73.1
4.5	11.8	18.6	25.4	32.3	39.1	46.0	52.2	58.8	66.3	73.8
5.0	12.5	19.2	26.0	32.8	39.6	46.4	52.5	59.0	66.5	74.6
6	13.8	20.5	27.2	33.9	40.6	47.2	53.2	59.6	67.2	76.1
7	15.2	21.8	28.4	35.1	41.6	48.2	54.1	60.3	68.0	77.0
8	16.5	23.0	29.7	36.2	42.7	49.2	55.1	61.2	68.8	77.4
9	17.8	24.3	30.9	37.4	43.8	50.3	56.1	62.2	69.6	77.4
10	19.0	25.6	32.1	38.7	45.0	51.4	57.2	63.3	70.5	77.2
15	25.5	32.1	38.4	44.7	50.9	57.0	63.1	69.1	75.0	81.0
20	31.7	38.3	44.6	50.8	56.9	62.9	68.9	74.8	80.7	86.6
30	—	50.4	56.7	62.8	68.8	74.7	80.5	86.3	92.1	97.9
40	—	62.0	68.2	74.3	80.3	86.1	92.0	97.7	103	109
50	—	—	79.4	85.4	91.4	97.2	103	109	114	120
60	—	—	90.3	96.3	102	108	114	120	125	131
70	—	—	—	107	113	118	124	130	136	141
80	—	—	—	117	123	129	134	140	146	151
90	—	—	—	127	133	139	144	150	156	161
100	—	—	—	137	143	149	154	160	165	171

[a] Standard state: $H^{\circ} = 0$ at 25°C and 1 bar. The tabulated values may be referred to a standard state $H^{*\circ} = 0$ at the triple point of water ($P^{*\circ} = 6.1$

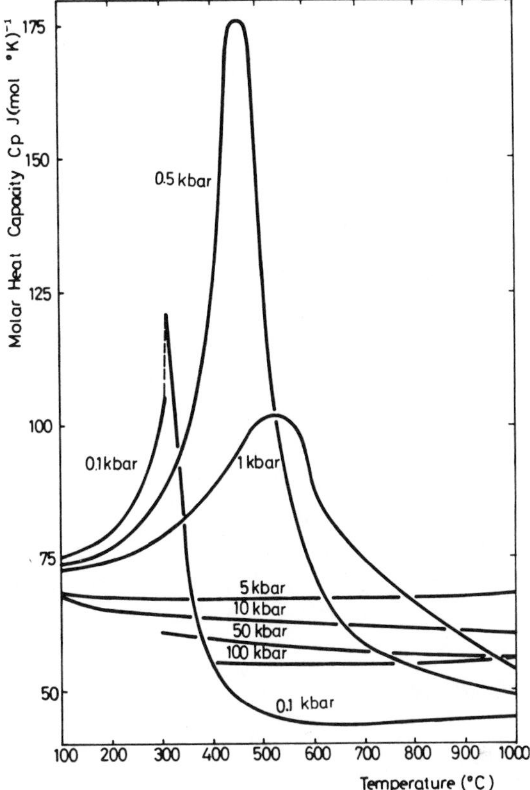

Fig. 4. Molar heat capacity at constant pressure C_P of water as a function of temperature for various pressures.[566,953]

10 kbar were calculated from the Gibbs free energies and entropies given in Tables V and VI, respectively. The values of Jůza *et al.*[566] are tabulated above 10 kbar.

At temperatures above 300°C, the enthalpy exhibits a shallow minimum as a function of pressure. This minimum shifts from about 600 bars at 300°C to about 3 kbar at 1000°C. The enthalpy increases with temperature according to the requirement of thermal stability, $0 < C_P = (\partial H/\partial T)_P$.

The behavior of the molar heat capacity at constant pressure C_P is shown in Fig. 4, where it is plotted as a function of temperature at various constant pressures. The isobars for pressures lower than the critical pressure show the usual discontinuity at the boiling temperatures. On the 0.1-kbar isobar, the molar heat capacity of the saturated vapor is already greater than that of the liquid. On the critical isobar, not shown in Fig. 4, C_P

approaches infinity at the critical temperature. At higher pressures, the isobars are continuous. For moderate pressures, they still show a pronounced maximum, which shifts to higher temperatures and becomes less pronounced at higher pressures. Above 3 kbar the maximum finally vanishes and C_P becomes almost independent of temperature. It decreases slightly with further increasing pressure. The values of the heat capacity in this range are 10–25% lower than for water at normal conditions.

The temperature dependence of the molar heat capacity at constant volume C_V gives some insight into the nature of dense, hot water as compared with water at normal conditions. At a constant specific volume of 1 $cm^3 g^{-1}$, C_V drops to 60% of its original value in going from 0 to 1000°C.[153] More than half of this decrease occurs below 200°C and more than three-quarters below 400°C. Moreover, the value of C_V at 1000°C and 1 $cm^3 g^{-1}$ is only 30% higher than the ideal gas value[367] for that temperature, whereas at 0°C, C_V of water is three times greater at 1 $cm^3 g^{-1}$ than the ideal gas value. Therefore, one may conclude that water loses most of the abnormal features, responsible for its unusually high specific heat, at temperatures below 200°C even if the density is held constant. In this respect it becomes a normal liquid at high temperatures and pressures.

2.2.4. *Fugacity Coefficient*

The fugacity coefficient φ is defined as

$$\varphi = f/p \tag{11}$$

where the fugacity f is given by the relation

$$\ln(f/f^\circ) = [G(P) - G(P^\circ)]/RT \tag{12}$$

f° and $G(P^\circ)$ being the fugacity and Gibbs free energy in an arbitrarily chosen reference state. Combining eqns. (11) and (12) yields the result

$$\ln \varphi = \{[G(P) - G(P^\circ)]/RT\} + \ln f^\circ - \ln P \tag{13}$$

The values for the fugacity coefficient of water above 100°C and 1 kbar tabulated in Table VIII were calculated by Burnham *et al.*[153] using eqn. (13). The reference pressure chosen was $P^\circ = 0.01$ bar. At such a low pressure, the relation $f^\circ/P^\circ = 1$ holds for all temperatures in Table VIII. Equation (13) can thus be written in the form

$$\ln \varphi = \{[G(P) - G(P^\circ)]/RT\} + \ln P^\circ - \ln P \tag{14}$$

TABLE VIII. Fugacity Coefficients of Water above 100°C and 1 kbar[a]

P, kbar	Fugacity coefficients									
	at T=100°C	200°C	300°C	400°C	500°C	600°C	700°C	800°C	900°C	1000°C
1.0	0.002	0.024	0.105	0.263	0.459	0.632	0.762	0.848	0.905	0.940
1.5	0.002	0.020	0.088	0.219	0.389	0.557	0.700	0.807	0.883	0.930
2.0	0.002	0.019	0.082	0.202	0.361	0.523	0.671	0.788	0.875	0.932
2.5	0.002	0.020	0.081	0.197	0.351	0.510	0.660	0.786	0.880	0.945
3.0	0.002	0.020	0.083	0.199	0.353	0.512	0.664	0.795	0.895	0.964
3.5	0.002	0.022	0.087	0.206	0.361	0.523	0.678	0.813	0.917	0.990
4.0	0.003	0.024	0.092	0.216	0.376	0.540	0.700	0.838	0.947	1.020
4.5	0.003	0.027	0.100	0.229	0.395	0.564	0.728	0.871	0.982	1.055
5.0	0.003	0.030	0.109	0.245	0.418	0.594	0.762	0.909	1.024	1.096
6	0.005	0.038	0.132	0.287	0.478	0.668	0.848	1.004	1.124	1.193
7	0.007	0.050	0.163	0.343	0.556	0.763	0.956	1.121	1.246	1.315
8	0.009	0.066	0.205	0.414	0.654	0.881	1.088	1.262	1.391	1.460
9	0.014	0.088	0.259	0.504	0.776	1.024	1.246	1.428	1.561	1.631
10	0.020	0.119	0.330	0.618	0.925	1.195	1.432	1.621	1.755	1.826

[a] Standard state: $f^{\circ}/P^{\circ} = 1$ at $P^{\circ} = 0.01$ bar for each temperature.

The G(0.01 bar) values were taken from the NEL Steam Tables.[45] No values are tabulated above 10 kbar, where the fugacity coefficient increases very rapidly with pressure. At 100 kbar, for example, its order of magnitude is 10^7 and 10^4 at 400 and 1000°C, respectively.

3. TRANSPORT PROPERTIES

For a discussion of the transport properties of water at high temperatures and high pressures, the behavior of the viscosity, self-diffusion coefficient, and thermal conductivity of normal substances should briefly be recalled. At low densities, that is, in the dilute gas state, where the mean free path of the molecules is large compared with their diameter, transport of energy, momentum, and mass is mainly due to the translational movement of the molecules. For this range, the kinetic theory of gases gives the result[516] that the viscosity η, the coefficient of thermal conductivity λ, and the product ϱD of density ϱ and self-diffusion coefficient D are about proportional to one another and to the square root of the absolute temperature. In liquids, however, where the average distance between molecules is comparable with their diameters, collisional transfer of energy and momentum becomes important. Collisional transfer means the instantaneous transport from the center of one molecule to the center of a second one during a collision. Collisional transfer does not contribute to mass transport. Consequently, in a dense phase, viscosity and thermal conductivity are about one or two orders of magnitude greater than in the corresponding dilute gas and increase with increasing density or pressure at constant temperature, whereas the self-diffusion coefficient is about five orders of magnitude smaller and decreases with increasing density or pressure.

The isochoric temperature dependence of the viscosity of a dense fluid phase is best described by the transition-state theory expression

$$\eta = \eta_0 \exp[(E_\eta)_V/RT] \tag{15}$$

where $(E_\eta)_V > 0$ is the isochoric activation energy and η_0 is a constant. Hence, the viscosity decreases with temperature, contrary to the dilute gas case, where $(\partial\eta/\partial T)_V > 0$. Consequently, at some intermediate density, an inversion of the isochoric temperature dependence of the viscosity must occur.

Since in dense phases, according to the transition-state theory, the self-diffusion coefficient is normally related to the viscosity by $D \sim kT/\eta$, it increases with temperature.

The isochoric temperature dependence of the thermal conductivity also remains positive over the whole density range, whereas the isobaric temperature dependence $(\partial\lambda/\partial T)_P$ due to thermal expansion is usually negative at high densities.

It will be shown in the next three sections that the transport properties of water deviate considerably from what is regarded as normal.

3.1. Viscosity

Bridgman[119] found that the viscosity of water at low temperatures first decreases upon compression, passes through a minimum, and then increases with further increasing pressure. The pressure of the minimum shifts from about 1.3 kbar at 0°C toward lower values with increasing temperature. At about 30°C, the minimum disappears. This result has been confirmed by several investigations.[80,464,530] In the last decade, the viscosity of water has been measured over a considerable range of temperatures and pressures. Various techniques for viscosity measurements at ordinary conditions were adapted for use at high pressures by the construction of viscometers in such a way that they could be operated in a high pressure vessel. At high temperatures and moderate pressures (below 1.2 kbar), capillary viscometers have usually been employed. At low temperatures and high pressures, i.e., at high densities, falling cylinder[80,464] and rolling ball[530] viscometers are better suited. In the intermediate range (to 560°C and and 3.5 kbar), Dudziak and Franck[279] used an oscillating disk method. The viscosity values above 100°C and 100 bars reported in the literature[2,3,279,464,555,597,613,737,769,893,1054,1068,1169] are collected in Table IX.

For the temperatures listed in Table IX, the viscosity increases monotonically with pressure, as it does for normal substances. Its temperature dependence is quite complex. Neither the isochoric nor the isobaric temperature dependence follows a simple Arrhenius relation in the range of high densities. The isobaric activation energy $(E_\eta)_P$, defined as $(E_\eta)_P = R \times [\partial(\ln\eta)/\partial(1/T)]_P$, does not only depend on pressure, but also varies with temperature. At 1 bar (or saturation pressure), $(E_\eta)_P$ falls from 23.0 kJ mol^{-1} at 0°C to 8.8 kJ mol^{-1} at 150°C,[301] with a tendency to become constant at higher temperatures.[464] A similar behavior is observed at higher pressures, where the decrease of $(E_\eta)_P$ with temperature is slightly smaller and the high-temperature limiting values are slightly higher than at lower pressures.[464] For given temperatures in the range below 100°C, the activation energy decreases with pressure.

The isochoric activation energy $(E_\eta)_V$, defined by eqn. (15), is approxi-

TABLE IX. Viscosity of Water above 100°C and 0.1 kbar[2,3,279,464,555,597,613,737,769,893,1054,1068,1169]

P, kbar	Viscosity,[a] μN sec m⁻²														
	at T =100°C	150°C	200°C	250°C	300°C	350°C	400°C	450°C	500°C	550°C	600°C	650°C	700°C	750°C	800°C
0.1	287	185	137	110	89.5[b]	23.6	25.4	27.2	29.2	31.2	33.2	35.2	37.2	39.2	41.2
0.5	296	194	148	121	101	84.8	68.8	52.0	42.0	39.8	40.0	41.0	42.1	43.5	44.9
1.0	308	205	162	132	113	98.6	86.7	76.0	66.9	60.4	56.1	53.3	52.1	51.9	52.3
1.5	320	216	175	145	124	109	98.0	87.0	79.0	72.5	—	—	—	—	—
2.0	333	227	184	155	133	117	105	95.0	86.0	81.0	—	—	—	—	—
2.5	346	238	192	163	141	124	111	100	92.0	87.0	—	—	—	—	—
3.0	359	249	199	171	148	130	116	105	96.0	91.0	—	—	—	—	—
3.5	372	260	206	178	155	135	121	109	100	95.0	—	—	—	—	—
4.0	385	—	—	—	—	—	—	—	—	—	—	—	—	—	—
5.0	415	—	—	—	—	—	—	—	—	—	—	—	—	—	—
6.0	446	—	—	—	—	—	—	—	—	—	—	—	—	—	—
7.0	477	—	—	—	—	—	—	—	—	—	—	—	—	—	—
8.0	510	—	—	—	—	—	—	—	—	—	—	—	—	—	—
9.0	544	—	—	—	—	—	—	—	—	—	—	—	—	—	—

[a] To convert the viscosity values into centipoise, multiply by 10^{-3}.
[b] Phase transition; do not interpolate.

mately equal to $(E_\eta)_P$. It also decreases with increasing temperature along an isochore[279,464] and becomes almost constant at higher temperatures. At low temperatures, $(E_\eta)_V$ decreases with decreasing volume.

From the behavior of the viscosity, it follows that at low temperatures, compression of water tends to facilitate displacements of the water molecules. This result is supported by Hertz and Rädle,[495,882] who measured the proton spin–lattice relaxation time T_1 in water as a function of pressure for several temperatures. At 0°C, they found an increase of T_1 when pressure was applied up to 1.4 kbar, where T_1 passes through a maximum. At 3 kbar, T_1 is about 6% less than at 1 bar. At 30°C, T_1 is almost independent of pressure (also found by Benedek and Purcell[62]), and at higher temperatures, it decreases continually with pressure. Since the spin-rotational contribution is negligible,

$$\begin{aligned} 1/T_1 &= (1/T_1)_{\text{intra}} + (1/T_1)_{\text{inter}} \\ &= \gamma^4\hbar^2[(3/2b^6)\tau_r + (2/5)\pi(N_s/aD)] \end{aligned} \tag{16}$$

where γ is the gyromagnetic ratio of the proton; b is the intramolecular proton–proton distance; a is the closest distance of two protons on different water molecules; N_s is the number of spins per cm^3; $\hbar$ is Planck's constant divided by 2π; τ_r is the rotational correlation time of the water molecules, and D is the self-diffusion coefficient. From this relation and the pressure dependence of T_1 and of the self-diffusion coefficient (see Section 3.2), it follows that at temperatures close to the melting temperature, the rotational correlation time of the water molecule is smaller under the pressure of maximal T_1 than at 1 bar. The rotational diffusive motion of the water molecule thus becomes faster with increasing pressure in this range.

Dudziak and Franck[279] plotted the viscosity of water as a function of density for various constant temperatures and found that the isotherms intersect at a density of about 0.8 g cm^{-3}, that is, at 80% of the normal liquid density. This density seems to be surprisingly high for the transition of the isochoric temperature dependence of the viscosity from the dilute gas type (positive sign) to the liquid type (negative sign) of behavior, but a comparison with the temperature dependences of the viscosities of argon and krypton, which also change sign at about 80% of the respective normal liquid density,[747,1072] indicates that water does not behave abnormally in this respect.

The viscosity of heavy water shows the same general behavior over the whole range of temperature and pressure.[3,464] Harlow[464] calculated the ratio η_{D_2O}/η_{H_2O} as a function of pressure for several temperatures. He found that the viscosity ratio does not depend on pressure, except at temperatures below 75°C, where it decreases with increasing pressure up to 4 kbar, because of the difference in the pressure effects on the structures of water and heavy water. At constant pressure, the ratio decreases with increasing temperature from a value slightly smaller than $(I_{D_2O}/I_{H_2O})^{1/2} = 1.34$, and remaining always greater than $(m_{D_2O}/m_{H_2O})^{1/2} = 1.054$ where I and m are the moment of inertia and the mass of the molecules respectively. Harlow concluded that the rotational contribution to momentum transfer in water is large, independent of pressure, and decreases with rising temperature.

Mineev and Zaidel'[(755,756)] carried out measurements of the viscosity of water in the pressure range from 40 to 150 kbar. They propagated a sinusoidally corrugated shock front into water and observed the change of periodic perturbation as the shock front proceeded. From the rate of that change, and some other measurable quantities, Mineev and Zaidel' calculated the viscosity of water behind the shock front. They arrived at the result that the viscosity increases rapidly with increasing pressure. At 80 kbar, it reaches a value of about 2×10^3 N sec m^{-2} and remains constant up to 150 kbar, the highest pressure in the experiments. This result means that the viscosity of water under shock conditions is about six orders of magnitude higher than under normal conditions. As Hamann and Linton[(451)] pointed out, this result seems to be very improbable for several reasons:

1. A calculation of the viscosity under shock conditions (1.5 to 1.9 g cm^{-3} and 700 to 1700°K) using Enskog's formula[(516)] for the viscosity of a hard-sphere fluid yields values of the order 10^{-3}–10^{-2} N sec m^{-2}. These values may be regarded as upper limits, since Dudziak and Franck[(279)] demonstrated that Enskog's formula fits their experimental viscosities at 773°K up to a density of 1 g cm^{-3} to within 25% with a tendency to overestimate the viscosity at higher densities.

2. The molar conductance Λ of strong electrolytes in shock-compressed aqueous solutions measured by Hamann and Linton[(451)] is only slightly different from that in unshocked solutions under ordinary conditions. If Walden's rule ($\Lambda \propto 1/\eta$) holds approximately, the viscosity cannot change by six orders of magnitude.

3. According to the increase in viscosity, the relaxation time of the water molecules should also increase by six orders of magnitude to a value of about 10^{-5} sec in the shock front. Such a long relaxation time is incompatible with the high values found for the excess proton mobility under shock conditions (see Section 4.2) and with the observation of Kormer *et al.*[(624)] that the phase transition from water to ice VII between 40 and 100 kbar produced by consecutive shocks is accomplished in 10^{-6}–10^{-7} sec.

Thus, it seems that the viscosity of water under shock conditions may not be greatly different from that under normal conditions.

3.2. Self-Diffusion

Very little experimental information on the self-diffusion coefficient of water at high temperatures and pressures is available so far. This is mainly due to the fact that conventional tracer diffusion experiments are difficult to perform in fluid phases under high pressures and high temperatures because of the problems relating to the continuous measurement of the tracer concentration in a pressure vessel and the long time involved. In addition, due to the uncertainty of the initial conditions, the accuracy of such experiments is very poor.

Cuddeback *et al.*[222] measured the self-diffusion coefficient of water at 0, 25, and 50°C in the pressure range to 10 kbar using THO as tracer. Although their results suffer from a considerable scatter, they clearly show that the self-diffusion coefficient D at 0 and 25°C first increases with rising pressure. After passing through a maximum, D decreases with further increasing pressure. At 50°C, it immediately drops to smaller values upon compression.

Benedek and Purcell[62] measured the self-diffusion coefficient of water at 28.8°C, also in the range to 10 kbar, using the nuclear magnetic resonance spin-echo technique. They found that D decreases when the pressure is raised at this temperature. At 10 kbar, it arrives at about 40% of its initial value. The authors plotted the normalized self-diffusion coefficient $D(P)/D(1)$, spin–lattice relaxation time $T_1(P)/T_1(1)$, which they also measured, and reciprocal viscosity $\eta(1)/\eta(P)$ as a function of pressure. The ratios $D(P)/D(1)$ and $\eta(1)/\eta(P)$ show almost the same pressure dependence. Thus, the relation $D \propto 1/\eta$ applies. The decrease of $T_1(P)/T_1(1)$ with pressure is only half that for D and $1/\eta$. From these observations, Benedek and Purcell concluded that the frequently used relation

$$\tau_r \propto \eta/T \tag{17}$$

which links the rotational correlation time of the water molecule and the bulk viscosity, does not hold for density changes at constant temperature. If eqn. (17) is valid, it follows from eqn. (16) and the experimentally confirmed relation $D \propto T/\eta$ that

$$1/T_1 = (A + BN_s)\eta/T \tag{18}$$

That is, the pressure dependence of T_1 and of $1/\eta$ should be the same, since the spin density varies only slightly with pressure. This is ruled out by the experimental results, and consequently eqn. (17) is not valid, possibly because the translational motion of the molecules is influenced more by density increases than is the rotational motion.

Hertz and Rädle[495,882] also measured the self-diffusion coefficient of water at 0 and 25°C up to 3 kbar using the spin-echo technique. At 0°C, they found a maximum of D at about 1.5 kbar, where it is about 20% higher than at 1 bar. At 25°C, D decreases as a function of pressure.

Measurements of the self-diffusion coefficient of water under pressure, at temperatures exceeding 100°C, were made by Hausser *et al.*,[474] who applied the spin-echo technique to water under its saturation pressure up to the critical point. They found that D increases from 2.3×10^{-5} cm² sec⁻¹ at 25°C to about 6×10^{-4} cm² sec⁻¹ at 370°C. Its temperature dependence cannot be described by a simple Arrhenius relation with a constant activation energy E_D. As found for the viscosity, the activation energy decreases with increasing temperature and becomes almost constant above 170°C, where it has the value 6.7 kJ mol⁻¹. The relation $D \propto kT/\eta$ provides an excellent fit to the experimental data over the whole temperature range.

From the few experimental results on the self-diffusion coefficient of compressed water, one can draw the conclusion that it is roughly proportional to the reciprocal viscosity and also shows the anomalous features discussed in the preceding subsection.

3.3. Thermal Conductivity

The thermal conductivity λ of water is anomalous in two respects: (1) its value is about a factor of four higher than for other liquids; and (2) its isobaric temperature dependence is positive, whereas the thermal conductivities of most liquids decrease linearly when the temperature is raised at constant pressure. As can be seen from Table *X*, in which the experimental thermal conductivity data of water above 100°C and 0.1 kbar are listed, the positive isobaric temperature dependence persists up to about 150–200°C, where a maximum is passed. At pressures above 1 kbar, the temperature of maximal thermal conductivity has not been reached in the experiments. However, Bridgman's[118] values at 30 and 75°C and those at 100°C in Table X indicate that even at a pressure of 10 kbar, λ increases with temperature.

The behavior of the thermal conductivity of water in the temperature range below about 300°C is very similar to that of dissociating gases under

TABLE X. Coefficient of Thermal Conductivity of Water above 100°C and 0.1 kbar[131,661,666]

P, kbar	Coefficient of thermal conductivity, mJ m^{-1} sec^{-1} $°K^{-1}$ at T = 100°C	150°C	200°C	250°C	300°C	350°C	400°C	450°C	500°C	550°C	600°C	650°C	700°C	750°C	800°C
0.1	689	697	677	629	554	67.5[a]	63.9	67.6	74.2	82.0	90.4	99.1	108	117	126
0.25	694	706	690	650	582	483	162	108	99.5	104	110	110	123	129	136
0.6	703	719	713	684	628	538	437	307	212	166	155	143	149	152	155
0.75	712	733	736	718	674	594	—	—	—	—	—	—	—	—	—
1.0	726	—	—	—	—	—	—	—	—	—	—	—	—	—	—
1.5	747	—	—	—	—	—	—	—	—	—	—	—	—	—	—
2.0	767	—	—	—	—	—	—	—	—	—	—	—	—	—	—
2.5	786	—	—	—	—	—	—	—	—	—	—	—	—	—	—
3.0	804	—	—	—	—	—	—	—	—	—	—	—	—	—	—
3.5	822	—	—	—	—	—	—	—	—	—	—	—	—	—	—
4.0	840	—	—	—	—	—	—	—	—	—	—	—	—	—	—
4.5	856	—	—	—	—	—	—	—	—	—	—	—	—	—	—
5.0	872	—	—	—	—	—	—	—	—	—	—	—	—	—	—
6.0	903	—	—	—	—	—	—	—	—	—	—	—	—	—	—
7.0	935	—	—	—	—	—	—	—	—	—	—	—	—	—	—
8.0	964	—	—	—	—	—	—	—	—	—	—	—	—	—	—

[a] Phase transition; do not interpolate.

conditions of pressures and temperatures where the equilibrium of the dissociation reaction is strongly shifted by temperature changes. Under such conditions, dissociation occurs at the hotter end of a temperature gradient, where the gas takes up the dissociation energy. At the cold end, the gas reassociates and delivers the dissociation energy, which thus has been transported in addition to the amount of energy transferred if no dissociation occurs. Eigen[293] treated water as a dissociating substance and was able to calculate the thermal conductivity within the experimental limits of error. His calculation also provided the correct isobaric temperature dependence of the thermal conductivity.

Heavy water has a slighlty lower thermal conductivity than ordinary water, but the general behavior is the same in both cases.[667] The ratio $\lambda_{D_2O}/\lambda_{H_2O}$ is almost independent of pressure and decreases with increasing temperature.

4. ELECTRICAL PROPERTIES

Experimental information on the electrical properties of water at high temperatures and high pressures is scarce. Since measurements of the pressure dependence of the dielectric constant in the dispersion region have not yet been made, nothing is known about the effects of pressure on the dielectric relaxation time. It was only recently that the static dielectric constant, the specific conductance, and the ionic conductances of protons and hydroxyl ions in water have been studied over wide ranges of temperature and pressure.

4.1. Static Dielectric Constant

4.1.1. *Experimental Methods and Results*

Measurements of the static dielectric constant of water are complicated by the appreciable electrical conductivity of water at high temperatures and pressures (see Section 4.2). So far, only two of the various experimental techniques for the determination of dielectric constants have been applied to hot, dense water. Akerlof and Oshry[4] determined the frequencies $\omega_R(P, T)$ and $\omega_R(0)$ for which the impedance $Z = [R^2 + (\omega L - 1/(\omega C))^2]^{1/2}$ of a resonant circuit in a pressure vessel has its minimum, when filled with water at pressure P and temperature T, and when filled with air under normal conditions, respectively. If the inductance L is kept constant, the relations $\omega_R^2(P, T) = 1/[LC(P, T)]$ and $\omega_R^2(0) = 1/[LC(0)]$ hold at the minimum of the impedance Z. Here, $C(P, T)$ and $C(0)$ are the capacitances of the resonant circuit filled with water and air, respectively. Since the dielectric constant is defined as $\varepsilon_0(P, T) = C(P, T)/C(0)$, it may be calculated from the relation $\varepsilon_0(P, T) = \omega_R^2(P, T)/\omega_R^2(0)$. In order to obtain good results for ε_0, the attenuation of the resonant circuit must be kept small, that is, the condition $1/R \ll \omega_R C$ must be satisfied. Since the specific conductance of water in the range covered is of the order 10^{-4} mho cm^{-1}, a frequency of about 100 MHz is necessary. Measurements of the static dielectric constant at such a high frequency are still feasible because the dispersion of the dielectric constant of water does not start below 1 GHz.

Applying the technique described, Akerlof and Oshry measured the dielectric constant of water along the saturation line between 100 and 370°C. Their results may be expressed by the relation

$$\varepsilon_0 = (5321/T) + 233.76 - 0.9297T + 1.417 \times 10^{-3}T^2 - 8.292 \times 10^{-7}T^3 \quad (19)$$

where T is the temperature in °K.

Heger[(485)] measured the capacitance of a condenser in a pressure vessel using a high-frequency bridge. Since this method allows $1/R$ and ωC to be of equal size, frequencies in the range 100 kHz to 1 MHz can be used. The disadvantage of this method is that only the sum of the capacitances of the condenser and of the leads can be measured. With experiments in high-pressure vessels, the capacitance of the leads C_L may be comparable to that of the condenser and also often varies with pressure and temperature in an irreproducible manner. In order to eliminate the capacitance of the leads, Heger used a geometrically variable condenser consisting of two coaxial half-cylinders made from noble metals. The inner half-cylinder could be rotated against the rigidly mounted, and electrically insulated, outer one from outside the pressure vessel by rotating the stem of a high-pressure valve. With this arrangement, Heger measured the capacitances

$$C(P, T, \varphi_1) = C_c(P, T, \varphi_1) + C_L(P, T)$$

and

$$C(P, T, \varphi_2) = C_c(P, T, \varphi_2) + C_L(P, T)$$

where $C_c(P, T, \varphi_1)$ and $C_c(P, T, \varphi_2)$ are the capacitances of the condenser filled with water at a pressure P and a temperature T for two different angles of rotation φ_1 and φ_2, chosen in such a way that $C(P, T, \varphi)$ had its maximum at φ_1 and its minimum at φ_2. Heger also measured the corresponding values under vacuum $C(0, \varphi_1) = C_c(0, \varphi_1) + C_L(0)$ and $C(0, \varphi_2) = C_c(0, \varphi_2) + C_L(0)$ and calculated ε_0 from the relation

$$\varepsilon_0(P, T) = [C(P, T, \varphi_1) - C(P, T, \varphi_2)]/[C(0, \varphi_1) - C(0, \varphi_2)]$$

into which the capacitance of the leads does not enter.

Table XI gives the results obtained by Heger up to 550°C and 5 kbar. The accuracy is claimed to be better than 1% for most of the values given. For ε_0 values below 10, the accuracy drops to about 3%.

Calculations of the static dielectric constant of water based on the Onsager–Kirkwood formula made by Franck(350) and Quist and Marshall(877) yield values which agree with the experimental results to within <4%, Franck's values generally being higher and Quist and Marshall's being lower than the experimental ones. Experimental values obtained by Gier and Young (published by Lawson and Hughes(660)) deviate considerably from the values in Table XI. As Quist and Marshall(877) pointed out, extrapolation of Gier and Young's results to a density of 1 g cm^{-3} yields a higher dielectric constant for 250°C than for 200°C, which seems to indicate some inconsistency in the results.

TABLE XI. Static Dielectric Constant of Water above 100°C and 0.1 kbar(485)

P, kbar	Static dielectric constant								
	at T = 100°C	200°C	250°C	300°C	350°C	400°C	450°C	500°C	550°C
0.1	55.7	34.8	27.1	20.2	—	—	—	—	—
0.25	56.2	35.4	28.1	21.5	14.8	2.7	1.9	1.7	1.5
0.5	57.0	36.4	29.4	23.0	17.7	12.5	6.5	3.7	2.7
0.75	57.8	37.3	30.5	24.2	19.3	14.9	10.2	7.0	4.9
1.0	58.4	38.0	31.4	25.3	20.5	16.5	12.5	9.3	7.0
1.5	59.6	39.5	32.8	27.0	22.4	18.7	15.0	12.0	9.9
2.0	60.8	40.8	34.1	28.4	23.8	20.0	16.6	13.0	11.6
2.5	61.9	41.9	35.1	29.5	24.9	21.1	17.7	15.1	12.9
3.0	62.9	42.8	36.0	30.5	25.9	22.1	18.7	16.1	13.9
3.5	63.8	43.7	36.8	31.3	26.7	22.9	19.5	17.0	14.8
4.0	64.7	44.5	37.5	32.0	27.4	23.5	20.3	17.7	15.5
4.5	65.5	45.2	38.2	32.7	28.0	24.1	20.9	18.3	16.1
5.0	66.3	45.8	38.7	33.2	28.5	24.6	21.5	18.8	16.6

4.1.2. *Dielectric Theory and Experimental Results* (see Chapter 7)

For the static dielectric constant ε_0 of a liquid consisting of polarizable molecules with permanent dipoles, Onsager[808] and Kirkwood[608] derived the equation

$$(\varepsilon_0 - \varepsilon_\infty)(2\varepsilon_0 + \varepsilon_\infty)/\varepsilon_0 = (4\pi N_A/V)[(\varepsilon_\infty + 2)/3]^2 g\mu_0{}^2/kT \qquad (20)$$

where ε_∞ is the dielectric constant at a frequency high enough that dipole orientation does not contribute, N_A is Avogadro's number, V is the molar volume, μ_0 is the dipole moment of a molecule in its local environment without the contribution due to the reaction field, and g is a correlation parameter that accounts for local ordering of neighboring molecules. The correlation parameter may be expressed by the equation

$$g = 1 + \sum_{i=1}^{\infty} N_i \langle \cos \gamma_i \rangle \qquad (21)$$

where N_i is the number of molecules in the ith shell around a central molecule and $\langle \cos \gamma_i \rangle$ is the average cosine of angles between the dipole moments of molecules in the ith shell and that of the central molecule. If there is no local ordering in the liquid, $\langle \cos \gamma_i \rangle$ is equal to zero and the correlation parameter g is equal to unity.

It is somewhat uncertain whether the square of the refractive index n^2 or the dielectric constant at infrared frequencies should be introduced for ε_∞ into eqn. (20). The difference between these two quantities for water is appreciable, the former being 1.78, the latter about 5 (see e.g., Ref. 888). Whalley[1163] pointed out "that contributions to the dielectric constant due to intermolecular motion are not to be included in calculating the enhancement of the dipole moment of a molecule by its own reaction field" and that the polarizability due to intramolecular vibrations contributes only little to the total polarizability. Thus, the value of n^2 gives a better approximation than the infrared dielectric constant. Introducing n^2 for ε_∞ into eqn. (20), we arrive at

$$(\varepsilon_0 - n^2)(2\varepsilon_0 + n^2)/\varepsilon_0 = (4\pi N_A/V)[(n^2 + 2)/3]^2 g\mu_0{}^2/kT \qquad (22)$$

To calculate the product $g\mu_0{}^2$ as a function of temperature and density from the experimental values of ε_0, the refractive index must be known as a function of pressure and temperature. Experimental determinations of the refractive index of water at high pressures have been made below 55°C in the range to 1.1 kbar by Waxler and Weir[1155] and under shock conditions (33–150 kbar, 185–875°C, 1.4–1.7 g cm^{-3}) by Zeldovich *et al.*[1196] For most of the temperatures and pressures covered by Table XI, no experimental values for n are available. Since the temperature dependence of n at constant volume was found to be very small [$(\partial n/\partial T)_V = -1.9 \times 10^{-5}$ °K^{-1} (886)], it seems reasonable to assume that n is independent of temperature and that its density dependence is described by the Lorentz–Lorenz formula

$$[(n^2 - 1)/(n^2 + 2)]V = \text{const} \tag{23}$$

This implies that the polarizability of the water molecule is independent of pressure. Values of n for shock conditions calculated from the refractive index of water at ordinary conditions using eqn. (23) agree with the experimental values[886] to better than 5%, which may be regarded as proof for the approximate validity of eqn. (23).

The product $g\mu_0^2$, calculated from eqn. (22) with the aid of eqn. (23) and the ε_0 values from Table XI, is plotted in the form of $g\mu_0^2/\mu_g^2$ as a function of density for various constant temperatures in Fig. 5. μ_g is the dipole moment of the water molecule in the dilute vapor state. The density dependence of $g\mu_0^2/\mu_g^2$ is quite complex. At supercritical temperatures, it has a value of unity for zero density, as expected, since both g and μ_0^2/μ_g^2

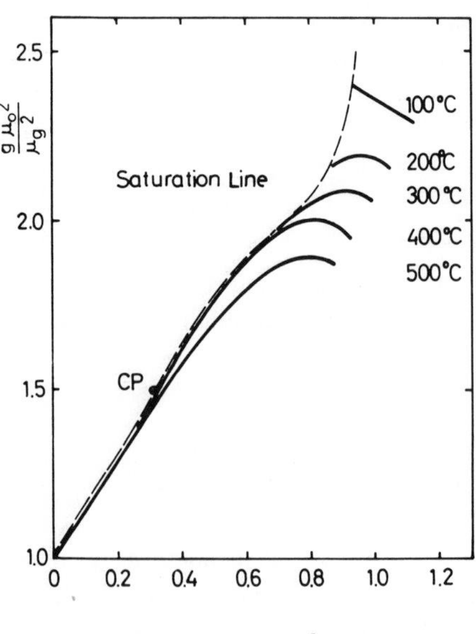

Fig. 5. Correlation parameter g times ratio of dipole moments squared as a function of density for various temperatures.[485]

should equal unity. It increases rapidly with increasing density, most probably due to a small increase in μ_0^2/μ_g^2 and a considerable increase in g. At higher densities, it passes through a maximum where, even at 500°C, it has a value close to two. The maximum is shifted slightly toward lower densities with increasing temperature. No distinction can be made from the experimental results whether the small decrease beyond the maximum is caused by a decrease of g or of μ_0^2/μ_g^2. At subcritical temperatures, the density range covered is small because of the limits set by the two-phase region. At 100°C, no maximum is observed and the product decreases monotonically from the saturation line to higher densities.

At constant density, $g\mu_0^2/\mu_g^2$ decreases with increasing temperature. Heger[485] calculated the correlation parameter g at a constant density of 1 g cm^{-3} as a function of temperature using the expression derived by Pople[855] for his distorted-hydrogen-bond model of water:

$$g = 1 + \sum_{i=1}^{3} 3^{1-i} N_i (\cos^{2i}\alpha)\{\coth(f/kT) - (kT/f)\}^{2i} \tag{24}$$

where 2α is the H—O—H angle of the water molecule and f is the hydrogen bond bending force constant. Heger used the values $N_1 = 4$, $N_2 = 11$, $N_3 = 22$, and $f = 10\,kT_0$, with $T_0 = 273$°K, which Pople derived from the experimental X-ray radial distribution function. The result of Heger's calculation was that the temperature dependence of g alone, as calculated from eqn. (24), accounts for the temperature dependence of $g\mu_0^2/\mu_g^2$ above 100°C.

Although the contributions of g and μ_0^2/μ_g^2 cannot clearly be separated, it may be concluded that the experimental ε_0 values indicate that at moderate and high densities, considerable local ordering persists in water up to supercritical temperatures.

4.2. Specific Conductance and Ionic Conductances

4.2.1. *Specific Conductance*

The specific conductance of water $\varkappa$ and the ionic conductances of hydrated protons and hydroxyl ions in water, λ_{H^+} and λ_{OH^-}, to be discussed in this and the following subsection, are related by the equation

$$10^3\varkappa = \lambda_{H^+}c_{H^+} + \lambda_{OH^-}c_{OH^-} \tag{25}$$

where c_{H^+} and c_{OH^-} are the concentration of protons and hydroxyl ions, respectively. Since in pure water, $c_{H^+} = c_{OH^-}$, it follows that $10^3\varkappa = (\lambda_{H^+}$

$+ \lambda_{OH^-})c_{H^+}$. Assuming that corrections for finite ion concentration may be neglected, because of the small ion concentration in pure water, the sum $\lambda_{H^+} + \lambda_{OH^-}$ may be replaced by $\lambda^{\circ}_{H^+} + \lambda^{\circ}_{OH^-} = \Lambda^{\circ}_{H_2O}$, the equivalent conductance at infinite dilution, and c_{H^+} by $K_w^{1/2}$, the square root of the ionic product of water. This leads to

$$\log \varkappa = \log(\Lambda^{\circ}_{H_2O}/10^3) + \tfrac{1}{2} \log K_w \tag{26}$$

Thus, information on $\varkappa$ may be obtained either from direct experimental determinations, or from measurements of both the sum of the ionic conductances and the ionic product. At low temperatures and pressures, where the specific conductance of water is small, direct measurements are complicated by the effects of small amounts of dissociating impurities. In this range, most of the information on $\varkappa$ comes from measurements of $\lambda^{\circ}_{H^+}$, $\lambda^{\circ}_{OH^-}$, and K_w. Experimental determinations of these quantities at very high temperatures and pressures raise serious problems. Fortunately, $\varkappa$ of water is large enough in this range not to be greatly affected by small amounts of impurities.

Direct experimental results for the specific conductance of water have been obtained up to 1000°C and about 130 kbar. David and Hamann[(236)] found that in strong shock waves, $\varkappa$ rises to the order of 1 mho cm^{-1} at the highest temperatures and pressures reached in the experiments. This result has been confirmed by other shock wave experiments[(126,237,450,1193)] and by the static measurements of Holzapfel and Franck,[(525)] who studied the electrical conductance of water to 1000°C and 100 kbar in an opposed anvil apparatus. Figure 6 shows two perpendicular cross sections of the conductivity cell. This consists of a disk with a central hole and an annular groove, both filled with water, and a flat cover. The two wires of a thermocouple, insulated by an alumina tube and cast in epoxy resin, enter the cell through radial bores. To obtain accurate temperature measurement, the hot junction of the thermocouple is placed inside the central hole of the cell. Here, the alumina tube has a narrow slit, which provides electrical contact between the water and the hot junction so that it can also be used as one of the electrodes. The cell itself, which is insulated from the tungsten carbide pistons by mica sheets, serves as the second electrode. The water is heated by two graphite resistance heaters above and below the central hole. Since the cell is cooled from the outside, it is possible to keep the water in the annular groove frozen while the water in the central hole is heated to 1000°C. Thus, a good seal is obtained when the cell is squeezed between the tungsten carbide pistons to about 100 kbar by means of a hydraulic ram.

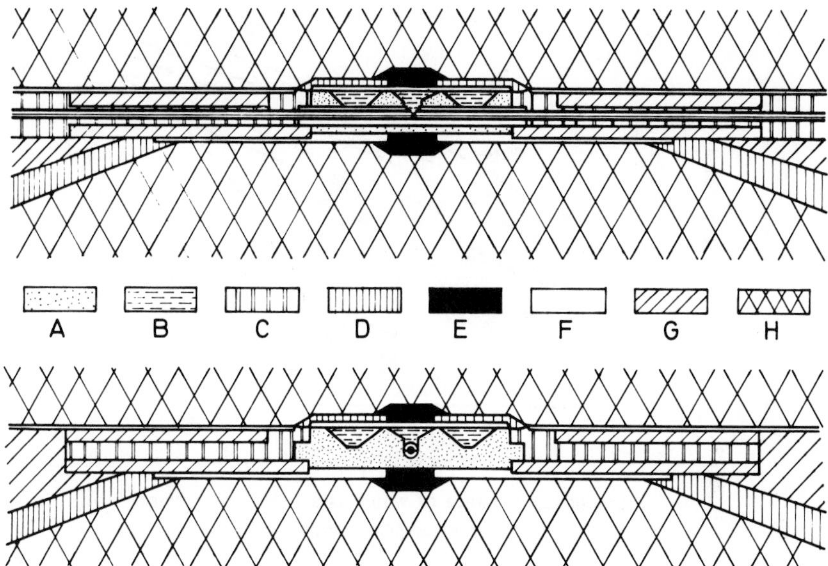

Fig. 6. Conductance cell for measurements to 1000°C and 100 kbar.[525] A: Cell; B: water; C: epoxy resin; D: pyrophyllite; E: graphite heaters; F: mica sheets; G: steel rings; H: tungsten carbide pistons.

In the pressure range below 10 kbar measurements of the specific conductance have been made by Gier and Young (published in Ref. 660), Mangold,* and v. Osten.[817] All the information on $\varkappa$, including that from measurements of the ionic product and ionic conductances, has been collected by Holzapfel.[523] The results are shown in Fig. 7, where the specific conductance of water is plotted on a logarithmic scale against density for various constant temperatures. Several isobars are indicated by dashed lines. It may be seen from Fig. 7 that $\varkappa$ increases with increasing density and increasing temperature. At the highest temperatures and pressures, it becomes comparable with the specific conductance of molten salts. The reasons for the high specific conductance will be discussed in the following subsections.

Holzapfel found that the specific conductance may be described by the equation

$$\log \varkappa(\varrho, T) = [7.2 + 2.5(\varrho/\varrho_0)] \log(\varrho/\varrho_0) + \log \varkappa(\varrho_0, T) \qquad (27)$$

where $\varrho_0 = 1 \text{ g cm}^{-3}$ and $\log \varkappa(\varrho_0, T)$ may be calculated from eqn. (26)

* Private communication.

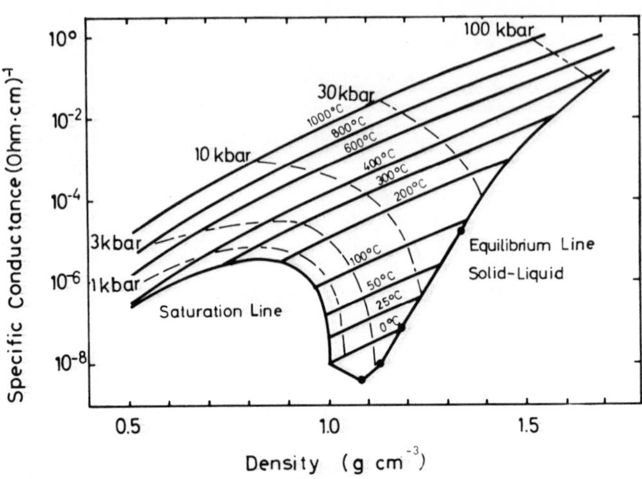

Fig. 7. Specific conductance of water as a function of density for various temperatures.[523]

with the aid of eqns. (33) and (35). Equation (27) should be regarded as an interpolation formula for $\varkappa$ over wide ranges rather than a close fit of experimental data at about ordinary conditions.

4.2.2. *Ionic Conductances*

A determination of the ionic conductances of the hydrated proton and hydroxyl ion in water from measurements of the specific conductance, the ionic product, and the transference numbers is not feasible. As already mentioned, measurements of the very small intrinsic conductance of pure water at low temperatures and pressures suffer from impurity effects. Consequently, the ionic conductances must be evaluated from measurements of the equivalent conductances and transference numbers in dilute aqueous solutions of electrolytes containing protons and hydroxyl ions, respectively, and subsequent extrapolation to infinite dilution. Thus, $\lambda^{\circ}_{\mathrm{H}^+}$ and $\lambda^{\circ}_{\mathrm{OH}^-}$ may be calculated from the relations

$$\begin{aligned} \lambda^{\circ}_{\mathrm{H}^+} &= \lim_{c\to 0}(t_{\mathrm{H}^+}\Lambda_{\mathrm{HCl}}) \\ \lambda^{\circ}_{\mathrm{OH}^-} &= \lim_{c\to 0}(t_{\mathrm{OH}^-}\Lambda_{\mathrm{KOH}}) \end{aligned} \tag{28}$$

The transference numbers t_{H^+} and t_{OH^-} in eqns. (28) are strongly dependent on pressure and temperature.[511] Since their temperature and pressure dependence is not known at high pressures and temperatures, eqns. (28) are not suited for practical use in that range. However, because Walden's rule ($\lambda \propto 1/\eta$) is approximately obeyed by simple ions,[350,493,1071] the transference numbers of ions in solutions containing no protons or hydroxyl ions are nearly independent of pressure and temperature. A possibility is thus provided for the elimination of t_{H^+} and t_{OH^-} and eqns. (28) may be replaced by

$$\begin{aligned} \lambda^{\circ}_{H^+} &= \lim_{c\to 0} \Lambda_{HCl} - \lim_{c\to 0} t_{Cl^-}\Lambda_{KCl} \\ \lambda^{\circ}_{OH^-} &= \lim_{c\to 0} \Lambda_{KOH} - \lim_{c\to 0} t_{K^+}\Lambda_{KCl} \end{aligned} \tag{29}$$

Since t_{Cl^-} and t_{K^+} in KCl solutions are almost equal to 0.5, eqns. (29) reduce to

$$\begin{aligned} \lambda^{\circ}_{H^+} &= \Lambda^{\circ}_{HCl} - \tfrac{1}{2}\Lambda^{\circ}_{KCl} \\ \lambda^{\circ}_{OH^-} &= \Lambda^{\circ}_{KOH} - \tfrac{1}{2}\Lambda^{\circ}_{KCl} \end{aligned} \tag{30}$$

and for the equivalent conductance of water, the expression

$$\Lambda^{\circ}_{H_2O} = \lambda^{\circ}_{H^+} + \lambda^{\circ}_{OH^-} = \Lambda^{\circ}_{HCl} + \Lambda^{\circ}_{KOH} - \Lambda^{\circ}_{KCl} \tag{31}$$

is obtained.

The equivalent conductances of dilute aqueous solutions of KCl, HCl, and KOH have been studied as a function of concentration up to 1000°C and 12 kbar by several investigators.[349–352,730,800,817,878,899 905,1075] Using the limiting equivalent conductances from these references, $\lambda^{\circ}_{H^+}$ and $\lambda^{\circ}_{OH^-}$ have been calculated for the density range 0.7–1.15 g cm^{-3} up to 400°C and 300°C, respectively. Although the results show considerable scatter, the following features are observed:

1. $\lambda^{\circ}_{H^+}$ and $\lambda^{\circ}_{OH^-}$ show roughly the same behavior in the density and temperature ranges covered.

2. They are both nearly independent of density.

3. At low temperatures, they increase with increasing temperature. At about 300°C, a maximum is passed and at higher temperatures, $\lambda^{\circ}_{H^+}$ decreases slightly with further rise in temperature. No experimental information on $\lambda^{\circ}_{OH^-}$ is yet available for this range.

4. The absolute values are high. In the range close to the maximum (300°C, 0.9 g cm^{-3}), $\lambda^{\circ}_{H^+} = 950$ cm^2 ohm^{-1} equiv.$^{-1}$ compared with 350 cm^2 ohm^{-1} equiv.$^{-1}$ in water at 25°C and 1 bar; the corresponding values for the K^+ ion are 470 and 74 in the same units. The ratio $\lambda^{\circ}_{H^+}/\lambda^{\circ}_{OH^-}$, which is

equal to 1.76 under ordinary conditions, decreases with increasing temperature, but remains greater than unity.

For a further discussion of the temperature and density effects on the ionic conductances of protons and hydroxyl ions, it is useful to regard the total conductances as the sum of two separate contributions: (1) the normal contributions, $\lambda^{\circ N}_{H^+}$ and $\lambda^{\circ N}_{OH^-}$, which arise from the movement of the hydrated ions in the direction of the applied field, approximately considered as the movement of charged spheres in a viscous medium; the normal contributions thus follow Walden's rule, that is, they increase with temperature, and decrease with increasing density in nearly the same way as the reciprocal viscosity; (2) the excess contributions, $\lambda^{\circ E}_{H^+}$ and $\lambda^{\circ E}_{OH^-}$, which are due to the ability of the proton and defect proton to move very fast along the line joining the oxygen atoms of two adjacent hydrogen-bonded water molecules. With the reasonable assumption that the normal contributions are of the same order as the conductances of the K^+ and Cl^- ions ($\lambda^{\circ N}_{H^+} = \lambda^{\circ}_{K^+}$, $\lambda^{\circ N}_{OH^-} = \lambda^{\circ}_{Cl^-}$), the following expressions are obtained for the excess conductances:

$$\lambda^{\circ E}_{H^+} = \Lambda^{\circ}_{HCl} - \Lambda^{\circ}_{KCl}, \qquad \lambda^{\circ E}_{OH^-} = \Lambda^{\circ}_{KOH} - \Lambda^{\circ}_{KCl} \tag{32}$$

The values for the excess conductances obtained from eqns. (32) are shown in Fig. 8, where they are plotted against temperature for various constant

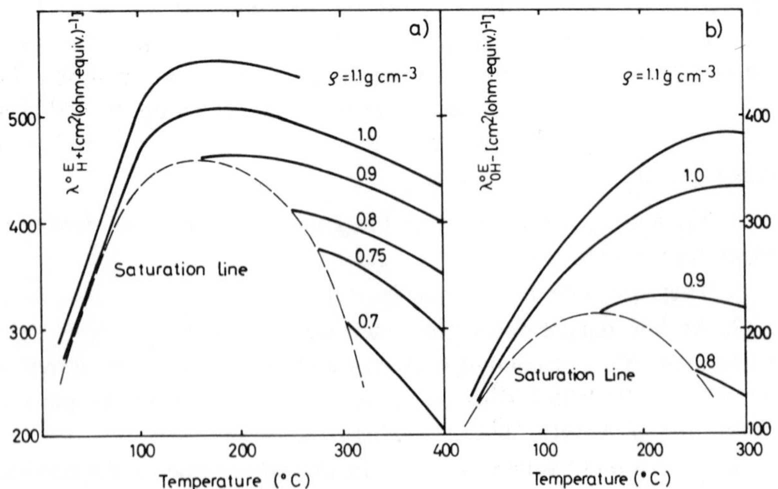

Fig. 8. Excess conductances of (a) proton and (b) hydroxyl ion as functions of temperature for various densities (Refs. 351, 352, 730, 800, 817, 878, 899, 905, 1075; also Mangold, private communication).

densities. It may be seen from Fig. 8 that at low temperatures, the excess conductances increase with rising temperature. At about 150–300°C, a broad maximum is passed and at higher temperatures, a decrease is observed. This result is in agreement with Eigen and De Mayer's[294–297] model for proton and hydroxyl ion conductances in water. In this model, it is assumed that the high conductance of the proton is due to relatively rigid $H_9O_4^+$ complexes formed by a central H_3O^+ ion with three water molecules in its first hydration shell. The rigidity of the complex favors fast proton tunneling along the hydrogen bonds, so that the excess proton can fluctuate quickly over the whole complex. In the case of the hydroxyl ion, the defect proton can fluctuate over an $H_7O_4^-$ complex, formed by the hydroxyl ion and three water molecules. Charge transport in the direction of the applied field occurs when a water molecule at the periphery of the $H_9O_4^+$ or $H_7O_4^-$ complex is oriented in such a way that a proton can transfer to it. After the proton jump is completed, a new $H_9O_4^+$ ($H_7O_4^-$) complex is formed around the new H_3O^+ (OH^-) ion. This process may be regarded as "structural" diffusion. From the absolute values of the ionic conductances and from the ratio of the conductances of H^+ and D^+ (which is about 1.4), it was concluded that the orientation of water molecules at the periphery of the complexes is the rate-determining step in the conductance mechanism. This step is accelerated as the temperature rises. Assuming that the average lifetime of the H_3O^+ ion, $\tau_{H_3O^+}$, about 2×10^{-12} sec at 25°C,[739] is not very much influenced by temperature changes at low temperatures, and that the temperature dependence at constant density of the rotational orientation time τ_r ($4\text{–}8 \times 10^{-12}$ sec at 25°C[494]), is sufficiently well described by relation (17), one would expect that τ_r becomes shorter than $\tau_{H_3O^+}$ above 150°C. Consequently, the orientation of water molecules at the periphery of the complexes is no longer rate-determining. The movement of the proton in the complex now determines the rate of proton transfer. This is slowed down with further increasing temperature, because the complex itself becomes less rigid and the rate of proton transfer is strongly dependent on the orientation of the molecules in the complex. At very high temperature, the excess conductance goes to zero, again in agreement with the model.

Contrary to the normal contribution, the excess conductance increases with increasing density. Near the temperature of maximum excess conductance, a change in the density dependence of $\lambda^{\circ E}_{H^+}$ and $\lambda^{\circ E}_{OH^-}$ seems to appear. At low temperatures, the change of excess conductance with density is small, possibly because the increase in the number of water molecules at the periphery of the $H_9O_4^+$ or $H_7O_4^-$ complexes more than compensates for the relatively small effect of density increase on τ_r (see Section 3.3). At higher temperatures, especially at low densities, the density dependence increases, as one would expect, with the movement of the proton in the complex being the rate-determining step, since it is strongly dependent on changes in the O—O distance in a hydrogen bond.

Thus, all the experimental observations are in qualitative agreement with the model of Eigen and De Maeyer. Therefore, in order to account for the high values of the excess conductance, it may be concluded that at about 300–400°C and a density of 1 g cm^{-3}, a considerable degree of hydrogen bonding persists. Furthermore, the vanishing excess conductance at about 700°C and densities below 0.9 g cm^{-3} [351] indicates that very few, if any, hydrogen bonds providing fast proton jumps remain at such a high temperature.

The high values, as well as the temperature and density dependences, of the total ionic conductances $\lambda^\circ_{H^+}$ and $\lambda^\circ_{OH^-}$ can now be explained by considering the effects of temperature and density changes on both the normal and excess conductances. In particular, their independence of density is made plausible by the opposing effects of density changes on the normal and excess contributions.

The equivalent conductance of water $\Lambda^\circ_{H_2O}$ is also almost independent of density and therefore depends on temperature only. Between 0 and 300°C, the relation[523]

$$\log \Lambda^\circ_{H_2O}(T) = -[100/(T - 175)] + \log 3400 \tag{33}$$

where T is in °K and $\Lambda^\circ_{H_2O}$ in cm^2 ohm^{-1} equiv.$^{-1}$; provides a good fit to the experimental data. Although above 300°C, $\Lambda^\circ_{H_2O}$ may decrease slightly over a short temperature range since the excess contributions decrease, eqn. (33) may be used as a first approximation (upper limit) for an extrapolation of $\Lambda^\circ_{H_2O}$ to higher temperatures because of the positive temperature dependence of the normal contributions which become predominant at high temperatures.

5. IONIC PRODUCT

In dense, aqueous phases, some of the water molecules dissociate to form hydrated protons and hydroxyl ions. Under ordinary conditions, the equilibrium $H_2O \rightleftarrows (H^+)_{aq} + (OH^-)_{aq}$ is almost completely on the side of the water molecule. At 25°C and 1 bar the value of the ionic product $K_w = c_{H^+} \times c_{OH^-}$ is equal to 10^{-14} mol^2 liter^{-2}. Noyes *et al.*[799] found that the ionic product along the saturation line of water between 25 and 240°C increases by about three orders of magnitude. It exhibits a maximum at 240°C and decreases at still higher temperatures because the density decrease of liquid water on the saturation line then starts to become appreciable. Hamann[(449)] measured the pressure dependence of K_w at 25°C up to 2 kbar and found an increase by a factor of four.

For high temperatures and high pressures, no directly measured values of the ionic product are available. However, K_w may be calculated from measured values of the specific conductance $\varkappa$ and the limiting equivalent conductance $\Lambda^\circ_{H_2O}$. Combining eqns. (26) and (27) and assuming that $\Lambda^\circ_{H_2O}$ is independent of density for all temperatures (which has been proved experimentally only below 300°C—see Section 4.2.2), Holzapfel[(523)] obtained the following relation for the ionic product K_w as a function of density and temperature:

$$\log K_w(\varrho, T) = 2[7.2 + 2.5(\varrho/\varrho_0)] \log(\varrho/\varrho_0) + \log K_w(\varrho_0, T) \qquad (34)$$

where $\varrho_0 = 1$ g cm^{-3}. The temperature dependence of the ionic product at $\varrho_0 = 1$ g cm^{-3} may be approximated by the relation

$$\log K_w(\varrho_0, T) = -(3108/T) - 3.55 \qquad (35)$$

where T is in °K and K_w in mol^2 liter^{-2}. As Holzapfel demonstrated, eqn. (34), together with eqn. (35), fits the experimental results of K_w reasonably well, although it should be regarded as an interpolation formula for the ionic product over wide ranges of temperature and density rather than a close fit to experimental results under ordinary conditions. Quist[(876a)] obtained additional values for K_w to 800°C and 4 kbar, and derived an equation for $\log K_w(\varrho, T)$. Values thus calculated tend to be slightly more negative than those calculated from eqn. (34) at the highest temperatures and densities.

The values of $\log K_w$ calculated from eqn. (34) are plotted in Fig. 9 as a function of density for various constant temperatures. Some isobars are shown as dashed lines. It may be seen from Fig. 9 that K_w increases both with temperature and density. At about 1000°C and 100 kbar ($\varrho = 1.52$ g cm^{-3}), its value is of the order of 10^{-2}, that is, twelve orders of magnitude higher than at 25°C and 1 bar. Even if one takes into account that under such extreme conditions, K_w may be inaccurate by one or two orders of magnitude, it is clearly demonstrated that water is ionized to a considerable extent at such temperatures and pressures. The results indicate that one water molecule in about 840 is dissociated into ions, whereas at 25°C and 1 bar, the ratio is only one in 5.5×10^8. It may be concluded from the trends in Fig. 9 that at still higher temperatures and pressures, water will probably be almost completely dissociated and that its properties may then resemble those of molten hydroxides.

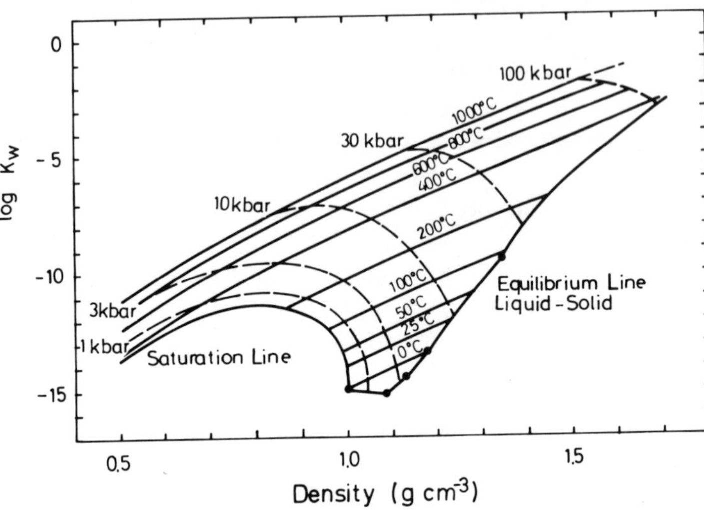

Fig. 9. Logarithm of ionic product of water as a function of density for various temperatures.[523]

Holzapfel[523] also calculated the changes in the standard thermodynamic functions associated with the dissociation reaction of water molecules, using eqns. (34) and (35) and *PVT* data from the literature. With the usual convention that the activity of water is set equal to unity and K_w is based on the molarity scale, he obtained the following results for 500°C and a density of 1 g cm^{-3} (the literature values for 25°C and 1 bar are given in brackets for comparison): $\Delta V^\circ = -(55 \pm 5)$ cm^3 mol^{-1} [−21.3]; $\Delta G^\circ = (111 \pm 4)$ kJ mol^{-1} [79.96]; $\Delta H^\circ = (25 \pm 12)$ kJ mol^{-1} [57.01]; $\Delta S^\circ = -(113 \pm 20)$ J mol^{-1} deg^{-1} [−77.00].

The most drastic change is in ΔV°, which becomes more negative by a factor of about 2.5. However, this result may be partly due to the fact that eqn. (34) does not allow for a change of $[\partial(\log K_w)/\partial\varrho]_T$ with temperature.

6. SPECTROSCOPIC STUDIES

Contrary to the properties discussed so far, from which information may be derived about the environment of a water molecule in hot, dense water averaged over a period of time comparable to or longer than the average time between diffusional motions (10^{-11}–10^{-12} sec), spectroscopic studies give some insight into the local structure around a water molecule averaged over a much shorter time (10^{-13}–10^{-14} sec). Unfortunately, spectroscopic measurements with hot, dense water raise serious problems, especially in the provision of suitable windows, strong enough to be used at high pressures, which do not dissolve in hot, dense water and are transparent in the frequency range of interest. Therefore, very few experimental studies of the spectroscopic properties of water at high temperatures and pressures have yet been undertaken.

6.1. Infrared Spectral Studies

6.1.1. *Experimental Technique*

Franck and Roth[354,919] measured the absorption of HDO in H_2O in the fundamental mode of the O—D stretching vibration at frequencies from 2200 to 3000 cm^{-1} up to 500°C and 4 kbar. In order to obtain a layer thickness of about 10–30 μm, not affected by the elastic deformation of the pressure vessel when the pressure is raised, an optical cell with a single window, shown in Fig. 10, had to be constructed. The body of the cell consists of two coaxial cylinders made from corrosion-resistant Ni–Cr alloys and is heated by an external resistance heater. A colorless sapphire single crystal with a 16 mm diameter and a thickness of 10 mm is used as a high-pressure window. A Pt–Ir mirror is pressed against the window by a spring in the pressure vessel. The layer thickness, being twice the distance between the window and the mirror, is determined by a spacer made from gold foil and is thus independent of the elastic deformation of the cell. The pressure vessel is connected to the pressure-generating system by high-pressure tubing and is completely filled with the fluid to be studied. Franck

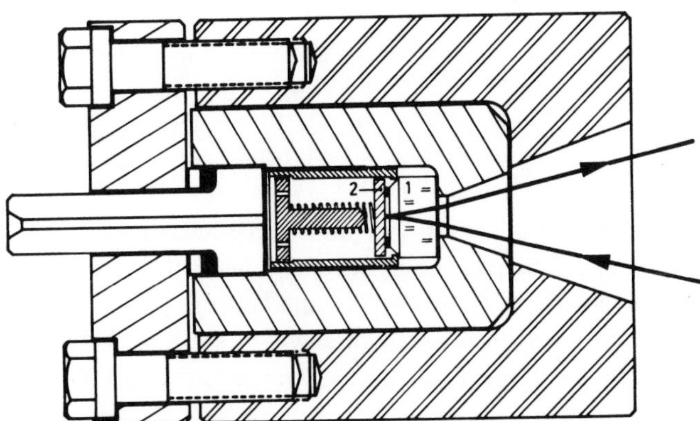

Fig. 10. High-pressure optical cell for measurements of infrared spectra.[919] 1: Sapphire window; 2: Pt–Ir mirror.

and Roth used a Perkin-Elmer model 521 grating spectrophotometer with a mirror specular reflectance accessory in the sample area to deflect the infrared beam through about 90°. The cell is mounted outside the sample area with its axis perpendicular to the light path in the spectrophotometer. The beam is focused on the Pt–Ir mirror, from which it is reflected back to the reflectance accessory. After a further deflection through 90°, it is refocused on the entrance slit of the detector unit of the spectrophotometer and compared with the reference beam in the usual way.

6.1.2. *Results*

With this technique, Franck and Roth measured the absorption of the O—D stretching vibration of the HDO molecule in a mixture of 5 mol % D_2O and 95 mol % H_2O (equilibrium concentrations in mole per cent: 9.5 HDO, 0.25 D_2O, 90.25 H_2O) as a function of temperature and density. As an example of the results obtained, a plot of the molar extinction coefficient ε as a function of frequency at 400°C and various densities is shown in Fig. 11. The measured spectra show the following features:

1. At very low densities, the rotational structure of the band is clearly discernible. At about 0.01 g cm^{-3}, the rotational lines are broadened to such an extent that only the contours of the *R*, *Q*, and *P* branches appear in the spectrum. With further increasing density, the intensity decreases in the *Q* branch and increases in the *P* branch, where a new, broad, and intense band appears. At densities above 0.05 g cm^{-3}, the *Q* branch becomes first a shoulder of this new band, and subsequently disappears. At still higher densities, the *R* branch also becomes a shoulder of the intense band and is totally incorporated above 0.3 g cm^{-3}.

2. At densities above 0.3 g cm^{-3}, only one broad, intense band exists for subcritical and supercritical temperatures. This band has no detectable shoulder, and at low temperatures, it is almost Gaussian in shape. However, with rising temperature at constant density, the band becomes increasingly asymmetric.

3. The integrated absorption of the band (defined as $B = \int_{\nu_1}^{\nu_2} \varepsilon(\nu)\, d\nu$, where ν_1 and ν_2 are frequencies lower and higher than ν_{max} respectively, and are chosen in such a way that $\varepsilon(\nu)$ is equal to zero in ν_1 and ν_2), increases with density, mainly in the range of low frequencies. It decreases with increasing temperature, markedly at low temperatures, and more gradually at high temperatures.

4. The frequency at maximal extinction ν_{max} decreases with increasing density due to the increase of intensity in the range of low frequency.

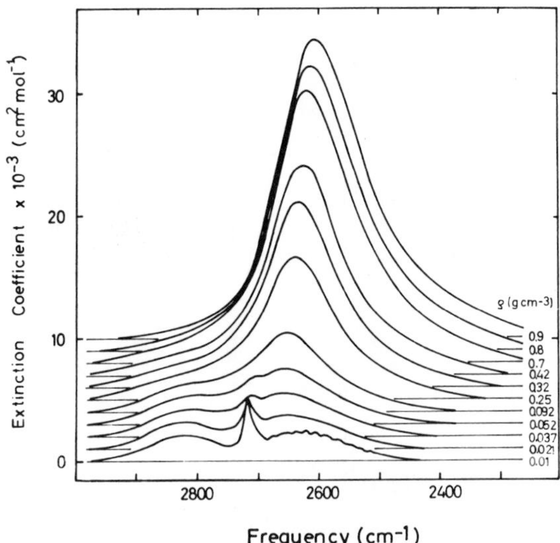

Fig. 11. Molar extinction coefficient of the O—D stretching vibration of the HDO molecule in water at 400°C and at various densities as a function of frequency.[354,919]

Rising temperature shifts ν_{max} toward higher values, again markedly at low temperatures and at a lesser rate at high temperatures.

5. If the integrated absorption B of the band is plotted against ν_{max}, the B values for all temperatures and densities fall on a single line. Such a correlation between B and ν_{max}, implying that the changes in these two quantities have a common source, has been observed for many hydrogen-bonded systems.[540]

6. The half-width of the band exhibits a maximum at about 100°C and also increases slightly with density.

7. No isosbestic point is observed.

An interpretation of these observations with respect to water structure will be given in Section 6.3.

6.2. Raman Spectral Studies

6.2.1. *Experimental Technique*

The Raman spectral technique has been applied by Lindner[686] to measurements on hot, dense water up to 400°C and 5 kbar. The high-pressure Raman cell used is shown in Fig. 12. It consists of two coaxial

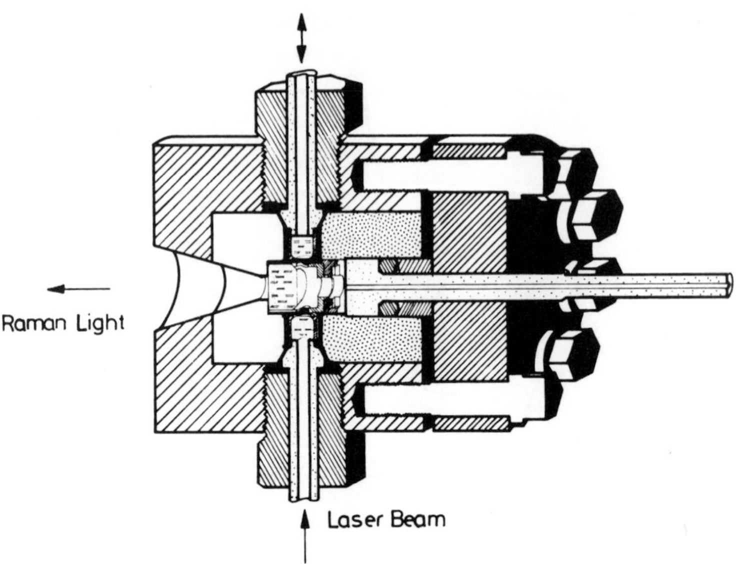

Fig. 12. High-pressure Raman cell.[686]

cylinders made from corrosion-resistant Ni–Cr alloys and has three sapphire windows. Two of the windows have a diameter and a thickness of 8 mm, and the third one has a diameter of 16 mm and is 10 mm thick. The vessel is externally heated and the temperature is measured by a sheathed thermocouple inside the cell. A Spectra Physics model 140 argon ion laser serves as a light source. Its spectral line at 4880 Å, having a power of 1 W, is used as the primary light. The laser beam is focused into the Raman cell through one of the smaller sapphire windows. The unscattered light leaves the cell through the opposite window and is reflected back into the cell by a mirror in order to increase the scattering intensity. The Raman-scattered light leaving the cell through the larger window is focused on the entrance slit of a Steinheil Raman spectrometer. In order to make measurements of the depolarization ratio, a polarization filter is placed where the light enters the spectrometer. A conventional arrangement consisting of two photomultipliers and a registration unit is used for the recordings of the Raman spectra relative to the intensity of the exciting light.

6.2.2. *Results*

Using this technique, Lindner[686] measured the Raman spectrum of the O—D stretching vibration of the HDO molecule in a 6.2 *M* solution of D_2O in H_2O to 400°C and 5 kbar. He also obtained results, not discussed here, for the O—D stretching vibration of HDO in the presence of various amounts of KI in this solution and for the O—H stretching vibration of the H_2O molecule in pure water. As an example of the results obtained, Fig. 13 shows a plot of intensity I'' against frequency at a density of 1 g cm^{-3} and for various temperatures. I'' is the fraction of the total intensity in the O—D Raman band which has a polarization parallel to that of the primary light. The measured O—D valence spectra show the following features:

1. Contrary to the infrared band, the Raman band has a shoulder on its high-frequency side which increases with temperature (Fig. 13).

2. The frequency of the band maximum ν_{max} decreases with increasing density for all temperatures, but it increases with temperature (Fig. 13). Below 100°C, this increase is about equal to that of the infrared band, but above 100°C, the shift of the Raman band is greater. This difference between Raman and infrared bands at 400°C and 0.9 g cm^{-3} amounts to 30 cm^{-1}.

3. The half-width of the band exhibits a maximum at about 100°C (Fig. 13). Above this temperature, the half-width of the Raman band is considerably smaller than that of the infrared band. The half-width increases with density.

4. The relative integrated intensity increases with density and decreases with temperature (Fig. 13) in the same way as the integrated absorption of the infrared band. However, above 200°C, the rate of decrease is much more pronounced for the Raman band.

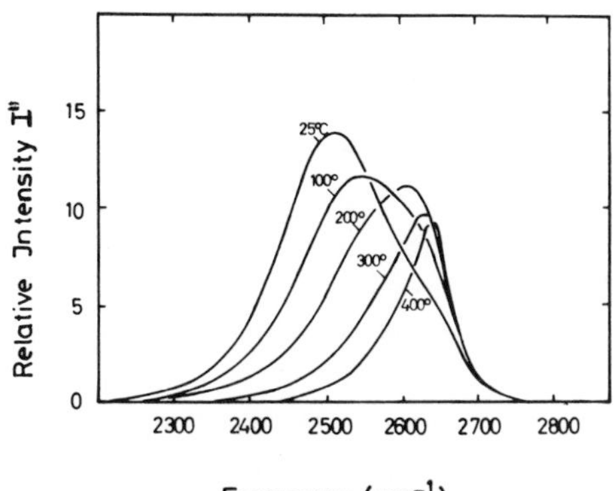

Fig. 13. Relative intensity of the Raman O—D stretching vibration of the HDO molecule in water as a function of frequency at a density of 1 g cm^{-3} and at various temperatures.[686]

5. The spectral depolarization ratio (defined as $\varrho_s(\nu) = I^{\perp}(\nu)/I^{\parallel}(\nu)$, where $I^{\perp}(\nu)$ and $I^{\parallel}(\nu)$ are the intensities of Raman scattered light at the frequency ν with a polarization perpendicular and parallel to that of the exciting light, respectively), has a minimum at $\nu = 2640\ cm^{-1}$, indicating that a strongly polarized subband is situated at this frequency. The minimum of ϱ_s becomes more pronounced at higher temperatures, but its position on the frequency scale is not shifted by temperature changes.

6.3. Interpretation of the Spectroscopic Results with Respect to Water Structure

The most obvious features of the infrared and Raman spectra of the O—D stretching vibration of HDO in hot, dense water, namely the strong dependences of ν_{max} and of the integrated absorption and intensity on temperature and density, may readily be explained by the existence of hydrogen bonds. If the O···D—O hydrogen bond length is reduced by a density increase at constant temperature, the vibrational force constant of the O—D vibration decreases, and consequently, the frequency of the vibration also decreases. For very low densities, the frequency ν_{max} approaches the value $\nu_{max} = 2727\ cm^{-1}$ of the isolated molecule in the dilute vapor state. The shift to higher frequencies with increasing temperature at constant density may be attributed to the increase of the vibrational force constant resulting from the increased distortion of the hydrogen bonds with increasing thermal motion of the molecules. Variations of the intensity of the infrared band are related to the changes in the polarity of the O—D bond resulting from changes in the average distance and the orientation of the neighboring molecules.

However, the treatment of the O—D valence band as a single broad band, reflecting a continuous distribution of O···D—O distances and angles, offers no explanation for the following observations: (1) the discrepancy in the shift of ν_{max} with temperature in the infrared and Raman spectra above 100°C; (2) the discrepancy in the decrease of the integrated absorption in the infrared and of the integrated intensity in the Raman spectrum when the temperature is raised above 200°C; (3) the maximum of the half-width at 100°C in both spectra; (4) the shoulder at high frequencies in the Raman spectrum; (5) the minimum in the depolarization ratio at 2640 cm^{-1} in the Raman band, and the temperature invariance of its frequency.

All these observations can be explained if one assumes that the total band is composed of two components the relative contributions of which to the total infrared band are different from those to the Raman band

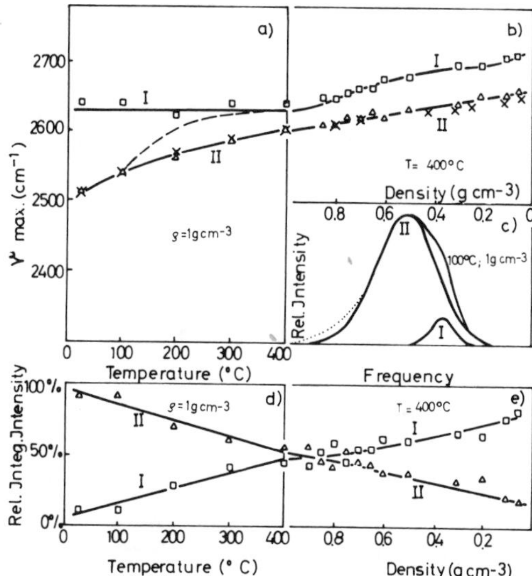

Fig. 14. Splitting of the Raman O—D stretching vibration band of HDO into two Gaussian component bands.[686] (I) High-frequency component; (II) low-frequency component. (a) Frequency of maximal intensity as a function of temperature at a density of 1 g cm^{-3}. Dashed line represents ν_{max} of the total band. Crosses represent ν_{max} of the total infrared band. (b) Frequency of maximal intensity as a function of density at 400°C. (c) Total band and component bands at 100°C and 1 g cm^{-3}. (d) Relative integrated intensity as a function of temperature at a density of 1 g cm^{-3}. (e) Relative integrated intensity as a function of density at 400°C.

under the same conditions of temperature and pressure. Figure 14 shows the result of splitting the parallel polarized O—D Raman band into the two Gaussian component bands which provide the best fit to the total band. Splitting into two Gaussian component bands is feasible over the whole range of temperature and density in such a way that the sum of the intensities of the two component bands accounts for 90–95% of the total intensity. As shown in Fig. 14c, at 100°C and 1 g cm^{-3} the low-frequency tail of the total band is not completely covered by the two Gaussian bands. It turns out that ν_{max} of the high-frequency component band, marked by I, has a value of about 2640 cm^{-1}, that is, almost exactly where the depolarization ratio has its minimum, and is independent of temperature (Fig. 14a). ν_{max} approaches the vapor value of 2727 cm^{-1} as the density drops to zero (Fig. 14b). The ν_{max} value of the low-frequency component band, designated by II, increases by about 100 cm^{-1} between 25 and 400°C and also increases with decreasing density.

The integrated intensity of the fairly narrow band I increases with temperature (Fig. 14d), and also with decreasing density (Fig. 14e). The reverse is true for the relatively broad band II. Because of these opposing temperature dependences of the intensities of bands I and II, ν_{max} of the total band is governed by band II at low temperatures and by band I at high temperatures, and consequently it changes from $\nu_{max,\,II}$ to $\nu_{max,\,I}$ with increasing temperature, as indicated by the dashed line in Fig. 14a.

Splitting of the infrared band into two Gaussian component bands yields the result that the high-frequency band is very weak and that the total band, therefore, is almost entirely governed by the intense low-frequency component band. This seems to be reasonable since hydrogen bonds normally produce intense infrared stretching vibration bands of the adjacent X—H groups. If one plots ν_{max} of the total infrared band against temperature, or density (crosses in Fig. 14a,b), it almost coincides with ν_{max} of the Raman subband II. This explains the discrepancy in the temperature dependences of ν_{max} of the total infrared and Raman bands.

The existence of two components also accounts for the maximum in the half-width. This increases when the high-frequency band I grows as a shoulder of the total band at low temperatures. Above the temperature of maximal half-width, the decrease of band II becomes predominant.

All the observed features of the total bands are compatible with the assumption that the infrared and Raman bands are each composed of two component bands. Therefore, the spectroscopic results obtained for the O—D vibration at high temperatures and high pressures favor a model of water in which the O—D (O—H) groups of the molecules can be present in two spectroscopically distinguishable states, rather than a continuous distribution of bond lengths and angles. The characteristics of the two Raman components in dense water lead to the conclusion that band II is due to O—D groups participating in a hydrogen bond, whereas band I is due to non-hydrogen-bonded O—D groups.

Since nothing is known about the absolute Raman scattering intensities of the O—D groups in the two different states, no quantitative calculation of the fraction of bonded O—D groups can be based on the spectroscopic results. However, the spectra clearly indicate that this fraction becomes appreciable at densities above 0.6 g cm^{-3} at temperatures of 400–500°C. If the temperature dependences of the relative integrated intensities per

O—D group in the Raman components I and II are assumed to be the same, the energy necessary to transfer an O—D group from the hydrogen-bonded into the nonhydrogen-bonded state may be derived from the slope of a plot of $\log(I^{\mathrm{I}}_{\mathrm{int}}/I^{\mathrm{II}}_{\mathrm{int}})$ against $1/T$. At a density of 1 g cm^{-3}, Lindner calculated a value of 9.8 kJ mol^{-1}, a value which is in good agreement with values from other sources (see, in particular, Chapter 5).

7. CONCLUDING REMARKS

As yet, neither the experimental information on the properties of hot, dense water nor a statistical mechanical theory relating the macroscopic properties of dense water to the properties of its molecules is sufficiently well-developed to enable a complete picture of the structure of water at high temperatures and high pressures to be deduced.

Although no striking anomalies, like the density maximum at 4°C or the viscosity minimum with increasing pressure in water at low temperatures, occur above 100°C and 100 bar, water is far from being a normal polar fluid in this range. The peculiarities of hot, dense water, like those of water at ordinary conditions, stem from the ability of the water molecule to form hydrogen bonds. This ability, combined with the low thermal energy of the molecules below room temperature, leads to a relatively rigid arrangement of the water molecules in the open icelike structure of liquid water at its melting temperature. At high temperatures, the thermal energy of a fraction of the water molecules is high enough to distort the hydrogen bonds to such an extent that they may be regarded as being "broken." There is experimental evidence that at a constant density of 1 g cm^{-3}, marked structural changes occur between 100 and 200°C which become less pronounced at higher temperatures: the steep decrease of the molar heat capacity at constant volume, the maximum of the thermal conductivity at about 150°C, the decrease of the activation energies for viscosity and self-diffusion with temperature below about 200°C, and their approximate constancy above this temperature, the steeper decrease of the static dielectric constant below than above 200°C, the decrease of the proton excess conductance above 150–200°C, and the temperature dependence of the $\nu_{\max}$ of the O—D Raman valence band. Nevertheless, the influence of the hydrogen bonds on the properties of water at high densities and temperatures above 200°C remains significant, as indicated by the high static dielectric constant, which leads to a correlation parameter g closer to two than to one, and by the still high proton excess conductance.

Under these conditions, the number of molecules participating in hydrogen bonds may become equal to those with "free" O—H (O—D) groups. On a suitable time scale, one particular molecule may be regarded as spending comparable periods participating, or not participating, in a hydrogen bond; these periods are longer than the O—D or O—H stretching vibration periods.

Reduction of density at high (supercritical) temperatures results in a rapid decrease of the influence of hydrogen bonds, as may be seen from the density dependence of the static dielectric constant, of the proton excess conductance, and of the infrared and Raman O—D stretching vibration bands. As a consequence, water behaves nearly as a normal polar fluid at moderate and low densities.

To improve the present knowledge of hot, dense water, dielectric relaxation measurements and spectroscopic studies in the frequency range of intermolecular vibrations at high temperatures and high pressures would be particularly useful.

Properties which are affected by the steep increase of the ionic product with temperature and density in hot, dense water may be different by orders of magnitude in ordinary water and in water at high temperatures and high pressures. Examples are the specific conductance and the equilibrium positions of hydrolysis reactions in solutions of weak electrolytes in supercritical water.[(351)]

The properties reported here, especially the high dielectric constant, indicate that water at high temperatures and high pressures is an excellent ionizing solvent the properties of which may be adjusted to special requirements by temperature or pressure variations. Since dissolved particles have a very high mobility in hot, dense water, it may be particularly well suited for use as a solvent in future electrochemical processes at high temperatures.

CHAPTER 14

Structural Models

Henry S. Frank

Department of Chemistry
University of Pittsburgh
Pittsburgh, Pennsylvania

1. THE NATURE OF MODELS

The molecules of liquid water, the details of their mutual interactions, and the nature of the kinetic and transport processes in which they are involved are all inaccessible to direct observation or measurement. As things now stand, moreover, there is no early prospect that theoretical calculations will be able to give us, independently, information about water that could substitute for experimental observation. Therefore, insofar as we wish to "understand" water in a molecular sense or "account for" the results of observations made upon it, we must resort to the consideration of hypothetical models.

This is, of course, a common enough situation in physical science, but it is instructive to note that the methods by which models are appropriately employed in a given field are apt to change as the study of that field progresses. What is chiefly required at the outset, when a seemingly mysterious phenomenon is first encountered, may well be simply to show that physical assumptions do in fact exist which could account for it—that explanatory ideas can be formulated which have, so to speak, the same "shape" as the phenomenon to be explained. This was the case, for example, when Einstein[(299)] put forward his model of a crystal composed of oscillators, to establish that the quantum concept was capable of accounting for the falling off of heat capacities at low temperatures. A similar, if more homely, example is Röntgen's[(916)] interpretation of the maximum of density by postulating that cold water is a solution of ice in its truly fluid polymorph.

What follows upon such a beginning can be very different in different cases, depending on the nature of the questions at issue, on the kinds of information available, and on the rate of development of related areas of subject matter. Thus, understanding of the specific heats of solids was able to go forward rapidly and reach a kind of maturity within two or three decades of the appearance of Einstein's paper. The fact that progress in the study of water has been so much slower arises in part from the fact that liquids in general have remained less well understood. A major difficulty, also, has been that such interpretations as Röntgen's have generally turned out not to be unique, and that none of the experiments which could, until recently, be done on water were able to yield results which seemed to support unambiguous choices among competing models or model features. The result is that, while there has been no dearth of ingenious suggestions about what water *might* be like in order to display this or that set of properties, it is only recently, by use of new experimental techniques, and by combining data and interpretations from several fields, that it has begun to be possible to draw useful inferences about what water *must* be like in this or that respect. Recent developments do, however, seem to portend real progress toward a comprehensive, self-consistent model of water structure upon which we will be justified in expecting successive approximations to converge.

2. REQUIREMENTS A WATER MODEL MUST ATTEMPT TO SATISFY

Until approximately 40 years ago, the goal of a "convergent" model could not realistically have been entertained, because the framework within which a model could be formulated had not yet achieved any claim to finality. Such a claim is now made in principle for the Schrödinger equation and for the fundamental laws of statistical mechanics, and it is therefore necessary to require of a water model that it be compatible with any valid inference from these, regarding molecular interactions or statistical thermodynamics.* Indeed, we must go farther and require that a water model be considered as a prediction about the details of the result that can in principle be expected to emerge when (in some future century) the mathematical diffi-

* The earliest publication about water in which such a requirement was imposed seems to be the pioneering paper of Bernal and Fowler,[71] published in 1933.

culties have been overcome and a solution obtained for the Schrödinger equation, and the sums and integrals evaluated which will occur in the partition function for a system of N water molecules in a volume V at a temperature T. We know, for example, that if 6×10^{23} H_2O molecules are confined in a volume of 25 cm^3 at 300°K, the mathematics will have to tell us that all but about 1 in 10^5 of the molecules are to be found in a volume of 18 cm^3, the balance filling the remaining space at a density of about 8.65×10^{17} molecules cm^{-3}. Correspondingly, the adherent of a "mixture" model for liquid water is compelled, if he is serious, to predict that the partition function of a sample of 6×10^{23} H_2O molecules in 18 cm^3 at 300°K will eventually turn out to be dominated by configurations and states which correspond to the separation of the molecules into groups of appropriate sizes, which constitute the distinct populations in respect of energy, modes of motion, etc., which he takes as the "constituents" of the mixture.

What the foregoing means in practice is that before any water model can be considered acceptable, it must show itself capable of being paraphrased in statistical thermodynamic terms (unless it was proposed in such terms to begin with[(788)]) and the resulting statement found not to contradict any reasonable expectation from quantum mechanics or statistical thermodynamics. For this purpose, when details of the partition function must remain the subjects of hypothesis or of speculative estimate, it is often more convenient, and is entirely equivalent, to conduct the discussion, and assess the acceptability of the model, in terms of entropies of mixing, standard free energies, etc., in the first instance.[(360)]

On a still more practical level, one of the principal requirements upon a water model is to fit data and, as more and more new kinds of data become available, it is important to emphasize that an acceptable model must fit all of the (reliable) data that can be adduced from any source. As Narten and Levy[(785)] have pointed out, however, not all models can be tested against all kinds of data. Thus, the model of Röntgen could not be tested against the results of X-ray scattering experiments, not so much because Röntgen had not yet discovered X-rays[(917)] as because the model is a "limited-purpose" one, designed only to give a qualitative account of the anomalous *PVT*, heat capacity, and viscosity–pressure properties of water. A similar limitation is observed in the model suggested by Hall[(444)] to account for the "anomalous" acoustic absorption of water. This model also contains nothing that could be tested against X-ray scattering data, or, indeed, against any save the acoustic data. The discussion of the present chapter is not primarily historical in purpose, and will therefore give relatively little attention to models which are too limited in this sense.

3. SOME PROPERTIES OF WATER OF SPECIAL CURRENT INTEREST

As just discussed, the eventually acceptable model will, by definition, account for all the properties of water. In the present stage, however, when so many questions about water structure are still without generally agreed answers, there are certain kinds of information which seem especially likely to be useful, both separately and when taken together, as pointers in the search for such answers. The purpose of this section is to identify and comment on some of these. Not included, for categorical reasons, are various properties of aqueous solutions, both of ionic and of nonionic solutes, which have seemed to various workers to be determined by, or to have implications for, the structuredness of water and the alteration of its structure when solutes are added (e.g., Refs. 359, 361, 364). It may, however, be permissible to note that the requirement that an acceptable model be consistent with (i.e., that it have the right "shape" to accommodate) the way water is known to behave as a solvent subjects it to a set of conditions which are quite as stringent as those imposed by the properties of water as a pure substance.

3.1. Thermodynamic and Mechanical Properties; Structure and Fluctuations

The maximum of density and certain associated properties have long been known[916] to call for a structural explanation in which the distinction is drawn between greater and lesser degrees of bulkiness that must somehow exist in water. The various ways in which such a distinction has been introduced are discussed at greater length in a later section; suffice it here to say that all ascribe a fairly high degree of generalized "icelikeness" to cold water at low pressures, and find ways of accounting for the gradual disappearance of this property as either temperature or pressure is raised. Included among the phenomena often discussed in this connection is the *decrease* in viscosity with increasing pressure which water displays up to 500 bars at the ice point and up to 30°C at low pressures.[80]

The unusually large heat capacity of water is usefully interpreted in statistical terms as a measure of an unusually large mean-square fluctuation in energy $\overline{E^2} - \bar{E}^2 = kT^2C_v$.[509] When numerical values are inserted and the calculation is made on a hypothetical single-molecule basis, it turns out that the root-mean- square fluctuation in energy in water equals some 1.7 kcal mol^{-1}, slightly less than 20% of the energy of vaporization and over 50% of the energy content relative to ice at 0°K. In considering the meaning of this figure, it is to be remembered that the vibration frequencies of the internal degrees of freedom in water are so high that the whole of this fluctuation is of intermolecular energy.

The mean-square fluctuation in density is also unusually large in the case of water, as may be seen by inserting experimental values in the equation $(\overline{\varrho^2} - \bar{\varrho}^2)/\bar{\varrho}^2 = kT\varkappa/V$, where $\varkappa$ is the coefficient of compressibility. It turns out that at 20°C, the relative mean-square fluctuation in density for water is between two and three times as large ($\varkappa/V = 2.54 \times 10^{-6}$ $\text{cm}^{-3}\ \text{bar}^{-1}\ \text{mol}^{-1}$) as for benzene (1.05), chloroform (1.25), or carbon tetrachloride (1.07), in spite of the fact that these are at much higher reduced temperatures than is water.* Thus, the compressibility of water seems to be just as clearly "too big" as is its heat capacity. On a hypothetical single-molecule basis, the root-mean-square fluctuation in relative density amounts to something like 25%.

It is of interest that these two fluctuations, in energy and in density, respond very differently to application of external pressure. From a two-dimensional table of V as a function of P and T,[120] it is straightforward to estimate $\varkappa = -(1/V)(\partial V/\partial P)_T$ and $(\partial C_P/\partial P)_T = -T(\partial^2 V/\partial T^2)_P$ and conclude that at 20°C, increasing the pressure to 6 kbar reduces the volume by about 15%, reduces the mean-square fluctuation in enthalpy in approximately the same proportion, but causes the mean-square fluctuation in relative density to fall off three or four times as much (by about 55%).

3.2. X-Ray Scattering; Structure and Fluctuations

The scattering of X-rays by liquid water has been comprehensively discussed in Chapter 8 of this volume; what is intended here is simply to point out some special aspects of its relation to the material of the preceding paragraph. The finding from X-ray scattering that a large fraction of the O—O—O angles in water have equilibrium values which are nearly tetrahedral confirms the inference from PVT data that there is a special "bulkiness" in water, and identifies it as "icelikeness" in the general sense of relative openness of packing. On the other hand, the shallowness of the

* Methanol and ethanol both have room-temperature $\varkappa/V$ ratios greater than that of water. The reason for this is doubtless also their much higher reduced temperatures, i.e., it seems safe to assume that if they could be considered at $\Theta = 0.46$ (the reduced temperature of water at 25°C), they would also be found considerably less compressible than is room-temperature water.

trough between the first and second maxima in the pair correlation function[785] means that there also exist a certain proportion of smaller second-nearest-neighbor distances, or of regions where there are a larger number of nearest neighbors—there is a certain amount of semantic overlap between these two descriptive statements.

Of particular interest in this last connection is the shape (or lack of shape) of the low-angle scattering curve which, following Narten and Levy,[785] places limits on the kinds of fluctuations in density which can occur in the liquid. That there are sizable density fluctuations of some sort was seen above to follow from the numerical value of the compressibility, and local fluctuations must, in any case, be expected from the shape of the pair correlation curve between the two maxima near 2.9 and 4.5 Å. The fact that water seems to contain no significantly "dense" or "bulky" domains larger than the range of short-range order is thus of great importance, as will be further discussed in later sections.

Two ingenious calculations had previously been published by Fisher and Adamovich[333] in which an assumed form of the pair correlation function was used to draw inferences regarding local density fluctuations. The openness of the structure of water finds expression in both results, but it is difficult to assess their reliability, in part because the details of the pair correlation function $g^{(2)}(r)$ need to be restudied,[785] and in part because the asymptotic form employed for that function for large r was derived by use of the superposition approximation, and, therefore, contains an error of unknown seriousness.[509]

3.3. Hydrogen Bonds and Structure

As described in Chapter 2 of this book, several groups of workers, exploiting recent advances in the techniques of *ab initio* quantum mechanical calculations and in computer capabilities, have made contributions of first-rate importance to our understanding of what goes on when hydrogen bonds are formed between water molecules. Three qualitative results stand out: (a) the great sensitivity of bond energy to bond configuration, (b) the fact that there is some actual transfer of electron density from the "Lewis-base" partner to the Lewis-acid partner, and (c) the fact that in a chain of three molecules connected by two hydrogen bonds, the latter reinforce each other in strength—i.e., are stronger than "separate" bonds—if the charge separations reinforce each other, but are weaker than separate bonds if the charge separations buck each other—i.e., if the central H_2O is "donating" either electrons or protons toward both end molecules.[251,454] Moreover, the mutual reinforcement of the bonds in a chain of n favorably oriented molecules increases with n, at least up to $n = 6$,[251] the geometry being such that for $n = 5$ or $n = 6$, the chains can easily undergo joining of their ends so as to form closed rings, which are then particularly stable. Smaller rings can also be stable in an absolute sense, but are relatively destabilized by "bond bending," so that, by one calculation,[251] 16 H_2O molecules bonded as 4 ring tetramers could obtain an energetic advantage by disproportionating into 3 ring pentamers and 1 monomer.

The likelihood that charge separation in the hydrogen bond would lead to mutual stabilization of adjacent bonds which were favorably oriented had been commented upon[355,361] and the suggestion made that it might produce a cooperativeness in bond formation and bond breaking (the meaning of "breaking" a bond is discussed below) which, interacting with thermal fluctuation, would result in the existence in liquid water of "flickering clusters" of structure. This speculation is seen to receive some possible support from the quantum mechanical calculations, which, moreover, suggest an elaboration. If a thermal fluctuation permits the local formation of a group of mutually reinforcing hydrogen bonds, the detailed geometry of the cluster which is produced will of course be different from case to case depending on the momentary locations (and motions? see below) of the molecules in the volume element in question. In favorable situations, the lowest energy that can be achieved will correspond to the formation of the largest possible number of bonds, which will be at advantageous angles to each other. This will result in there being an open, tetrahedrally bonded region of low ("icelike") density. At the opposite extreme will be situations in which the locations of the molecules are so "unfavorable" that the maximum number of bonds could be formed only if the angles between them were grossly distorted. In such cases, it seems thinkable that a three-dimensional analog of the two-dimensional disproportionation mentioned above might result in the leaving of an unbonded molecule surrounded by a polyhedral cage, which had been made stable by the favorable bond angles which the "freezing out" of the monomer had made possible. The presence of the monomer inside the cage would of course cause the local density and effective coordination number to be higher than average.

Between the two extremes just described, it is possible to visualize a large number of intermediate geometries, including regions completely bonded but with more or less drastically bent bonds, and regions not quite completely bonded, with "pseudointerstitials" which filled vacancies but were also bonded to one or two "wall" molecules, and so had two or three unused bonding functions. The kind of distribution to be expected among the various possibilities will of course depend on the degree of cooperativeness in bond formation which is assumed. Such assumptions about cooperativeness may be explicit or implied, and it is worth noting that the recent statistical mechanical treatments of water which start with the assumption that the total N-molecule potential may be represented as the sum of pair potentials contain the implication that cooperativeness of the sort here discussed does not exist.

3.4. Spectroscopic Properties: Broken Hydrogen Bonds

The term bond breaking and bond making used in the preceding section require further discussion, for, while this usage has long been current, it has never been easy to state just what a broken hydrogen bond is, physically, nor what is involved in breaking it. It is clear that bending a hydrogen bond (rotating either the proton or the lone-pair electrons out of the O—O line) must weaken it—i.e., raise the potential energy of the configuration—but there seems to be no logical basis on which some degree of bending might be assigned beyond which the bond is said to be broken. It is therefore of great interest that recent spectroscopic observations seem to have provided both an experimental definition of a broken bond and some insight into its physical nature.

These observations are of the "uncoupled" O—H and O—D valence-stretching frequencies of the HOD molecule (in D_2O and in H_2O media, respectively) and were made in the Raman spectra by Wall and Hornig[(1132)] Walrafen,[(1138)] Bernstein,[(73)] and Franck and Lindner,[(353)] and in the infrared by Falk and Ford,[(323)] Hartmann,[(469)] Senior and Verrall,[(962)] and Franck and Roth.[(354)] Experimental and other details are to be found in the original papers and in Chapters 5 and 13 of the present volume. What is of interest here is exemplified in the fact that the broad O—D stretching band in H_2O is analyzable into component bands centered near 2620 and 2520 cm^{-1}, respectively. This stretching motion is represented in the dilute vapor by a very narrow band[(63)] of frequency 2720 cm^{-1}, and in ice by a broader, more intense band of frequency 2420 cm^{-1}.* Both downward shift in frequency and the broadening and increase in infrared intensity upon condensation from vapor to ice are well-known accompaniments of hydro-

* Some of these frequencies have been variously estimated to be slightly larger or slightly smaller than those here quoted. These differences in no case call the assignment in question, and are, for the present purpose, immaterial.

gen bond formation.[844] The 2520-cm^{-1} band in liquid water is also assigned to the hydrogen-bonded O—D stretch, its smaller lowering in frequency and greater breadth as compared with the ice band agreeing with the expectation that the hydrogen bond should be less strong (on average) and that there should be irregularity in the liquid which would produce the extra broadening.

The principal novelty here is in the assignment of the 2620-cm^{-1} band in room-temperature water to the stretching motion of non-hydrogen-bonded O—D groups. This is strongly supported by two, experimentally quite independent, considerations. The first is that a rise in temperature is accompanied by the expected change in relative intensities of the 2520- and 2620-cm^{-1} bands, the former decreasing and the latter increasing as they should do if heating breaks bonds. In the limit, the 2620-cm^{-1} frequency is strongly in evidence, and the 2520-cm^{-1} frequency has disappeared, in high-density supercritical water substance[353] under conditions which preclude the existence of hydrogen bonds. As the second, and very important fact, 2620 cm^{-1} is the asymptote approached by the frequency of maximum infrared absorption in this region when the temperature of HOD in H_2O, at a density of 0.9 g cm^{-3}, is raised without limit.[354] Such a temperature regime corresponds to random orientation of the HOD molecule with respect to its neighbors, and this means that the lowering in frequency from 2720 cm^{-1} (the dilute-vapor value) to 2620 cm^{-1} must have been produced by some interaction which, while imitating the hydrogen bond in this regard, is qualitatively (because dynamically and geometrically) very different from it. A plausible interpretation is that in the dense, disordered, aqueous medium, dispersion interactions involving both electronic and rotational motions cause each O—D to be, effectively, vibrating outward toward the wall of a cavity lined with negative charge, and that this produces the effect in question.

This interpretation of the physical state of an unbonded O—D in dense, supercritical water also implies how an unbonded O—D can be thought of at room temperature, namely, as an OD which is not pointing toward a free lone-pair, but is stabilized by some sort of mutual polarization interaction with what it does "see." Moreover, aside from its being spectroscopically less subject to irrelevant complications, there is nothing about the O—D stretch of an HOD in H_2O which makes it different, for the present purpose, from the O—H stretch of an H_2O molecule. It therefore follows that if the above interpretation is accepted, of the spectroscopic data in terms of bonded and unbonded O—D functions, it will be necessary in future to require of a room-temperature water model that a significant fraction of its O—H functions be unbonded in the sense just described.*

Brief mention should also be made of the lower-frequency spectra in liquid water which have been ascribed to librational, or hydrogen-bond-bending (1000–300 cm^{-1}) and to translational, or hydrogen-bond-stretching ($\sim$180 cm^{-1}) motions of whole water molecules. These have been observed in the infrared by Draegert *et al.*[273] and studied by means of Raman

* In stating this conclusion, which will from this point be accepted as requiring no further discussion, reference should be made to several spectroscopic arguments which purport, with varying degrees of cogency, to prove that liquid water at room temperature cannot contain an appreciable percentage of "unbonded" molecules. The basis of one of these claims[1022] is that the long-wavelength tail of the absorption band in the vacuum-ultraviolet spectrum of liquid water displays only a fraction of a per cent of the intensity of the vapor absorption at the same wavelengths. Since that claim was made, however, two groups of workers[1104] have succeeded in getting complete profiles of the liquid absorption band in question, and although both these refer to very thin layers of water, it seems unlikely that this fact can alter the following interpretations. The liquid absorption band has its maximum at about 1470 Å, as compared with the maximum of the vapor band at 1650 Å, i.e., there is a blue shift on condensation of some 7000 cm^{-1} or 20 kcal mol^{-1}. But the difference in energy of the ground states in the two cases averages less than 10 kcal mol^{-1} (ΔE of vaporization), so that the upper states in the two cases must be separated by an additional 10 kcal mol^{-1}, i.e., the electronically excited state in liquid water is 10 kcal mol^{-1} higher-lying (referred to a common baseline) than that in the vapor, presumably as a result of there not being enough room in the condensed phase for the expanded oxygen orbital. Whatever the cause of this difference in excited-state energies, it is clear that no grounds exist in these spectra, as now known, for drawing inferences regarding the proportion of H_2O molecules in the liquid which may be "unbonded." A similar conclusion, that very few unbonded O—H functions could be present in HOD dissolved in D_2O, was drawn[469] from the fact that the O—H valence-stretching band showed very little absorption at the frequency characteristic of the O—H stretch of HOD dissolved in deutrochloroform. This argument is also seen not to stand up, this time because of the obviously different "medium effects" to be expected in $CDCl_3$ and in D_2O. Finally, it should be mentioned that the argument[323,1132] that the great breadth of the O—D and O—H stretching bands arises from the wide range of variation in environments of the absorbing or scattering molecules is not challenged by the finding that these environments are to be classified into two categories, bonded and unbonded. In each category, there is indeed a range of O—O distances, and of degrees of vibrational excitation of the motions of the molecules with respect to each other. This causes both bonded and unbonded bands to be very broad, and results in so pronounced an overlap that a considerable range of frequencies exists (around 2550 cm^{-1} in the O—D band in H_2O) which can belong, almost equally well, in either category. There is, however, no more logical difficulty in this than in the fact that a human being, of whom nothing more is specified than a height of 140 cm, may almost equally well be a man or a woman.

scattering, particularly by Walrafen.[1134,1135] An item of special interest regarding these spectra is that they seem to be counterparts to corresponding spectra in ice, which are characterized by somewhat better-resolved bands at slightly higher frequencies. Walrafen has found that the integrated intensities of the Raman bands in both frequency ranges decrease sharply with temperature, in perfect phase with each other, amounting at 40°C to about half of their values at 0°C, and to only about 15% at 100°C. The positions and shapes of the bands, on the other hand, do not change greatly over this temperature interval. There thus seems to be "something icelike" in cold water which disappears as the water is heated.

This "something," according to Walrafen (see Chapter 5), consists of centers of C_{2V} symmetry, each being a water molecule hydrogen-bonded to four others in tetrahedral positions around it, and he suggests that the disappearance of these centers arises from the breaking of the hydrogen bonds. An alternative, or supplemental, suggestion has also been made,[358] namely that if the bonds bend sufficiently at high temperatures, the local symmetry might be destroyed, with the same disappearance of scattering intensity. That this may indeed be a part of what happens receives some support from the fact that the effect of increase in temperature on the infrared absorption between 300 and 1000 cm^{-1} [273] seems to be very different, consisting of a marked shifting of the band to lower frequencies, but with relatively little apparent change in integrated intensity. If the only effect of raising the temperature were the breaking of hydrogen bonds, it might be expected that the Raman and infrared bands would decrease in intensity in more nearly the same way, whereas a change in local symmetry might well leave the infrared intensity almost unchanged, affecting instead its frequency of maximum absorption, as observed.*

3.5. Vapor-Pressure Isotope Effect

The striking fact has long been known[612] and commented upon[356] that whereas H_2O is more volatile than D_2O as ice and as cold water, the reverse is true above 200°C, so that D_2O has the lower critical temperature

* A considerable literature exists[701] in which the alterations which change of temperature produces on various overtone and combination spectra of water are used to draw conclusions about the percentage of hydrogen bonds which are formed or broken. Inferences drawn from these spectra can be expected to provide important information when the general outline of water structure has become somewhat clearer. As present-day "proofs" of what water must be like, however, they suffer from the uncertainty which still surrounds the laws of addition of intensities in combination bands, etc. It is for this reason that no use has been made of such data in the argument of this section.

(370.9°C as compared with 374.15°C for H_2O). The theory of the vapor-pressure isotope effect (VPIE) has been given in general form by Bigeleisen[82] and has been applied to ice and to cold water by several workers,[881,1086,1178] but without emphasis upon the drawing of inferences about the water model. Important inferences can, however, be drawn.

As is made clear in the general theory,[82] the VPIE is an intrinsically quantal phenomenon and arises from the existence of differences between zero-point energies of vibrational motions in condensed and in vapor phases. Using water as an example, the antisymmetric O—H stretching frequency in the vapor is 3756 cm^{-1} and, less precisely, about 3490 cm^{-1} in the liquid.[57] This means zero-point energies ($\sim\frac{1}{2}hc\tilde{\nu}$) of 5.37 kcal mol^{-1} ascribable to this motion in the vapor and 4.99 kcal mol^{-1} in the liquid, so that, of the approximately 10,000 cal mol^{-1} heat of vaporization of water at room temperature, some 380 cal mol^{-1} consists of zero-point energy of the antisymmetric valence-stretching motion which must be restored to the H_2O molecule as part of the vaporization process. The fact that in D_2O the corresponding frequencies are 2788 and 2540 cm^{-1}, respectively, means that here only 350 cal mol^{-1} needs to be supplied in the vaporization process, so that, were this the only difference between the two liquids, H_2O would have the higher heat of vaporization by 30 cal mol^{-1}. This is (in this case) also the difference in the standard free energies of vaporization and therefore produces a factor of about 1.05 by which the vapor pressure of D_2O should be higher than that of H_2O. The symmetric valence stretch makes a similar multiplicative contribution, and these are only triflingly offset by a contribution in the opposite direction arising from the valence bending frequency, which is slightly *raised* on condensation. The net effect ascribable to the intramolecular frequencies of the H_2O and D_2O molecules is thus in the direction of making the latter considerably the more volatile. Corresponding effects are at work in the cases of benzene and cyclohexane and, since only feeble intermolecular influences are present in those cases, the perdeutero compound is in fact the more volatile in each of the pairs, and has the lower boiling point.[242,546]

It is thus to the existence of the strong "intermolecular" interactions, which make themselves felt through the librational (and "hindered translation") oscillations discussed in the preceding section, that appeal must be made if the reversal of sign of VPIE as between water and, say, benzene is to be understood. Taking 685 cm^{-1} [301] as a representative librational frequency in liquid H_2O, with a zero-point energy of about 980 cal mol^{-1}, we note that no librational oscillation, and hence no corresponding zero-point energy, is present in the vapor, so that its existence in the liquid decreases (the absolute amount of) the heat of vaporization by the full 980 cal mol^{-1}. Since, again, the D_2O frequencies are lower (in this case, because of the larger moment of inertia of the D_2O molecule, so that $\tilde{\nu}_L \approx 505$ cm^{-1}), the corresponding negative contribution to the heat of vaporization is only 720 cal mol^{-1}, with the result that there is an isotopic difference of 260 cal mol^{-1}, arising from this one librational mode in the liquid, in the direction of making H_2O the more volatile. Considering that in the liquid there can be librational oscillations about three axes, and that a considerable proportion of the molecules in cold water are expected to be fully hydrogen-bonded, it is not surprising that at room temperature, H_2O should, in net result, be more volatile than D_2O. The experimental fact[(612)] that P_{H_2O}/P_{D_2O} decreases from about 1.15 at 25° to 1.05 at 100° and to 1.005 at 200°C is explicable as arising in part from a gradual breaking of hydrogen bonds as temperature is raised, with a resulting conversion of the torsional oscillations into "unbound" rotations, unaccompanied by zero-point energies. That the ratio, however, should go below unity at still higher temperatures shows something else as well, namely that the disappearance of the librational oscillations is not accompanied by a disappearance of the lowering of the intramolecular frequencies. That is, the lowering of valence-stretching frequencies as a result of non-hydrogen-bonded interactions which was inferred from spectroscopic evidence receives strong confirmation from the VPIE above 225°C. Except, in fact, for the existence of this non-hydrogen-bonded effect on the *spectra*, it would seem very difficult, within the framework of the Bigeleisen theory, to account for the difference in critical temperatures between H_2O and D_2O, whereas with it, this isotope effect is just what must be expected.*

3.6. Molecular Motions

One inference which must be drawn from the molecular explanation of the VPIE is that there are limits to how completely liquid water can be

* It seems that more new insights are likely to result from more systematic comparisons between the properties of H_2O and D_2O. The widely held view that, at any given temperature, D_2O is "more structured" than H_2O[(790)] may, for example, require reinterpretation in the light of the finding of Narten *et al.*[(784)] that the X-ray scattering curves for H_2O and D_2O, both at 4°C, are virtually indistinguishable. This, together with the long-known fact that their molal volumes are only triflingly different, makes it hard to ascribe to D_2O more of the bulkiness ordinarily associated with structuredness. An alternative suggestion is that the two liquids are closely similar in structure and that it is the greater gain in entropy accompanying structure-breaking in D_2O, because of its larger moment of inertia, which has given the other impression.

characterized in purely classical terms and that, in one respect at least, the relative positions and orientations of the water molecules cannot fully be understood unless their relative motions are also taken into account. Although Pople, in his 1951 paper,[855] recognized the possibility that statistical-mechanical treatments of water might require quantum corrections, it has been customary to preceed in the hope that these could be neglected. To the extent that this hope must now be called in question, the possibility must be considered that a full understanding of the structure of water from the standpoint of bonding and geometry will be impossible until its molecular dynamics is also well understood.

There are a number of experimental fields, some of them relatively new, from which microdynamic information about water can in principle be obtained. Many of the most important of these are discussed in earlier chapters of this book dealing with nuclear resonance, dielectric relaxation, neutron scattering, and acoustic relaxation, in Chapters 6, 7, 9, and 12, respectively. Study of these chapters reveals important agreements among the findings from the different kinds of measurements, but some apparent discrepancies also, which can be expected to be illuminating. In addition, some recent inferences from the band contours of infrared absorption and Raman scattering in other liquids[922] seem to be relevant to one of the important questions at issue.

The drawing of inferences about molecular dynamics from infrared or Raman band contours employs a theorem[413] which shows the time-dependent correlation function, $C(t) \equiv \langle \mathbf{r}(t) \cdot \mathbf{r}(0) \rangle$, of the transition moment vector, i.e., the averaged (over the ensemble) cosine of the angle between the direction in which the vector points at time t and the direction it pointed at time 0, to be given by the real part of the (normalized) Fourier transform of the band envelope (intensity versus frequency in radians) $I(\omega)$. That is,

$$C(t) = \left[\int_{\text{band}} I(\omega) \cos(\omega t) \, d\omega \right] \Big/ \int_{\text{band}} I(\omega) \, d\omega \qquad (1)$$

When the indicated integrations are carried out with one of the infrared bands of methly iodide[326] or of cyclohexane,[920] it turns out that a plot of $\ln C(t)$ against t resembles a parabola for several tenths of a picosecond and then straightens out to a rather long, nearly linear segment in which $C(t)$ falls to about 0.1 in perhaps 2 psec. A linear segment in such a plot corresponds to the exponential form of the time correlation function which is so often assumed in relaxation calculations, partly on intuitive grounds and partly because it is relatively tractable for mathematical purposes.

Physically, it can be understood as arising from the cumulative effect of numerous (small) rotations following upon numerous collisions with neighbors—a "Brownian rotational diffusion" such as Debye assumed in developing his formalism for dielectric relaxation.[(245)]

The initial "parabolic" portion of the plot, on the other hand, has been claimed to be indistinguishable from what would be produced by free rotation in the vapor phase,[(921)] and it persists, typically, until the vector has rotated through some 30–40°. That is, when first "observed," the molecule seems to make an uninterrupted rotational jump of a considerable fraction of a radian before the first "collision" with a neighbor begins to introduce the stochastic element that characterizes Brownian motion. As pointed out by Rothschild,[(922)] this tends to discredit the numerical values (except for orders of magnitude) of correlation times as calculated from relaxation times by use of the exponential assumption.

Wall[(1131)] has explored the application of such an analysis to the envelopes of the Raman scattering bands of the O—D and O—H stretches of HOD in H_2O and D_2O. He remarks that this is not fully justified, because the HOD molecule is strongly perturbed by its neighbors and is, in any case, asymmetric. If the conclusion reached in Section 3.4 is justified and these bands are of complex origin, the justification for drawing inferences from their Fourier transforms is still weaker. Furthermore, it appears that the application of this theorem to Raman scattering should be carried out in a somewhat different way.[(413)] It is, nevertheless, of interest that the transforms Wall obtains look very much like those described above, the chief difference being that the time scales are shorter by perhaps half an order of magnitude, though the O—H and D—O times bear the "proper" relation to each other.

Whether or not these observations on the HOD spectra are in themselves significant, it seems extremely likely on intuitive grounds that if, in such liquids as methyl iodide and cyclohexane, a molecule, when it starts to rotate, makes a big rotational jump before "settling down" to Brownian rotational diffusion, then *a fortiori* the same should be true in water. This notion, moreover, tends to confirm the view expressed in Chapter 7 that in water the phenomenological dielectric relaxation time and the molecular reorientation time are not greatly different from each other, for this is what theory says would result if the molecules independently rotated in big jumps. Accepting this, we get $\tau_c = 8 \times 10^{-12}$ sec (see Chapter 7) as estimated from dielectric dispersion.

On the other hand, the "big jump" mechanism seems very unlikely to make either the neutron scattering or the nuclear resonance data predict

τ_c values bigger than, say, $2\text{–}3 \times 10^{-12}$ sec and it would appear quite incapable of making the time constant for structural collapse (acoustic relaxation)[(1000)] exceed about 1×10^{-12} sec. There is thus, on the face of it, a clear contradiction among the values of τ_c derived by customary procedures from the data from different fields. One possibility that suggests itself as a tentative explanation for this as well as for the "jump-and-wait" character that analysis of neutron scattering ascribes to the self-diffusion mechanism is that the motions of adjacent molecules may be more strongly coupled than the conventional formalisms assume—i.e., that the "flickering cluster" previously speculated about might somehow be involved. Neither this nor any other suggestion will be very useful, of course, until some way is found of making it the basis of more detailed predictions (see Section 5.4).

4. BRIEF SURVEY OF MODELS

As already remarked, the fact that water shrinks upon warming from the ice point implies the presence in it of some special bulkiness which can be produced and destroyed by some mechanism not found in "ordinary" liquids. It was natural to identify this with the peculiar bulkiness which makes ice float, and it is therefore not surprising that suggestions based on this notion had become current* hardly less than a century ago. Indeed, the general idea is still accepted, and the chief differences among the various water models that have been put forward over the years are in what they assume about the nature of the other, denser, constituent of water, and about its relationship to the generalized icelikeness which preponderates at the lowest temperatures.

The models that have been proposed may be classified in terms of what they visualize as happening to ice when it melts. Chronologically, the first suggestion was that part of the ice was converted into a fluid in which the remainder dissolved. The fraction of icelikeness remaining was described by Röntgen[(916)] as subject to the influence of changes in temperature and pressure through the operation of the le Chatelier principle,

* It seems first to have been spelled out by Röntgen.[(916)] Earlier statements of essentially this idea were made by Rowland[(924)] and Whiting,[(1170)] but in each of these cases, the observation was incidental to a discussion that was not primarily concerned with water. Röntgen referred to the idea he was setting forth as "by no means new," and it would not be surprising if additional early reference to it should someday come to light.

and this representation is the prototype of the class of "mixture models" of water.

The second suggestion was more specifically structural, having been made after the hydrogen bond had been identified[(657)] and the structure of ice I established,[(56)] and describes the melting of ice as consisting in the bending of the hydrogen bonds, which remain unbroken. In this concept, every molecule in the liquid is pictured as being surrounded in the same way, on average, by neighbors in distorted positions and orientations, and the average degree of bond bending is the parameter which is sensitive to changes in temperature and pressure. These general outlines were proposed in 1933 by Bernal and Fowler[(71)] and have become the basis of the class of models now often known as "uniformist."

A third way in which the density could increase when ice melts was pointed out by Samoilov,[(939)] namely that individual molecules could be "shaken loose" from their positions in the lattice (which would then "heal" itself) and take up positions in the cavities which the lattice encloses. Samoilov's is, therefore, the prototype of the so-called interstitial models, in which the temperature- and pressure-sensitive parameter is taken as the fraction of the interstitial sites which are occupied. Always implied, and sometimes explicit, in the formulation of an interstitial model is the recognition that the structure of a liquid cannot display long-range order. This feature may be taken care of by assuming that the random irregularities in the local geometry are great enough to produce a cumulative disorder which is complete beyond 10 Å or so[(785)] or by picturing the framework clusters as regular enough, but limited in instantaneous extent, requiring, therefore, a "third state" of water (disordered) between the ordered patches.[(360)] Because an interstitial model assigns different instantaneous roles to different groups of water molecules, it should, strictly, also be classed as a "mixture model." For present purposes, however, the "mixture" label will be restricted to the simpler noninterstitial representations.

Because of the rapid rate at which new data and insights about water are becoming available, many of the detailed models which have been proposed have been overtaken by events and are now of hardly more than historical interest. Perhaps the classic example of such obsolescence is found in the theory put forward in 1900 by Sutherland[(1038)] representing water as a mixture of monohydrol (monomeric, steamlike), dihydrol (dimeric, intrinsically liquid), and trihydrol (trimeric, icelike) molecules. This was a particularization of the Röntgen mixture concept, but was attempted too soon, so that, although it dominated the scene for more than a decade, it ended up being completely discredited by fundamental developments in valence and structure theories. In view of this situation, the surveys in the following sections make no attempt to be exhaustive, the hope, instead, being to make observations of present-day usefulness. Fuller discussions of the older theories are to be found, for example, in the review by Chadwell,[173] the book by Dorsey,[269] or the introductory paragraphs of papers by Feates and Ives[327] and Nemethy and Scheraga.[788]

4.1. Mixture Models

None of the mixture models which have been proposed seems now to be immune against serious criticism of one kind or another. For one thing, none of them can claim to be unique—i.e., none purports to answer the question "What must water be like?." This is consistent with the fact that most of the mixture models were introduced for special, and limited, purposes—e.g., that of Hall[444] to show how the "anomalous" acoustic absorption of water might be accounted for. In many cases, the special purpose was, at least in part, to lay groundwork for suggested explanations of some striking properties of aqueous solutions. This was true of the elaborately quantitative model of Eucken[317] and of the largely qualitative one assumed, partly by implication, by Frank and Evans.[359] Those of Grjötheim and Krogh-Moe[429] and of Wada[1125] brought a few, and that of Davis and Litovitz[240] rather more, properties together, but none of these seem likely to be able to satisfy the criterion apparently established by the low-angle X-ray scattering in the matter of density fluctuations (Section 3.2). Nor does any of them seem likely to provide a plausible environment for the hydrogen bonds which are "broken" in the sense of Sections 3.4 and 3.5, which the spectroscopic evidence requires to be present.

The mixture model of Jhon *et al.*[560] is unique in applying to water a special theory of the liquid state, the "significant structure" theory,[322] and succeeds in fitting a large variety of data, some with high accuracy. Apart, however, from questions regarding the implications for water of the general theory, the use of such approximations as a single Θ to represent three degrees of translational and three degrees of librational freedom in Einstein oscillator functions places limits on how far this formalism can be counted on to tell us about the details of what the molecules are actually doing.

A somewhat simpler application of the significant structure method[322] leads to a special kind of "mixture" model which pictures water as a mixture, not of 2 kinds of species, "bonded" and "unbonded," but rather of 5 kinds, each defined as a water molecule of which 0, 1, 2, 3, or all 4 of its bonding functions are actually bonded. This concept was employed by Haggis *et al.*[443] in an attempt to account for the dielectric dispersion properties of water: a 3- or 4-bonded molecule could not rotate, a 0-bonded one could rotate about three axes, etc. Their calculation of the rates of making or breaking bonds, and therefore of the relative numbers of molecules of different degrees of bondedness, is, however, now seen to be vitiated by containing the simplifying assumption that the several bonds a molecule forms are energetically independent of each other—i.e., that the chance that a water molecule will form a new bond is independent of the number of bonds in which it may already be a partner. (Section 3.3)

A more elaborate treatment of the (0, 1, 2, 3, 4) mixture is that of Nemethy and Scheraga,[788] who constructed a significant structure partition function on this basis and derived a number of model details, such as cluster sizes and relative numbers of molecules of different bondedness. Unfortunately, there are now known to be logical inconsistencies in this theory also, among them an erroneous combinatory expression for the number of ways of distributing N molecules among the five categories, and, again, the assumed energetic interaction of bonds involving the same molecule. That there should be such interaction[361] is, indeed, allowed for, but in a strictly formal manner, of which there is no way to judge the adequacy.

Finally, special mention should be made of the very recent 0, 1, 2, 3, 4 representation of Walrafen[1139] (see also Chapter 5), for it, while still a special-purpose model, differs from all those previously mentioned in claiming, on the strength of data which ought to be reliable, to say something definite about what water *must* be like. The data are the Raman scattering spectra in the valence-stretching region, and the analyses of these into bands which represent bonded and unbonded O—H and O—D motions seems, as set forth in Section 3.4, to be sound. The assumption of a linear relation between the integrated intensity of a band and the concentration of the scattering species is also straightforward. Taking these all together, Walrafen can obtain, for each temperature, a pair of numbers which are proportional to the concentrations of the bonded and the unbonded O—H functions. Proceeding beyond this, however—e.g., to the actual concentrations, or to the translation into a ΔH of bond breaking derived from the temperature dependence of a number proportional to the concentration ratio—is a chancier business. Either of the latter calculations requires the making of some additional assumptions, for which neither the actual experimental data nor the theory of Raman scattering gives any hints, and making them thus means leaving the realm of "must be" and reentering that of "might be." Indeed, there seems to be nothing in the experiments or in spectroscopic theory which could distinguish between the acceptability of a 0, 1, 2, 3, 4, mixture and that of an interstitial representation in accounting for these data.

4.2. Uniformist Models

The best-known sequel to the classic work of Bernal and Fowler[(71)] is the 1951 paper by Pople,[(855)] in which the methods of statistical mechanics were employed to obtain an average degree of hydrogen bond bending in terms of an assumed harmonic restoring force constant and the further assumption that each bond bends independently of all others. When this formalism was used to calculate a radial distribution function for the relative positions of the molecules, the result was consistent with what had been deduced by Morgan and Warren[(761)] from X-ray scattering. Calculated values for the dielectric constant could also be obtained, and, while these came out somewhat low in absolute value, their temperature dependence was almost exactly right.

More recently, Bernal[(69)] has proposed a uniformist picture of water based on his concept of a liquid as an intrinsically irregular structure, and some detailed machine calculations of classical models for water have been undertaken by Rahman and Stillinger[(884)] using a molecular dynamics approach and by Barker and Watts[(51)] using the Monte Carlo procedure. All of these calculations, however, assume the potential energy of the water to be the simple sum of the potentials of pairs of molecules. Therefore, they cannot include effects arising from the energetic interaction between the several hydrogen bonds into which a single H_2O can enter (Section 3.3) nor can they, on the face of it, take any cognizance of a broken bond, since a continuous gradation of energy with degree of bending is built into their pair potentials. They should thus be able to display no feature corresponding to the fact that spectroscopy gives two, and only two, categories of environment for the O—D vibration of HOD in an H_2O medium (Section 3.4). The zero-point energy differences between H_2O and D_2O which were seen in Section 3.5 to play a decisive role in the VPIE are also foreign to these formalisms. While, therefore, it seems certain that continued exploration of theoretical calculations within this uniformist framework will provide further insights into the nature of water, it also seems likely that these will be of an indirect character rather than supplying first instance descriptions of what the molecules are actually doing.

4.3. Interstitial Models

The idea that the increase in density when ice melts arises from the invasion by interstitial molecules of some of the empty spaces in the ice lattice occurred independently in at least three laboratories. The first publication putting it forward seems to have been that of Samoilov,[939] and those of Forslind[344] and Danford and Levy[226] were apparently written in ignorance of Samoilov's work. The somewhat different notion that liquid water could be regarded as a "water hydrate" was put forward by Pauling,[827] also independently. His suggestion was that a framework resembling that in a clathrate hydrate might contain monomeric water as the "guest molecules" and thus attain the necessary density, radial distribution, etc. The possibility of accounting for the peculiar *PVT* behavior of water by allowing the degree of occupancy of interstitial sites to vary, and making this variation follow changes in P and T in such a way as to minimize free energy, was pointed out by Frank and Quist.[360] Their discussion is of the Pauling model, but the statistical thermodynamic formalism amployed applies equally well to any interstitial representation.

Samoilov's model was developed partly with a view to understanding water itself and partly in order to make possible a discussion within a consistent framework of the nature of hydration and the structure of ionic solutions. This is clear from his monograph, "The Structure of Aqueous Electrolyte Solutions and the Hydration of Ions" (published in 1957 in Russian, in German translation in 1961, and in English translation in 1965, with a 1964 author's preface[940]), which has been followed by a large number of papers by Samoilov and others, devoted to developments of, or variations upon, the interstitial concept. These are mostly published in the *Journal of Structural Chemistry*. Inasmuch as these studies deal almost exclusively with questions of what water *might* be like, no attempt will be made here to enumerate or discuss them as a group. Moreover, the march of events has created a situation in which not all of them are to be accepted without further consideration—e.g., in one,[435] a distinction between centrosymmetric and mirror-symmetric hydrogen bonds in water is supported by ascribing to them the different far infrared vibration frequencies 60 cm^{-1} and 180 cm^{-1}, whereas there is now reason[1134] to believe that the 60-cm^{-1} frequency arises from the bending, and the 180-cm^{-1} frequency from the stretching of a configuration O—H···O—H···O—H; or, again, in another,[1101] the thermodynamic differences between H_2O and D_2O are discussed without reference to the part played by zero-point energy differences. On the other hand, there are a number of physical ideas which may well reappear at a later time in discussions of what water *must* be like and may, therefore, be mentioned here. Among these is that interstitial molecules may enter into hydrogen bond formation with framework molecules[437,438]; an application to liquid water of the mechanism proposed by Haas[439] for self-diffusion in ice through exchange of roles between interstitial and framework molecules, involving what are called DI defects[436]; and some suggestions regarding the mechanism of self-diffusion which involve the motion *en masse* of "regions of short-range order."[1189]

Another elaboration of the interstitial concept, of an entirely different nature, arose out of the success of Danford and Levy[226] in employing the computerized least-squares methods of the X-ray crystallographer to "refine" an interstitial water model so that the calculated scattering of X-rays from "model water" agreed with the experimentally measured scattering. This development is described and explained in Chapter 8 of the present volume, but deserves special mention here because it contains one of the first systematic attempts to formulate, and then give experimental answers to, questions of what water *must* be like. These questions, and their answers are once more taken up in the next section.

5. THE PRESENT "BEST GUESS"

There are two reasons why any attempt, in the present state of our knowledge, to say anything about what water structure *must* be like may well seem to be premature. For one thing, the pieces of evidence which are most compelling for this purpose have not yet been long enough exposed to technical scrutiny, or are, in a psychological sense, too recently adduced, as yet to be sure of universal acceptance. Second, the knowledge that we so far have at our disposal contains known gaps of such size and of such a nature that the things we will learn in filling them may well cast quite different lights on things that we already know. For both of these reasons, the tentative conclusions expressed in this section are "subject to change without notice," and are put forward at all only because the importance of the subject requires that, at each stage, we try to squeeze out of the information at our disposal all of the meaning that it contains.

Another limitation which must be kept in mind is that the structure which it is here desired to describe is, in the first instance, that of "cold water" at atmospheric pressure. Excursions must be made into other ranges of temperature and pressure, for the purpose of observing contrasts or of obtaining certain other kinds of information, but it is in relating these contrasts and this information to "ordinary cold water" that our immediate interest lies.

5.1. Bonding

With these reservations and limitations, it seems appropriate to begin by repeating that the long-disputed question whether water does or does not contain broken hydrogen bonds seems now to have been decided in the affirmative. When the "best" procedures are followed for drawing baselines so as to obtain reliable band contours, and the "best" techniques of curve analyzing (both analog and digital) are employed, it appears from the Raman and from the infrared spectra that there really are two physically distinguishable environments into which an O—D stretching vibration can "look," and that all of the expectations are fulfilled which follow from assigning to one of these environments a hydrogen-bonded and to the other a nonbonded nature, and observing the effects of changing temperature and density. Moreover, the physical description of a broken bond as deduced from the spectroscopic studies is in full agreement with what is required to account for the effect upon the VPIE of raising the temperature to the critical point.

5.2. Structure Pattern

Acceptance of this point of view makes it necessary, as pointed out in Section 4.2, to rule out the uniformist class of models from further consideration, for these are, by hypothesis, unable to describe a liquid which contains broken bonds. The next question, therefore, is whether there is any similar experimental criterion for choosing among the remaining possibilities, and it will be remembered that the possible existence of such a criterion is indeed implied in the interpretation placed by Narten and Levy on the low-angle X-ray scattering data, presented in Chapter 8, and referred to in Section 3.2. Its spectroscopic and VPIE behavior entitle us to say that cold water contains bondedness and nonbondedness, and it seems straightforward to associate the former with the bulkiness and the latter with the denseness, of which the presence was inferred in Section 3.1 from *PVT* properties. How the bonded-bulkiness and the nonbonded-denseness are organized, and how related to each other, is the question to which Narten and Levy offer an answer. It is (a) that the denseness and bulkiness can probably not be segregated into patches which are large enough and still different enough from each other to satisfy the presumable requirements of the "simple" mixture models, but (b) that there is no difficulty if the denseness is, so to speak, subdivided to molecular dimensions by ascribing it to the occupied sites, and the bulkiness to the vacant sites, of an interstitial structure.

While the low-angle X-ray scattering data thus seem to place the interstitial class of models in a preferred position as compared with the mixture category, they do not, by themselves, justify an unqualified statement that cold water must be interstitial in structure. For one thing, the corrections that must be made to raw X-ray scattering data, in order to sort out the desired components, are of significant magnitude[782] so that until, for instance, the scattering in question has been studied over a range of temperatures and its penultimate limiting behavior accounted for theoretically, there may still be something hypothetical about the results under discussion. The possible compatibility with these data of alternative structural possibilities (e.g., the existence of larger patches of less pronounced bulkiness and denseness) may also require further exploration. There is, however, a quite different point of view from which the interstitial representation also seems to have special advantages.

What are referred to above either as "simple" or as 0, 1, 2, 3, 4 mixture models seem somewhat less attractive from the physical and geometric standpoints than from the purely spectroscopic and energetic ones. The difficulty has already been pointed out (Section 4.1) on quantum mechanical grounds of considering the several bonds on a given water molecule as independently breakable or formable. To this must now be added the connection between bondedness and bulkiness, which is established by the correlation between the ways the spectroscopic and the *PVT* properties of water respond to temperature changes. Now, for bondedness to produce bulkiness to best effect requires the formation of three-dimensional structures —i.e., in the first instance, of four-bonded molecules which possess the C_{2V} symmetry emphasized by Walrafen as characterizing cold water (Section 3.4). But to the extent that bondedness is thus concentrated on a few molecules, conservation rules require that nonbondedness also be concentrated. In addition, physical difficulties begin ro appear unless the nonbonded protons (lone-pairs) are suitably separated from the lone-pairs (protons) with which they could recombine. The obvious automatic way of protecting a nonbonded bonding function from such attack is to encapsulate it in an interstitial site, and we thus have another argument favoring the interstitial class of models.

Provisional acceptance of the idea that cold water is interstitial in nature still leaves open many questions about the probable details of the interstitial structure. From the discussion of Section 3.3, it appears that accepting an interstitial structure at all requires the assumption of a high degree of cooperativity in hydrogen bond formation, and this assumption will presumably become testable as the sizes increase of the groups of water molecules that can be studied by the methods of quantum mechanics (Chapter 3, and Section 3.3). In the meantime, there are those to whom the idea of a thoroughgoing cooperativity of the kind here under discussion will seem implausible, and they, and others, may well find merit in picturing a rather irregular framework with many semiinterstitial molecules in the sense of Section 3.3, in preference to a more highly regular one with fully interstitial guests. A future development which might help in deciding such questions would be a clarification of the relationship between the configurations and the motions of a group of molecules, and the extent to which this relationship enters into the specification of the group's instantaneous state. Also to be considered must be the topological question of how well such a semiinterstitial model could reproduce the *PVT* properties of real water.

Even now, however, further experimental suggestions regarding the degree of order in the cold-water framework are to be found in the temperature dependence of the integrated intensity of Raman scattering by the hydrogen-bond-stretching and hydrogen-bond-bending motions (Section 3.4). As previously mentioned, this suggests that the concentration of centers of C_{2V} symmetry is high at low temperatures, and this would be compatible with the existence near the icepoint of a framework which exhibited a relatively high degree of local order, and therefore, probably, a truly interstitial character.

Whatever the degree of local order, its geometry becomes a matter of interest, and the lack of success that Narten and Levy report[(785)] for their efforts to refine an interstitial structure based on the Pauling notion[(827)] of a pentagonal dodecahedral framework must be taken as support for their (and Samoilov's) assumption of an icelike framework geometry, for which they did succeed in refining a structure. It would be possible, however, to overestimate the amount of reliable detail, even regarding local order, that can be inferred from the Narten–Danford–Levy calculations. They employ some half dozen adjustable parameters in obtaining a least-squares fit to the scattering-intensity data, and the possibility exists that the adjustments which these parameters undergo in adapting to the changes, for instance, that occur in water as the temperature rises, may represent the latter in a sort of code language rather than as real physical alterations in the quantities the parameters purport to represent. The variety of things that can be accounted for as "adjustments in the parameters" is brought out in an exchange of reports in *Science*,[(570,786)] and attention may also be called to the fact that real water under its saturation pressure at 200°C must surely differ more strikingly from room-temperature water (witness the change in VPIE) than would appear from the relatively trifling changes which the Narten–Danford–Levy parameters undergo over this range.[784] The actual framework in cold water thus doubtless contains a less strict local order than the formalism of the Oak Ridge model specifies, but is presumably still closer to an icelike than to a pentagonal dodecahedral geometry.

5.3. Equilibrium Relations

A mixture model has not been adequately characterized until attention has been given not only to the nature of the species of which it is composed, but also to the proportions in which these constituents are mixed. This question is more easily discussed for some models than for others; for example, in the interstitial representation as quantified by the Oak Ridge group, cold water must have approximately 20% of its molecules in the interstitial form. This is the straightforward conclusion from the assumed geometry (in which there is one interstitial site for each two framework molecules), the parametric specifications of the framework (an O—O distance of 2.94 Å in the *a* plane and of 2.76 Å along the *c* axis), and the density of the liquid, which, together, require that about half of the interstitial sites be occupied.[784] In contrast, the analysis of the Raman bands in the valence-stretching region can, as indicated in Section 4.2, only give numbers proportional, at any temperature, to the concentration of "made" and "broken" hydrogen bonds, for which the proportionality constants are different in the two cases. This is also true of the oscillator strengths of the corresponding band components in the infrared. In these cases, calibration must be sought from some other source and it is therefore gratifying that there appears to be nothing about the expectations regarding scattering power and oscillator strengths that makes the estimate of some 20% of unbonded molecules at room temperature seem inherently implausible.

When it comes to formulating an equilibrium constant, however, and inferring a ΔH of bond breaking by use of the van't Hoff formalism $\partial(\ln K)/\partial T = \Delta H/RT^2$, additional assumptions are required, and it is not always clear on what grounds these can be justified. For the simple form of K to be valid, namely, $K = c_A/c_B$, where c_A and c_B are the concentrations either of bound and of free molecules or of formed and broken hydrogen bonds, it is necessary that the entropy of mixing of the entities referred to be ideal—i.e., that an A and a B be able to be interchanged without making any more difference than the redirecting of attention to a different, but equivalent, microscopic configuration. But this is clearly impossible when A is bonded and B is not, since bondedness means being bonded to a neighbor, and this neighbor cannot remain in an unaltered state when the exchange in question is performed. In special cases, the problem can be solved of deriving the alternative form which the equilibrium constant should take,[360,784] but for the general 0, 1, 2, 3, 4 mixture, this has not yet been accomplished.

Another difficulty besets the use of the van't Hoff equation to obtain ΔH in a system in which cooperativeness is at work. If, effectively, n bonds form at once and break at once, then the ΔH obtained can be expected* to refer to the making or breaking of n bonds. This raises the question whether some of the reported van't Hoff heats of hydrogen bond breaking are not several times too large.

5.4. Molecular Motions

The model tentatively settled upon in the preceding paragraphs must still be tested against what is known and what can be inferred about the nature of molecular motions in water, for, as indicated at the end of Section 3.6, there seems to be here the possibility of making another claim about what water must be like. That is, although many questions remain unanswered, it seems clear that a discrepancy exists among time scales for molecular reorientation as inferred from different experiments, and this appears susceptible of resolution only if the molecules do not reorient independently but, instead, by some concerted or correlated mechanism. It is thus of interest

* This is by analogy to a topological one-dimensional case, the helix-coil transition in protein denaturation, for which a complete theory exists. Zimm and Bragg[1202] define a cooperativeness parameter σ, the factor by which the formation of a new bond is favored if the immediately preceding bond has not been formed, and show that for small σ (small chance of forming a new bond if the precondition has not been met), the equilibrium percentage of "formed" bonds changes very rapidly with temperature. For this formalism, Applequist[25,26] has shown both that the van't Hoff heat of bond formation $RT^2\, \partial(\ln K)/\partial T$ is σ^{-1} times the true, or calorimetric, ΔH of bond formation, and that, on average, σ^{-1} adjacent bonds are formed or broken together, i.e., that the van't Hoff heat is the true heat for however many bonds form and break "at the same time." That the same thing is true in the three-dimensional case of hydrogen bonding in water is an attractive theorem (B. Zimm, private communication) and can hardly help being closer to the truth than the contrary assumption, namely that, regardless of cooperativity, the van't Hoff heat is the same as the calorimetric heat of forming or breaking an individual bond.

that the interstitial model seems to be compatible with this requirement also. Indeed, if the model is taken seriously, the energetic interaction between interstitial and framework molecules must be expected to produce some coupling of this nature.

As has previously been pointed out,[357,1161] some free rotational states of the H_2O molecule have only small Stark splitting constants—i.e., would be pulled only weakly into an inhomogeneous electric field and, while a molecule in an interstitial site could hardly be called a "free" rotator, it seems not impossible that it might be, on a picosecond time scale, in a state which was well enough characterized to possess some measure of this "hydrophobicity."[360] On the other hand, a sample of liquid water must be the seat of a rapidly fluctuating electromagnetic field, coupled with mechanical fluctuations (phonons), so that, whatever the instantaneous state of an interstitial molecule, it must, before long, change to one of more pronounced interaction with its framework neighbors. This would then, presumably, result in a local collapse of stability, to be followed by a new structural arrangement in which, not improbably, framework and interstitial molecules might have exchanged roles. Such a flickering of clusters would be characterized by (and shaped and guided by the requirements of) the exchange of angular momentum, the local intensity and extent of fluctuations of thermal energy, and whatever combination of configurations and motions sufficed, at any given time and place, to constitute a (conditionally) stable "state." In any case, assuming, as this picture does, the rather high degree of cooperativity in bonding discussed in Section 3.3, neither the occasions upon which, nor the angles through which, neighboring molecules altered their orientations in space could be independent of each other, and the difficulty which, in current analyses of relaxation times, results from the assumptions of small and independent rotational jumps is seen as an expected outcome of what the interstitial representation would give reason to expect.

5.5. Heat Capacity

As already noted, if a model such as that just sketched is to reach the point of being regarded as a serious possibility, there must be the implied assumption that it will be able to provide explanations not only for the phenomena on which its formulation was based, but for the rest of the outstanding water puzzles as well. Even though it is still too soon to expect, for example, an accounting in detail for the heat capacity behavior of water, one must believe that the direction, at least, can be specified in which an explanation is likely to lie.

The heat capacity in particular needs to be considered, since in an earlier thermodynamic discussion of an interstitial model,[360] it was found that this was the property which was least easily accounted for, and heavy reliance was placed upon the "state III" of water which was pictured as coming between (either geometrically or temporally) interstitial clusters, to allow for mismatches between the edges of the latter, and to furnish extra heat capacity contributions. In the meantime, an ingenious discussion of "configurational" heat capacity has been given by Eisenberg and Kauzmann[301] as arising, so to speak, from an upward movement with rising temperature, of the potential energy "floor" from which translational and vibrational thermal energies are measured. This is of special interest in the present context since what has been said in Sections 3.4 and 5.2 about the disappearance with rising temperature of centers of C_{2V} symmetry prepares us to expect that the straining of angles when one of these is converted to a center of lower symmetry must be accompanied by an increase in potential energy of just the sort Eisenberg and Kauzmann envisage. An alternative suggestion is that the centers of different symmetry be identified, as Kamb has done,[569] with structural patterns similar to those known to exist in high-pressure ices[569] which, because they are not greatly different in energy from ice I, can be expected to coexist to some extent in cold water. In that case, a rise in temperature should shift this equilibrium with an accompanying increase in potential energy of the sample, thus, again, producing the necessary heat capacity contribution. There seem, however, to be two important conditions with which such configurational heat capacity notions must be required to comply. (a) Such a shift in potential energy must be of an "equilibrium" potential energy about which librational oscillations still take place, so as to continue to make their VPIE contribution. (b) There must still also be some semantic equivalent of interstitial character in the model, so that both requirements can still be satisfied simultaneously, of librational oscillations and the broken bonds which are needed to account for the unbonded spectra. This also means that an appropriate heat capacity contribution must also be expected, arising from the transformation, with rising temperature, of bonded to nonbonded species. This, it may be noted, is a return to the earliest known published speculation on this subject.[924,1170]

References

1. A. Abragam, "The Principles of Nuclear Magnetism," Oxford University Press, London (1961).
2. N. A. Agaev and A. D. Yusibova, *Dokl. Akad. Nauk SSSR* **180**, 334 (1968).
3. N. A. Agaev and A. D. Yusibova, *Soviet Phys.—Doklady* **13**, 472 (1968).
4. G. C. Akerlof and H. J. Oshry, *J. Am. Chem. Soc.* **72**, 2844 (1950).
5. W. J. Albery and M. H. Davies, *Trans. Faraday Soc.* **65**, 1059 (1969).
6. P. S. Albright and L. J. Gosting, *J. Am. Chem. Soc.* **68**, 1061 (1946).
7. B. J. Alder and W. G. Hoover, *in* "Physics of Simple Liquids," Chapter 4, North-Holland Publishing Co., Amsterdam (1968).
8. I. V. Alexandrov, N. N. Korst, and N. D. Sokolov, *Opt. i Spectroskopiya* **8**, 575 (1960).

8a. J. D. Allen and H. Shull, *Chem. Phys. Letters* **9**, 339 (1971).

9. L. C. Allen, *Ann. Rev. Phys. Chem.* **20**, 315 (1969).
10. L. C. Allen and P. A. Kollman, *Science* **169**, 1443 (1970).
11. E.-H. Amagat, *Ann. Chim.* **29**, 543 (1893).
12. E. Amaldi, *Phys. Z.* **32**, 914 (1931).
13. Kh. I. Amirkhanov and A. M. Kerimov, *Teploenergetika* **1963**(8), 64.
14. Kh. I. Amirkhanov and A. M. Kerimov, *Teploenergetika* **1963**(9), 61.
15. I. S. Andrianova, O. Ya. Samoilov, and I. Z. Fisher, *J. Struct. Chem.* **8**, 736 (1967).
16. I. S. Andrianova, O. Ya. Samoilov, and I. Z. Fisher, *Zh. Strukt. Khim.* **8**, 813 (1967).
17. C. A. Angell and E. J. Sare, *Science* **168**, 280 (1970).
18. S. Angus and D. M. Newitt, *Phil. Trans. Roy. Soc.* **A259**, 107 (1966).
19. Anonymous, *Chem. Eng. News* **47**, 39 (1969).
20. Anonymous, *Nature* **224**, 937 (1969).
21. Anonymous, *New Scientist* **43**, 55 (1969).
22. Anonymous, *Chem. Eng. News* **48**, 48 (1970).
23. Anonymous, *New Scientist*, p. 599, 26 March 1970.
24. Anonymous, *The Times* (*London*), 14 July 1970.
25. J. Applequist, *J. Chem. Phys.* **38**, 934 (1963).
26. J. Applequist, *J. Chem. Phys.* **49**, 3327 (1968).
27. R. Aris, "Vectors, Tensors, and the Basic Equations of Fluid Mechanics," Prentice-Hall, Englewood Cliffs, N.J. (1962).
28. G. P. Arrighini and C. Guidotti, *Chem. Phys. Letters* **6**, 435 (1970).
29. G. P. Arrighini, C. Guidotti, and O. Salvetti, *J. Chem. Phys.* **52**, 1037 (1970).
30. G. P. Arrighini, M. Maestro, and R. Moccia, *Chem. Phys. Letters* **1**, 242 (1967).
31. G. P. Arrighini, M. Maestro, and R. Moccia, *J. Chem. Phys.* **49**, 882 (1968).

32. G. P. Arrighini, M. Maestro, and R. Moccia, *Chem. Phys. Letters* **7**, 351 (1970).
33. G. P. Arrighini, M. Maestro, and R. Moccia, *J. Chem. Phys.* **52**, 6411 (1970).
34. P. Ascarelli and G. Caglioti, *Nuovo Cimento* **XLIIIB**, 375 (1966).
36. S. Aung, R. M. Pitzer, and S. I. Chan, *J. Chem. Phys.* **49**, 2071 (1968).
37. R. P. Auty and R. H. Cole, *J. Chem. Phys.* **20**, 1309 (1952).
38. A. Azman, J. Koller, and D. Hadzi, *Chem. Phys. Letters* **5**, 157 (1970).
39. G. Bacon, "Neutron Diffraction," Clarendon Press, Oxford (1962).
40. R. F. W. Bader, *J. Am. Chem. Soc.* **86**, 5070 (1964).
40a. R. F. W. Bader and R. A. Gangi, *Chem. Phys. Letters* **6**, 312 (1970).
40b. R. F. W. Bader and R. A. Gangi, *J. Am. Chem. Soc.* **93**, 1831 (1971).
41. R. F. W. Bader and G. A. Jones, *Can. J. Chem.* **41**, 586 (1963).
42. W. A. Bagdade, doctoral dissertation, University of California, Berkeley, 1966.
43. B. J. Bailey, *J. Phys.* **D3**, 550 (1970).
44. R. W. Bain, *J. Mech. Eng. Sci.* **3**, 289 (1961).
45. R. W. Bain, "National Engineering Laboratory Steam Tables 1964," Her Majesty's Stationery Office, Edinburgh (1964).
46. G. Bancroft, *The Morning Call* (Allentown, Pa.), 22 June 1970.
47. K. E. Banyard and N. H. March, *Acta Cryst.* **9**, 385 (1956).
48. K. E. Banyard and N. H. March, *J. Chem. Phys.* **26**, 1416 (1957).
49. C. R. Barber, *Nature* **222**, 929 (1969).
50. J. A. Barker, *Proc. Roy. Soc.* **A259**, 442 (1961).
51. J. A. Barker and R. O. Watts, *Chem. Phys. Letters* **3**, 144 (1969).
52. A. J. Barlow and E. Yazgan, *Brit. J. Appl. Phys.* **17**, 807 (1966).
53. J. Barna, *Rheol. Acta* **5**, 46 (1966).
54. D. E. Barnaal and I. J. Lowe, *J. Chem. Phys.* **48**, 4614 (1968).
55. W. H. Barnes, *Proc. Roy. Soc.* **A125**, 670 (1929).
56. A. K. Barua and A. Das Gupta, *Trans. Faraday Soc.* **59**, 2243 (1963).
57. J. G. Bayly, V. B. Kartha, and W. H. Stevens, *Infrared Physics* **3**, 211 (1963).
58. G. Bäz, *Phys. Z.* **40**, 394 (1939).
59. R. E. Bedford and G. M. Kirby, *Metrologia* **5**, 83 (1969).
60. C. L. Bell and G. M. Barrow, *J. Chem. Phys.* **31**, 300 (1959).
60a. S. Bell, *J. Mol. Spect.* **16**, 205 (1965).
61. L. J. Bellamy, A. R. Osborn, E. R. Lippincott, and A. R. Bandy, *Chem. Ind.* (*London*) **1969**, 686.
62. G. B. Benedek and E. M. Purcell, *J. Chem. Phys.* **22**, 2003 (1954).
63. W. S. Benedict, N. Gailar, and E. K. Plyler, *J. Chem. Phys.* **24**, 1139 (1956).
63a. W. S. Benedict, S. A. Clough, L. Frenkel, and T. E. Sullivan, *J. Chem. Phys.* **53**, 2565 (1970).
64. C. Bénézech, "L'eau, base structurelle et fonctionelle des êtres vivants," Masson & Co., Paris (1962).
65. A. Ben-Naim, *J. Chem. Phys.* **50**, 404 (1969).
66. A. Ben-Naim, *J. Chem. Phys.* **52**, 5531 (1970).
66a. A. Ben-Naim, *J. Chem. Phys.* **54**, 3682 (1971).
67. A. Ben-Naim and F. H. Stillinger, *in* "Structure and Transport Processes in Water and Aqueous Solutions" (R. A. Horne, ed.), John Wiley and Sons, New York (1971).
68. J. D. Bernal, *Sci. Am.* **203**, 124 (1960).
69. J. D. Bernal, *Proc. Roy. Soc.* (*London*) **A280**, 299 (1964).

70. J. D. Bernal, P. Barnes, I. A. Cherry, and J. L. Finney, *Nature* **224**, 393 (1969).
71. J. D. Bernal and R. H. Fowler, *J. Chem. Phys.* **1**, 515 (1933).
72. U. Bernini, F. Fittipaldi, and E. Ragozzino, *Nature* **224**, 910 (1969).
73. H. J. Bernstein, *Raman Newsletter* **1968** (November), No. 1.
74. R. Bersohn, *J. Chem. Phys.* **32**, 85 (1960).
75. J. E. Bertie, L. D. Calvert, and E. Whalley, *J. Chem. Phys.* **38**, 840 (1963).
76. J. E. Bertie, H. J. Labbé, and E. Whalley, *J. Chem. Phys.* **50**, 4501 (1969).
77. J. E. Bertie and E. Whalley, *J. Chem. Phys.* **40**, 1637 (1964).
78. J. E. Bertie and E. Whalley, *J. Chem. Phys.* **40**, 1646 (1964).
79. J. E. Bertie and E. Whalley, *J. Chem. Phys.* **46**, 1271 (1967).
80. K. E. Bett and J. B. Cappi, *Nature* **207**, 620 (1965).
81. J. Bigeleisen, *J. Chem. Phys.* **23**, 2264 (1955).
82. J. Bigeleisen, *J. Chem. Phys.* **34**, 1485 (1961).
83. P. H. Bigg, *Brit. J. Appl. Phys.* **18**, 521 (1967).
84. E. C. Bingham and R. F. Jackson, *Bull. Bur. Stand.* **14**, 59 (1918).
85. J. B. Birks, *Proc. Phys. Soc.* (*London*) **60**, 282 (1948).
86. D. M. Bishop and A. A. Wu, *J. Chem. Phys.* **54**, 2917 (1971).
86a. D. M. Bishop and A. A. Wu, *Theoret. Chim. Acta* **21**, 287 (1971).
87. N. Bjerrum, *Kgl. Danske Videnskab. Selskab, Skr.* **27**, 1 (1951).
88. N. Bjerrum, *Kgl. Danske Videnskab. Selskab, Skr.* **27**, 41 (1951).
89. M. Blackman and N. D. Lisgarten, *Proc. Roy. Soc.* **A239**, 93 (1957).
90. G. Blank, *Wärme-Stoffübertragung* **2**, 53 (1969).
91. L. A. Blatz and P. Waldstein, *J. Phys. Chem.* **72**, 2614 (1968).
92. H. Blicks, O. Dengel, and N. Riehl, *Phys. Kondens. Materie* **4**, 375 (1966).
93. R. Blinc, G. Lahajnar, I. Zupanic, and H. Gränicher, *Chem. Phys. Letters* **4**, 363 (1969).
94. N. Bloembergen, *Am. J. Phys.* **35**, 989 (1967).
95. R. W. Blue, *J. Chem. Phys.* **22**, 280 (1954).
96. L. Blum, *J. Comput. Phys.* **7**, 592 (1971).
97. H. Bluyssen, A. Dynamus, J. Reuss, and J. Verhoeven, *Phys. Letters* **25A**, 584 (1967).
98. W. Bol, *J. Appl. Cryst.* **1**, 234 (1968).
99. R. W. Bolander, J. L. Kassner, and J. T. Zung, *J. Chem. Phys.* **50**, 4402 (1969).
100. N. F. Bonarenko, *Russ. J. Phys. Chem.* **42**, 116 (1968).
100a. L. G. Bonner, *Phys. Rev.* **46**, 458 (1934).
101. O. D. Bonner and G. B. Woolsey, *J. Phys. Chem.* **72**, 899 (1968).
102. F. Booth, *J. Chem. Phys.* **19**, 391, 1327, 1615 (1951).
103. M. Born and R. Oppenheimer, *Ann. Physik* **84**, 457 (1927).
104. A. A. Bothner-By, *J. Mol. Spectry.* **5**, 52 (1960).
105. C. J. F. Böttcher, "Theory of Electric Polarization," Elsevier, New York (1952).
105a. M. J. T. Bowers and R. M. Pitzer (to be published).
106. R. S. Bradley, *Trans. Faraday Soc.* **53**, 687 (1957).
107. G. W. Brady and W. J. Romanow, *J. Chem. Phys.* **32**, 306 (1960).
108. T. J. S. Brain, *Int. J. Heat Mass Transfer* **10**, 737 (1967).
109. T. J. S. Brain, The Thermal Conductivity of Steam at Atmospheric Pressure, University of Glasgow, Mechanical Engineering Department, Technical Report No. 26 (1968).
110. T. J. S. Brain, *J. Mech. Eng. Sci.* **11**, 392 (1969).

111. J. Braundmüller and H. Moser, "Einführung in die Ramanspektroskopie," Dr. Dietrich Steinkopf Verlag, Darmstadt, (1962), p. 387.
112. S. Bratoz, *Adv. Quantum Chem.* **3**, 209 (1967).
113. O. C. Bridgeman and E. W. Aldrich, A Note on Vapor-Pressure Equations for Water, American Society of Mechanical Engineers, Paper 63-HT-33 (1963).
114. O. C. Bridgeman and E. W. Aldrich, *J. Heat Transfer* **86C**, 279 (1964).
115. O. C. Bridgeman and E. W. Aldrich, *J. Heat Transfer* **87**, 266 (1965).
116. P. W. Bridgman, *Proc. Am. Acad. Arts Sci.* **47**, 441 (1912).
117. P. W. Bridgman, *Proc. Am. Acad. Arts Sci.* **48**, 309 (1912).
118. P. W. Bridgman, *Proc. Am. Acad. Arts Sci.* **59**, 141 (1923).
119. P. W. Bridgman, *Proc. Am. Acad. Arts Sci.* **61**, 57 (1926).
120. P. W. Bridgman, *J. Chem. Phys.* **3**, 597 (1935).
121. P. W. Bridgman, *J. Chem. Phys.* **5**, 964 (1937).
122. P. W. Bridgman, *Proc. Am. Acad. Arts Sci.* **74**, 399 (1942).
123. P. W. Bridgman, "The Physics of High Pressure," G. Bell and Sons, London (1949).
124. L. J. Briggs, *J. Appl. Phys.* **21**, 721 (1950).
125. R. Brill and A. Tippe, *Acta Cryst.* **23**, 343 (1967).
126. A. A. Brish, M. S. Tarasov, and V. A. Tsukerman, *Zh. Eksperim. i Teor. Fiz.* **38**, 22 (1960); English transl.: *Soviet Phys.—JETP* **11**, 15 (1960).
127. B. Brockamp and H. Querfurth, *Z. Polarforsch.* **5**, 253 (1964).
128. B. N. Brockhouse, *Nuovo Cimento* **9** (Suppl.) 45 (1958).
129. B. N. Brockhouse, *Phys. Rev. Letters* **2**, 287 (1959).
130. A. J. Brown and E. Whalley, *J. Chem. Phys.* **45**, 4360 (1966).
131. E. A. Bruges, Transport Properties of Water and Steam—Thermal Conductivity, Technical Report 9(B), Mechanical Engineering Department, University of Glasgow (1963).
132. E. A. Bruges, *Engineer* **224**, 411 (1967).
133. E. A. Bruges and M. R. Gibson, The Dynamic Viscosity of Compressed Water to 10 kbar and Steam to 1500°C, Mechanical Engineering Department, University of Glasgow, TR 25 (1968).
134. E. A. Bruges and M. R. Gibson, *J. Mech. Eng. Sci.* **11**, 189 (1969).
135. E. A. Bruges and M. R. Gibson, *J. Mech. Eng. Sci.* **12**, 232 (1970).
136. E. A. Bruges, B. Latto, and A. K. Ray, *J. Heat Mass Transfer* **9**, 465 (1966).
137. R. M. Brugger, *in* "Thermal Neutron Scattering," (P. A. Egelstaff, ed.), Academic Press, London (1965), p. 53.
138. R. M. Brugger, *Nucl. Sci. Eng.* **33**, 187 (1968).
139. M. Bruma, *Compt. Rend.* **232**, 42, 219 (1951).
139a. C. R. Brundle and D. W. Turner, *Proc. Roy. Soc.* (*London*) **A307**, 27 (1968).
140. D. G. Bruscato and V. A. Genusa, *J. Chem. Phys.* **49**, 4462 (1968).
141. J. B. Bryan, doctoral dissertation, Kansas State University, Manhattan, Kansas, 1969.
142. T. J. Buchanan, *Proc. IEE* **99**, 61 (1952).
143. T. J. Buchanan and E. H. Grant, *Brit. J. Appl. Phys.* **6**, 64 (1955).
144. A. D. Buckingham and J. A. Pople, *Trans. Faraday Soc.* **51**, 1173 (1955).
145. A. D. Buckingham, T. Shaefer, and W. G. Schneider, *J. Chem. Phys.* **32**, 1227 (1960).
146. K. Buijs and G. R. Choppin, *J. Chem. Phys.* **39**, 2035 (1963).
147. K. Buijs and G. R. Choppin, *J. Chem. Phys.* **40**, 3120 (1964).

148. H. B. Bull and K. Breese, *Arch. Biochem. Biophys.* **137**, 299 (1970).
149. B. Bullemer, I. Eisele, H. Engelhardt, N. Riehl, and P. Seige, *Solid State Commun.* **6**, 663 (1968).
150. B. Bullemer and N. Riehl, *Phys. kondens. Materie* **7**, 248 (1968).
151. J. O. Burgman, J. Sciensinski, and K. Skold, *Phys. Rev.* **170**, 808.
152. L. Burnelle and C. A. Coulson, *Trans. Faraday Soc.* **53**, 403 (1957).
153. C. W. Burnham, J. R. Holloway, and N. F. Davis, Thermodynamic Properties of Water to 1000°C and 10,000 bars, Department of Geochemistry and Mineralogy, College of Earth and Mineral Sciences, Pennsylvania State University, University Park, Pennsylvania (1968).
154. C. W. Burnham, J. R. Holloway, and N. F. Davis, *Am. J. Sci.* **267A**, 70 (1969).
155. C. W. Burnham, J. R. Holloway, and N. F. Davis, Thermodynamic Properties of Water to 1000°C and 10,000 bars, The Geological Society of America, Special Paper No. 132 (1969).
156. E. F. Burton and W. F. Oliver, *Proc. Roy. Soc.* **A153**, 166 (1935).
157. T. R. Butkovich and J. K. Landauer, U.G.G.I., Assn. Internat. d'Hydrologie Scientifique, Symp. de Chamonix, publication No. 47, 318 (1958).
158. J. A. V. Butler, *Trans. Faraday Soc.* **33**, 229 (1937).
159. J. A. V. Butler, C. N. Rachmandi, and D. W. Thomson, *J. Chem. Soc.* **1935**, 28.
160. R. Camille, Dissertation, University of London, 1970.
161. E. S. Campbell, G. Gelernter, H. Heinen, and V. R. G. Moorti, *J. Chem. Phys.* **46**, 2690 (1967) and the references therein.
162. C. G. Cannon, *Spectrochim. Acta* **10**, 341 (1958).
163. J. Cappi, The Viscosity of Water at High Pressure (unpublished Doctoral Dissertation), The University of London, 1964.
164. A. Carnvale, P. Bowen, M. Basileo, and J. Sprenke, *J. Acoust. Soc. Am.* **44**, 1098 (1968).
165. H. Y. Carr and E. M. Purcell, *Phys. Rev.* **94**, 630 (1954).
166. S. Carrà and C. Zanderrighi, *Chem. Ind.* (*Milan*) **52**, 249 (1970).
167. J. Carrington, *J. Chem. Phys.* **41**, 2012 (1964).
168. C. H. Cartwright, *Nature* **135**, 872 (1935), **136**, 181 (1935).
169. C. H. Cartwright and I. Errera, *Phys. Rev.* **49**, 470 (1936).
170. C. H. Cartwright and I. Errera, *Proc. Roy. Soc.* **A154**, 138 (1936).
171. J. A. Caskey and W. B. Barlage, Jr., *Ind. Eng. Chem. Fundamentals* **9**, 495 (1970).
172. F. Cernushi and H. Eyring, *J. Chem. Phys.* **7**, 547 (1939).
173. H. M. Chadwell, *Chem. Revs.* **4**, 375 (1927).
174. A. R. Challoner and R. W. Powell, *Proc. Roy. Soc.* (*London*) **A238**, 90 (1956).
175. J. E. Chamberlain, G. W. Chantry, H. A. Gebbie, N. W. B. Stone, T. B. Taylor, and G. Wyllie, *Nature* **210**, 790 (1966).
176. J. E. Chamberlain, M. S. Zafar, and J. B. Hasted (in course of publication).
177. M.-Y. Chan, L. Wilardjo, and P. M. Parker, *J. Mol. Spectry.* **40**, 473 (1971).
178. R. K. Chan, D. W. Davidson, and E. Whalley, *J. Chem. Phys.* **43**, 2376 (1965).
179. T. Y. Chang and M. Karplus (to be published).
180. P. Chappuis, *Trav. Mém. Bur. Intern. Poids et Mesures* **13**, D1 (1907).
181. S. K. Chetterjee, *J. Indian Inst. Sci.* **34**, 43 (1952).
182. R. Chidenbaram, *Acta Cryst.* **14**, 467 (1961).
183. R. Chidambaram and S. K. Sikka, *Chem. Phys. Letters* **2**, 162 (1968).
184. T. F. Child, N. G. Pryce, M. J. Tait, and S. Ablett, *Chem. Commun.* **1970**, 1214.

185. N. J. Chonacky and W. H. Beeman, *Acta Cryst.* **A25**, 564 (1969).
186. H. A. Christ, *Helv. Phys. Acta* **33**, 572 (1960).
187. W. N. Christiansen, R. W. Crabtree, and T. H. Laby, *Nature* **125**, 870 (1935).
188. W. F. Claussen, *Science* **156**, 1226 (1967).
189. C. R. Claydon, G. A. Segal, and H. S. Taylor, *J. Chem. Phys.* **54**, 3799 (1971).
190. C. J. Clemett, *J. Chem. Soc.* **A1969**, 458, 761.
191. A. H. Cockett and A. Ferguson, *Phil. Mag.* **29**, 185 (1940).
192. S. J. Cocking, A.E.R.E. Report R-5867, H.M.S.O., London (1968).
193. J. R. Coe and T. B. Godfrey, *J. Appl. Phys.* **15**, 625 (1944).
194. N. V. Cohan, M. Cotti, J. V. Uribarne, and M. Weissmann, *Trans. Faraday Soc.* **58**, 490 (1962).
195. N. V. Cohan, M. Cotti, J. V. Uribarne, and M. Weissmann, *Nature* **201**, 490 (1964).
196. K. S. Cole and R. H. Cole, *J. Chem. Phys.* **9**, 341 (1941).
197. R. H. Cole, *J. Chem. Phys.* **27**, 33 (1957).
198. R. H. Cole and O. Wörz, *in* "Physics of Ice" (N. Riehl, B. Bullemer, and H. Engelhardt, eds.), Plenum Press, New York (1969).
199. M. J. Colles, G. E. Walrafen, and K. W. Wecht, *Chem. Phys. Letters* **4**, 621 (1970).
200. C. H. Collie, J. B. Hasted, and D. M. Ritson, *Proc. Phys. Soc.* (*London*) **60**, 71 (1948).
201. C. H. Collie, J. B. Hasted, and D. M. Ritson, *Proc. Phys. Soc.* (*London*) **60**, 145 (1948).
202. C. H. Collie, D. M. Ritson, and J. B. Hasted, *Trans. Faraday Soc.* **42A**, 129 (1946).
203. S. C. Collins and F. G. Keyes, *Proc. Am. Acad. Arts Sci.* **72**, 283 (1938).
204. Comité International des Poids et Mesures, *Metrologia* **5**, 35 (1969).
205. Compton Scattering Factors for Spherically Symmetric Free Atoms, LA-3689 (1967).
206. H. Conroy, *J. Chem. Phys.* **47**, 5307 (1967).
207. H. F. Cook, *Brit. J. Appl. Phys.* **3**, 249 (1952).
208. J. M. Costello and S. T. Bowden, *Research* **10**, 329 (1957).
209. C. A. Coulson, *Research* (*London*) **10**, 149 (1957).
210. C. A. Coulson, *Spectrochim. Acta* **14**, 161 (1959).
211. C. A. Coulson, "Valence," Oxford University Press (1961).
212. C. A. Coulson and V. Danielsson, *Ark. Fys.* **8**, 239, 245 (1954).
213. C. A. Coulson and B. M. Deb, *Int. J. Quantum Chem.* **5**, 411 (1971).
214. C. A. Coulson and D. Eisenberg, *Proc. Roy. Soc.* **A291**, 445, 454 (1966).
215. J. Cox, *Trans. Roy. Soc. Canada*, **III** 3 (1904).
216. R. A. Cox, M. J. McCartney, and F. Culkin, *Deep-Sea Res.* **15**, 319 (1968).
217. H. Craig, *Science* **133**, 1833 (1961).
218. B. Crawford and R. E. Frech, doctoral dissertation of the latter, University of Minnesota, 1968.
219. D. Cribier and B. Jacrot, "Inelastic Scattering of Neutrons in Solids and Liquids, I.A.E.A., Vienna (1961).
220. F. G. Cripwell and G. B. Sutherland, *Trans. Faraday Soc.* **42**, 149 (1946).
221. D. T. Cromer and J. B. Mann, X-Ray Scattering Factors Computed from Numerical Hartree–Fock Wave Functions, LA-3816 (1968).
222. R. B. Cuddeback, R. C. Koeller, and H. G. Drickamer, *J. Chem. Phys.* **21**, 589 (1953).

223. C. F. Curtiss and J. O. Hirschfelder, *J. Chem. Phys.* **10**, 491 (1942).
223a. S. J. Cyvin, "Molecular Vibrations and Mean Square Amplitudes," Elsevier, Amsterdam (1968), p. 205.
224. S. J. Cyvin, J. E. Rauch, and J. C. Decius, *J. Chem. Phys.* **43**, 4083 (1965).
225. T. W. Dakin and C. N. Works, *J. Appl. Phys.* **18**, 789 (1947).
226. M. D. Danford and H. A. Levy, *J. Am. Chem. Soc.* **84**, 3965 (1962).
227. G. Dantl, *Z. Phys.* **166**, 115 (1962).
228. G. Dantl, *Phys. kondens. Materie* **7**, 390 (1968).
229. G. Dantl, *in* "Physics of Ice" (N. Riehl, B. Bullemer, and H. Engelhardt, eds.), Plenum Press, New York (1969).
230. G. Dantl and I. Gregora, *Naturwiss.* **55**, 176 (1968).
231. R. L. d'Arcy and J. D. Leeder, *Trans. Faraday Soc.* **66**, 1236 (1970).
232. B. T. Darling and D. M. Dennison, *Phys. Rev.* **57**, 128 (1940).
233. G. Das and A. C. Wahl, *Phys. Rev. Letters* **24**, 440 (1970).
234. T. P. Das and T. Ghose, *J. Chem. Phys.* **31**, 42 (1959).
235. N. Dass, *J. Geophys. Res.* **74**, 5557 (1969).
236. H. G. David and S. D. Hamann, *Trans. Faraday Soc.* **55**, 72 (1959).
237. H. G. David and S. D. Hamann, *Trans. Faraday Soc.* **56**, 1043 (1960).
238. C. M. Davis and J. Jarzynski, *Adv. Mol. Relax. Processes* **1**, 155 (1967–68).
239. C. M. Davis and J. Jarzynski, Measurement of the Pressure and Temperature Dependence of the Volume Viscosity in Water, Final Report of Office of Saline Water, Grant No. 14-01-0001-684 (1968).
240. C. M. Davis and T. A. Litovitz, *J. Chem. Phys.* **42**, 2563 (1965).
241. J. C. Davis, K. S. Pitzer, and C. N. R. Rao, *J. Phys. Chem.* **64**, 1744 (1960).
242. R. T. Davis, Jr., and R. W. Schiessler, *J. Phys. Chem.* **57**, 966 (1953).
243. R. A. Dawe and E. B. Smith, *Science* **163**, 675 (1969).
244. P. Debye, *Ann. Physik* **46**, 809 (1915).
245. P. Debye, "Polar Molecules," Chemical Catalog, New York, 1929.
246. P. Debye and M. Mencke, *Phys. Z.* **31**, 797 (1930).
247. W. J. de Haas, *Procès-Verbaux Comité International Poids et Mesures* **22**, 85 (1950).
248. R. E. Dehl and C. A. Hoeve, *J. Chem. Phys.* **50**, 3245 (1969).
249. P. Dehyde, "Polare Molekeln," S. Hirzel, Leipzig (1929); English Transl., Dover, New York, Chapter V.
250. J. Del Bene and J. A. Pople, *Chem. Phys. Letters* **4**, 426 (1969).
251. J. Del Bene and J. A. Pople, *J. Chem. Phys.* **52**, 4858 (1970).
252. V. A. Del Grosso, *J. Acoust. Soc. Am.* **47**, 947 (1970).
253. P. Delibaltas, O. Dengel, D. Helmreich, N. Riehl, and H. Simon, *Phys. kondens. Materie* **5**, 166 (1966).
253a. F. C. De Lucia, R. L. Cook, P. Helminger, and W. Gordy, *J. Chem. Phys.* **55**, 5334 (1971).
254. O. Dengel, E. Jacobs, and N. Riehl, *Phys. kondens. Materie* **5**, 58 (1966).
255. M. De Paz, S. Ehrenson, and L. Friedman, *J. Chem. Phys.* **52**, 3362 (1970).
256. B. V. Deryagin, *Disc. Faraday Soc.* **42**, 109 (1966).
257. B. V. Deryagin and N. V. Churaev, *Priroda* (*Moskva*) **1968**(4), 16.
258. A. Deubner, R. Reise, and K. Wenzel, *Naturwiss.* **47**, 600 (1960).
259. L. Devell, *Acta Chem. Scand.* **16**, 2177 (1962).
260. S. S. Dharmatti, K. J. Sundara Rao, and R. Vijayaraghavan, *Nuovo Cimento* **11**, 656 (1959).

261. M. Diaz-Peña and M. L. McGlashan, *Trans. Faraday Soc.* **55**, 2018 (1959).
262. W. C. Dickenson, *Phys. Rev.* **80**, 563 (1950).
263. G. Diercksen, *Chem. Phys. Letters* **4**, 373 (1969); *Theoret. Chim. Acta* **21**, 335 (1971).
264. P. Dirac, *Proc. Roy. Soc.* **A112**, 661 (1926).
265. J. Dirian, "Constantes Physiques des Variétés Isotopiques de l'Eau (^{17}O–^{18}O–D–T)," Commissariat à l'Energie Atomique, France, Sér. Bibliographies No. 15 (1962).
266. F. J. Donahoe, *Nature* **224**, 198 (1969).
267. J. Donohue, *in* "Structural Chemistry and Molecular Biology" (A. Rich and N. Davidson, eds.), Freeman & Co., San Francisco (1968).
268. J. Donohue, *Science* **166**, 1000 (1969).
269. N. E. Dorsey, "Properties of Ordinary Water Substance," Reinhold Publishing Corporation, New York (1940); reprinted, Hafner Publishing Co., New York (1968).
270. D. C. Douglass and D. W. McCall, *J. Chem. Phys.* **31**, 569 (1959).
271. J. G. Dowell and A. P. Rinfret, *Nature* **188**, 1144 (1960).
272. D. A. Draegert, doctoral dissertation, Kansas State University, Manhattan, Kansas, 1967.
273. D. A. Draegert, N. W. B. Stone, B. Curnutte, and D. Williams, *J. Opt. Soc. Am.* **56**, 64 (1966).
274. D. A. Draegert and D. Williams, *J. Chem. Phys.* **48**, 401 (1968).
275. F. H. Drake, G. W. Pierce, and M. T. Dow, *Phys. Rev.* **35**, 613 (1930).
276. M. Dreyfus and A. Pullman, *Theor. Chim. Acta* **17**, 109 (1970).
277. W. Drost-Hansen, *Ann. N.Y. Acad. Sci.* **125**, 471 (1965).
278. W. Drost-Hansen, *Ind. Eng. Chem.* **61**, 10 (1969).
279. K. H. Dudziak and E. U. Franck, *Ber. Bunsenges.* **70**, 1120 (1966).
280. A. B. F. Duncan and J. A. Pople, *Trans. Faraday Soc.* **49**, 217 (1953).
281. J. D. Dunitz, *Nature* **197**, 860 (1963).
281a. T. H. Dunning, Jr., *J. Chem. Phys.* **53**, 2823 (1970).
281b. T. H. Dunning, Jr., *J. Chem. Phys.* **55**, 716 (1971).
281c. T. H. Dunning, Jr., *J. Chem. Phys.* **55**, 3958 (1971).
282. R. Dunsmuir and J. G. Powles, *Phil. Mag.* **37**, 747 (1946).
283. H. Eck, *Phys. Z.* **40**, 3 (1939).
284. C. Eckart, *Am. J. Sci.* **256**, 225 (1958).
285. G. Eckhardt, R. W. Hellwarth, F. J. McClung, S. E. Schwarz, D. Weiner, and E. J. Woodbury, *Phys. Rev. Letters* **9**, 455 (1962).
286. C. Edmiston and K. Ruedenberg, *Rev. Mod. Phys.* **35**, 457 (1963).
287. P. A. Egelstaff, *in* "Inelastic Scattering of Neutrons in Solids and Liquids," I.A.E.A., Vienna (1961), p. 25.
288. P. A. Egelstaff, "An Introduction to the Liquid State," Academic Press, London (1967).
289. P. A. Egelstaff and D. H. C. Harris, *Disc. Faraday Soc.* **49**, 193 (1970).
290. P. A. Egelstaff, B. C. Haywood, and I. M. Thorson, *in* "Inelastic Scattering of Neutrons in Solids and Liquids, I.A.E.A., Vienna (1963), Vol. 1, p. 356.
291. P. A. Egelstaff and J. W. Ring, *in* "Physics of Simple Liquids," (H. N. V. Temperley, J. S. Rowlinson, and G. S. Rushbrook, eds.), North-Holland Publishing Company, Amsterdam (1968), p. 253.
292. P. A. Egelstaff and P. Schofield, *Nucl. Sci. Eng.* **12**, 260 (1962).

293. M. Eigen, *Z. Elektrochem.* **56**, 176 (1952).
294. M. Eigen, *Angew. Chem.* **75**, 489 (1963).
295. M. Eigen, *Angew. Chem.* (*Int. Ed.*) **3**, 1 (1964).
296. M. Eigen and L. De Maeyer, *Proc. Roy. Soc.* (*London*) **A247**, 505 (1958).
297. M. Eigen and L. De Maeyer, *in* "The Structure of Electrolytic Solutions (W. J. Hamer, ed.), John Wiley and Sons, New York (1959), pp. 64–85.
298. M. Eigen, L. de Mayer, and H. C. Spatz, *Ber. Bunsenges. Phys. Chem.* **68**, 19 (1964).
299. A. Einstein, *Ann. Phys.* **22**, 180 (1907).
300. D. Eisenberg and C. A. Coulson, *Nature* **199**, 368 (1963).
301. D. Eisenberg and W. Kauzmann, "The Structure and Properties of Water," Oxford University Press (1969).
302. M. Eisenstadt and A. G. Redfield, *Phys. Rev.* **132**, 635 (1963).
303. Electrical Research Association, "1967 Steam Tables," Edward Arnold, London (1967).
304. D. D. Eley, *Trans. Faraday Soc.* **35**, 1281, 1421 (1939).
305. J. N. Elliott, "Tables of the Thermodynamic Properties of Heavy Water" (AECL-1673), Atomic Energy of Canada Limited (1963).
306. F. O. Ellison, *J. Chem. Phys.* **23**, 2358 (1955).
307. F. O. Ellison and H. Shull, *J. Chem. Phys.* **23**, 2348 (1955).
308. J. W. Emsley, J. Freeney, and L. H. Sutcliffe, "High-Resolution Nuclear Magnetic Resonance Spectroscopy," Pergamon Press, New York (1965).
309. J. E. Enderby, D. M. North, and P. A. Egelstaff, *Phil. Mag.* **14**, 961 (1966).
310. L. Endon, H. G. Hertz, B. Thul, and M. D. Zeidler, *Ber. Bunsenges-Phys. Chem.* **71**, 1008 (1967).
311. G. Engel and H. G. Hertz, *Ber. Bunsenges. Phys. Chem.*, to be published.
312. H. Engelhardt, B. Bullemer, and N. Riehl, "Physics of Ice," Plenum Press, New York (1969).
313. S. R. Erlander, *Phys. Rev. Letters* **22**, 177 (1969).
314. S. R. Erlander, *Sci. J.* **1969** (November), 80.
315. W. C. Ermler and C. W. Kern, *J. Chem. Phys.* **55**, 4851 (1971).
316. A. Eucken, *Physik. Z.* **14**, 324 (1913).
317. A. Eucken, *Nachr. Ges. Wiss. Göttingen* **1946**, 38.
318. A. Eucken, *Z. Elektrochem.* **52**, 255 (1948).
319. A. Eucken, *Z. Elektrochem.* **53**, 102 (1949).
320. D. H. Everett, J. M. Haynes, and P. J. McElroy, *Nature* **224**, 394 (1969).
321. H. E. Eyring, *J. Chem. Phys.* **4**, 283 (1936).
322. H. Eyring, T. Ree, and N. Hirai, *Proc. Natl. Acad. Sci.* (*U.S.*) **44**, 683 (1958).
323. M. Falk and T. A. Ford, *Can. J. Chem.* **44**, 1699 (1966).
324. E. Fatuzzo and P. R. Mason, *Proc. Phys. Soc.* **19**, 729 (1967).
325. E. Fatuzzo and P. R. Mason, *Proc. Phys. Soc.* **19**, 741 (1967).
326. C. E. Favelukes, A. A. Clifford, and B. Crawford, Jr., *J. Phys. Chem.* **72**, 962 (1968).
327. F. S. Feates and D. J. G. Ives, *J. Chem. Soc.* (*London*) **1956**, 2798.
328. N. N. Fedyakin, *Colloid J. USSR* **24**, 425 (1962).
329. H. Fellner-Feldegg, *J. Phys. Chem.* **73**, 616 (1969).
330. W. Fieggen, *Rec. Trav. Chim. Pays-Bas* **89**, 625 (1970).
331. C. Finbak and H. Viervoll, *Tidsskr. Kjemi, Bergvesen Met.* **3**, 36 (1943).
332. I. Z. Fisher, "Statistical Theory of Liquids," University of Chicago Press, Chicago (1964).

333. I. Z. Fisher and V. I. Adamovich, *J. Struct. Chem.* **4**, 759 (1963).
334. M. E. Fisher, *J. Math. Phys.* **5**, 944 (1964).
335. M. E. Fisher, *in* "Critical Phenomena," (M. S. Green and J. V. Sengers, eds.), U.S. National Bureau of Stardards, Special Publication 273 (1966), p. 21.
336. F. Fister and H. G. Hertz, *Ber. Bunsenges. Phys. Chem.* **71**, 1032 (1967).
337. N. H. Fletcher, "The Chemical Physics of Ice," Cambridge University Press (1970).
338. A. E. Florin and M. Alei, *J. Chem. Phys.* **47**, 5268 (1967).
339. P. Flubacher, A. J. Leadbetter, and J. A. Morrison, *J. Chem. Phys.* **33**, 1751 (1960).
340. W. H. Flygare, *J. Chem. Phys.* **41**, 793 (1964).
341. T. A. Ford and M. Falk, *Can. J. Chem.* **46**, 3579 (1968).
342. T. A. Ford and M. Falk, *J. Mol. Struct.* **3**, 445 (1969).
343. "Forms of Water in Biologic Systems," Conference Proceedings reported in *Ann. N.Y. Acad. Sci.* **125**(2) (1965).
344. E. Forslind, *Acta Polytech. Scand.* **115**, 9 (1952).
345. M. Forte and S. Menardi, *Nuovo Cimento* **XLVIB**, 7 (1966).
346. P. E. Fraley and K. N. Rao, *J. Mol. Spectry.* **29**, 348 (1969).
347. P. E. Fraley, K. N. Rao, and L. H. Jones, *J. Mol. Spectry.* **29**, 312 (1969).
348. P. F. Franchini, R. Moccia, and M. Zandomeneghi, *Intl. J. Quantum Chem.* **4**, 487 (1970).
349. E. U. Franck, *Z. Phys. Chem. (Frankfurt)* **8**, 92 (1956).
350. E. U. Franck, *Z. Phys. Chem. (Frankfurt)* **8**, 107 (1956).
351. E. U. Franck, *Z. Phys. Chem. (Frankfurt)* **8**, 192 (1956).
352. E. U. Franck, D. Hartmann, and F. Hensel, *Disc. Faraday Soc.* **39**, 200 (1965).
353. E. U. Franck and H. Lindner (in press).
354. E. U. Franck and K. Roth, *Disc. Faraday Soc.* **43**, 108 (1967).
355. H. S. Frank, *Proc. Roy. Soc. (London)* **A247**, 481 (1958).
356. H. S. Frank, "Desalination Research Conference Proceedings," Nat. Acad. Sci., Nat. Res. Council Publ. **942** (1963), p. 141.
357. H. S. Frank, *Fed. Proc.* **24**, No. 2, Part III, P. S-1 (1965).
358. H. S. Frank, *Disc. Faraday Soc.* **43**, 147 (1967).
359. H. S. Frank and M. W. Evans, *J. Chem. Phys.* **13**, 507 (1945).
360. H. S. Frank and A. S. Quist, *J. Chem. Phys.* **34**, 604 (1961).
361. H. S. Frank and W. Y. Wen, *Disc. Faraday Soc.* **24**, 133 (1957).
362. F. Franks, "Physico-Chemical Processes in Mixed Aqueous Solvents," (F. Franks, ed.), Heinemann, London (1967).
363. F. Franks, "Hydrogen-Bonded Solvent Systems," (A. K. Covington and P. Jones, eds.), Taylor and Francis, London (1968), pp. 31–48.
364. F. Franks and D. J. G. Ives, *Quart. Rev. (London)* **20**, 1 (1966).
365. F. Franks, J. Ravenhill, P. A. Egelstaff, and D. I. Page, *Proc. Roy. Soc.* **A319**, 189 (1970).
366. J. F. Frenkel, "Kinetic Theory of Liquids," Dover, New York (1943), p. 93.
367. A. S. Friedman and L. Haar, *J. Chem. Phys.* **22**, 2051 (1954).
368. L. Friedman and V. J. Shiner, *J. Chem. Phys.* **44**, 4639 (1966).
369. H. L. Frisch and J. L. Lebowitz, "The Equilibrium Theory of Classical Fluids," W. A. Benjamin, New York (1964).
370. H. Fröhlich, *Physica* **22**, 898 (1956).
371. H. Fröhlich, "Theory of Dielectrics," Oxford University Press (1949); (2nd ed., 1958).

372. M. E. Fuller and W. B. Brey, *J. Biol. Chem.* **243**, 274 (1968).
372a. R. A. Gangi and R. F. W. Bader, *J. Chem. Phys.* **55**, 5369 (1971).
373. B. B. Garrett, A. B. Denison, and S. W. Rabideau, *J. Phys. Chem.* **71**, 2606 (1967).
374. J. A. Ghormely, *J. Chem. Phys.* **25**, 599 (1956).
375. J. A. Ghormely, *J. Chem. Phys.* **48**, 503 (1968).
376. W. F. Giauque and J. W. Stout, *J. Am. Chem. Soc.* **58**, 1144 (1936).
377. M. R. Gibson, *Engineer* **224**, 417 (1967).
378. M. R. Gibson, *J. Mech. Eng. Sci.* **11**, 521 (1969).
379. M. R. Gibson and E. A. Bruges, *J. Mech. Eng. Sci.* **9**, 24 (1967).
380. M. R. Gibson and E. A. Bruges, *J. Mech. Eng. Sci.* **10**, 319 (1968).
381. P. A. Giguère, *Phys. Chem. Solids* **11**, 249 (1959).
382. D. C. Ginnings and H. F. Stimson, *in* "Experimental Thermodynamics" (J. P. McCullough and D. W. Scott, eds.), Plenum Press, New York (1968), Vol. I, p. 395.
383. G. Girard and M. Menaché, *Compt. Rend.* **265B**, 709 (1967).
384. G. Girard and M. Menaché, *Compt. Rend.* **270B**, 1513 (1970).
385. W. Gissler, *Z. Naturforsch.* **19A**, 1422 (1964).
386. G. J. Gittens, *J. Colloid Interface Sci.* **30**, 406 (1969).
387. R. M. Glaeser and C. A. Coulson, *Trans. Faraday Soc.* **61**, 389 (1965).
388. S. H. Glarum, *J. Chem. Phys.* **33**, 1371 (1960).
390. J. A. Glasel, *Proc. Natl. Acad. Sci.* (*U.S.*) **58**, 27 (1967).
391. J. A. Glasel, *J. Sci. Instr.* **2**, 963 (1968).
392. J. A. Glasel, *in* "Developments in Applied Spectroscopy," Vol. 6, Plenum Press, New York (1968), p. 241.
393. J. A. Glasel, *J. Am. Chem. Soc.* **92**, 372 (1970).
394. J. A. Glasel, *J. Am. Chem. Soc.* **92**, 375 (1970).
395. S. Glasstone, K. J. Laidler, and H. Eyring, "The Theory of Rate Processes," McGraw-Hill, New York (1941).
396. J. W. Glen, *Adv. Physics* **7**, 254 (1958).
397. R. E. Glick and K. C. Tewari, *J. Chem. Phys.* **44**, 546 (1966).
398. W. A. Goddard, III, *Phys. Rev.* **157**, 73 (1967).
399. W. A. Goddard, III, *Phys. Rev.* **157**, 81 (1967).
400. W. A. Goddard, III, *J. Chem. Phys.* **48**, 450 (1968).
401. W. A. Goddard, III, *J. Chem. Phys.* **48**, 5337 (1968).
402. A. Goel, A. S. N. Murthy, and C. N. R. Rao, *Chem. Commun.* **148**, 423 (1970).
403. A. Goel, A. S. N. Murthy, and C. N. R. Rao, *Indian J. Chem.* **9**, 56 (1971).
404. A. Goel, A. S. N. Murthy, and C. N. R. Rao, *J. Chem. Soc.* (*London*) **A1971**, 190.
405. J. A. Goff and S. Gratch, *Trans. Am. Soc. Heating Ventillating Eng.* **52**, 95 (1946).
406. E. v. Goldammer and M. D. Zeidler, *Ber. Bunsenges. Phys. Chem.* **73**, 4 (1969).
407. H. Goldstein, "Classical Mechanics," Addison-Wesley, Reading, Mass. (1950).
408. V. V. Golikov, I. Zubovska, F. L. Shapiro, A. Szkatula, and J. Janik, *in* "Inelastic Scattering of Neutrons," I.A.E.A., Vienna (1965), Vol. II, p. 210.
409. R. D. Goodwin, *J. Res. Natl. Bur. Stand.* **73A**, 585 (1969).
410. R. W. Goranson, *Am. J. Sci.* (*5th Series*) **35A**, 71 (1938).
411. A. R. Gordon, *J. Chem. Phys.* **2**, 65, 549 (1934).

412. R. G. Gordon, *J. Chem. Phys.* **42**, 3658 (1965).
413. R. G. Gordon, *J. Chem. Phys.* **43**, 1307 (1965).
414. R. G. Gordon, *Adv. Magnetic Resonance* **3**, 1 (1968).
415. A. Grandsard, *Onde Electrique* **30**, 245 (1950).
416. H. Gränicher, *Z. Krist.* **110**, 432 (1958).
417. H. Gränicher, *Phys. kondens. Materie* **1**, 1 (1963).
418. E. H. Grant, *J. Chem. Phys.* **26**, 1575 (1957).
419. E. H. Grant, *J. Phys. Chem.* **73**, 4386 (1969).
420. E. H. Grant, T. J. Buchanan, and H. F. Cook, *J. Chem. Phys.* **26**, 156 (1957).
421. E. H. Grant and R. Shack, *Brit. J. Appl. Phys.* **18**, 1807 (1967).
422. E. H. Grant and R. Shack, *Trans. Faraday Soc.* **65**, 1519 (1969).
423. F. T. Green, T. A. Milne, A. E. Vandergrift, and J. Beachey, An Experimental Study of the Structure, Thermodynamics and Kinetic Behavior of Water, U.S. Office of Saline Water, RDPR No. 493 (1969).
424. P. Gregorio and C. Merlini, *Termotechnica* **23**(5) (Suppl.), 41 (1969).
425. R. B. Griffiths, *Phys. Rev.* **158**, 176 (1967).
426. U. Grigull and J. Bach, *Brennst.-Wärme-Kraft* **18**, 73 (1966).
427. U. Grigull, F. Mayinger, and J. Bach, *Wärme-und Stoffübertragung* **1**, 15 (1968).
428. T. Grindley and J. E. Lind, Jr., *J. Chem. Phys.* **54**, 3983 (1971).
429. K. Grjotheim and J. Krogh-Moe, *Acta Chem. Scand.* **8**, 1193 (1954).
430. S. L. Guberman and W. A. Goddard, III, *J. Chem. Phys.* **53**, 1803 (1970).
431. E. A. Guggenheim, "Thermodynamics," 3rd ed., North-Holland Publishing Co., Amsterdam (1957), p. 28.
432. A. Guinier, "X-Ray Diffraction," W. H. Freeman Co., San Francisco (1963).
433. A. Guinier and G. Fournet, "Small-Angle Scattering of X-Rays," John Wiley and Sons, New York (1955).
434. S. Gulliksen, *J. Geophys. Res.* **75**, 2247 (1970).
435. Yu. V. Gurikov, *J. Struct. Chem.* **4**, 763 (1963).
436. Yu. V. Gurikov, *J. Struct. Chem.* **5**, 173 (1964).
437. Yu V. Gurikov, *J. Struct. Chem.* **6**, 786 (1965).
438. Yu. V. Gurikov, *J. Struct. Chem.* **7**, 6 (1966).
439. C. Haas, *Phys. Lett.*, **3**, 126 (1962).
440. C. Haas and D. F. Hornig, *J. Chem. Phys.* **32**, 1763 (1960).
441. P. T. Hacker, Experimental Values of the Surface Tension of Supercooled Water, National Advisory Committee for Aeronautics, Washington, NACA TN 2510 (1951).
442. D. Hadzi and H. W. Thompson (eds.), "Hydrogen Bonding," Pergamon Press, Oxford (1959).
443. G. H. Haggis, J. B. Hasted and T. J. Buchanan, *J. Chem. Phys.* **20**, 1452 (1952).
444. L. Hall, *Phys. Rev.* **73**, 775 (1948).
445. R. T. Hall and J. M. Dowling, *J. Chem. Phys.* **47**, 2454 (1967).
446. R. T. Hall and J. M. Dowling, *J. Chem. Phys.* **52**, 1161 (1970).
447. W. H. Hall, I. C. Holliday, W. A. Johnson, and S. Walker, *Trans. Faraday Soc.* **42**, 136 (1946).
448. J. Hallett, *Proc. Phys. Soc.* **82**, 1046 (1963).
449. S. D. Hamann, *J. Phys. Chem.* **67**, 2233 (1963).
450. S. D. Hamann and M. Linton, *Trans. Faraday Soc.* **62**, 2234 (1966).
451. S. D. Hamann and M. Linton, *J. Appl. Phys.* **40**, 913 (1969).

452. S. Hameed, S. S. Hui, J. I. Musher, and J. M. Schulman, *J. Chem. Phys.* **51**, 502 (1969).
453. W. C. Hamilton and J. Ibers, "Hydrogen Bonding in Solids," Benjamin, New York (1968).
454. D. Hankins, J. W. Moskowitz, and F. H. Stillinger, *Chem. Phys. Letters* **4**, 527 (1970).
455. D. Hankins, J. W. Moskowitz, and F. H. Stillinger, *Chem. Phys. Letters* **4**, 581 (1970).
456. H. J. M. Hanley and G. E. Childs, *Science* **159**, 1114 (1968).
457. H. J. M. Hanley and G. E. Childs, *in* "Thermal Conductivity," (D. R. Flynn and B. A. Peavy, eds.), U.S. National Bureau of Standards, Special Publication 302 (1968), p. 597.
458. Y. Harada and J. N. Murrell, *Mol. Phys.* **14**, 153 (1968).
459. R. C. Hardy and R. L. Cottington, *J. Chem. Phys.* **17**, 509 (1949).
460. R. C. Hardy and R. L. Cottington, *J. Res. Natl. Bur. Stand.* **42**, 573 (1949).
461. W. D. Harkins and A. E. Alexander, *in* "Physical Methods of Organic Chemistry," (A. Weissberger, ed.), 3rd. ed., Interscience, New York (1959), Vol. I, Part I, p. 757.
462. O. K. Harling, Report BNWL 436 (1967).
463. O. K. Harling, *in* "Neutron Inelastic Scattering," Vol. 1, I.A.E.A., Vienna (1968), p. 507.
464. A. Harlow, Thesis, Imperial College, London (1967).
465. F. E. Harris, *J. Chem. Phys.* **23**, 1663 (1955).
466. F. E. Harris, *Adv. Quantum Chem.* **3**, 61 (1966).
467. F. E. Harris and B. J. Alder, *J. Chem. Phys.* **21**, 1031 (1953).
468. J. F. Harrison, *J. Chem. Phys.* **47**, 2990 (1967).
469. K. A. Hartman, *J. Phys. Chem.* **70**, 270 (1966).
470. C. B. Haselgrove, *Math. Computation* **15**, 323 (1961).
471. J. B. Hasted, *Prog. Dielectric* **3**, 101 (1961).
472. J. B. Hasted and S. H. M. El-Sabeh, *Trans. Faraday Soc.* **49**, 1003 (1953).
473. L. Hausen, *Ann. Phys.* **5**, 597 (1901).
474. R. Hausser, G. Maier, and F. Noack, *Z. Naturforsch.* **21a**, 1410 (1966).
475. R. Hausser and F. Noack, *Z. Naturforsch.* **20a**, 1668 (1965).
476. J. Havlicek and L. Miskovský, *Helv. Phys. Acta* **9**, 161 (1936).
478. S. Hawley, J. Allegra, and G. Holton, *J. Acoust. Soc. Am.* **47**, 137 (1970).
479. A. T. J. Hayward, *Brit. J. Appl. Phys.* **18**, 965 (1967).
480. A. T. J. Hayward, "Compressibility Equations for Liquids—A Comprehensive Study," National Engineering Laboratory, East Kilbride, Glasgow, NEL Report No. 295 (1967).
481. A. T. J. Hayward and J. D. Isdale, *J. Phys.* **D2**, 251 (1969).
482. B. C. Haywood, *J. Nucl. Eng.* **21**, 249 (1967).
483. B. C. Haywood and D. I. Page, *in* "Neutron Thermalization and Reactor Spectra," I.A.E.A., Vienna (1968), p. 361.
484. J. Heemskerk, *Rec. Trav. Chim.* **81**, 904 (1962).
485. K. Heger, Dissertation, Karlsruhe, 1969.
486. J. R. Heiks, M. K. Barnett, L. V. Jones, and E. Orban, *J. Phys. Chem.* **58**, 488 (1954).
487. F. Heinmets and R. Blum, *Trans. Faraday Soc.* **59**, 1141 (1963).

488. D. Helmreich and B. Bullemer, *Phys. kondens. Materie* **8**, 384 (1969).
489. D. Henderson, *J. Chem. Phys.* **37**, 631 (1962).
490. R. W. Hendricks, presented at the American Chemical Society Meeting, New York City, September 1969; to be submitted to *J. Chem. Phys.*
491. R. W. Hendricks, *J. Appl. Cryst.* **3**, 348 (1970).
492. R. W. Hendricks and P. W. Schmidt, *Acta Phys. Austriaca* **26**, 97 (1967).
493. F. Hensel and E. U. Franck, *Z. Naturforsch.* **19a**, 127 (1964).
494. H. G. Hertz, *in* "Progress in Nuclear Magnetic Resonance Spectroscopy," (J. W. Emsley, J. Feeney, and L. H. Sutcliffe, eds.), Vol. 3, Pergamon Press, Oxford (1967), pp. 159–230.
495. H. G. Hertz and C. Rädle, *Z. Phys. Chem. (Frankfurt)* **68**, 324 (1969).
496. H. G. Hertz and C. Rädle, *Ber. Bunsenges.* (in press).
497. H. G. Hertz and M. D. Zeidler, *Ber. Bunsenges. Phys. Chem.* **67**, 774 (1968).
498. G. Herzberg, "Molecular Spectra and Molecular Structure. II. Infrared and Raman Spectra of Polyatomic Molecules," Van Nostrand, Princeton, New Jersey (1945).
499. G. Herzberg, "Molecular Spectra and Molecular Structure. I. Spectra of Diatomic Molecules," Van Nostrand, Princeton, New Jersey (1950).
500. G. Herzberg, "Molecular Spectra and Molecular Structure. III. Electronic Spectra and Electronic Structure of Polyatomic Molecules," Van Nostrand, Princeton, New Jersey (1966), Chapter III.
501. K. F. Herzfeld, *in* "Proceedings of the Fourth Symposium on Thermophysical Properties" (J. R. Moszynski, ed.), American Society of Mechanical Engineers, New York (1968), p. 1.
502. K. Herzfeld and T. Litovitz, "Absorption and Dispersion of Ultrasonic Waves," Academic Press, New York (1959).
503. A. Higashi, S. Koinuma, and S. Mae, *J. Appl. Phys.* **3**, 610 (1964).
504. A. Higashi, S. Koinuma, and S. Mae, *J. Appl. Phys.* **4**, 575 (1965).
505. J. H. Hildebrand and R. L. Scott, "The Solubility of Non-Electrolytes," 3rd ed., Reinhold, New York (1950).
506. N. E. Hill, *Trans. Faraday Soc.* **59**, 344 (1963).
507. N. E. Hill, *J. Phys. C. (Solid State Phys.)* (in press).
508. N. E. Hill, W. E. Vaughan, A. H. Price, and Mansel Davies, "Dielectric Properties and Molecular Behavior," van Nostrand, New York (1970).
509. T. L. Hill, "Statistical Mechanics," McGraw-Hill Book Co., New York (1956).
510. T. L. Hill, *J. Chem. Phys.* **28**, 1179 (1958).
511. G. J. Hills, P. J. Ovenden, and D. R. Whitehouse, *Disc. Faraday Soc.* **39**, 207 (1965).
512. J. C. Hindman, *J. Chem. Phys.* **44**, 4582 (1966).
513. J. C. Hindman, A. Svirmickas, and M. Wood, *J. Phys. Chem.* **72**, 4188, (1968).
514. J. C. Hindman, A. Svirmickas, and M. Wood, *J. Phys. Chem.* **74**, 1266, (1970).
515. J. O. Hirschfelder, W. Byers Brown, and S. T. Epstein, *Adv. Quantum Chem.* **1**, 255 (1964).
516. J. O. Hirschfelder, C. F. Curtiss, R. B. Bird, "Molecular Theory of Gases and Liquids," John Wiley and Sons, New York (1954, 1966).
517. J. O. Hirschfelder, F. T. McClure, and I. F. Weeks, *J. Chem. Phys.* **10**, 201 (1942).
518. R. Hoffmann, *J. Chem. Phys.* **39**, 1397 (1963).
519. C. Hollister and O. Sinanoğlu, *J. Am. Chem. Soc.* **88**, 13 (1966).

520. J. R. Holmes, D. Kivenson, and W. C. Drinkard, *J. Chem. Phys.* **37**, 150 (1962).
521. W. T. Holser and G. C. Kennedy, *Am. J. Sci.* **256**, 744 (1958).
522. W. T. Holser and G. C. Kennedy, *Am. J. Sci.* **257**, 71 (1959).
523. W. Holzapfel, *J. Chem. Phys.* **50**, 4424 (1969).
524. W. Holzapfel and H. G. Drickamer, *J. Chem. Phys.* **48**, 4798 (1968).
525. W. Holzapfel and E. U. Franck, *Ber. Bunsenges.* **70**, 1105 (1966).
526. G. Honjo and K. Shimaoka, *Acta Cryst.* **10**, 710 (1957).
527. A. Höpfner, *Angew. Chem.* (*Internat. Ed.*) **8**, 689 (1969).
528. R. A. Horne, "Marine Chemistry," Wiley–Interscience, New York (1969), p. 11.
529. R. A. Horne, R. A. Courant, D. S. Johnson, and F. F. Margosian, *J. Phys. Chem.* **69**, 3988 (1965).
530. R. A. Horne and D. S. Johnson, *J. Phys. Chem.* **70**, 2182 (1966).
531. F. Horner, T. A. Taylor, R. Dunsmuir, J. Lamb, and W. Jackson, *JIEE* **93** (III), 53 (1946).
532. D. F. Hornig, H. F. White, and F. P. Reding, *Spectrochim. Acta* **12**, 338 (1958).
533. J. A. Horsley and W. H. Fink, *J. Chem. Phys.* **50**, 750 (1969).
534. J. A. Horsley and F. Flouquet, *Chem. Phys. Letters* **5**, 165 (1970).
535. R. P. Hosteny, R. R. Gilman, T. H. Dunning, A. Pipano, and I. Shavitt, *Chem. Phys. Letters* **7**, 325 (1970).
536. B. B. Howard, B. Linden, and M. T. Emerson, *J. Chem. Phys.* **36**, 485 (1962).
537. B. J. Howard and R. E. Moss, *Mol. Phys.* **19**, 433 (1970).
538. J. R. Hoyland and L. B. Kier, *Theor. Chim. Acta* **15**, 1 (1969).
539. P. S. Hubbard, *Phys. Rev.* **131**, 275 (1963).
540. C. M. Huggins and G. C. Pimentel, *J. Phys. Chem.* **60**, 1615 (1956).
541. D. J. Hughes, H. Palevesky, W. Kley, and E. Tunkelo, *Phys. Rev.* **119**, 872 (1960).
542. F. Humbel, F. Jona, and P. Scherrer, *Helv. Phys. Acta* **26**, 17 (1953).
543. W. J. Hunt and W. A. Goddard, *Chem. Phys. Letters* **3**, 414 (1969).
544. A. C. Hurley, J. Lennard-Jones, and J. A. Pople, *Proc. Roy. Soc.* (*London*) **A220**, 446 (1953).
545. J. A. Ibers and D. P. Stevenson, *J. Chem. Phys.* **41**, 122 (1964).
546. C. K. Ingold, C. G. Rasin, C. L. Wilson, and C. R. Bailey, *J. Chem. Soc.* (*London*) **1936**, 915.
547. International Formulation Committee, The 1967 IFC Formulation for Industrial Use, IFC Secretariat, Verein Deutscher Ingenieure, Düsseldorf, Germany, 1967.
548. International Formulation Committee, The 1968 IFC Formulation for Scientific and General Use, Secretariat of the International Conference on the Properties of Steam, American Society of Mechanical Engineers, New York, 1968.
549. K. Itagaki, *J. Phys. Soc. Japan* **19**, 1081 (1964).
550. D. J. G. Ives and T. Lemon, *RIC Rev.* **1**, 62 (1968).
551. C. J. Jaccard, *Hevl. Phys. Acta* **32**, 89 (1959).
552. C. J. Jaccard, *N.Y. Acad. Ann. Sci.* **125**, 391 (1965).
553. J. D. Jackson, "Classical Electrodynamics," John Wiley and Sons, New York (1962), p. 102.
554. J. A. Jackson and S. W. Rabideau, *J. Chem. Phys.* **41**, 4008 (1964).
555. T. W. Jackson, B. Latto, C. E. Willbanks, J. W. Hodgson, and H. H. Y. Yen, *in* "Advances in Thermophysical Properties at Extreme Temperatures and Pressures," Proceedings of the Third Symposium on Thermophysical Properties (S. Gratch, ed.), American Society of Mechanical Engineers, New York (1965), p. 221.

556. W. Jackson, *Trans. Faraday Soc.* **42A**, 91 (1946).
557. W. Jaeger and H. von Steinwehr, *Ann. Physik* **369**, 305 (1921).
557a. H. A. Jahn and E. Teller, *Proc. Roy. Soc.* (*London*) **A161**, 220 (1937).
558. G. Jancso, J. Pupezin, and W. A. van Hook, *Nature* **225**, 723 (1970).
559. G. A. Jeffrey and R. K. McMullan, *Progr. Inorg. Chem.* **8**, 43 (1967).
560. M. S. Jhon, J. Grosh, T. Ree, and H. Eyring, *J. Chem. Phys.* **44**, 1465 (1966).
561. J. W. C. Johns, *Can. J. Phys.* **49**, 944 (1971).
562. G. Jones and W. A. Ray, *J. Chem. Phys.* **5**, 505 (1937).
563. J. R. Jones, D. L. G. Rowlands, and C. B. Monk, *Trans. Faraday Soc.* **61**, 1384 (1965).
564. S. J. Jones and J. W. Glen, *Phil. Mag.* **19**, 13 (1969).
565. W. M. Jones, Vapor Pressures of Tritium Oxide and Deuterium Oxide; Interpretation of the Isotope Effect, Los Alamos Scientific Laboratories, Los Alamos, New Mexico, LADC-5905 (1964).
566. J. Jůza, V. Kmoníček, and O. Šifner, *in* "An Equation of State for Water and Steam," Academia, Nakladatelstvi Československé Akademie VĚD, Praha (1966), pp. 131–142.
566a. U. Kaldor and F. E. Harris, *Phys. Rev.* **183**, 1 (1969).
567. B. Kamb, *Acta Cryst.* **17**, 1437 (1964).
568. B. Kamb, *Science* **150**, 205 (1965).
569. B. Kamb, *in* "Structural Chemistry and Molecular Biology," (A. Rich and N. Davidson, eds.), Freeman, San Francisco (1968), p. 507.
570. B. Kamb, *Science* **167**, 1520 (1970).
571. B. Kamb and S. K. Datta, *Acta Cryst.* **13**, 1029 (1960).
572. B. Kamb and S. K. Datta, *Nature* **187**, 140 (1960).
573. B. Kamb and B. L. Davis, *Proc. Natl. Acad. Sci.* **52**, 1433 (1964).
574. P. M. Kampmeyer, *J. Appl. Phys.* **23**, 99 (1952).
575. M. Karplus and D. H. Anderson, *J. Chem. Phys.* **30**, 6 (1959).
576. J. J. Katz, "Chemical and Biological Studies with Deuterium" (Thirty-Ninth Annual Priestley Lecture), Pennsylvania State University (1965).
577. S. Katzoff, *J. Chem. Phys.* **2**, 841 (1934).
578. J. J. Kaufmann, L. M. Sachs, and M. Geller, *J. Chem. Phys.* **49**, 4369 (1968).
579. W. Kauzmann, *Adv. Protein Chem.* **14**, 1 (1959).
580. J. L. Kavanau, "Water and Solute–Water Interactions," Holden-Day, San Francisco (1964).
581. R. L. Kay, G. A. Vidulich, and K. S. Pribadi, *J. Phys. Chem.* **73**, 445 (1969).
582. L. P. Kayushin, *in* "Water in Biological Systems," (L. P. Kayushin, ed.), Consultants Bureau, New York (1969).
583. Ya. Z. Kazavchinskii, P. M. Kessel'man, V. A. Kirillin, S. L. Rivkin, V. V. Sychev, D. L. Timrot, A. E. Sheindlin, and E. E. Shpil'rain, *in* "Tyazhelaya Voda, Teplofizicheskie Svoistva" (V. A. Kirillin, ed.), Gosenergoizdat, Moscow (1963).
584. P. Kebarle, R. N. Haynes, and J. C. Collins, *J. Am. Chem. Soc.* **89**, 5753 (1967).
585. P. Kebarle, S. K. Searles, A. Zolla, J. Scarborough, and M. Arshadi, *J. Am. Chem. Soc.* **89**, 6393 (1967).
586. J. H. Keenan, F. G. Keyes, P. G. Hill, and J. G. Moore, "Steam Tables (English Units)," John Wiley and Sons, New York (1969).
587. J. H. Keenan, F. G. Keyes, P. G. Hill, and J. G. Moore, "Steam Tables (Internat. Ed.—Metric Units)," John Wiley and Sons, New York (1969).

588. G. S. Kell, *J. Chem. Eng. Data* **12**, 66 (1967).
589. G. S. Kell, *J. Chem. Eng. Data* **15**, 119 (1970).
590. G. S. Kell, *in* "Water and Aqueous Solutions" (R. A. Horne, ed.), John Wiley and Sons, New York (1971), p. 331.
591. G. S. Kell and G. E. McLaurin, *J. Chem. Phys.* **51**, 4345 (1969).
592. G. S. Kell, G. E. McLaurin, and E. Whalley, *J. Chem. Phys.* **49**, 2839 (1968).
593. G. S. Kell, G. E. McLaurin, and E. Whalley, *J. Chem. Phys.* **48**, 3805 (1968).
594. G. S. Kell and E. Whalley, *Phil. Trans. Royal Soc.* **258A**, 565 (1965).
595. G. C. Kennedy and W. T. Holser, *in* "Handbook of Physical Constants—Revised Edition" (S. P. Clark, Jr., ed.), Geological Society of America, Memoir 97 (1966), p. 371.
596. G. C. Kennedy, W. L. Knight, and W. T. Holser, *Am. J. Sci.* **256**, 590 (1958).
597. A. M. Kerimov, N. A. Agaev, and A. A. Abaszade, *Teploenergetika* **1969** (11), 87.
598. A. M. Kerimov, N. A. Agaev, and A. A. Abaszade, *Thermal Eng.* **1969**(11), 126.
599. C. W. Kern and M. Karplus, *J. Chem. Phys.* **40**, 1374 (1964).
600. C. W. Kern and R. L. Matcha, *J. Chem. Phys.* **49**, 2081 (1968).
601. J. Kestin and J. H. Whitelaw, *J. Eng. Power* **88**, 82 (1966).
602. F. G. Keyes, *Intern. J. Heat Mass Transfer* **5**, 137 (1962).
603. F. G. Keyes, L. B. Smith, and H. T. Gerry, *Proc. Am. Acad. Arts Sci.* **70**, 319 (1936).
604. F. G. Keyes and R. G. Vines, *Int. J. Heat Mass Transfer* **7**, 33 (1964).
605. G. W. King, R. M. Hainer, and P. C. Cross, *J. Chem. Phys.* **11**, 27 (1943).
606. J. P. Kintzinger and J. M. Lehn, *Mol. Phys.* **14**, 133 (1968).
607. J. G. Kirkwood, *J. Chem. Phys.* **4**, 592 (1936).
608. J. G. Kirkwood, *J. Chem. Phys.* **7**, 911 (1939).
609. J. G. Kirkwood and F. P. Buff, *J. Chem. Phys.* **19**, 774 (1951).
610. J. G. Kirkwood, E. K. Mann, and B. J. Alder, *J. Chem. Phys.* **18**, 1040 (1950).
611. G. T. Kirouac, W. E. Moore, and K. W. Seemann, Report KAPL-M-6536 (1966).
612. I. Kirshenbaum, "Physical Properties and Analysis of Heavy Water," McGraw-Hill Book Co., New York (1951).
613. E. Kjelland-Forsterud, *J. Mech. Eng. Sci.* **1**, 30 (1959).
614. M. Klein and H. J. M. Hanley, *Trans. Faraday Soc.* **64**, 2927 (1968); H. J. M. Hanley and M. Klein, On the Selection of the Intermolecular Potential Function: Application of Statistical Mechanical Theory to Experiment, U. S. National Bureau of Standards, NBS TN 360 (1967).
615. G. Klopman, *J. Am. Chem. Soc.* **86**, 4550 (1964).
616. W. Koch, *Forsch. Geb. Ingenieurwesens* **5**, 138 (1934).
617. P. G. Kohn, *Symp. Soc. Experimental Biology* **19**, 3 (1965).
618. P. A. Kollman, Ph.D. thesis, Princeton University, 1970.
619. P. A. Kollman and L. C. Allen, *J. Am. Chem. Soc.* **92**, 753 (1970).
620. P. A. Kollman and L. C. Allen, *J. Chem. Phys.* **51**, 3286 (1969).
621. P. A. Kollman and L. C. Allen, *J. Chem. Phys.* **52**, 5085 (1970).
622. P. A. Kollman and L. C. Allen, *Theor. Chim. Acta* **18**, 399 (1970).
623. M. Kopp, D. E. Barnaal, and I. J. Lowe, *J. Chem. Phys.* **43**, 2965 (1965).
624. S. B. Kormer, K. B. Yushko, and G. V. Krishkevich, *Zh. Eksperim. i Teor. Fiz.* **54**, 1640 (1968); English transl.: *Soviet Phys.—JETP* **27**, 879 (1968).
625. V. S. Korobkov, *J. Struct. Chem.* **6**, 460 (1965).
626. A. Korosi and B. M. Fabuss, Second Report on the Properties of Saline Water Systems, U. S. Office of Saline Water, RDPR 249 (1967).

627. A. Korosi and B. M. Fabuss, *Anal. Chem.* **40**, 157 (1968).
628. L. Korson, W. Drost-Hansen, and F. J. Millero, *J. Phys. Chem.* **73**, 34 (1969).
629. R. Kosfield and L. Oehlmann, *Naturwiss.* **53**, 357 (1966).
630. H. Köster and E. U. Franck, *Ber. Bunsenges. Phys. Chem.* **73**, 716 (1969).
631. M. Kotani, A. Amemiya, E. Ishiguro, and T. Kimuro, "Tables of Molecular Integrals," Maruzen Co., Tokyo (1955), Chapter 1.
632. D. A. Kottwitz and B. R. Leonard, "Inelastic Scattering of Neutrons in Solids and Liquids," I.A.E.A., Vienna (1963).
633. J. J. Kozak, W. S. Knight, and W. Kauzman, *J. Chem. Phys.* **48**, 675 (1968).
634. W. Kraemers and G. Diercksen, *Chem. Phys. Letters* **5**, 463 (1970).
635. O. Kratky, *Z. Elektrochemie* **62**, 66 (1958).
636. V. S. Kravchenko, *Atom. Energiya* **20**, 168 (1966).
637. R. F. Kruh, *Chem. Rev.* **62**, 319 (1962).
638. K. Krynicki, *Physica* **32**, 167 (1966).
639. R. Kubo, *in* "Fluctuation, Relaxation and Resonance in Magnetic Systems" (D. Jer Haar, ed.), Oliver and Boyd, London (1961), p. 23.
639a. K. Kuchitsu, *Bull. Chem. Soc. Japan* **44**, 96 (1971).
640. K. Kuchitsu and L. S. Bartell, *J. Chem. Phys.* **36**, 2460 (1962).
641. K. Kuchitsu and Y. Morino, *Bull. Chem. Soc. Japan* **38**, 814 (1965).
642. W. Kuhn and M. Türkauf, *Helv. Chim. Acta* **41**, 938 (1958).
643. K. J. Kume, *J. Phys. Soc. Japan* **15**, 1493 (1960).
644. S. L. Kurtin, C. A. Mead, W. A. Mueller, B. C. Kurtin, and E. D. Wolf, *Science* **167**, 1720 (1970).
645. E. F. Labuda, E. I. Gordon, and R. C. Miller, *IEEE Trans. Quantum Electronics* **1**, 273 (1967).
646. T. H. Laby and E. O. Hercus, *Phil. Trans. Roy. Soc.* **A227**, 63 (1927).
647. R. C. Ladner and W. A. Goddard, III, *J. Chem. Phys.* **51**, 1073 (1969).
648. J. Lamb, *Trans. Faraday Soc.* **42A**, 238 (1946).
649. J. Lamb, in "Physical Acoustics," Vol. IIA, Academic Press, New York (1965), Chapter 4.
650. J. A. Lane and J. A. Saxton, *Proc. Roy. Soc.* **A214**, 531 (1952).
651. P. W. Langhoff, M. Karplus, and R. P. Hurst, *J. Chem. Phys.* **44**, 505 (1966).
652. S. R. La Paglia, *J. Chem. Phys.* **41**, 1427 (1964).
653. S. LaPlaca and B. Post, *Acta Cryst.* **13**, 503 (1960).
654. K. E. Larsson, *in* "Thermal Neutron Scattering" (P. A. Egelstaff, ed.), Academic Press, London (1965), p. 350.
655. K. E. Larsson and U. Dahlborg, *Reactor Sci. Tech.* (*J. Nucl. Energy*) **16**, 81 (1962).
656. K. E. Larsson, U. Dahlborg, and S. Holmryd, *Arkiv Fysik* **17**, 369 (1960).
657. W. M. Latimer and W. H. Rodebush, *J. Am. Chem. Soc.* **42**, 1419 (1920).
658. R. T. Lattey, O. Gatty, and W. G. Davies, *Phil. Mag.* **12**, 1019 (1931).
659. B. Latto, The Viscosity of Steam at Atmospheric Pressure, Mechanical Engineering Department, University of Glasgow, TR 16 (1965).
660. A. W. Lawson and A. J. Hughes, *in* "High Pressure Physics and Chemistry" (R. S. Bradley, ed.), Vol. 1, Academic Press, London (1963), pp. 207–225.
661. A. W. Lawson, R. Lovell, and A. L. Jain, *J. Chem. Phys.* **30**, 643 (1959).
662. A. J. Leadbetter, *Proc. Roy. Soc.* **A287**, 403 (1965).
663. J. C. R. le Bot, *Compt. Rend.* **228**, 749 (1949).

A. S. N. Murthy and C. N. R. Rao, *Chem. Phys. Letters* **2**, 123 (1968).
A. S. N. Murthy and C. N. R. Rao, *J. Mol. Struct.* **6**, 253 (1970).
A. S. N. Murthy, K. G. Rao, and C. N. R. Rao, *J. Am. Chem. Soc.* **92**, 3544 (1970).
U. Nakaya, *SIPRE Res. Report* **28** (1958).
A. H. Narten, X-Ray Diffraction Data on Liquid Water in the Temperature Range 4–200°C, ORNL-4578 (1970).
A. H. Narten, M. D. Danford, and H. A. Levy, X-Ray Diffraction Data on Liquid Water in the Temperature Range 4–200°C, ORNL-3997 (1966).
A. H. Narten, M. D. Danford, and H. A. Levy, *Disc. Faraday Soc.* **43**, 97 (1967).
A. H. Narten and H. A. Levy, *Science* **165**, 447 (1969).
A. H. Narten and H. A. Levy, *Science* **167**, 1521 (1970).
G. Nemethy and H. A. Scheraga, *J. Phys. Chem.* **66**, 1773 (1962).
G. Nemethy and H. A. Scheraga, *J. Chem. Phys.* **36**, 3382 (1962).
G. Nemethy and H. A. Scheraga, *J. Chem. Phys.* **36**, 3401 (1962).
G. Nemethy and H. A. Scheraga, *J. Chem. Phys.* **41**, 680 (1964).
D. Neumann and J. W. Moscowitz, *J. Chem. Phys.* **49**, 2056 (1968).
Neutron Diffraction Commission, *Acta Cryst.* **A25**, 391 (1969).
G. F. Newell, *Phys. Rev.* **80**, 476 (1950).
. D. M. Newitt, S. Angus, and M. B. Hooper, The Experimental Determination of the Specific Enthalpy of Steam in the Superheat and Supercritical Regions, Electrical Research Association, Report No. 5050, Leatherhead (1964).
. H. H. Nielsen, *Rev. Mod. Phys.* **23**, 90 (1951).
. E. S. Nowak and R. J. Grosh, An Investigation of Certain Thermodynamic and Transport Properties of Water and Water Vapor in the Critical Region, Argonne National Laboratory, ANL-6064 (1959).
. E. S. Nowak and R. J. Grosh, Atomic Energy Commission Technical Report ANL-6508 (1961).
3. E. S. Nowak, R. J. Grosh, and P. E. Liley, *J. Heat Transfer* **83**, 1, 14 (1961).
9. A. A. Noyes, Y. Kato, and R. B. Sosman, *Z. Phys. Chem.* **73**, 1 (1910).
0. A. A. Noyes, A. C. Melcher, H. C. Cooper, and G. W. Eastman, *Z. Phys. Chem.* **70**, 335 (1910).
1. A. J. Nozik and M. Kaplan, *Chem. Phys. Letters* **1**, 391 (1967).
2. N. Ockman, *Adv. Phys.* **7**, 199 (1958).
3. J. P. O'Connell and J. M. Prausnitz, *Ind. Eng. Chem. Fundamentals* **8**, 453 (1969).
4. J. J. O'Dwyer, *J. Chem. Phys.* **26**, 878 (1957).
5. G. D. Oliver and J. W. Grisard, *J. Am. Chem. Soc.* **78**, 561 (1956).
6. P. C. Olympia, Jr., and B. M. Fung, *J. Chem. Phys.* **51**, 2976 (1969).
7. J. R. O'Neil and L. H. Adami, *J. Phys. Chem.* **73**, 1553 (1969).
8. L. Onsager, *J. Am. Chem. Soc.* **58**, 1486 (1936).
9. L. Onsager and L. K. Runnels, *J. Chem. Phys.* **50**, 1089 (1969).
0. N. S. Osborne, *J. Res. Natl. Bur. Std.* **4**, 609 (1930).
1. N. S. Osborne and C. H. Meyers, *J. Res. Natl. Bur. Std.* **13**, 1 (1934).
2. N. S. Osborne, H. F. Stimson, E. F. Fiock, and D. C. Ginnings, *J. Res. Natl. Bur. Std.* **10**, 155 (1933).
13. N. S. Osborne, H. F. Stimson, and D. C. Ginnings, *J. Res. Natl. Bur. Std.* **18**, 389 (1937).
14. N. S. Osborne, H. F. Stimson, and D. C. Ginnings, *J. Res. Natl. Bur. Std.* **23**, 197 (1939).

664. R. E. Leckenby, E. J. Robbins, and P. A. Trevalion, *Proc. Roy. Soc.* (*London*) **A280**, 409 (1964).
665. B. le Neindre, P. Bury, R. Tufeu, P. Johannin, and B. Vodar, Résultats expérimentaux sur la conductivité thermique de l'eau et de l'eau lourde en phase liquide, jusqu'à une température de 370°C, et leurs discussions, Laboratoire des Hautes Pressions, Centre National de la Recherche, France (about 1968).
666. B. le Neindre, P. Johannin, and B. Vodar, *Compt. Rend.* **258**, 3277 (1964).
667. B. le Neindre, P. Johannin, and B. Vodar, *Compt. Rend.* **261**, 67 (1965).
668. J. Lennard-Jones and A. F. Devonshire, *Proc. Roy. Soc.* **A163**, 53 (1937).
669. J. Lennard-Jones and A. F. Devonshire, *Proc. Roy. Soc.* **A165**, 1 (1938).
670. J. Lennard-Jones and A. F. Devonshire, *Proc. Roy. Soc.* **A169**, 317 (1939).
671. J. Lennard-Jones and J. A. Pople, *Proc. Roy. Soc.* (*London*) **A202**, 166 (1950).
672. J. Lennard-Jones and J. A. Pople, *Proc. Roy. Soc.* (*London*) **A205**, 155 (1951).
673. J. M. H. Levelt Sengers, Thermodynamic Anomalies near the Critical Point of Steam, Paper communicated to the Seventh International Conference on the Properties of Steam, Tokyo, 1968.
674. J. M. H. Levelt Sengers, *Ind. Eng. Chem. Fundamentals* **9**, 470 (1970).
675. J. M. H. Levelt Sengers and M. Vicentini-Missoni, *in* "Proceedings of the Fourth Symposium on Thermophysical Properties," (J. R. Moszynski, ed.), American Society of Mechanical Engineers, New York (1968), p. 79.
676. A. M. Levelut and A. Guinier, *Bull. Soc. Franc. Mineral. Crystallogr.* **40**, 445 (1967).
677. S. Levine and J. W. Perram, *in* "Hydrogen-Bonded Solvent Systems," (A. K. Covington and P. Jones, eds.), Taylor and Francis, London (1968), p. 115.
678. H. A. Levy, M. Danford, and A. H. Narten, Data Collection and Evaluation with an X-Ray Diffractometer Designed for the Study of Liquid Structure, ORNL-3960 (1966).
679. G. N. Lewis and R. T. MacDonald, *J. Am. Chem. Soc.* **55**, 3057 (1933).
680. G. N. Lewis and R. T. MacDonald, *J. Am. Chem. Soc.* **55**, 4730 (1933).
681. Y.-H. Li, *J. Geophys. Res.* **72**, 2675 (1967).
682. J. H. Liang, Ph. D. Thesis, The Ohio State University, 1970.
683. M. Lichtenstein, V. E. Derr, and J. J. Gallagher, *J. Mol. Spectry.* **20**, 391 (1966).
684. T. F. Lin and A. B. F. Duncan, *J. Chem. Phys.* **48**, 866 (1968).
685. H.-M. Lin and R. L. Robinson, *J. Chem. Phys.* **52**, 3727 (1970).
686. H. Lindner, doctoral dissertation, University of Karlsruhe, 1970.
687. J. W. Linnett, *Science* **168**, 1719 (1970).
688. J. W. Linnett and A. Poe, *Trans. Faraday Soc.* **47**, 1033 (1951).
689. E. R. Lippincott, *J. Chem. Phys.* **23**, 603 (1955).
690. E. R. Lippincott and R. Schroeder, *J. Chem. Phys.* **23**, 1099 (1955).
691. E. R. Lippincott, R. Stromberg, W. Grant, and G. Cessac, *Science* **164**, 1482 (1969).
692. T. Litovitz and E. Carnevale, *J. Appl. Phys.* **26**, 816 (1955).
693. T. Litovitz and C. Davis, *in* "Physical Acoustics," Vol. IIA, Academic Press, New York (1965).
694. T. Litovitz and G. McDuffie, *J. Chem. Phys.* **39**, 729 (1963).
695. W. M. Lomer and G. G. Low, *in* "Thermal Neutron Scattering" (P. A. Egelstaff, ed.), Academic Press, London (1965), p. 1.
696. K. Lonsdale, *Proc. Roy. Soc.* **A247**, 424 (1958).
697. L. G. Longsworth, *J. Phys. Chem.* **64**, 1914 (1960).

698. H. A. Lorentz, "The Theory of Electrons," Dover Publications, New York (1952), p. 308.
699. P. O. Löwdin, *Phys. Rev.* **97**, 1509 (1955).
700. P. O. Löwdin and H. Shull, *Phys. Rev.* **101**, 1730 (1956).
701. W. A. P. Luck, *Disc. Faraday Soc.* **43**, 115 (1967).
702. W. A. P. Luck, *Disc. Faraday Soc.* **43**, 132 (1967).
703. W. A. P. Luck, *in* "Physico-Chemical Processes in Mixed Aqueous Solvents," (F. Franks, ed.), Heinemanns, London (1967).
704. W. A. P. Luck, Habilitationsschrift, "Spektroskopische Bestimmungen der Wasserstoffbrückenbindungen in Naher IR," University of Heidelberg, 1968.
705. W. A. P. Luck and W. Ditter, *Z. Naturforsch.* **24b**, 482 (1969).
706. B. Lundqvist and T. Persson, *Brennst.-Wärme-Kraft* **17**, 356 (1965).
707. Z. Luz and G. Yagil, *J. Phys. Chem.* **70**, 554 (1966).
708. B. J. McBride and S. Gordon, *J. Chem. Phys.* **35**, 2198 (1961).
709. D. W. McCall, D. C. Douglass, and E. W. Anderson, *J. Chem. Phys.* **31**, 1555 (1959).
710. M. McCarty, Jr., and S. V. K. Babu, *J. Phys. Chem.* **74**, 1113 (1970).
711. A. L. McClellan, "Dipole Moments," Freeman, San Francisco (1963).
712. J. P. McCullough, R. E. Pennington, and G. Waddington, *J. Am. Chem. Soc.* **74**, 4439 (1952).
713. J. R. Macdonald, *Rev. Mod. Phys.* **41**, 316 (1969).
714. P. C. McKinney and G. M. Barrow, *J. Chem. Phys.* **31**, 294 (1959).
715. E. McLaughlin, *Physica* **26**, 650 (1960).
716. E. McLaughlin, *in* "Thermal Conductivity" (R. P. Tye, ed.), Vol. 2, Academic Press, London (1969), p. 1.
717. J. A. McMillan and S. C. Los, *Nature* **206**, 806 (1965).
718. H. J. McSkimin, *J. Acoust. Soc. Am.* **37**, 325 (1965).
719. R. McWeeny and K. A. Ohno, *Proc. Roy. Soc.* (*London*) **A255**, 367 (1960).
720. M. Magat, *Ann. Phys.* (*Paris*) **6**, 109 (1936).
721. M. Magat, *J. Chem. Phys.* **45**, 91 (1948).
722. M. A. Maidique, A. von Hippel, and W. B. Westphal, *J. Chem. Phys.* **54**, 150 (1971).
723. S. Maier and E. U. Franck, *Ber. Bunsenges. Phys. Chem.* **70**, 639 (1966).
724. P. D. Maker, *Phys. Rev.* **A1**, 923 (1970).
725. C. G. Malmberg and A. A. Maryott, *J. Res. Natl. Bur. Std.* **56**, 1 (1956).
726. J. Malsch, *Ann. Phys. Leipzig* **84**, 841 (1927).
727. J. Malsch, *Phys. Z.* **29**, 770 (1928).
728. A. M. Mamedov, *Teploenergetika* **1960**(9), 71.
729. A. M. Mamedov, *Teploenergetika* **1963**(5), 69.
730. K. Mangold and E. U. Franck, *Ber. Bunsenges.* **73**, 21 (1969).
731. J. P. Marckmann and E. Whalley, *J. Chem. Phys.* **41**, 1450 (1964).
732. T. W. Marshall and J. A. Pople, *Mol. Phys.* **1**, 199 (1958).
733. E. A. Mason, *in* "Proceedings of the Fourth Symposium on Thermophysical Properties" (J. R. Moszynski, ed.), American Society of Mechanical Engineers, New York (1968), p. 21.
734. E. A. Mason and T. H. Spurling, "The Virial Equation of State," Pergamon Press, Oxford (1969), p. 1.
735. P. R. Mason, J. B. Hasted, and L. Moore, to be published.

736. F. A. Matsen, *Adv. Quantum Chem.* **1**, 60 (1964).
737. F. Mayinger, *Int. J. Heat Mass Transfer* **5**, 807 (19
738. H. D. Megaw, *Nature*, **134**, 900 (1934).
739. S. Meiboom, *J. Chem. Phys.* **34**, 375 (1961).
740. M. Menaché, *Cahiers Oceanographiques* **17**, 625 (19
741. M. Menaché, *Cahiers Oceanographiques* **18**, 477 (19
742. M. Menaché, *Metrologia* **3**, 58 (1967).
743. D. Mendeleef, *Phil. Mag.* **33**, 99 (1892).
744. R. P. Messmer, *Science* **168**, 679 (1970).
745. H. H. Meyer, *Ann. Physik* **5**, 701 (1930).
745a. W. Meyer, *Internat. J. Quantum Chem.* **5S**, 296 (1
746. A. Michels, B. Blaisse, C. A. Ten Seldam, and H. M (1943).
747. A. Michels, A. Botzen, and W. Schuurmann, *Physic*
748. C. Migchelsen, H. J. C. Berendsen, and A. Rupprecht,
749. F. T. Miles and A. W. C. Menzies, *J. Am. Chem. So*
750. A. A. Miller, *J. Chem. Phys.* **38**, 1568 (1963).
750a. J. H. Miller and H. P. Kelly, *Phys. Rev.* **A4**, 480 (
751. K. J. Miller, S. R. Mielczarek, and M. Krauss, *J. Che*
752. F. J. Millero, R. W. Curry, and W. Drost-Hansen, *J.* (1969).
753. F. J. Millero, R. Dexter, and E. Hoff, *J. Chem. Eng. L*
754. F. J. Millero and F. K. Lepple, *J. Chem. Phys.* **54**, 94
755. V. N. Mineev and R. M. Zaidel', *Zh. Eksperim. i Teor.*
756. V. N. Mineev and R. M. Zaidel', *Soviet Phys.—JETP*
757. A. P. Minton, *Nature* **226**, 151 (1970).
758. R. Moccia, *J. Chem. Phys.* **40**, 2186 (1964).
759. J. F. Mohler, *Phys. Rev.* **35**, 236 (1912).
760. L. Monchick and E. A. Mason, *J. Chem. Phys.* **35**, 1676
761. J. Morgan and B. E. Warren, *J. Chem. Phys.* **6**, 666 (
762. K. Morokuma, *Chem. Phys. Letters* **4**, 358 (1969).
763. K. Morokuma and L. Pederson, *J. Chem. Phys.* **48**, 32
764. K. Morokuma and J. R. Winick, *J. Chem. Phys.* **52**, 1
765. H. Moser, *Ann. Physik* **82**, 993 (1927).
766. H. Moser and A. Zmaczynski, *Physik. Z.* **40**, 221 (1939)
767. J. W. Moskowitz and M. C. Harrison, *J. Chem. Phys.* **4**
768. O. F. Mossotti, *Mem. de Mat. el Fis. Soc. Ital. de Sci.* 2
769. J. R. Moszynski, *J. Heat Transfer* **83**, 111 (1961).
770. W. G. Moulton and R. A. Kromhout, *J. Chem. Phys.* **2**
771. N. Muller and R. C. Reiter, *J. Chem. Phys.* **42**, 3265 (1
772. R. H. Muller and J. A. Shufle, *J. Geophys. Res.* **73**, 3345
773. R. S. Mulliken, *J. Chem. Phys.* **23**, 1833 (1955).
774. J. N. Murrell and F. B. van Duijneveldt, *J. Chem. Phys.*
775. A. S. N. Murthy, S. N. Bhat, and C. N. R. Rao, *J. Chen* 1251.
776. A. S. N. Murthy, R. E. Davis, and C. N. R. Rao, *The* (1969).
777. A. S. N. Murthy and C. N. R. Rao, *Appl. Spec. Revs.* **2**,

815. N. S. Osborne, H. F. Stimson, and D. C. Ginnings, *J. Res. Natl. Bur. Std.* **23**, 261 (1939).
816. W. H. Orttung and J. A. Meyers, *J. Phys. Chem. Ithaca* **67**, 1905 (1963).
817. W. v. Osten, Diplomarbeit, Karlsruhe, 1966.
818. G. Oster and J. G. Kirkwood, *J. Chem. Soc.* **11**, 175 (1943).
819. B. B. Owen, J. R. White, and J. S. Smith, *J. Am. Chem. Soc.* **78**, 3561 (1956).
820. D. I. Page, *Disc. Faraday Soc.* **43**, 143 (1967).
821. T. F. Page, R. J. Jakobsen, and E. R. Lippincott, *Science* **167**, 51 (1970).
822. L. R. Painter, R. D. Birkhoff, and E. T. Arakawa, *J. Chem. Phys.* **51**, 2113 (1969).
823. M. Pancholy, *J. Acoust. Soc. Am.* **25**, 1003 (1953).
824. V. Parkash, M. K. Dheer, and T. S. Jaseja, *Physics Letters* **29A**, 220 (1969).
825. F. Paschen, *Ann. Physik* **53**, 334 (1894).
826. L. Pauling, *J. Am. Chem. Soc.* **57**, 2680 (1935).
827. L. Pauling, *in* "Hydrogen Bonding" (D. Hadzi, ed.), Pergamon Press, New York (1959).
828. L. Pauling, "The Nature of the Chemical Bond," 3rd ed., Cornell University Press, Ithaca, New York (1960).
829. L. Pederson, *Chem. Phys. Letters* **4**, 280 (1969).
830. J. Pellam and J. Galt, *J. Chem. Phys.* **14**, 608 (1946).
831. R. P. Penrose, *Trans. Faraday Soc.* **42**, 108 (1946).
832. J. K. Percus and G. J. Yevick, *Phys. Rev.* **110**, 1 (1958).
833. G. Peschel and K. H. Adlfinger, *Naturwiss.* **56**, 558 (1969).
834. S. W. Peterson and H. A. Levy, *Acta Cryst.* **10**, 70 (1957).
835. S. W. Peterson and H. A. Levy, *Acta Cryst.* **10**, 344 (1957).
836. G. A. Petsko, *Science* **167**, 171 (1970).
837. S. D. Peyerimhoff, R. J. Buenker, and L. C. Allen, *J. Chem. Phys.* **45**, 734 (1966).
838. M. A. Pick, in "The Physics of Ice" (N. Riehl, B. Bullemer, and H. Engelhardt, eds.) Plenum Press, New York (1969).
839. A. Piekara and A. Chelkowski, *J. Chem. Phys.* **25**, 794 (1956).
840. A. Piekara, A. Chelkowski, and S. Kielich, *Z. Phys. Chem.* **206**, 375 (1957).
841. A. Piekara, A. Chelkowski, and S. Kielich, *Arch. Sci.* **12**, 61 (1959).
842. A. Piekara and S. Kielich, *J. Chem. Phys.* **29**, 1297 (1958).
843. F. Pilar, "Elementary Quantum Chemistry," McGraw-Hill Book Co., New York (1968).
844. G. C. Pimentel and A. L. McClellan, "The Hydrogen Bond," Freeman, San Francisco (1960).
845. C. J. Pings, *in* "Physics of Simple Liquids" (H. N. Temperley, J. S. Rowlinson, and G. S. Rushbrooke, eds.), North-Holland Publishing Co., Amsterdam (1968).
846. J. Pinkerton, *Nature* **160**, 128 (1947).
847. C. W. F. T. Pistorius, M. C. Pistorius, J. P. Blakey, and L. J. Admiraal, *J. Chem. Phys.* **38**, 600 (1963).
848. K. S. Pitzer, "Quantum Chemistry," Prentice Hall, Englewood Cliffs, New Jersey (1953).
849. R. M. Pitzer and D. P. Merrifield, Battelle Memorial Institute, Theoretical Chemistry Group Report No. 14 (1969).
850. R. M. Pitzer and D. P. Merrifield, *J. Chem. Phys.* **52**, 4782 (1970).
851. K. S. Pitzer and J. Polissar, *J. Phys. Chem.* **60**, 1140 (1956).

852. G. Placzek, *Phys. Rev.* **86**, 377 (1954).
853. N. K. Pope and R. Nation, *in* "Inelastic Scattering of Neutrons," Vol. II, I.A.E.A., Vienna (1965), p. 141.
854. J. A. Pople, *Proc. Roy. Soc.* **A202**, 323 (1950).
855. J. A. Pople, *Proc. Roy. Soc.* **A205**, 163 (1951).
856. J. A. Pople, D. L. Beveridge, and P. A. Dobosh, *J. Chem. Phys.* **47**, 2027 (1968).
857. J. A. Pople, D. P. Santry, and G. A. Segal, *J. Chem. Phys.* **43**, S129, S136 (1965).
858. J. A. Pople and G. A. Segal, *J. Chem. Phys.* **44**, 3289 (1966).
859. M. M. Popov and F. I. Tazetdinov, *Atom. Energiya* **8**, 420 (1960).
860. M. M. Popov and F. I. Tazetdinov, *Sov. J. Atom. Eng.* **8**, 353 (1960).
861. D. W. Posener, MIT Research Laboratory of Electronics Technical Report No. 255, 14 May 1953.
862. D. W. Posener and M. W. P. Strandberg, *Phys. Rev.* **95**, 374 (1954).
863. R. Pothier, *The Miami Herald*, 30 July 1969.
864. R. Pottel and O. Lossen, *Ber. Bunsenges. Phys. Chem.* **71**, 135 (1967).
865. F. X. Powell and D. R. Johnson, *Phys. Rev. Letters* **24**, 637 (1970).
866. R. W. Powell, *Adv. Phys.* **7**, 276 (1958).
867. J. G. Powles, *J. Chem. Phys.* **21**, 633 (1952).
868. J. G. Powles, M. Rhodes, and J. H. Strange, *Mol. Phys.* **11**, 515 (1966).
869. H. J. Prask and H. P. Boutin, Army Materials Research Agency, Tech. Rep. 3348 (1966).
870. H. Prask, S. Yip, and H. Boutin, *J. Chem. Phys.* **48**, 3367 (1968).
871. J. M. Prausnitz, Intermolecular Forces in Systems Containing Water, U.S. Office of Saline Water, RDPR No. 306 (1968).
872. A. H. Price, *Nature* **181**, 262 (1958).
873. I. Prigogine, "The Molecular Theory of Solutions," North-Holland Publishing Co., Amsterdam (1957).
874. J. A. Prins and W. Prins, *Physica* **23**, 253 (1957).
875. E. A. Pshenichnov and N. D. Sokolov, *Optics Spectry. USSR* **11**, 8 (1961).
876. M. R. Querry, B. Curnutte, and D. Williams, *J. Opt. Soc. Am.* **59**, 1299 (1969).
876a. A. S. Quist, *J. Phys. Chem.* **74**, 3396 (1970).
877. A. S. Quist and W. L. Marshall, *J. Phys. Chem.* **69**, 3165 (1965).
878. A. S. Quist and W. L. Marshall, *J. Phys. Chem.* **73**, 978 (1969).
879. S. W. Rabideau, E. D. Finch, and A. B. Denison, *J. Chem. Phys.* **49**, 4660 (1968).
880. S. W. Rabideau and H. G. Hecht, *J. Chem. Phys.* **47**, 544 (1967).
881. I. B. Rabinovich, "Influence of Isotopy on the Physicochemical Properties of Liquids," Consultants Bureau, New York (1970).
882. C. Rädle, Diplomarbeit, Universität Karlsruhe, 1969.
883. A. Rahman, *J. Chem. Phys.* **42**, 3540 (1965).
884. A. Rahman and F. H. Stillinger, Jr., *J. Chem. Phys.* **55**, 3336 (1971).
885. O. Rahn, M. Maier, and W. Kaiser, *Optics Commun.* **1**, 109 (1969).
886. V. Raman, K. S. Venkataraman, *Proc. Roy. Soc.* **A171**, 137 (1939).
887. M. B. Ramiah and D. A. I. Goring, *J. Polymer Sci.* **II** (Part C), 27 (1965).
888. R. W. Rampolla, R. C. Miller, and C. P. Smyth, *J. Chem. Phys.* **30**, 566 (1959).
889. R. O. Ramseier, *J. Appl. Phys.* **38**, 2553 (1967).
890. N. F. Ramsey, *Phys. Rev.* **78**, 699 (1950).
891. N. F. Ramsey, "Molecular Beams," Clarendon Press, Oxford (1956).

892. Yu. L. Rastorguev and V. V. Pugach, *Teploenergetika* **1970**(4), 77.
893. A. K. Ray, *J. Mech. Eng. Sci.* **6**, 137 (1964).
894. R. M. Redheffer, R. C. Wildman, and V. Gorman, *J. Appl. Phys.* **23**, 610 (1952).
895. R. M. Redheffer and E. D. Winkler, MIT Radiation Laboratory Report 483, 15, 18, 145–6.
896. C. Reid, *J. Chem. Phys.* **30**, 182 (1950).
897. R. Rein and F. Harris, *J. Mol. Struct.* **2**, 103 (1968).
898. H. Reiss, *Adv. Chem. Phys.* **9**, 1 (1966).
899. H. Renkert, Diplomarbeit, Universität Karlsruhe, 1965.
899a. R. Renner, *Z. Physik* **92**, 172 (1934).
900. J. Reuben, A. Tzalmona, and D. Samuel, *Proc. Chem. Soc.* **1962**, 353.
901. M. H. Rice and J. M. Walsh, *J. Chem. Phys.* **26**, 824 (1957).
902. E. H. Riesenfeld and T. L. Chang, *Z. Physik. Chem.* **B30**, 61 (1935).
903. E. H. Riesenfeld and T. L. Chang, *Z. Physik. Chem.* **B33**, 120 (1936).
904. E. H. Riesenfeld and T. L. Chang, *Z. Physik. Chem.* **B33**, 127 (1936).
905. G. Ritzert and E. U. Franck, *Ber. Bunsenges.* **72**, 798 (1968).
906. S. L. Rivkin and T. S. Akhundov, *Teploenergetika* **1962**(1), 57.
907. S. L. Rivkin and B. N. Egorov, *Atom. Energiya* **7**, 462 (1959).
908. S. L. Rivkin and A. Ya. Levin, *Teploenergetika* **1966**(4), 79.
909. S. L. Rivkin and A. Ya. Levin, *Thermal Eng.* **1966**(4), 104.
910. S. L. Rivkin, A. Ya. Levin, and L. B. Izrailevskii, *Teploenergetika* **1968**(12), 74.
911. S. L. Rivkin, A. Ya. Levin, and L. B. Izrailevskii, *Thermal Eng.* **1968**(12), 108.
912. H. A. Rizk and Y. M. Girgis, *Z. Physik. Chem.* (*Frankfurt*) **65**, 261 (1969).
913. S. Roberts and A. von Hippel, *J. Appl. Phys.* **17**, 610 (1946).
914. R. A. Robinson and R. H. Stokes, "Electrolyte Solutions," Butterworths, London (1959).
915. U. K. Rombusch, W. Barho, G. Ernst, A. Schaber, and D. Straub, *Atompraxis* **8**, 339 (1962).
916. W. K. Röntgen, *Ann. Phys.* **45**, 91 (1892).
917. W. K. Röntgen, *Ann. Phys.* **64**, 1 (1898).
918. C. C. J. Roothaan, *Rev. Mod. Phys.* **23**, 69 (1951).
919. K. Roth, Dissertation, Universität Karlsruhe, 1969.
920. W. G. Rothschild, *J. Chem. Phys.* **49**, 2250 (1968).
921. W. G. Rothschild, *J. Chem. Phys.* **52**, 6453 (1970).
922. W. G. Rothschild, *J. Chem. Phys.* **53**, 3265 (1970).
923. D. L. Rousseau and S. P. S. Porto, *Science* **167**, 1715 (1970).
924. H. A. Rowland, *Proc. Am. Acad. Arts Sci.* **15**, 75 (1880).
925. J. S. Rowlinson, *Trans. Faraday Soc.* **45**, 974 (1949).
926. J. A. Rowlinson, *Trans. Faraday Soc.* **47**, 120 (1951).
927. J. S. Rowlinson, *J. Chem. Phys.* **19**, 827 (1951).
928. J. S. Rowlinson, "Liquids and Liquid Mixtures," Butterworths, London (1969).
929. J. S. Rowlinson, Paper presented at 1st Internat. Conf. Calorimetry and Thermodynamics, Warsaw, 1969.
930. E. W. Rusche and W. B. Good, *J. Chem. Phys.* **45**, 4667 (1966).
931. J. W. Ryde and D. Ryde, Research Laboratories General Electric Co., Wembley, England, Report No. 8670 (1945).
932. J. R. Sabin, R. E. Harris, T. W. Archibald, P. A. Kollman, and L. C. Allen, *Theor. Chim. Acta* **18**, 235 (1970).

933. G. J. Safford, P. S. Leung, A. W. Naumann, and P. C. Schaffer, *J. Chem. Phys.* **50**, 4444 (1969).
934. G. J. Safford, A. W. Naumann, and P. C. Schaffer, A Neutron Scattering Study of Water and Ionic Solutions, Union Carbide Corporation Report (1967).
935. G. Safford, P. Schaffer, P. Leung, G. Doebbler, G. Brady, and E. Lyden, *J. Chem. Phys.* **50**, 2140 (1969).
936. M. Sakamoto, B. N. Brockhouse, R. G. Johnson, and N. K. Pope, *J. Japan Phys. Soc.* **17** (Suppl. B2), 370 (1962).
937. M. Salomon, Thermodynamic Properties of Liquid H_2O and D_2O and Their Mixtures, National Aeronautics and Space Administration, NASA TN D-5223 (1969).
938. O. Ya. Samoilov, *Zh. Fiz. Khim.* **20**, 10 (1946).
939. O. Ya. Samoilov, *Zh. Fiz. Khim.* **20**, 12 (1946).
940. O. Ya. Samoilov, Structure of Aqueous Electrolyte Solutions and the Hydration of Ions, (translated by D. J. G. Ives), Consultants Bureau, New York (1965).
941. S. I. Sandler, *Phys. Fluids* **11**, 2549 (1968).
942. O. Sandus and B. Lubitz, *J. Phys. Chem.* **65**, 881 (1961).
943. R. Sänger and O. Steiger, *Helv. Phys. Acta* **1**, 369 (1928).
944. S. C. Saxena and K. M. Joshi, *Phys. Fluids* **5**, 1217 (1962).
945. J. Saxton, *Proc. Roy. Soc.* (*London*) **A213**, 473 (1952).
946. J. A. Saxton and J. A. Lane, *Proc. Roy. Soc.* **A213**, 400 (1952).
947. B. K. P. Scaife, *in* "Molecular Relaxation Processes" (Chemical Society Special Publication No. 20), Academic Press, New York (1966), p. 15.
948. E. A. Scarzafava, Ph. D. Thesis, Indiana University, 1969.
949. H. F. Schaefer and C. F. Bender, *J. Chem. Phys.* **55**, 1720 (1971).
950. W. R. Schell, G. Sauzay, and B. R. Payne, *J. Geophys. Res.* **75**, 2251 (1970).
951. J. A. Schellman, *J. Chem. Phys.* **25**, 350 (1956).
952. J. Schiffer and D. F. Hornig, *J. Chem. Phys.* **49**, 4150 (1968).
953. E. Schmidt, "VDI–Steam Tables," Springer, Berlin–Heidelberg–New York (1968).
954. E. Schmidt, "Properties of Water and Steam in SI-Units," Springer Verlag, Berlin (1969).
955. E. Schmidt and W. Sellschopp, *Forsch. Geb. Ing.* **3**, 277 (1932).
956. W. G. Schneider, H. J. Bernstein, and J. A. Pople, *J. Chem. Phys.* **28**, 601 (1958).
957. R. Schroeder and E. R. Lippincott, *J. Phys. Chem.* **61**, 921 (1957).
958. G. Schwarz, European *J. Biochem.* **12**, 442 (1970).
959. R. A. Scott and H. A. Scheraga, *J. Chem. Phys.* **45**, 2091 (1966).
960. V. F. Sears, *Can. J. Phys.* **45**, 237 (1967).
961. V. Seidl, O. Knop, and M. Falk, *Can. J. Chem.* **47**, 1361 (1969).
962. W. A. Senior and R. E. Verrall, *J. Phys. Chem.* **73**, 4242 (1969).
963. L. B. Shaffer, Thesis, University of Wisconsin, 1964.
964. W. H. Shaffer and B. J. Krohn, Abstracts of the Symposium on Molecular Structure and Spectroscopy, The Ohio State University, 1970, Paper B1.
965. W. H. Shaffer and H. H. Nielsen, *Phys. Rev.* **56**, 188 (1939).
965a. W. H. Shaffer and R. P. Schuman, *J. Chem. Phys.* **12**, 504 (1944).
966. S. L. Shapiro, J. A. Giordmane, and K. W. Wecht, *Phys. Rev. Letters* **19**, 1093 (1967).
967. A. I. Shatenshtein, E. A. Yakovleva, E. N. Zvyagintseva, Ya. M. Varshavskii, E. A. Izrailevich, and N. M. Dykhno, "Isotopic Water Analysis," U.S. Atomic Energy Commission, AEC-tr-4136 (1960).

968. I. Shavitt, *J. Comp. Phys.* **6**, 124 (1970).
969. A. N. Shaw, *Trans. Roy. Soc. Canada* **III**, 187 (1924).
970. Y. R. Shen and N. Bloembergen, *Phys. Rev.* **137A**, 1787 (1965).
971. S. Shibata and L. S. Bartell, *J. Chem. Phys.* **42**, 1147 (1965).
972. K. Shimaoka, *J. Phys. Soc. Japan* **15**, 106 (1960).
973. J. A. Shufle, *Chem. Ind.* **1965**, 690.
974. J. A. Shufle, *J. Geophys. Res.* **74**, 5560 (1969).
975. J. A. Shufle and M. M. Venugopalan, *J. Geophys. Res.* **72**, 3271 (1967).
976. H. Shull, *J. Chem. Phys.* **30**, 1405 (1959).
977. H. Shull and P. O. Löwdin, *Phys. Rev.* **110**, 1466 (1958).
978. G. Siegle and M. Weithase, *Z. Phys.* **219**, 364 (1969).
979. M. G. Silk and B. O. Wade, *J. Nucl. Energy* **23**, 1 (1969).
980. L. Simons, Soc. Sci. Fennica, *Commentationes Phys. Math.* **10**(9) (1939).
981. J. H. Simpson and H. Y. Carr, *Phys. Rev.* **111**, 1201 (1958).
982. Y. Singh, S. K. Deb, and A. K. Barua, *J. Chem. Phys.* **46**, 4036 (1967).
983. K. S. Singwi and A. Sjolander, *Phys. Rev.* **119**, 863 (1960).
984. A. M. Sirota, *Teploenergetika* **1958**(7), 10.
985. A. M. Sirota, *Teploenergetika* **1963**(9), 57.
986. A. M. Sirota and B. K. Mal'tsev, *Teploenergetika* **1959**(6), 7.
987. A. M. Sirota and B. K. Mal'tsev, *Teploenergetika* **1960**(10), 67.
988. A. M. Sirota and B. K. Mal'tsev, *Teploenergetika* **1962**(1), 52.
989. A. M. Sirota and B. K. Mal'tsev, *Teploenergetika* **1962**(7), 70.
990. A. M. Sirota and Z. Kh. Shrago, *Teploenergetika* **1968**(3), 24.
991. A. M. Sirota and Z. Kh. Shrago, *Teploenergetika* **1968**(8), 86.
992. A. M. Sirota and Z. Kh. Shrago, *Thermal Engineering* **1968**(3), 32.
993. A. M. Sirota and Z. Kh. Shrago, *Thermal Engineering* **1968**(8), 114.
994. A. M. Sirota and D. L. Timrot, *Teploenergetika* **1956**(7), 16.
995. Sixth International Conference on the Properties of Steam, Official Release, Secretariat of the International Conference on the Properties of Steam, American Society of Mechanical Engineers, New York, 1963.
996. Sixth International Conference on the Properties of Steam, Supplementary Release on Transport Properties, Secretariat of the International Conference on the Properties of Steam, American Society of Mechanical Engineers, New York, 1964; also R. W. Haywood, *J. Eng. Power* **88**, 63 (1966).
997. A. Sjolander, *in* "Phonons and Phonon Interactions" (T. A. Bak, ed.), Benjamin, New York (1964).
998. A. Sjolander, *in* "Thermal Neutron Scattering" (P. A. Egelstaff, ed.), Academic Press, London (1965), p. 291.
998a. A. Skerbele, V. D. Meyer, and E. N. Lassettre, *J. Chem. Phys.* **43**, 817 (1965).
999. C. P. Slichter, "Principles of Magnetic Resonance," Harper and Row, New York (1963).
1000. W. Slie, A. Donfor, and T. Litovitz, *J. Chem. Phys.* **44**, 3712 (1966).
1001. L. B. Smith and F. G. Keyes, *Proc. Am. Acad. Arts Sci.* **69**, 285 (1934).
1002. L. B. Smith, F. G. Keyes, and H. T. Gerry, *Proc. Am. Acad. Arts Sci.* **69**, 137 (1934).
1003. A. Smith and A. Lawson, *J. Chem. Phys.* **22**, 351 (1954).
1004. G. Soda and T. Chiba, *J. Chem. Phys.* **50**, 439 (1969).
1005. N. D. Sokolov, *in* "Hydrogen Bonding" (D. Hadzi and H. W. Thompson, eds.), Pergamon Press, Oxford (1959).

1006. R. L. Somarjai and D. F. Hornig, *J. Chem. Phys.* **36**, 1980 (1962).
1007. C. J. Sparks, Metals and Ceramics Division Annual Progress Report for period ending 30 June 1966, ORNL-3970 (1966).
1008. H. W. Spiess, B. B. Garrett, R. K. Sheline, and S. W. Rabideau, *J. Chem. Phys.* **51**, 1201 (1969).
1009. T. Springer, C. Hofmeyr, S. Kornblicher, and H. D. Lemmel, Third International Conference on Peaceful Uses of Atomic Energy, A/CONF·28/P/763 (1964).
1010. A. E. Stanevich and N. G. Yaroslavskii, *Opt. Spectry.* (*USSR*) **10**, 278 (1961).
1011. E. M. Stanley and R. C. Batten, *J. Phys. Chem.* **73**, 1187 (1969).
1012. D. Staschewski, *Ber. Bunsenges. Phys. Chem.* **73**, 59 (1969).
1013. F. Steckel and S. Szapiro, *Trans. Faraday Soc.* **59**, 331 (1963).
1014. W. A. Steele, *J. Chem. Phys.* **38**, 2404, 2411 (1963).
1015. W. A. Steele and R. Pecora, *J. Chem. Phys.* **42**, 1863 (1965).
1016. A. Steinemann and H. Gränicher, *Helv. Phys. Acta* **30**, 553 (1957).
1017. R. M. Stevens and M. Karplus, *J. Chem. Phys.* **49**, 1094 (1968).
1018. R. M. Stevens and W. N. Lipscomb, *J. Chem. Phys.* **40**, 2238 (1964).
1019. R. M. Stevens and W. N. Lipscomb, *J. Chem. Phys.* **41**, 184, 3710 (1964).
1020. R. M. Stevens and W. N. Lipscomb, *J. Chem. Phys.* **42**, 3666, 4302 (1965).
1021. R. M. Stevens, R. M. Pitzer, and W. N. Lipscomb, *J. Chem. Phys.* **38**, 550 (1963).
1022. D. P. Stevenson, *J. Phys. Chem.* **69**, 2145 (1965).
1023. M. J. Stevenson and C. H. Townes, *Phys. Rev.* **107**, 635 (1957).
1024. G. W. Stewart, *Phys. Rev.* **37**, 9 (1931).
1025. H. H. Stiller and H. R. Danner, *in* "Inelastic Scattering of Neutrons," Vol. 1, International Atomic Energy Agency, Vienna (1961).
1026. F. H. Stillinger, Jr. (to be published).
1027. F. H. Stillinger and A. Ben-Naim, *J. Phys. Chem.* **73**, 900 (1969).
1028. H. F. Stimson, *Am. J. Phys.* **23**, 614 (1955).
1029. H. F. Stimson, *J. Res. Natl. Bur. Std.* **73A**, 493 (1969).
1030. W. H. Stockmayer, *J. Chem. Phys.* **9**, 398 (1941).
1031. G. Stokes, *Trans. Cambridge Soc.* **8**, 287 (1845).
1032. R. H. Stokes and R. Mills, "Viscosity of Electrolytes and Related Properties," Pergamon Press, Oxford (1965), p. 74.
1033. V. Stott and P. H. Bigg, *in* "International Critical Tables" (E. W. Washburn, ed.), Vol. III, McGraw-Hill Book Co., New York (1928), p. 24.
1034. A. Suggett, P. A. Mackness, M. J. Tait, H. W. Loeb, and G. M. Young, *Nature* **228**, 456 (1970).
1035. M. Sugisaki, H. Suga, and S. Seki, *Bull. Chem. Soc. Japan* **40**, 2984 (1967).
1036. M. Sugisaki, H. Suga, and S. Seki, *Bull. Chem. Soc. Japan* **41**, 2591 (1968).
1037. W. H. Surber and G. E. Crouch, *J. Appl. Phys.* **19**, 1130 (1948).
1038. W. Sutherland, *Phil. Mag.* **50**(5), 460 (1900).
1039. J. F. Swindells, J. R. Coe, and T. B. Godfrey, *J. Res. Natl. Bur. Std.* **48**, 1 (1952).
1040. J. F. Swindells, R. Ullman, and H. Mark, *in* "Techniques of Organic Chemistry," (A. Weissberger, ed.), 3rd ed., Interscience Publishers, New York (1959), Vol. I, Part I, p. 689.
1040a. E. Switkes, R. M. Stevens, and W. N. Lipscomb, *J. Chem. Phys.* **51**, 5229 (1969).
1041. S. Szapiro and F. Steckel, *Trans. Faraday Soc.* **63**, 883 (1967).
1042. M. J. Tait, U. R. L. Report, 1969.

1043. M. J. Tait and F. Franks, *Nature* **230**, 91 (1971).
1044. P. G. Tait, "Voyage of HMS Challenger" (Physics and Chemistry), (1889), Vol. II, Part IV, 76 pp.
1045. P. G. Tait, *Proc. Roy. Soc. (Edinburgh)* **20**, 63 (1893).
1046. P. G. Tait, "Scientific Papers," Cambridge Press, Cambridge (1898), Vol. I, p. 261.
1047. P. G. Tait, "Scientific Papers," Cambridge Press, Cambridge (1900), Vol. II, p. 334.
1048. R. Tait, *Acustica* **7**, 193 (1957).
1049. T. E. Talpey, *Onde Electrique* **33**, 561 (1953).
1050. G. Tammann, *Z. Physik. Chem.* **17**, 620 (1895).
1051. G. Tammann, *Ann. Phys.* **2**, 1 (1900).
1052. G. Tammann and A. Rühenbeck, *Ann. Physik* **5**. Folge **13**, 63 (1932).
1053. G. Tammann and E. Schwarzkopf, *Z. Anorg. Allgem. Chem.* **174**, 216 (1928).
1054. K. Tanaka, M. Sasaki, H. Hattori, Y. Kobayashi, K. Haishima, K. Sato, and M. Tashiro, Viscosity of Steam at High Pressures and High Temperatures, J.C.P.S. Report No. 10, Resources Research Institute, Kawaguchi, Saitama, Japan, 1964.
1055. I. Tanishita and A. Nagashima, *Bull. Japan. Soc. Mech. Eng.* **11**, 1161 (1968).
1056. W. J. Taylor, *J. Chem. Phys.* **48**, 2385 (1968).
1057. M. J. Taylor and E. Whalley, *J. Chem. Phys.* **40**, 1660 (1964).
1058. B. Ya. Teitel'baum, T. A. Gortalova, and E. E. Sidorova, *Zh. Fiz. Khim.* **25**, 911 (1951).
1059. H. N. V. Temperley, J. S. Rowlinson, and G. S. Rushbrooke, "Physics of Simple Liquids," North-Holland Publishing Co., Amsterdam (1968).
1060. R. W. Terhune, P. D. Maker, and C. M. Savage, *Phys. Rev. Letters* **14**, 681 (1965).
1061. M. Thiesen, *Wiss. Abhandl. Physik.-Techn. Reichsanstalt* **4**, 1 (1904).
1062. M. Thiesen, K. Scheel, and H. Diesselhorst, *Wiss. Abhandl. Physik.-Techn. Reichsanstalt* **3**, 1 (1900).
1063. A. Thorhaug, Ph. D. thesis, University of Miami, 1969.
1064. A. Thorhaug and W. Drost-Hansen, *Nature* **215**, 506 (1967).
1065. P. Thomas, *Z. Physik* **208**, 338 (1968).
1066. R. N. Thurston, *in* "Physical Acoustics," (W. P. Mason, ed.), Academic Press, New York (1964), Vol. 1, Part A, pp. 1, 11.
1067. L. W. Tilton and J. K. Taylor, *J. Res. Natl. Bur. Std.* **18**, 205 (1937).
1068. D. L. Timrot and A. V. Khlopkina, *Teploenergetika* **1963** (**7**), 64.
1069. D. L. Timrot and K. F. Shuiskaya, *Atom. Energiya* **7**, 459 (1959).
1070. D. L. Timrot and N. B. Vargaftik, *Zh. Tekh. Fiz.* **10**, 1063 (1940).
1071. K. Tödheide, *Ber. Bunsenges.* **70**, 1022 (1966).
1071a. S. Trajmar, W. Williams, and A. Kuppermann, *J. Chem. Phys.* **54**, 2274 (1971).
1072. N. J. Trappeniers, A. Botzen, J. van Oosten and H. R. van den Berg, *Physica* **31**, 945 (1965).
1073. N. J. Trappeniers, C. J. Gerritsma, and P. H. Oosting, *Phys. Letters* **18**, 256 (1965).
1074. E. B. Treacy, and Y. Beers, *J. Chem. Phys.* **36**, 1473 (1962).
1075. G. Treiber, Diplomarbeit, Universität Karlsruhe, 1967.
1075a. D. G. Truhlar, Ph. D. thesis, California Institute of Technology, Proposition V (1970).
1076. H. Tsubomura, *Bull. Chem. Soc. Japan* **27**, 445 (1954).
1077. O. Tumlirz, *Sitzber. Akad. Wiss. Wien, Math.-Naturw. Kl., Abt.* **IIa**, **118**, 203 (1909).
1078. V. F. Turchin, "Slow Neutrons," Oldbourne Press, London (1965).

1079. A. Tursi and E. Nixon, *J. Chem. Phys.* **52**, 1521 (1970).
1080. G. M. Uddin, Ph. D. thesis, University of London, 1968.
1081. O. V. Uvarov, N. M. Sokolov, and N. M. Zhavoronkov, *Kernergie* **5**, 323 (1962).
1082. V. Vand and W. A. Senior, *J. Chem. Phys.* **43**, 1878 (1965).
1083. F. B. van Duijneveldt, *J. Chem. Phys.* **49**, 1424 (1968).
1084. J. C. V. M. van Duijneveldt-van de Rijdt and F. B. van Duijneveldt, *Chem. Phys. Letters* **2**, 565 (1968).
1085. A. L. van Geet and D. N. Hume, *Anal. Chem.* **37**, 979 (1965).
1086. W. A. van Hook, *J. Phys. Chem.* **72**, 1234 (1968).
1087. L. van Hove, *Phys. Rev.* **95**, 249 (1954).
1088. C. L. van Panthaleon van Eck, H. Mendel, and W. Boog, *Disc. Faraday Soc.* **24**, 200 (1957).
1089. C. L. van Panthaleon van Eck, H. Mendel, and W. Boog, *Proc. Roy. Soc.* **A247**, 472 (1958).
1090. M. van Thiel, E. D. Becker, and G. C. Pimentel, *J. Chem. Phys.* **27**, 486 (1957).
1091. J. H. van Vleck, "The Theory of Electric and Magnetic Susceptibilities," Oxford University Press, London (1932).
1092. J. H. van Vleck, *J. Chem. Phys.* **5**, 556 (1937).
1093. J. H. van Vleck, *Phys. Rev.* **74**, 1168 (1948).
1094. N. B. Vargaftik, *Zh. Tekh. Fiz.* **4**, 341 (1937).
1095. N. B. Vargaftik and O. N. Oleshchuk, *Izv. Vses. Teplotekh. Inst.* **6**, 7 (1946).
1096. N. B. Vargaftik and O. N. Oleshchuk, *Teploenergetika* **1959**(10), 70.
1097. N. B. Vargaftik, O. N. Oleshchuk, and P. E. Belyakova, *Soviet J. Atomic Energy* **7**, 931 (1959).
1098. N. B. Vargaftik and E. V. Smirnova, *Soviet Phys.—Tech. Phys.* **1**, 1221 (1956).
1099. N. B. Vargaftik and E. V. Smirnova, *Zh. Tekh. Fiz.* **26**, 1251 (1956).
1100. N. B. Vargaftik and A. Zimina, *Teploenergetika* **1964**(12), 84.
1101. V. M. Vdovenko, Yu. V. Gurikov, and E. K. Legin, *J. Struct. Chem.* **7**, 765 (1966).
1102. R. Vedam and G. Holton, *J. Acoust. Soc. Am.* **43**, 108 (1968).
1103. J. E. S. Venart, *in* "Advances in Thermophysical Properties at Extreme Temperatures and Pressures—Proceedings of the Third Symposium on Thermophysical Properties," (S. Gratch, ed.), American Society of Mechanical Engineers, New York (1965), p. 237.
1104. R. E. Verall and W. A. Senior, *J. Chem. Phys.* **50**, 2746 (1969).
1105. J. Verhoeven, H. Bluyssen, and A. Dymanus, *Phys. Letters* **26A**, 424 (1968).
1106. J. Verhoeven and A. Dymanus, *J. Chem. Phys.* **52**, 3222 (1970).
1107. J. Verhoeven, A. Dymanus, and H. Bluyssen, *J. Chem. Phys.* **50**, 3330 (1969).
1108. H. H. V. Vernon, *Phil. Mag.* **31**, 387 (1891).
1109. E. J. W. Verwey, *Rec. Trav. Chim.* **60**, 887 (1941).
1110. M. Vincentini-Missoni, R. I. Joseph, M. S. Green, and J. M. H. Levelt Sengers, *Phys. Rev.* **B1**, 2312 (1970).
1111. M. Vincentini-Missoni, J. M. H. Levelt Sengers, and M. S. Green, *J. Res. Natl. Bur. Std.* **73A**, 563 (1969).
1112. G. Vineyard, *Phys. Rev.* **110**, 999 (1958).
1113. H. Vogel, *Physik. Z.* **22**, 645 (1921).
1114. H. Vogel, *Z. Angew. Chem.* **35**, 561 (1922).
1115. A. von Hippel, Technical Reports 1 and 2, Contract NOOO14-67A-0204-0003, Massachusetts Institute of Technology (1967).

1116. A. von Hippel and co-workers, Technical Report 6, Contract N00014-67A-0204-0003, Massachusetts Institute of Technology (1969).
1117. A. von Hippel, *J. Chem. Phys.* **54**, 145 (1971).
1118. A. von Hippel, D. B. Knoll, and W. B. Westphal, *J. Chem. Phys.* **54**, 134 (1971).
1119. M. P. Vukalovich, "Thermodynamic Properties of Water and Steam," VEB Verlag Technik, Berlin (1958).
1120. M. P. Vukalovich, M. S. Trakhtengerts, and G. A. Spiridonov, *Teploenergetika* **1967**(7), 65; tabulated by J. H. Dymond and E. B. Smith, "The Virial Coefficients of Gases," Oxford (1969), p. 168.
1121. M. P. Vukalovich, V. N. Zubarev, A. A. Aleksandrov, and Yu. Ya. Kalinin, *Teploenergetika* **1959**(10), 74.
1122. M. P. Vukalovich, V. N. Zubarev, A. A. Aleksandrov, and Yu. Ya. Kalinin, *Teploenergetika* **1961**(4), 76.
1123. M. P. Vukalovich, V. N. Zubarev, and P. G. Prusakov, *Teploenergetika* **1958**(7), 22.
1124. M. P. Vukalovich, V. N. Zubarev, and P. G. Prusakov, *Teploenergetika* **1962**(3), 56.
1125. G. Wada, *Bull. Chem. Soc. Japan* **34**, 604 (1961).
1126. G. Wada, *Bull. Chem. Soc. Japan* **34**, 955 (1961).
1127. P. Waldstein and S. W. Rabideau, *J. Chem. Phys.* **47**, 5338 (1967).
1128. P. Waldstein, S. W. Rabideau, and J. A. Jackson, *J. Chem. Phys.* **41**, 3407 (1964).
1129. P. Waldstein, S. W. Rabideau, and J. A. Jackson, *J. Chem. Phys.* **41**, 4008 (1964).
1130. W. A. Walker, Thermodynamic Properties of Liquid Water to 1 kbar, U.S. Naval Ordnance Laboratory, NOLTR 66-217 (1967).
1131. T. T. Wall, *J. Chem. Phys.* **51**, 113 (1969).
1132. T. T. Wall and D. F. Hornig, *J. Chem. Phys.* **43**, 2079 (1965).
1133. G. E. Walrafen, *J. Chem. Phys.* **36**, 1035 (1962).
1134. G. E. Walrafen, *J. Chem. Phys.* **40**, 3249 (1964).
1135. G. E. Walrafen, *J. Chem. Phys.* **44**, 1546 (1966).
1136. G. E. Walrafen, *J. Chem. Phys.* **46**, 1870 (1967).
1137. G. E. Walrafen, *J. Chem. Phys.* **47**, 114 (1967).
1138. G. E. Walrafen, *J. Chem. Phys.* **48**, 244 (1968).
1139. G. E. Walrafen, *in* "Hydrogen-Bonded Solvent Systems" (A. K. Covington and P. Jones, eds.), Taylor and Francis, London (1968).
1140. G. E. Walrafen, *J. Chem. Phys.* **50**, 560 (1969).
1141. G. E. Walrafen, *J. Chem. Phys.* **50**, 567 (1969).
1142. G. E. Walrafen, *J. Chem. Phys.* **52**, 4176 (1970).
1143. J. M. Walsh and M. H. Rice, *J. Chem. Phys.* **26**, 815 (1957).
1144. J. H. Wang, *J. Am. Chem. Soc.* **73**, 510 (1951).
1145. J. H. Wang, *J. Am. Chem. Soc.* **73**, 4181 (1951).
1146. J. H. Wang, *J. Am. Chem. Soc.* **76**, 4755 (1954).
1147. J. H. Wang, *J. Phys. Chem.* **69**, 4412 (1965).
1148. J. H. Wang, C. B. Anfinsen, and F. M. Polestra, *J. Am. Chem. Soc.* **76**, 4763 (1954).
1149. J. H. Wang, C. V. Robinson, and I. S. Edelman, *J. Am. Chem. Soc.* **75**, 466 (1953).
1150. J. Waser and V. Schomaker, *Rev. Mod. Phys.* **25**, 671 (1953).
1151. K. Watanabe and M. Zelikoff, *J. Opt. Soc. Amer.* **43**, 753 (1953).
1151a. J. K. G. Watson, *J. Chem. Phys.* **45**, 1360 (1966); **46**, 1935 (1967); **48**, 4517 (1968).
1152. J. K. G. Watson, *Mol. Phys.* **15**, 479 (1968).
1153. J. K. G. Watson, *Mol. Phys.* **19**, 465 (1970).
1154. I. C. Watt and J. D. Leeder, *J. Text. Inst.* **59**, 358 (1968).

1155. R. M. Waxler and C. E. Weir, *J. Res. Natl. Bur. Std.* **67A**, 163 (1963).
1156. W. Weber, *Z. Angew. Phys.* **7**, 96 (1955).
1157. M. Weissman, *J. Chem. Phys.* **44**, 422 (1966).
1158. M. Weissmann and L. Blum, *Trans. Faraday Soc.* **64**, 2606 (1968).
1159. M. Weissman, L. Blum, and M. Cohen, *Chem. Phys. Letters* **1**, 95 (1967).
1160. K. H. Welge and F. Stuhl, *J. Chem. Phys.* **46**, 2440 (1967).
1161. B. G. West and M. Mizushima, *J. Chem. Phys.* **38**, 251 (1963).
1162. E. Whalley, "Proc. Conf. Thermodynamics Transport Properties Fluids, London, 1957," Institution of Mechanical Engineers, London (1958), p. 15.
1163. E. Whalley, *in* "Advances in High-Pressure Research" (R. S. Bradley, ed.) Vol. 1, Academic Press, London–New York (1966), pp. 143–193.
1164. E. Whalley, "Developments in Applied Spectroscopy, Vol. 6, Plenum Press, New York (1968), p. 277.
1165. E. Whalley, D. W. Davidson, and J. B. R. Heath, *J. Chem. Phys.* **45**, 3976 (1966).
1166. E. Whalley and J. B. R. Heath, *J. Chem. Phys.* **45**, 3976 (1966).
1167. E. Whalley, J. B. R. Heath, and D. W. Davidson, *J. Chem. Phys.* **48**, 2362 (1968).
1168. D. H. Whiffen and H. W. Thompson, *Trans. Faraday Soc.* **42**, 114 (1946).
1169. J. H. Whitelaw, *J. Mech. Eng. Sci.* **2**, 288 (1960).
1170. H. Whiting, Thesis, Harvard University, 1884.
1171. B. Widom, *J. Chem. Phys.* **43**, 3898 (1965).
1172. E. Willis, G. K. Rennie, C. Smart, and B. A. Pethica, *Nature* **222**, 159 (1969).
1173. E. B. Wilson, J. C. Decius, and P. C. Cross, "Molecular Vibrations," McGraw-Hill Book Co., New York (1955).
1174. E. B. Wilson and J. B. Howard, *J. Chem. Phys.* **4**, 260 (1936).
1175. G. J. Wilson, R. K. Chan, D. W. Davidson, and E. Whalley, *J. Chem. Phys.* **43**, 2384 (1965).
1176. W. Wilson and D. Bradley, *Deep-Sea Res.* **15**, 355 (1968).
1177. D. E. Woessner, *J. Chem. Phys.* **40**, 2341 (1964).
1178. H. Wolff, *in* "Physics of Ice," (N. Riehl, B. Bullemer, and H. Engelhardt, eds.), Plenum Press, New York (1969), p. 305.
1179. H. Wolff and E. Wolff, *Ber. Bunseng. Phys. Chem.* **73**, 393 (1969).
1180. E. O. Wollan, W. L. Davidson, and C. G. Shull, *Phys. Rev.* **75**, 1348 (1949).
1181. J. Wonham, *Nature* **215**, 1053 (1967).
1182. W. W. Wood, "Physics of Simple Liquids," North-Holland Publishing Co., Amsterdam (1968), Chapter 5.
1183. E. J. Woodbury and W. K. Ng, *Proc. IRE* **50**, 2347 (1962).
1184. H. W. Woolley, *J. Chem. Phys.* **21**, 236 (1953).
1185. E. J. Workman and S. E. Reynolds, *Phys. Rev.* **78**, 254 (1950).
1186. J. D. Worley and I. M. Klotz, *J. Chem. Phys.* **45**, 2868 (1966).
1187. J. Wyman and E. N. Ingalls, *J. Amer. Chem. Soc.* **60**, 1182 (1938).
1188. K. K. Yallabandi and P. M. Parker, *J. Chem. Phys.* **49**, 410 (1968).
1189. V. I. Yashkichev, *J. Struct. Chem.* **10**, 676 (1969).
1190. M. Yasumi, *Bull. Chem. Soc. Japan* **24**, 53, 60 (1951).
1191. W. Yellin and W. L. Courchene, *Nature* **219**, 852 (1968).
1192. T. F. Young, L. F. Maranville, and H. M. Smith, *in* "The Structure of Electrolytic Solutions" (W. J. Hamer, ed.), John Wiley and Sons, New York (1959), pp. 35–63.
1193. M. M. Yuknavech, Ballistic Research Laboratories, Aberdeen Proving Ground, Md., Memorandum Rept. No. 1563 (1964).

1194. W. H. Zachariasen, *Phys. Rev.* **47**, 277 (1935).
1195. A. Zajac, *J. Chem. Phys.* **29**, 324 (1958).
1196. Y. B. Zeldovich, S. B. Kormer, M. V. Sinitsyn, and K. B. Yushko, *Dokl. Akad. Nauk SSSR* **138**, 1333 (1961); English transl.: *Soviet Phys.—Doklady* **6**, 494 (1961).
1197. F. Zernike and J. Prins, *Z. Physik* **41**, 184 (1927).
1198. F. Zernike and J. Prins, *Z. Physik* **56**, 617 (1929).
1199. B. V. Zheleznii, *Zh. Fiz. Khim.* **43**, 2343 (1969).
1200. H. Ziebland and J. T. A. Burton, *Int. J. Heat Mass Transfer* **1**, 242 (1960).
1201. K. Zieborak, *Z. Physik. Chem.* (*Leipzig*) **231**, 248 (1966).
1202. B. H. Zimm and J. K. Bragg, *J. Chem. Phys.* **31**, 526 (1959).
1203. R. Zimmermann and G. C. Pimental, *in* "Advances in Molecular Spectroscopy," (A. Mangini, ed.), Macmillan, New York (1962).
1204. V. M. Zolotarev, B. A. Mikhailov, L. I. Alperovich, and S. I. Popov, *Opt. Spectry.* (*USSR*) **1969**, 430.

Index

664. R. E. Leckenby, E. J. Robbins, and P. A. Trevalion, *Proc. Roy. Soc.* (*London*) **A280**, 409 (1964).

665. B. le Neindre, P. Bury, R. Tufeu, P. Johannin, and B. Vodar, Résultats expérimentaux sur la conductivité thermique de l'eau et de l'eau lourde en phase liquide, jusqu'à une température de 370°C, et leurs discussions, Laboratoire des Hautes Pressions, Centre National de la Recherche, France (about 1968).

666. B. le Neindre, P. Johannin, and B. Vodar, *Compt. Rend.* **258**, 3277 (1964).

667. B. le Neindre, P. Johannin, and B. Vodar, *Compt. Rend.* **261**, 67 (1965).

668. J. Lennard-Jones and A. F. Devonshire, *Proc. Roy. Soc.* **A163**, 53 (1937).

669. J. Lennard-Jones and A. F. Devonshire, *Proc. Roy. Soc.* **A165**, 1 (1938).

670. J. Lennard-Jones and A. F. Devonshire, *Proc. Roy. Soc.* **A169**, 317 (1939).

671. J. Lennard-Jones and J. A. Pople, *Proc. Roy. Soc.* (*London*) **A202**, 166 (1950).

672. J. Lennard-Jones and J. A. Pople, *Proc. Roy. Soc.* (*London*) **A205**, 155 (1951).

673. J. M. H. Levelt Sengers, Thermodynamic Anomalies near the Critical Point of Steam, Paper communicated to the Seventh International Conference on the Properties of Steam, Tokyo, 1968.

674. J. M. H. Levelt Sengers, *Ind. Eng. Chem. Fundamentals* **9**, 470 (1970).

675. J. M. H. Levelt Sengers and M. Vicentini-Missoni, *in* "Proceedings of the Fourth Symposium on Thermophysical Properties," (J. R. Moszynski, ed.), American Society of Mechanical Engineers, New York (1968), p. 79.

676. A. M. Levelut and A. Guinier, *Bull. Soc. Franc. Mineral. Crystallogr.* **40**, 445 (1967).

677. S. Levine and J. W. Perram, *in* "Hydrogen-Bonded Solvent Systems," (A. K. Covington and P. Jones, eds.), Taylor and Francis, London (1968), p. 115.

678. H. A. Levy, M. Danford, and A. H. Narten, Data Collection and Evaluation with an X-Ray Diffractometer Designed for the Study of Liquid Structure, ORNL-3960 (1966).

679. G. N. Lewis and R. T. MacDonald, *J. Am. Chem. Soc.* **55**, 3057 (1933).

680. G. N. Lewis and R. T. MacDonald, *J. Am. Chem. Soc.* **55**, 4730 (1933).

681. Y.-H. Li, *J. Geophys. Res.* **72**, 2675 (1967).

682. J. H. Liang, Ph. D. Thesis, The Ohio State University, 1970.

683. M. Lichtenstein, V. E. Derr, and J. J. Gallagher, *J. Mol. Spectry.* **20**, 391 (1966).

684. T. F. Lin and A. B. F. Duncan, *J. Chem. Phys.* **48**, 866 (1968).

685. H.-M. Lin and R. L. Robinson, *J. Chem. Phys.* **52**, 3727 (1970).

686. H. Lindner, doctoral dissertation, University of Karlsruhe, 1970.

687. J. W. Linnett, *Science* **168**, 1719 (1970).

688. J. W. Linnett and A. Poe, *Trans. Faraday Soc.* **47**, 1033 (1951).

689. E. R. Lippincott, *J. Chem. Phys.* **23**, 603 (1955).

690. E. R. Lippincott and R. Schroeder, *J. Chem. Phys.* **23**, 1099 (1955).

691. E. R. Lippincott, R. Stromberg, W. Grant, and G. Cessac, *Science* **164**, 1482 (1969).

692. T. Litovitz and E. Carnevale, *J. Appl. Phys.* **26**, 816 (1955).

693. T. Litovitz and C. Davis, *in* "Physical Acoustics," Vol. IIA, Academic Press, New York (1965).

694. T. Litovitz and G. McDuffie, *J. Chem. Phys.* **39**, 729 (1963).

695. W. M. Lomer and G. G. Low, *in* "Thermal Neutron Scattering" (P. A. Egelstaff, ed.), Academic Press, London (1965), p. 1.

696. K. Lonsdale, *Proc. Roy. Soc.* **A247**, 424 (1958).

697. L. G. Longsworth, *J. Phys. Chem.* **64**, 1914 (1960).

698. H. A. Lorentz, "The Theory of Electrons," Dover Publications, New York (1952), p. 308.
699. P. O. Löwdin, *Phys. Rev.* **97**, 1509 (1955).
700. P. O. Löwdin and H. Shull, *Phys. Rev.* **101**, 1730 (1956).
701. W. A. P. Luck, *Disc. Faraday Soc.* **43**, 115 (1967).
702. W. A. P. Luck, *Disc. Faraday Soc.* **43**, 132 (1967).
703. W. A. P. Luck, *in* "Physico-Chemical Processes in Mixed Aqueous Solvents," (F. Franks, ed.), Heinemanns, London (1967).
704. W. A. P. Luck, Habilitationsschrift, "Spektroskopische Bestimmungen der Wasserstoffbrückenbindungen in Naher IR," University of Heidelberg, 1968.
705. W. A. P. Luck and W. Ditter, *Z. Naturforsch.* **24b**, 482 (1969).
706. B. Lundqvist and T. Persson, *Brennst.-Wärme-Kraft* **17**, 356 (1965).
707. Z. Luz and G. Yagil, *J. Phys. Chem.* **70**, 554 (1966).
708. B. J. McBride and S. Gordon, *J. Chem. Phys.* **35**, 2198 (1961).
709. D. W. McCall, D. C. Douglass, and E. W. Anderson, *J. Chem. Phys.* **31**, 1555 (1959).
710. M. McCarty, Jr., and S. V. K. Babu, *J. Phys. Chem.* **74**, 1113 (1970).
711. A. L. McClellan, "Dipole Moments," Freeman, San Francisco (1963).
712. J. P. McCullough, R. E. Pennington, and G. Waddington, *J. Am. Chem. Soc.* **74**, 4439 (1952).
713. J. R. Macdonald, *Rev. Mod. Phys.* **41**, 316 (1969).
714. P. C. McKinney and G. M. Barrow, *J. Chem. Phys.* **31**, 294 (1959).
715. E. McLaughlin, *Physica* **26**, 650 (1960).
716. E. McLaughlin, *in* "Thermal Conductivity" (R. P. Tye, ed.), Vol. 2, Academic Press, London (1969), p. 1.
717. J. A. McMillan and S. C. Los, *Nature* **206**, 806 (1965).
718. H. J. McSkimin, *J. Acoust. Soc. Am.* **37**, 325 (1965).
719. R. McWeeny and K. A. Ohno, *Proc. Roy. Soc.* (*London*) **A255**, 367 (1960).
720. M. Magat, *Ann. Phys.* (*Paris*) **6**, 109 (1936).
721. M. Magat, *J. Chem. Phys.* **45**, 91 (1948).
722. M. A. Maidique, A. von Hippel, and W. B. Westphal, *J. Chem. Phys.* **54**, 150 (1971).
723. S. Maier and E. U. Franck, *Ber. Bunsenges. Phys. Chem.* **70**, 639 (1966).
724. P. D. Maker, *Phys. Rev.* **A1**, 923 (1970).
725. C. G. Malmberg and A. A. Maryott, *J. Res. Natl. Bur. Std.* **56**, 1 (1956).
726. J. Malsch, *Ann. Phys. Leipzig* **84**, 841 (1927).
727. J. Malsch, *Phys. Z.* **29**, 770 (1928).
728. A. M. Mamedov, *Teploenergetika* **1960**(9), 71.
729. A. M. Mamedov, *Teploenergetika* **1963**(5), 69.
730. K. Mangold and E. U. Franck, *Ber. Bunsenges.* **73**, 21 (1969).
731. J. P. Marckmann and E. Whalley, *J. Chem. Phys.* **41**, 1450 (1964).
732. T. W. Marshall and J. A. Pople, *Mol. Phys.* **1**, 199 (1958).
733. E. A. Mason, *in* "Proceedings of the Fourth Symposium on Thermophysical Properties" (J. R. Moszynski, ed.), American Society of Mechanical Engineers, New York (1968), p. 21.
734. E. A. Mason and T. H. Spurling, "The Virial Equation of State," Pergamon Press, Oxford (1969), p. 1.
735. P. R. Mason, J. B. Hasted, and L. Moore, to be published.

736. F. A. Matsen, *Adv. Quantum Chem.* **1**, 60 (1964).
737. F. Mayinger, *Int. J. Heat Mass Transfer* **5**, 807 (1962).
738. H. D. Megaw, *Nature*, **134**, 900 (1934).
739. S. Meiboom, *J. Chem. Phys.* **34**, 375 (1961).
740. M. Menaché, *Cahiers Oceanographiques* **17**, 625 (1965).
741. M. Menaché, *Cahiers Oceanographiques* **18**, 477 (1966).
742. M. Menaché, *Metrologia* **3**, 58 (1967).
743. D. Mendeleef, *Phil. Mag.* **33**, 99 (1892).
744. R. P. Messmer, *Science* **168**, 679 (1970).
745. H. H. Meyer, *Ann. Physik* **5**, 701 (1930).
745a. W. Meyer, *Internat. J. Quantum Chem.* **5S**, 296 (1971).
746. A. Michels, B. Blaisse, C. A. Ten Seldam, and H. M. Wouters, *Physica* **10**, 613 (1943).
747. A. Michels, A. Botzen, and W. Schuurmann, *Physica* **20**, 1141 (1954).
748. C. Migchelsen, H. J. C. Berendsen, and A. Rupprecht, *J. Mol. Biol.* **37**, 235 (1968).
749. F. T. Miles and A. W. C. Menzies, *J. Am. Chem. Soc.* **58**, 1067 (1936).
750. A. A. Miller, *J. Chem. Phys.* **38**, 1568 (1963).
750a. J. H. Miller and H. P. Kelly, *Phys. Rev.* **A4**, 480 (1971).
751. K. J. Miller, S. R. Mielczarek, and M. Krauss, *J. Chem. Phys.* **51**, 26 (1969).
752. F. J. Millero, R. W. Curry, and W. Drost-Hansen, *J. Chem. Eng. Data* **14**, 422 (1969).
753. F. J. Millero, R. Dexter, and E. Hoff, *J. Chem. Eng. Data* **16**, 85 (1971).
754. F. J. Millero and F. K. Lepple, *J. Chem. Phys.* **54**, 946 (1971).
755. V. N. Mineev and R. M. Zaidel', *Zh. Eksperim. i Teor. Fiz.* **54**, 1633 (1968).
756. V. N. Mineev and R. M. Zaidel', *Soviet Phys.—JETP* **27**, 874 (1968).
757. A. P. Minton, *Nature* **226**, 151 (1970).
758. R. Moccia, *J. Chem. Phys.* **40**, 2186 (1964).
759. J. F. Mohler, *Phys. Rev.* **35**, 236 (1912).
760. L. Monchick and E. A. Mason, *J. Chem. Phys.* **35**, 1676 (1961).
761. J. Morgan and B. E. Warren, *J. Chem. Phys.* **6**, 666 (1938).
762. K. Morokuma, *Chem. Phys. Letters* **4**, 358 (1969).
763. K. Morokuma and L. Pederson, *J. Chem. Phys.* **48**, 3275 (1968).
764. K. Morokuma and J. R. Winick, *J. Chem. Phys.* **52**, 1301 (1970).
765. H. Moser, *Ann. Physik* **82**, 993 (1927).
766. H. Moser and A. Zmaczynski, *Physik. Z.* **40**, 221 (1939).
767. J. W. Moskowitz and M. C. Harrison, *J. Chem. Phys.* **43**, 3550 (1965).
768. O. F. Mossotti, *Mem. de Mat. el Fis. Soc. Ital. de Sci.* **24**, 49 (1850).
769. J. R. Moszynski, *J. Heat Transfer* **83**, 111 (1961).
770. W. G. Moulton and R. A. Kromhout, *J. Chem. Phys.* **25**, 34 (1956).
771. N. Muller and R. C. Reiter, *J. Chem. Phys.* **42**, 3265 (1965).
772. R. H. Muller and J. A. Shufle, *J. Geophys. Res.* **73**, 3345 (1968).
773. R. S. Mulliken, *J. Chem. Phys.* **23**, 1833 (1955).
774. J. N. Murrell and F. B. van Duijneveldt, *J. Chem. Phys.* **46**, 1759 (1967).
775. A. S. N. Murthy, S. N. Bhat, and C. N. R. Rao, *J. Chem. Soc.* (*London*) **A1970**, 1251.
776. A. S. N. Murthy, R. E. Davis, and C. N. R. Rao, *Theor. Chim. Acta* **13**, 81 (1969).
777. A. S. N. Murthy and C. N. R. Rao, *Appl. Spec. Revs.* **2**, 69 (1968).

778. A. S. N. Murthy and C. N. R. Rao, *Chem. Phys. Letters* **2**, 123 (1968).
779. A. S. N. Murthy and C. N. R. Rao, *J. Mol. Struct.* **6**, 253 (1970).
780. A. S. N. Murthy, K. G. Rao, and C. N. R. Rao, *J. Am. Chem. Soc.* **92**, 3544 (1970).
781. U. Nakaya, *SIPRE Res. Report* **28** (1958).
782. A. H. Narten, X-Ray Diffraction Data on Liquid Water in the Temperature Range 4–200°C, ORNL-4578 (1970).
783. A. H. Narten, M. D. Danford, and H. A. Levy, X-Ray Diffraction Data on Liquid Water in the Temperature Range 4–200°C, ORNL-3997 (1966).
784. A. H. Narten, M. D. Danford, and H. A. Levy, *Disc. Faraday Soc.* **43**, 97 (1967).
785. A. H. Narten and H. A. Levy, *Science* **165**, 447 (1969).
786. A. H. Narten and H. A. Levy, *Science* **167**, 1521 (1970).
787. G. Nemethy and H. A. Scheraga, *J. Phys. Chem.* **66**, 1773 (1962).
788. G. Nemethy and H. A. Scheraga, *J. Chem. Phys.* **36**, 3382 (1962).
789. G. Nemethy and H. A. Scheraga, *J. Chem. Phys.* **36**, 3401 (1962).
790. G. Nemethy and H. A. Scheraga, *J. Chem. Phys.* **41**, 680 (1964).
791. D. Neumann and J. W. Moscowitz, *J. Chem. Phys.* **49**, 2056 (1968).
792. Neutron Diffraction Commission, *Acta Cryst.* **A25**, 391 (1969).
793. G. F. Newell, *Phys. Rev.* **80**, 476 (1950).
794. D. M. Newitt, S. Angus, and M. B. Hooper, The Experimental Determination of the Specific Enthalpy of Steam in the Superheat and Supercritical Regions, Electrical Research Association, Report No. 5050, Leatherhead (1964).
795. H. H. Nielsen, *Rev. Mod. Phys.* **23**, 90 (1951).
796. E. S. Nowak and R. J. Grosh, An Investigation of Certain Thermodynamic and Transport Properties of Water and Water Vapor in the Critical Region, Argonne National Laboratory, ANL-6064 (1959).
797. E. S. Nowak and R. J. Grosh, Atomic Energy Commission Technical Report ANL-6508 (1961).
798. E. S. Nowak, R. J. Grosh, and P. E. Liley, *J. Heat Transfer* **83**, 1, 14 (1961).
799. A. A. Noyes, Y. Kato, and R. B. Sosman, *Z. Phys. Chem.* **73**, 1 (1910).
800. A. A. Noyes, A. C. Melcher, H. C. Cooper, and G. W. Eastman, *Z. Phys. Chem.* **70**, 335 (1910).
801. A. J. Nozik and M. Kaplan, *Chem. Phys. Letters* **1**, 391 (1967).
802. N. Ockman, *Adv. Phys.* **7**, 199 (1958).
803. J. P. O'Connell and J. M. Prausnitz, *Ind. Eng. Chem. Fundamentals* **8**, 453 (1969).
804. J. J. O'Dwyer, *J. Chem. Phys.* **26**, 878 (1957).
805. G. D. Oliver and J. W. Grisard, *J. Am. Chem. Soc.* **78**, 561 (1956).
806. P. C. Olympia, Jr., and B. M. Fung, *J. Chem. Phys.* **51**, 2976 (1969).
807. J. R. O'Neil and L. H. Adami, *J. Phys. Chem.* **73**, 1553 (1969).
808. L. Onsager, *J. Am. Chem. Soc.* **58**, 1486 (1936).
809. L. Onsager and L. K. Runnels, *J. Chem. Phys.* **50**, 1089 (1969).
810. N. S. Osborne, *J. Res. Natl. Bur. Std.* **4**, 609 (1930).
811. N. S. Osborne and C. H. Meyers, *J. Res. Natl. Bur. Std.* **13**, 1 (1934).
812. N. S. Osborne, H. F. Stimson, E. F. Fiock, and D. C. Ginnings, *J. Res. Natl. Bur. Std.* **10**, 155 (1933).
813. N. S. Osborne, H. F. Stimson, and D. C. Ginnings, *J. Res. Natl. Bur. Std.* **18**, 389 (1937).
814. N. S. Osborne, H. F. Stimson, and D. C. Ginnings, *J. Res. Natl. Bur. Std.* **23**, 197 (1939).